小林 弘 珪藻図鑑

第1巻

小林　弘
出井 雅彦・真山 茂樹
南雲　保・長田 敬五
共　著

内田老鶴圃

H. Kobayasi's *Atlas of Japanese Diatoms* based on electron microscopy

Volume 1

Hiromu KOBAYASI
Masahiko IDEI Shigeki MAYAMA
Tamotsu NAGUMO Keigo OSADA

Uchida Rokakuho Publishing Co., Ltd.
Tokyo

ISBN4-7536-4046-9

序

　本書は分類学を中心に据えた日本初の珪藻図鑑であり，扱われたすべての分類群の説明に電子顕微鏡写真が加えられている点では世界初の図鑑である．本書は当初，日本淡水藻図鑑（内田老鶴圃発行）の姉妹編として企画され，小林弘ひとりの執筆により光学顕微鏡写真を列挙した図鑑になる予定であった．1980年代前半の出版を目指し作業が進められ，70年代後半には中心珪藻，羽状無縦溝珪藻を主とする数十枚の写真図版が完成していた．80年代になると，日本珪藻学会が設立され，国内の珪藻研究者の交流が深まると共に研究者も増加し，珪藻研究は活発化した．また，この頃から走査電子顕微鏡による殻の微細構造が解明されるようになり，分類学は格段の進歩を遂げるようになった．そして，1985年には日本珪藻学会から会誌Diatomが，翌年には国際珪藻学会誌Diatom Researchが発刊され，珪藻学の研究は国内外を問わず一層活発なものとなった．この流れの中で，世界の珪藻分類学の主流が，電子顕微鏡観察に基づく研究に移行することをいち早く感じ取った小林は，それまでの図版を一切破棄し，新たに電子顕微鏡写真を主体とする図鑑を一から作り直すことを決意した．電子顕微鏡写真による図鑑は1種あたりに使われるページ分量が大幅に増加するため，当初は河川産珪藻に限定した図鑑を作成しようとしたが，次第に河川産以外の重要な種も含むようになった．

　小林は図鑑作成のため，大学近くの武蔵小金井のマンションに専用の居を構え（後に東京珪藻研究所と命名)，そこに膨大な文献資料と写真資料を蓄積した．そして，それらを丹念に吟味し，種類ごとに整理することから始めた．当初，珪藻1種につき1図版の計画で進めたが，その緻密さゆえ途中から2図版に変更された種も多々あった．また小林は多くの種に，電子顕微鏡写真をトレースして描いた図を加えることで，一層わかりやすい図鑑の作成を心掛けた．そして1996年夏には400枚を超す図版が完成し，さらに殻構造と用語の解説案が作成されるに至った．この年の9月には第14回国際珪藻学会大会（14th International Diatom Symposium)が東京で開催されることになっており，同学会の副会長であった小林はその職務に加え，この大会の会長として，準備に多忙な毎日を過ごしていた．しかし，開催直前の7月12日に無念の急逝を遂げ，大会の成功を見ることもなく，また図鑑作成も志半ばにて終わることになった．

　その後，小林玲子夫人から依頼をうけた我々4人の弟子が，師の遺志を引き継ぎ図鑑の完成を目指すことになり，翌年から作業を再開した．小林は生前，写真図版は作成していたが，記載文の執筆をしていなかったため，これを4人で分担して進めることになった．また，不足していた写真の追加や構造と用語解説の改訂も行うことになった．折しも，この頃より多くの珪藻で属名の変更や新提案が行われたため，世界の潮流の中でそれらの学名が定着し，図鑑として妥当な学名を採用するために多少の時間が必要となった．また，内容の増加に伴い本図鑑は2部に分けて出版することが決定された．

　ここに第1巻を発行するに至ったことは我々一同にとって喜びに堪えないところである．しかし，多くの人の期待にも関わらず，殊の外時間を費やしてしまったことは我々の不徳の致すところである．また，内外の文献をもとに，できるだけ最新の分類学的知見に基づいて執筆を心掛けたつもりであるが，筆者らの浅学ゆえに誤りや遺漏がないとはいえない．また，4人の分担による執筆のため多少とも表現に個性が出たが，ご容赦願いたい．本書が日本の珪藻学，

そして世界の珪藻学の発展に少しでも寄与し，多くの読者に利用されるなら，著者一同望外の喜びである．今後，読者諸氏のご指摘やご叱責を頂き，それらを糧に早急な第 2 巻の刊行を目指す所存である．

　本図鑑では小林弘の直接，および間接の弟子により撮影された写真や資料が随所に使用されている．これら協力者には，東京教育大学在職時の内地留学生であった三友清史，原口和夫，安藤一男，山下不二子，東京学芸大学在職時の内地留学生の吉田稔，小原一基，北沢星磁，海外からの留学生の Eduardo A. Lobo, そして，研究室の学生・院生等であった阿蘇一博，上山敏，長谷川恵子，大淵晴美，田中勝子，渡辺真由美，梶原敦，加藤久美子，高井淳子，上村孝之，野沢美智子，山田裕子，宮坂裕子，赤星えりか，猿渡厚史，伊東布佐子，江原真理，柳下信，木村いずみ，須永智，高橋宗弘，服部彩，小林浩司，寺嶋剛，片岡宏文，小林秀明，沢田明美，田中俊二，井上裕喜，森純子，広田和幸，浦野浩二，石代俊則，徳田欣之，米村好朗，石井俊治，小沢淳子，内山喜代子，久保あゆみ，河崎優子，小堀晋爾，平康博，石原三正，高橋美佳，柿木孝文，勝本英嗣，作田浩美，大桃純子，真山なぎさ，さらに外部より師事した田中宏之，永沼治，飯島敏雄，浜篤，定年退官後に師事した河島綾子，佐竹俊子，寺尾和明，洲澤多美枝がいる．また，大島海一，加藤季夫の両氏には資料の提供を受けた．さらに，小林弘逝去後も図鑑編纂のため小林玲子夫人，令嬢の小林美咲氏および令息の小林理氏には，東京珪藻研究所の使用および所蔵されたすべての文献，資料の自由な閲覧を許可して頂いた．本書はご遺族の力添えがあって初めて完成に至ったものであり，ここに心からの感謝の意を表したい．また，筑波大学名誉教授の千原光雄氏には終始指導と激励を頂き感謝に堪えない．最後に，本書の企画に始まり出版までの長きにわたりご苦労頂いた，内田老鶴圃社長内田悟氏並びに内田学氏，笠井千代樹氏に深く御礼を申し上げる．

　　2006 年 10 月　　　　　　　　　　　　　　　　　　　　　　　　　　執筆者一同

Preface

This is the first atlas of Japanese diatoms compiled from a taxonomic viewpoint as well as the first monograph, in which all species are described with electron micrograms. This volume was intended to be a sequel to Illustrations of the Japanese Fresh-water Algae (Hirose & Yamagishi (eds) 1977) at first. Originally, it was planned to be published as a monograph composed of only light micrograms by the late Professor Dr. Hiromu Kobayasi in the early 1980s. He had, in fact, been preparing dozens of photo-plates of centric and araphid diatoms till the end of the 1970s. In 1980 the Japanese Society of Diatomolgy was established, and the research of diatoms was promoted domestically, around the time when the study of valve structures using scanning electron microscopy (SEM) was increasing worldwide and the taxonomy of diatoms was more vigorously pursued.

Sensing the new current in taxonomic studies, Kobayasi relinquished the plan to make an atlas of light micrograms and decided to make a new monograph composed of detailed electron micrograms. As the total number of pages increased inevitably in the atlas with electron micrograms, he planned to make a volume with only river diatoms at the beginning; however, as it progressed, it began to contain important species from other habitats also. Although one plate had been prepared per one species, later it was changed to two plates for many species to reveal structures in detail. Throughout the procedure of the study, all specimens were observed by SEMs equipped with an electric gun of field emission type, and a huge number of photographs was taken from various directions and at several magnifications. As almost all species had been originally described from outside of Japan, he took as much time as possible to compare his specimens with the types housed abroad. He had completed more than 400 plates by the summer of 1996. In September of that year, the 14th International Diatom Symposium was held in Tokyo, and Kobayasi spent hard days and nights for its preparation as a combiner during the summer. However, shortly before the congress he suddenly passed away due to myocardial infarction on 12 July.

After his death, Reiko, the widow of Kobayasi, who understood the value of the publication of the volume very well, offered his private rooms, literature and plenty of photo-data as a facility for his older students to continue the work. With her support, work on the atlas was restarted by us the next year. In the past sixteen years, many taxonomic changes were proposed in the rank of diatom genera by many researchers. As one mission of the volume is to be used by not only taxonomists but by researchers in other fields for a long time, we were very careful to adopt such new and revived names, and spent several years for re-examination. In this volume, about half of the taxa, which Kobayasi originally prepared for the publication, were included, and the rest will be published as volume 2.

We are grateful to Reiko Kobayasi, her daughter Misaki and son Osamu for their kind support. Thanks are due to all the old students of Kobayasi and to the many people concerned with preparing materials and taking photographs. We are also thankful to Prof. Mitsuo Chihara for his special encouragement of this posthumous edition and to Satoru Uchida, a publisher, who has helped with editing since the book was first planned.

IDEI, Masahiko
MAYAMA, Shigeki
NAGUMO, Tamotsu
OSADA, Keigo

凡　例

● 新名と新用語
　本書では 4 新種，1 新変種ならびに 6 分類群の新組み合わせを行った．国際植物命名規約に従った新分類群の記載および英文記述，並びに属の組み替えは，凡例の後にまとめて行った．新用語として，殻肩（valve shoulder）を提案した．

● 用語の解説
　珪藻の被殻には，他の微小藻類とは比べものにならないほど多くの構造が存在し，それぞれに特殊な用語が与えられている．ここではそれぞれの構造と用語をわかりやすく説明するため，電子顕微鏡写真を添え解説した．また，これらの用語の日本語，英語，ラテン語表記対照表を添え，利用者の便宜を図った．

● 分類体系
　これまで様々な分類体系が提唱されてきたが，形態情報と遺伝子情報に基づく，現時点で最新かつ妥当と思われる体系を示した．

● 分類群の説明
　本書で扱った分類群の記述順は，前述の分類体系に従った．ただし，同じ科内では，属名のアルファベット順，さらに同じ属内では，種形容語のアルファベット順に記載した．すべての説明は，学名，原典，必要に応じて基礎異名（Basionym）もしくは異名（Synonym）の引用，計測値，本文の順に記述した．

　＜学　名＞　近年，珪藻の分類学では，新属の設立や復活された属による細分化が著しいが，国際植物命名規約上合法であり，分類学的に妥当と思われる新学名を積極的に採用した．このため，従来親しまれてきた学名と異なるものも多くなったが，今後，これらの学名が世界的に普及すると思われる．

　＜命名者＞　学名の命名者の記述は，その姓のみを省略せずに用い，同姓の分類学者がいる場合は，識別のため名等の頭文字を付記した．また，用語の解説などでは "Authors of plant names"（Brummitt & Powell 編　1992）の提唱する短縮名を用いた．

　＜引　用＞　表題の学名（正名）にはすべて原典を引用し，それらが初出の名前と異なる場合は，基礎異名（Basionym）も引用した．また，必要に応じて異名（Synonym）を付加した．

　＜計測値＞　分類群の解説文に先立ち，便宜のため，計測値を次の短縮語とともに記した．
　　　L.：殻長，　W.：殻幅，　D.：殻径，　H.：殻套高，　Str.：条線数，　Ar.：胞紋数，
　　　F.：条線束数，T.C.：横走肋数．
　なお，ここに記した値は我々が実際に観察した個体から得られた計測値である．

　＜本　文＞　より特徴を捉えやすくするため，分類形質となる被殻の主な構造などをゴシッ

ク体の小見出しで表し，その後に具体的な説明を行った．また，被殻形質の他に，生体からの情報も可能な限り記述した．汚濁耐性の項に示したものは，河川の水質判定方法の1つである識別珪藻群法（Kobayasi & Mayama 1989）で用いられる3つのカテゴリーのいずれかである．出現地の項には，我々がその種類を実際に観察記録した地点のみを列記した．なお，特記事項がある場合は，ノートとして本文の最後に付記した．

● 図　　　版

　各図版には，原則として右側頁に当該分類群の光学顕微鏡写真，スケッチ，走査電子顕微鏡写真（SEM）および透過電子顕微鏡写真（TEM）を，左側頁にその英語説明文を載せた．光学顕微鏡写真の倍率は，特殊な場合を除き2000倍に統一した．電子顕微鏡写真の倍率は各図の説明に付した．また，各図の撮影個体の産地をLocalityの項に記した．

● 属和名一覧

　珪藻には1000を超える属が記載されているが，それらの属に対して和名が付けられているものはほんの一部に過ぎない．本書では，従来から与えられていた和名に加え，ここで新たに提唱する和名を合わせ，計373属について和名と学名を並記した．

New Taxa

Tabellaria pseudoflocculosa H. Kobayasi ex Mayama sp. nov.

Plate 112

Coloniae "zigzag" vel stellae ad instar. Frustula matura 3-6 dissepimmentis planis absque dissepimmento rudimentalis formato. Valvae fusiformes modice inflatae ad centrum, apicibus leniter capitatis, 35-45 μm longae, 5-6.5 μm latae. Area axialis anguste linearis. Striae transapicales parallelae circiter 16 in 10 μm. Plastidia discoidales multa.

Zigzag or starlike colonies. Mature frustules with 3-6 flat septa but without rudimentary septum. Valves fusiform with light inflation centrally, with weakly capitate ends, 35-45 μm long, 5-6.5 μm wide. Axial area is narrowly linear. Transapical striae are parallel, ca. 16 in 10 μm. Plastids are discoid, numerous.

Holotype: TNS-AL-55565 in TNS (Department of Botany, The National Science Museum, Tokyo).

Isotype: TNS-AL-55566.

Type locality: An irrigation pond (20×40 m) in Kamikomatsuki, Hino, Shiga, Japan. Collected by S. Mayama, on 31 March 1988.

Etymology: The species epithet refers to its morphological resemblance to *Tabellaria flocculosa* and was prepared by the late Hiromu Kobayasi in his life time.

SEM observations:

The outer surface of the valve is smooth and bears spinula on the interstria along the valve shoulder. Areolae in the stria are poroidal, ca. 50 in 10 μm. Both ends have apical pore fields, through which mucilage is secreted and links neighboring cells together. A labiate process is placed near the inward terminal of the stria in the valve center. The external opening of the labiate process is a transversely expanded slit. All bands have a single row of areolae along the pars media. Occasionally double or multiple rows of areolae are observed on the closed ends of several bands. These bands are complete bands up to the 2nd or the 3rd, and further bands are split bands. The septa are formed in the complete bands. The number of the complete or open bands is not stable among individual cells, but the cells with four septa are observed to be dominant in the mature cells.

This species resembles *Tabellaria flocculosa* (Ross) Kützing in some morphological features but differs in the flatness of the septa, the lesser number of septa, the lack of a subseptum in any band and the slenderer girdle view. *T. pseudoflocculosa* is also similar to *Tabellaria quadriseptata*, but differs in that the latter has the valve with more parallel sides and the bands with rudimentary septa. Moreover, this species usually occurs in eutrophic ponds of neutral to alkaline waters together frequently with *Tabellaria fenestrata*, which does not occur in peat areas, whereas *T. quadriseptata* never occurs in waters over pH 6 (Flower & Battarbee 1985).

Gomphonema curvipedatum H. Kobayasi ex K. Osada sp. nov.

Plate 122

Valvae claviformes, lineari-lanceolatae, plerumque modice cymbelloideae ad instar, polis curvatis versus later valvae sine stigmate, 21.5-44.5 µm longae, 4.5-8 µm latae. Curvatura poli basalis illa verticis fortior in speciminibus majoribus. Raphe lateralis, distincte undulata. Area axialis lata, lanceolatae. Stigma unicum in latere uno noduli centralis. Aperturae externae et internae stigmatis parvae, rotundae (in SEM). Striae transapicales, uniseriata, breves, modice radiatae omnino, 13-17 in 10 µm. Aperturae areolarum externae C-, I- vel 3-formes (in SEM).

Valves claviform, linear-lanceolate, mostly moderately cymbelloid in outline, with both poles curved towards the valve side containing no stigma, 21.5-44.5 µm long, 4.5-8 µm broad. Curvature of the footpole stronger than that of the headpole in larger specimens. Raphe lateral, distinctly undulate. Axial area broad, lanceolate. A single stigma on one side of the central nodule. Both external and internal openings of the stigma small, round (in SEM). Transapical striae uniseriate, short, moderately radiate throughout, 13-17 in 10 µm. External openings of the areolae slitlike, C-, I- or 3-shaped (in SEM).

Holotype: TNS-AL-53991sa in TNS (Department of Botany, The National Science Museum, Tokyo).

Isotype: TNS-AL-53991sb.

Type locality: A small concrete tank in Tokyo Gakugei University, Tokyo, Japan. Collected by M. Takahasi, on 1 September 1988.

Etymology: The species epithet is derived from the curved footpoles and was prepared by the late Hiromu Kobayasi in his lifetime. The Latin *curvipedatum* means "possessing curved foot".

This species resembles *G. clevei* in the short striae and the wide axial area, but is distinguishable from it by having curved footpoles, the small, round openings of the stigma and the C-, I- or 3-shaped external openings of the areolae.

Gomphonema kinokawaensis H. Kobayasi ex K. Osada sp. nov.

Plate 125

Valvae claviforme rhombicae, vertice cuneato et polo basali obtusato, 17-46 µm longae, 6.5-9 µm latae. Raphe modice lateralis, distincte undulata. Area axialis lata, rhombica. Stigma unicum in latere uno noduli centralis, apertura externa parva et rotunda, et apertura interna elliptica intra depressionem laterale elongatam (in SEM). Striae transapicales breves, modice radiatae omnino, 14-16 in 10 µm. Lineae longitudinalis visibiles in area striae. Areolae in quoque alveolo uniseriata, Aperturae areolarum externae C-formes in areolis juxta aream axialem, et longitudinaliter elongatae in areolis ceteris (in SEM).

Valves clavately rhombic, with cuneate headpole and obtuse footpole, 17-46 µm long, 6.5-9 µm broad. Raphe slightly lateral, distinctly undulate. Axial area broad, rhombic. A single stigma on one side of the central nodule, with the small, round external opening and the internal opening lying within a laterally elongated depression (in SEM). Transapical striae short, moderately radiate throughout, 14-16 in 10 µm. Longitudinal lines visible, crossing the stria. Areolae in each alveolus uniseriate. External openings of the areolae C-shaped in the areolae adjacent to the axial area, and longitudinally elongated in the other areolae (in SEM).

Holotype: TNS-AL-53992sa in TNS (Department of Botany, The National Science Museum, Tokyo).

Isotype: TNS-AL-53992sb.

Type locality: Near Inabe-bashi (Inabe Bridge), Kino-kawa (Kino River), Wakayama, Japan. Collected by S. Nakai, on 17 September 1983.

Etymology: The species epithet is derived from the name of the river, Kino-kawa, which is the type locality. The name was prepared by the late Hiromu Kobayasi in his lifetime.

This species resembles *G. clevei*, *G. curvipedatum* and *G. inaequilongum* in having relatively short striae, but is distinguished from the first two species in partially, internally occluded alveoli, and from the last in the valve outline and the shapes of both the striae and the areolar openings.

Gomphonema parvulum var. *neosaprophilum* H. Kobayasi ex K. Osada var. nov.
Plates 129, 130

Differt a var. *parvulum* valvis comparate late lanceolatis, apicibus obtusatis, polis vix productis striis transapicalibus sparsis 10-16 in 10 μm, alveolis extensis ad marginem valvae, tigillis in pagina alveolum interna. Longitudo 12-36 μm, latitudo 4.5-8 μm.

This variety is distinguished from var. *parvulum* by comparatively broadly lanceolate valves, barely produced poles, sparser transapical striae 10-16 in 10 μm, alveoli extending to the valve margin, and struts on the internal surface of the alveolus. Length 12-36 μm, breadth 4.5-8 μm.

Holotype: TNS-AL-53990sa in TMS (Department of Botany, The National Science Museum, Tokyo).

Isotype: TNS-AL-53990sb.

Raw type material: TNS-AL-53990m.

Type locality: Negishi-bashi (Negishi Bridge), Sakai-gawa (Sakai River), Tokyo, Japan. Collected by Eduardo A. Lobo, on 20 August 1992.

Hiromu Kobayasi had named this taxon *Gomphonema parvulum* var. *saprophilum* during his lifetime, however, the epithet "*saprophilum*" is a later homonym. Therefore, the taxon is here described as var. *neosaprophilum*.

Gomphonema yamatoensis H. Kobayasi ex K. Osada sp. nov.
Plate 135

Valvae lineares ad lineari-lanceolatae, apicibus obtusatis, 15-33 μm longae, 4-5 μm latae. Poli ambus obtusis cum latitudinibus fere aequalibus, vel polus basalis vertice leviter angustior. Raphe levier lateralis, modice undulata. Area axialis angusta, linealis. Area centralis relative angusta, unilateralis. Stigma unicum in latere uno noduli centralis, apertura externa rotunda cum margine incrassato, et apertura interna elliptica (in SEM). Striae transapicales uniseriatae, modice radiatae omnino, 11-12 in 10 μm. Puncta striarum invisibilia in LM. Aperturae areolarum externae (40-50 in 10 μm) C-formes (raro S-formes) in fronte et limbo (in SEM).

Valves linear, linear-lanceolate, with obtuse apices, 15-33 μm long, 4-5 μm broad. Both poles with nearly equal breadth, or the footpole slightly narrower than the headpole. Raphe slightly lateral, moderately undulate. Axial area narrow, linear. Central area relatively narrow, unilateral. A single stigma on one side of the central nodule, with the round external opening with thickened margin and the elliptical internal opening (in SEM). Transapical striae uniseriate, lightly radiate throughout, 11-12 in 10 μm. Puncta of striae invisible in LM. External openings of areolae (40-50 in 10 μm) C-shaped (rarely S-shaped) in both valve face and mantle (in SEM).

Holotype: TNS-AL-53993sa in TNS (Department of Botany, The National Science Museum, Tokyo).

Isotype: TNS-AL-53993sb.

Type locality: Inabe-gawa (Inabe River), Mie, Japan. Collected by N. Ishida, on 8 December 1989.

Etymology: The species epithet is derived from the classical name of Japan "Yamato" and was prepared by the late Hiromu Kobayasi in his lifetime.

This species is similar to *Gomphonema pumilum* (Grunow) Reichardt & Lange-Bertalot (1991) and *G. procerum* Reichardt & Lange-Bertalot (l.c.), but different from the former in the smaller central area and from the latter in the internal opening of the stigma in not becoming linear. Hiromu Kobayasi had named this taxon *Gomphonema yamatoensis* in his lifetime, however, he prepared no description for it.

New Combinations

Puncticulata shanxiensis **(S. Q. Xie & Y. Z. Qi) Nagumo comb. nov.**
Basionym: *Cyclotella shanxiensis* S. Q. Xie & Y. Z. Qi 1984. In: Mann, D. G. (ed.) Proc. 7th Intern. Diat. Symp. p. 188. pl. 1-4.

Hannaea arcus* var. *recta **(Cleve) M. Idei comb. nov.**
Basionym: *Fragilaria arcus* var. *recta* Cleve 1898. Bih Kongl. Svenska Vet.-Akad. Handl. **24**: 9.

Ulnaria inaequalis **(H. Kobayasi) M. Idei comb. nov.**
Basionym: *Synedra inaequalis* H. Kobayasi in Kobayasi, H., Idei, M., Kobori, S. & Tanaka, H. 1987. Diatom **3**: 9 ; Kobayasi, H. 1965. Journ. Jap. Bot. **40**: 347. f. 1a-d.

Ulnaria pseudogaillonii **(H. Kobayasi & M. Idei) M. Idei comb. nov.**
Basionym: *Fragilaria pseudogaillonii* H. Kobayasi & M. Idei 1979. Jap. J. Phycol. **27**: 196. f. 1-3.

Achnanthidium gracillimum **(Meister) Mayama comb. nov.**

Basionym: *Microneis gracillima* Meister 1912. Die Kieselalgen der Schweiz. p. 97. f. 12.

Achnanthidium subhudsonis **(Hustedt) H. Kobayasi comb. nov.**

Basionym: *Achnanthes subhudsonis* Hustedt 1921. Hedwigia **63**: 144. f. 9-12.

Newly Proposed Term

Valve shoulder

  When valve face mantle and are clearly distinguishable, "valve shoulder" is proposed as a term indicating the boundary between the two parts. This is equivalent to the longer terms "valve face and mantle juncture" or "valve face and mantle junction," which have been used by several authors. There are lots of species, in which the valve shoulder is distinct; moreover, various elements are frequently located on it, e.g., spine, marginal ridge, hyaline area and labiate process. Therefore, the use of this shorter name will offer convenience for taxonomists, who often describe this area in many diatoms.

Abbreviation

L.: valve length,   W.: valve width,   D.: valve diameter,   H.: mantle height,
Str.: striae number,   Ar.: areolae number,   F.: fascicles number,
T. C.: transapical costae number.

目　次

序 ·· **3**
Preface ·· **5**
凡　例 ·· **7**
新分類群・新組み合わせ・新用語
　New Taxa, New Combinations, Newly Proposed Term, Abbreviation ············ **9**
収録分類群一覧 ·· **17**
珪藻の殻構造と用語 ··· **21**
珪藻用語対照表 ·· **47**
珪藻分類体系 ··· **53**

和文解説 ··· **1〜145**
欧文解説と図版 ··· **147〜507**
　Plates and Legends
属の学名-和名対照表 ··· **509〜518**
引用文献 ··· **519〜526**
学名索引 ··· **527〜531**

収録分類群一覧

Melosira moniliformis (O. Müller) C. Agardh　Plates 1, 2 ··· 2, 148, 150
Melosira nummuloides C. Agardh　Plates 3, 4 ··· 3, 152, 154
Melosira varians C. Agardh　Plates 5, 6 ··· 4, 156, 158
Ellerbeckia arenaria f. *teres* (Brun) R. M. Crawford　Plate 7 ···································· 5, 170
Aulacoseira ambigua (Grunow) Simonsen　Plates 8, 9 ··· 6, 162, 164
Aulacoseira distans (Ehrenberg) Simonsen　Plate 10 ·· 7, 166
Aulacoseira granulata (Ehrenberg) Simonsen　Plates 11, 12 ·································· 8, 168, 170
Aulacoseira italica (Ehrenberg) Simonsen　Plates 13, 14 ····································· 9, 172, 174
Aulacoseira longispina (Hustedt) Simonsen　Plates 15, 16 ·································· 10, 176, 178
Aulacoseira nipponica (Skvortsov) Tuji　Plates 17, 18 ·· 11, 180, 182
Aulacoseira subarctica (O. Müller) Haworth　Plates 19, 20 ·································· 12, 184, 186
Aulacoseira tenuis (Hustedt) H. Kobayasi　Plates 21, 22 ···································· 13, 188, 190
Aulacoseira valida (Grunow) Krammer　Plates 23, 24 ··· 14, 192, 194
Orthoseira asiatica (Skvortsov) H. Kobayasi　Plate 25 ··· 15, 196
Orthoseira epidendron (Ehrenberg) H. Kobayasi　Plate 26 ·· 16, 198
Brebisira arentii (Kolbe) Krammer　Plate 27 ·· 17, 200
Thalassiosira allenii Takano　Plate 28 ·· 18, 202
Thalassiosira eccentrica (Ehrenberg) Cleve　Plate 29 ·· 19, 204
Thalassiosira faurii (Gasse) Hasle　Plate 30 ··· 20, 206
Thalassiosira guillardii Hasle　Plate 31 ··· 21, 208
Thalassiosira lacustris (Grunow) Hasle　Plate 32 ·· 22, 210
Thalassiosira nordenskioeldii Cleve　Plate 33 ·· 23, 212
Thalassiosira pseudonana Hasle & Heimdal　Plate 34 ··· 24, 214
Thalassiosira tenera Proshkina-Lavrenko　Plate 35 ··· 25, 216
Thalassiosira weissflogii (Grunow) G. A. Fryxell & Hasle　Plate 36 ···························· 26, 218
Cyclostephanos dubius (Fricke) Round　Plates 37, 38 ······································· 27, 220, 222
Cyclostephanos invisitatus (Hohn & Hellerman) Theriot *et al.*　Plates 39, 40
 ··· 28, 224, 226
Cyclotella atomus Hustedt　Plate 41 ·· 29, 228
Cyclotella atomus Hustedt var. *gracilis* Genkal & Kiss　Plate 42 ······························· 30, 230
Cyclotella criptica Reimann *et al.*　Plate 43 ··· 31, 232
Cyclotella litoralis Lange & Syvertsen　Plates 44, 45 ·· 32, 234, 236
Cyclotella meduanae Germain　Plate 46 ··· 33, 238
Cyclotella meneghiniana Kützing　Plate 47 ··· 34, 240
Cyclotella ocellata Pantocsek　Plate 48 ·· 35, 242
Cyclotella pantaneliana Castracane　Plate 49 ··· 36, 244
Cyclotella striata (Kützing) Grunow　Plate 50 ·· 37, 246
Discostella asterocostata (Lin *et al.*) Houk & Klee　Plate 51 ·································· 38, 248
Discostella pseudostelligera (Hustedt) Houk & Klee　Plate 52 ·································· 39, 250

Discostella stelligera (Ehrenberg) Houk & Klee   Plates 53, 54 .... 40, 252, 254
Puncticulata praetermissa (Lund) Håkansson   Plate 55 .... 41, 256
Puncticulata shanxiensis (S. Q. Xie & Y. Z. Qi) Nagumo comb. nov.   Plate 56 .... 42, 258
Stephanodiscus hantzshii f. *tenuis* (Hustedt) Håkansson & Stoermer   Plates 57, 58
.... 43, 260, 262
Stephanodiscus minutulus (Kützing) Round   Plates 59, 60 .... 44, 264, 266
Stephanodiscus rotula (Kützing) Hendey   Plate 61 .... 45, 268
Eucampia zodiacus Ehrenberg   Plate 62 .... 46, 270
Asterionella formosa Hassall   Plate 63 (Figs 1a-e) 47, 272
Asterionella gracillima (Hantzsch) Heiberg   Plate 63 (Figs 2a-g) 48, 272
Asterionella ralfsii W. Smith   Plate 64 .... 49, 274
Asterionellopsis glacialis (Castracane) Round   Plate 65 .... 50, 276
Catacombas obtusa (Pantocsek) Snoeijs   Plate 66 .... 51, 278
Ctenophora pulchella (Rakfs ex Kützing) D. M. Williams & Round   Plate 67 .... 52, 280
Diatoma mesodon (Ehrenberg) Kützing   Plate 68 .... 53, 282
Diatoma tenuis C. Agardh   Plate 69 .... 54, 284
Diatoma vulgaris Bory   Plate 70 .... 55, 286
Fragilaria capitellata (Grunow) J. B. Petersen   Plates 71, 72 .... 56, 288, 290
Fragilaria crotonensis Kitton   Plates 73, 74 .... 57, 292, 294
Fragilaria mesolepta Rabenhorst   Plate 75 .... 58, 296
Fragilaria neoproducta Lange-Bertalot   Plate 76 .... 59, 298
Fragilaria perminuta (Grunow) Lange-Bertalot   Plate 77 .... 60, 300
Fragilaria vaucheriae (Kützing) J. B. Petersen   Plate 78 .... 61, 302
Fragilariforma bicapitata (A. Mayer) D. M. Williams & Round   Plate 79 .... 62, 304
Hannaea arcus (Ehrenberg) R. M. Patrick   Plate 80 .... 63, 306
Hannaea arcus var. *recta* (Cleve) M. Idei comb. nov.   Plate 81 .... 64, 308
Martyana martyi (Héribaud) Round   Plates 82, 83 .... 65, 310, 312
Meridion circulare (Greville) C. Agardh   Plate 84 .... 66, 314
Pseudostaurosira brevistriata (Grunow) D. M. Williams & Round   Plate 85 .... 67, 316
Pseudostaurosira brevistriata var. *nipponica* (Skvortsov) H. Kobayasi   Plate 86 .... 68, 318
Pseudostaurosira robusta (Fusey) D. M. Williams & Round   Plate 87 .... 69, 320
Punctastriata linearis D. M. Williams & Round   Plate 88 .... 70, 322
Staurosira construens Ehrenberg var. *construens*   Plate 89 .... 71, 324
Staurosira construens var. *binodis* (Ehrenberg) P. B. Hamilton   Plate 90 .... 72, 326
Staurosira construens var. *exigua* (W. Smith) H. Kobayasi   Plate 91 .... 73, 328
Staurosira construens var. *triundulata* (H. Reichelt) H. Kobayasi   Plate 92 .... 74, 330
Staurosira elliptica (Schumann) D. M. Williams & Round   Plate 93 .... 75, 332
Staurosira venter (Ehrenberg) H. Kobayasi   Plate 94 .... 76, 334
Staurosira venter (Ehrenberg) var. *binodis* H. Kobayasi   Plate 95 .... 77, 336
Staurosirella lapponica (Grunow) D. M. Williams & Round   Plate 96 .... 78, 338
Staurosirella leptostauron (Ehrenberg) D. M. Williams & Round   Plate 97 .... 79, 340
Staurosirella pinnata (Ehrenberg) D. M. Williams & Round   Plate 98 .... 80, 342
Synedrella parasitica (W. Smith) Round & Maidana   Plate 99 .... 81, 344
Tabularia affinis (Kützing) Snoeijs   Plate 100 .... 82, 346

Ulnaria acus (Kützing) M. Aboal   Plate 101 ⋯⋯⋯⋯⋯⋯⋯⋯⋯⋯⋯⋯⋯⋯⋯⋯⋯⋯⋯⋯⋯ 83, 348
Ulnaria biceps (Kützing) Compère   Plate 102 ⋯⋯⋯⋯⋯⋯⋯⋯⋯⋯⋯⋯⋯⋯⋯⋯⋯⋯⋯⋯ 84, 350
Ulnaria capitata (Ehrenberg) Compère   Plate 103 ⋯⋯⋯⋯⋯⋯⋯⋯⋯⋯⋯⋯⋯⋯⋯⋯⋯ 85, 352
Ulnaria inaequalis (H. Kobayasi) M. Idei comb. nov.   Plate 104 ⋯⋯⋯⋯⋯⋯⋯⋯⋯ 86, 354
Ulnaria lanceolata (Kützing) Compère   Plate 105 ⋯⋯⋯⋯⋯⋯⋯⋯⋯⋯⋯⋯⋯⋯⋯⋯⋯ 87, 356
Ulnaria pseudogaillonii (H. Kobayasi & M. Idei) M. Idei comb. nov.   Plate 106 ⋯⋯⋯ 88, 358
Ulnaria ulna (Nitzsch) Compère   Plate 107 ⋯⋯⋯⋯⋯⋯⋯⋯⋯⋯⋯⋯⋯⋯⋯⋯⋯⋯⋯⋯⋯ 89, 360
Tabellaria fenestrata (Lyngbye) Kützing   Plates 108, 109 ⋯⋯⋯⋯⋯⋯⋯⋯⋯ 90, 362, 364
Tabellaria flocculosa (Roth) Kützing   Plates 110, 111 ⋯⋯⋯⋯⋯⋯⋯⋯⋯⋯ 91, 366, 368
Tabellaria pseudoflocculosa H. Kobayasi ex Mayama sp. nov.   Plate 112 ⋯⋯⋯⋯ 92, 370
Actinella brasiliensis Grunow   Plate 113 ⋯⋯⋯⋯⋯⋯⋯⋯⋯⋯⋯⋯⋯⋯⋯⋯⋯⋯⋯⋯⋯ 93, 372
Peronia fibula (Brébisson ex Kützing) R. Ross   Plate 114 ⋯⋯⋯⋯⋯⋯⋯⋯⋯⋯⋯ 94, 374
Rhoicosphenia abbreviata (C. Agardh) Lange-Bertalot   Plates 115-117
⋯⋯⋯⋯⋯⋯⋯⋯⋯⋯⋯⋯⋯⋯⋯⋯⋯⋯⋯⋯⋯⋯⋯⋯⋯⋯⋯⋯⋯⋯⋯⋯⋯⋯⋯ 95, 376, 378, 380
Gomphoneis heterominuta Mayama & Kawashima   Plate 118 ⋯⋯⋯⋯⋯⋯⋯⋯⋯ 96, 382
Gomphoneis rhombica (Fricke) V. Merino *et al.*   Plates 119, 120 ⋯⋯⋯⋯ 97, 384, 386
Gomphonema clevei Fricke   Plate 121 ⋯⋯⋯⋯⋯⋯⋯⋯⋯⋯⋯⋯⋯⋯⋯⋯⋯⋯⋯⋯⋯ 98, 388
Gomphonema curvipedatum H. Kobayasi ex K. Osada sp. nov.   Plate 122 ⋯⋯⋯⋯ 99, 390
Gomphonema gracile Ehrenberg   Plate 123 ⋯⋯⋯⋯⋯⋯⋯⋯⋯⋯⋯⋯⋯⋯⋯⋯⋯⋯ 100, 392
Gomphonema inaequilongum (H. Kobayasi) H. Kobayasi   Plate 124 ⋯⋯⋯⋯⋯ 101, 394
Gomphonema kinokawaensis H. Kobayasi ex K. Osada sp. nov.   Plate 125 ⋯⋯⋯⋯ 102, 396
Gomphonema micropus Kützing   Plate 126 ⋯⋯⋯⋯⋯⋯⋯⋯⋯⋯⋯⋯⋯⋯⋯⋯⋯⋯ 103, 398
Gomphonema nipponicum Skvortsov   Plate 127 ⋯⋯⋯⋯⋯⋯⋯⋯⋯⋯⋯⋯⋯⋯⋯⋯ 104, 400
Gomphonema parvulum var. *lagenula* (Kützing) Frenguelli   Plate 128 ⋯⋯⋯⋯⋯ 105, 402
Gomphonema parvulum var. *neosaprophilum* H. Kobayasi ex K. Osada var. nov.
   Plates 129, 130 ⋯⋯⋯⋯⋯⋯⋯⋯⋯⋯⋯⋯⋯⋯⋯⋯⋯⋯⋯⋯⋯⋯⋯⋯⋯⋯⋯ 106, 404, 406
Gomphonema parvulum (Kützing) Kützing var. *parvulum*   Plates 131, 132
⋯⋯⋯⋯⋯⋯⋯⋯⋯⋯⋯⋯⋯⋯⋯⋯⋯⋯⋯⋯⋯⋯⋯⋯⋯⋯⋯⋯⋯⋯⋯⋯⋯⋯⋯⋯ 107, 408, 410
Gomphonema pseudoaugur Lange-Bertalot   Plate 133 ⋯⋯⋯⋯⋯⋯⋯⋯⋯⋯⋯⋯ 108, 412
Gomphonema truncatum Ehrenberg   Plate 134 ⋯⋯⋯⋯⋯⋯⋯⋯⋯⋯⋯⋯⋯⋯⋯⋯ 109, 414
Gomphonema yamatoensis H. Kobayasi ex K. Osada sp. nov.   Plate 135 ⋯⋯⋯⋯⋯ 110, 416
Achnanthes coarctata (Brébisson) Grunow   Plates 136, 137 ⋯⋯⋯⋯⋯⋯ 111, 418, 420
Achnanthes crenulata Grunow   Plates 138, 139 ⋯⋯⋯⋯⋯⋯⋯⋯⋯⋯⋯ 112, 422, 424
Achnanthes inflata (Kützing) Grunow   Plates 140, 141 ⋯⋯⋯⋯⋯⋯⋯⋯ 113, 426, 428
Achnanthes kuwaitensis Hendey   Plate 142 ⋯⋯⋯⋯⋯⋯⋯⋯⋯⋯⋯⋯⋯⋯⋯⋯⋯ 114, 430
Cocconeis diminuta Pantocsek   Plate 143 ⋯⋯⋯⋯⋯⋯⋯⋯⋯⋯⋯⋯⋯⋯⋯⋯⋯⋯ 115, 432
Cocconeis pediculus Ehrenberg   Plates 144, 145 ⋯⋯⋯⋯⋯⋯⋯⋯⋯⋯⋯ 116, 434, 436
Cocconeis placentula Ehrenberg var. *placentula*   Plates 146, 147 ⋯⋯⋯ 117, 438, 440
Cocconeis placentula var. *lineata* (Ehrenberg) Van Heurck   Plates 148, 149 ⋯⋯ 118, 442, 444
Cocconeis scutellum Ehrenberg   Plate 150 ⋯⋯⋯⋯⋯⋯⋯⋯⋯⋯⋯⋯⋯⋯⋯⋯⋯⋯ 119, 446
Cocconeis stauroneiformis (Rabenhorst) Okuno   Plate 151 ⋯⋯⋯⋯⋯⋯⋯⋯⋯ 120, 448
Achnanthidium convergens (H. Kobayasi) H. Kobayasi   Plate 152 ⋯⋯⋯⋯⋯⋯ 121, 450
Achnanthidium exiguum (Grunow) Czarnecki   Plate 153 ⋯⋯⋯⋯⋯⋯⋯⋯⋯⋯ 122, 452
Achnanthidium gracillimum (Meister) Mayama comb. nov.   Plate 154 ⋯⋯⋯⋯⋯ 123, 454

Achnanthidium japonicum (H. Kobayasi) H. Kobayasi   Plate 155 ············· 124, 456
Achnanthidium minutissimum (Kützing) Czarnecki   Plates 156, 157 ············· 125, 458, 460
Achnanthidium pusillum (Grunow) Czarnecki   Plate 158 ············· 126, 462
Achnanthidium pyrenaicum (Hustedt) H. Kobayasi   Plates 159, 160 ············· 127, 464, 466
Achnanthidium saplophilum (H. Kobayasi & Mayama) Round & Bukhtiyarova
   Plate 161 ············· 128, 468
Achnanthidium subhudsonis (Hustedt) H. Kobayasi comb. nov.   Plates 162, 163
   ············· 129, 470, 472
Lemnicola hungarica (Grunow) Round & Basson   Plates 164, 165 ············· 130, 474, 476
Planothidium frequentissimum (Lange-Bertalot) Lange-Bertalot   Plate 166 ············· 131, 478
Planothidium lanceolatum (Brébisson ex Kützing) Lange-Bertalot   Plate 167 ············· 132, 480
Planothidium septentrionale (Østrup) Round & Bukhtiyarova ex U. Rumrich *et al.*
   Plate 168 ············· 133, 482
Psammothidium helveticum (Hustedt) Bukhtiyarova & Round   Plate 169 ············· 134, 484
Psammothidium hustedtii (Krasske) Mayama   Plate 170 ············· 135, 486
Psammothidium marginulata (Grunow) Bukhtiyarova & Round   Plate 171 ············· 136, 488
Psammothidium montanum (Krasske) Mayama   Plate 172 ············· 137, 490
Psammothidium subatomoides (Hustedt) Bukhtiyarova & Round   Plate 173 ············· 138, 492
Nupela lapidosum (Krasske) Lange-Bertalot   Plate 174 ············· 139, 494
Diatomella balfouriana Greville   Plate 175 ············· 140, 496
Diploneis elliptica (Kützing) Cleve   Plate 176 ············· 141, 498
Diploneis ovalis (Hilse) Cleve   Plate 177 ············· 142, 500
Diploneis smithii (Brébisson ex W. Smith) Cleve   Plate 178 ············· 143, 502
Entomoneis japonica (Cleve) K. Osada   Plate 179 ············· 144, 504
Entomoneis paludosa (W. Smith) Reimer   Plate 180 ············· 145, 506

珪藻の殻構造と用語

　電子顕微鏡，特に走査型電子顕微鏡 (SEM) が多用されるようになってから，珪藻の殻構造についての知見は飛躍的に増大し，同時にそれらを言い表すための用語も急激に増加した．そのため1つの構造に複数の名前が付くというような混乱も起こり始め，国際的に用語を統一する必要性が生じた．そこで，キール（独）で行われた第3回国際珪藻シンポジウムで G. R. Hasle を委員長とし，G. A. Fryxell, I. V. Makarova, R. Ross, R. Simonsen, H. A. von Stosch の6人の委員が選ばれ，用語統一のための委員会が発足した．その後，R. M. Crawford, J. Gerloff, P. E. Hargraves, N. I. Hendey の各氏が加わった．またこれと平行して，珪藻の記相の標準化のための委員会も設立され，G. W. Andrews（委員長），D. R. Hasle, R. Simonsen, N. I. Strelnikowa の4人が選ばれた．これら2つの委員会による提案は，早くも翌年に出版された (Anonymous 1975) が，これでもなお足りない部分があるということで，次の第4回シンポジウムにおいて，R. Ross, E. J. Cox, N. I. Karayeva, D. G. Mann, T. B. B. Paddock, R. Simonsen, P. A. Sims からなる委員会が発足し，改訂と補充が行われた (Ross *et al.* 1979)．

　この改訂版の用語は，英語で書かれた本文に加え，巻末に英語，フランス語，ドイツ語，ラテン語の対照表が付加されている．もちろん，日本語はない．そこで日本語訳をどうするかが大きな問題である．本書では上述の改訂版の英語の用語を忠実に紹介することを基本にし，収録されているすべての用語を取り上げ，さらに，足りない部分を補う意味で，個々の研究者によって新たに提案された用語も加えた．日本語訳については，幸いなことに，過去の文献で使われた用語の混乱を整理するために日本植物学会の協力でつくられた文部省学術用語集（文部省・日本植物学会1990，以下"用語集"と省略）があるので，本書ではこれに準拠することを基本とした．前述のように，SEM 時代に入ってから区別して表現しなければならない類似の構造が爆発的に増え，用語集に収録されていない独特の構造も多く，それらを区別するにふさわしい用語を，他の用語との整合性を考え，重複や曖昧さを避けながら，できる限り慣用されてきた用語を用いることに配慮した．また，その用語を選んだ理由などについては「注」欄を設けて，そのつど解説した．説明図は，原図にこだわらず，できる限り適切な SEM 写真を用い，それぞれの用語の右側に配置し，必要に応じて線画を付加した．

I. 細胞のタイプ

1 栄養細胞 — vegetative cell —

有糸分裂によってできた細胞をいう．細胞壁は，被殻を構成する珪酸質要素（図12参照）と有機質要素からできている．

被殻の形態は，次のような用語を使用すると説明しやすい．

被殻の軸と相称面（図1，2）
- 長軸 (apical axis) (aa')
- 貫殻軸 (pervalvar axis) (bb')
- 短軸 (transapical axis) (cc')
- 殻断面 (valvar plane) (A)
- 縦断面 (apical plane) (B)
- 横断面 (transapical plane) (C)

注：長軸は頂軸，また短軸は切頂軸と記されることもある．

なお，被殻が左右相称でない場合は，貫殻軸と殻断面のみ区別できる．

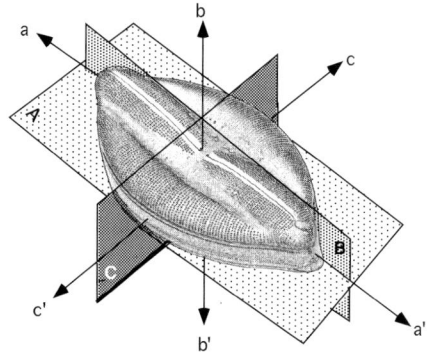

図1 羽状類被殻の軸と面

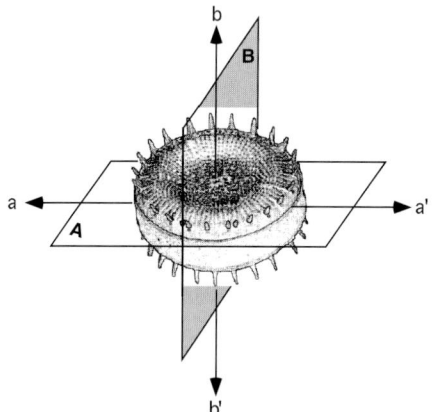

図2 中心類被殻の軸と面

2 増大胞子 — auxospore —

配偶子の有性的な融合または自家生殖の結果できる細胞，すなわち接合子．または純粋に栄養的な過程でつくられ，最大の大きさに膨潤した細胞をいう．その細胞壁は普通の細胞の細胞壁とは異なり，**鱗片** (scale) やペリゾニウム (perizonium) などの特殊化した構造をもつ．

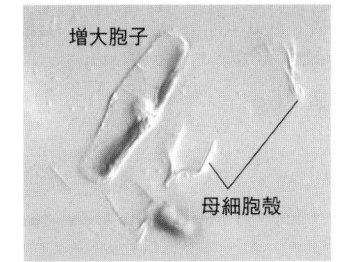

図 3　遊離増大胞子

1. 遊離増大胞子 — free auxospore —（図 3）
母細胞から離れてできる増大胞子をいう．
例：*Entomoneis decussata* (Grunow) K. Osada et H. Kobayasi

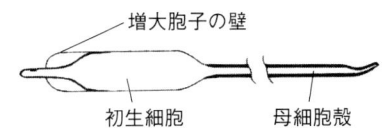

図 4　端生増大胞子

2. 端生増大胞子 — terminal auxospore —（図 4）
母細胞の末端にできる増大胞子をいう．
例：*Rhizosolenia alata* Brightw.（増大胞子の内側に，増大胞子の壁を貫いて第 1 初生殻がつくられている）

3. 側生増大胞子 — lateral auxospore —（図 5）
母細胞の殻帯部にでき，その貫殻軸（図 1 参照）が母細胞の貫殻軸と互いに直交または斜交する増大胞子をいう．
例：*Bacteriastrum hyalinum* Lauder

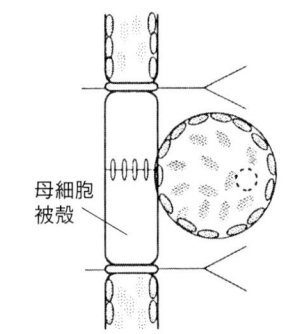

図 5　側生増大胞子

4. 介在増大胞子
　　　　― intercalary auxospore ―（図6）
　母細胞の上半被殻と下半被殻，またはその残骸が付着している増大胞子をいう（増大胞子が成長する間に，半殻帯は少なからず壊れる）．
例：*Melosira varians* C. Agardh

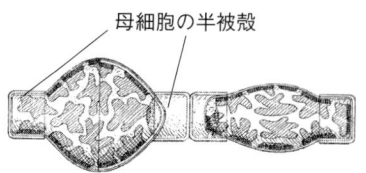

図6　介在増大胞子

5. 半介在増大胞子
　　　　― semi-intercalary auxospore ―（図7）
　片側で母細胞の半被殻と接し，反対側は離れているか，または隣の増大胞子と接している増大胞子をいう．
例：*Odontella regia* (M. Schultze) Simonsen

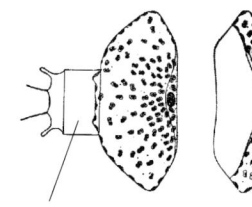

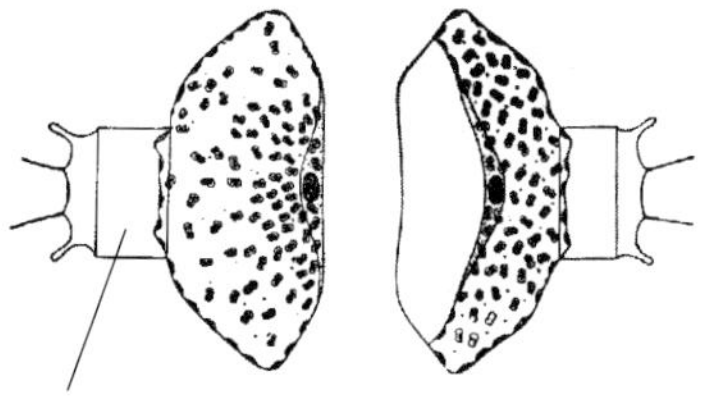

図7　半介在増大胞子

6. 初生細胞 ― initial cell ―（図8）
　増大胞子の中にできる殻をもった細胞をいう．通常この殻は普通の栄養細胞の殻とは形態的に異なっている．

3　休眠胞子 ― resting spore ―

　珪藻は土壌，岩石，コケなどの乾燥する基物上にも生育しているので，多くの種類は乾燥に耐えるしくみをもっていると考えられる．しかし，特殊な殻を母細胞の被殻内側につくったり，細胞壁が肥厚するなど，栄養細胞とは形態的に異なる殻ができる場合，これらを休眠胞子と呼ぶ．

　休眠胞子の知られている種類はごく限られており，羽状類では，*Eunotia soleirolli* (Kütz.) Rabenh. (von Stosch & Fecher 1979), *Craticula cuspidata* (Kütz.) D. G. Mann (=*Navicula cuspidata*) (Schmid 1979) などがある．また，中心類では形態的に栄養細胞とは大きく異なる場合も見られる．

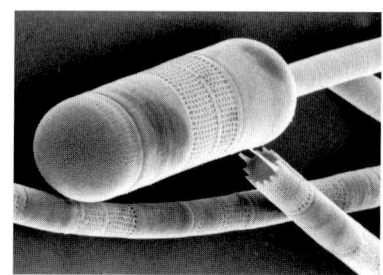

図8　*Aulacoseira* の初生細胞

たとえば，*Chaetoceros* 属では休眠胞子をつくる種類が多く，中でも化石種では母細胞の脆弱な殻が消失し休眠胞子殻だけが残ったため，*Goniothecium* 属，*Periptera* 属，*Ompharatheca* 属，*Xanthiopyxis* 属など，別属の種類に分類されたと考えられるものもある (Hargraves 1986).

休眠胞子は二分裂または不等分裂，細胞質分裂を伴わない有糸分裂によって，1つの細胞から4個，2個，または1個できるが，殻帯部を欠いている場合が多い．このとき，最初にできてくる殻を第1次殻 (primary valve)，次にできる殻を第2次殻 (secondary valve) と呼ぶ．

注：初生殻，後生殻(安達ら1982)という訳語もあるが，増大胞子内に最初につくられる初生殻 (initial valve) との重複を避けるため，第1次殻，第2次殻を使用した．

1. 外生休眠胞子
　　　　－ exogenous resting spore －（図9）
　母細胞の被殻によって包まれていない休眠胞子をいう．
例：*Detonula confervacea* (Cleve) Gran

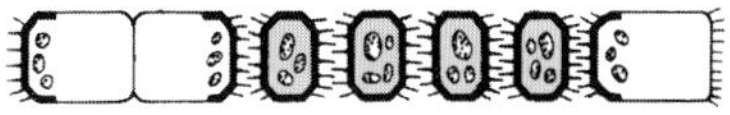

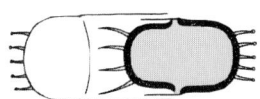

図9　外生休眠胞子

2. 半内生休眠胞子
　　　　－ semi-endogenous resting spore －（図10）
　一方の殻は母細胞の半被殻に包まれ，他方の殻は露出している休眠胞子をいう．
例：*Stephanopyxis turris* (Grev.) Ralfs

図10　半内生休眠胞子

3. 内生休眠胞子
　　　　－ endogenous resting spore －（図11）
　完全に母細胞の被殻で包まれている休眠胞子をいう．
例：*Chaetoceros compressus* Lauder

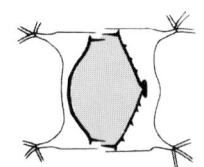

図11　内生休眠胞子

II. 細胞壁の珪酸質要素

1 被殻の基本的構成要素

1. 被殻 ― frustule ―（図 12）

被殻は，上半被殻 (epitheca) と下半被殻 (hypotheca) からなる．

注：上下の語は重力との関係で使うのが一般的である．珪藻の場合，上殻側がいつも上になっているとは限らないが，語源と"用語集"に従った．

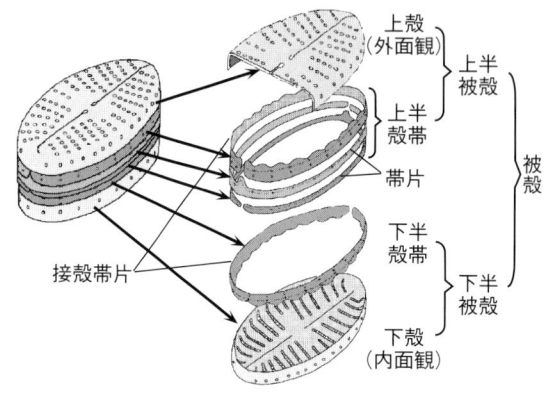

図 12　被殻の構成

　a. 上半被殻 ― epitheca ―（図 13）

上殻 (epivalve) と上半殻帯 (epicingulum) からなる．これらは親細胞から受け継いだものである．

　b. 下半被殻 ― hypotheca ―（図 13）

下殻 (hypovalve) と下半殻帯 (hypocingulum) からなる．これらは細胞分裂の結果，娘細胞につくられる．

注：下半殻帯のつくられ方については，なお不明な点が多いが，殻形成と同時に殻帯の全構成要素ができあがるわけではなく，次の分裂が起こる直前になってつくられる場合もある．上下を区別しない場合は，単に半被殻 (theca) の用語を用いる．

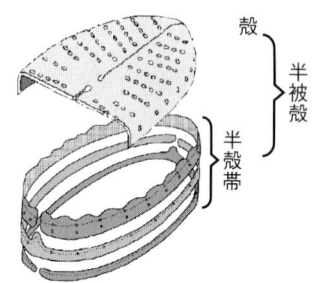

図 13　上(下)半被殻

　c. 殻面観 ― valve view ― と
　　　帯面観 ― girdle view ―（図 14）

光顕では細胞または被殻を正面（殻面）から見た状態を殻面観，側面（帯面）から見た状態を帯面観という．

2 殻 ― valve ―

被殻の構成要素の中で，両外に対置する平板状，または凸状の 2 枚の板のうちの 1 枚をいう（従来は *Rhizosolenia* 属のとんがり帽子状の殻を特別にカリプトラ (calyptra) と呼んでいたが，この語の使用はできるだけ避けたい）．

注："valve"は，殻とも蓋殻とも訳されてきたが，殻套，殻表面，殻面有基突起など，殻の字を冠した用語がしばしば出てくるので，字数を少なくするためにも殻という訳がよい．

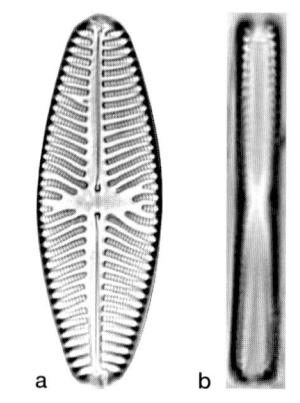

図 14　殻面観 (a) と帯面観 (b)

1. 殻の外形 — valve shapes —

いろいろな殻形のものが見られる．以下に代表的なものを示す．

　a．殻の全形—左右相称で両極が同形（同極）（図15）

　（1）円形 (circular)
　（2）楕円形 (elliptic)
　（3）狭楕円形 (narrow elliptic)
　（4）広皮針形 (lanceolate, wide)
　（5）狭皮針形 (lanceolate, narrow)
　（6）紡錘状皮針形 (lanceolate, fusiform)
　（7）ひし形 (rhombic)
　（8）長方形 (rectangular)
　（9）線形 (linear)
　(10) 中心部の膨れた線形
　　　 (linear with gibbous center)
　(11) 三波形 (triundulate)
　(12) S字形 (sigmoid)
　(13) S字状皮針形 (sigmoid lanceolate)
　(14) S字状ひし形 (sigmoid rhombic)
　(15) S字状線形 (sigmoid linear)
　(16) バイオリン形 (panduriform)
　(17) 弱狭さくバイオリン形
　　　 (panduriform, gently constricted)

注：高等植物ではササの葉のような形を皮針形というが，珪藻では凸レンズの断面のような形を皮針形と呼ぶ．

　b．殻の全形—左右相称で両極が異形（異極）および左右不相称で同極または異極（図16）

　(18) へら形 (spatulate)
　(19) 卵形 (ovate)
　(20) こん棒形 (clavate)
　(21) 二葉形 (bilobate)
　(22) ほこ形 (hastate)
　(23) 半皮針形 (semilanceolate)
　(24) 半円形 (semicircular)
　(25) 三日月形 (crescentic)
　(26) 弓形 (arcuate)

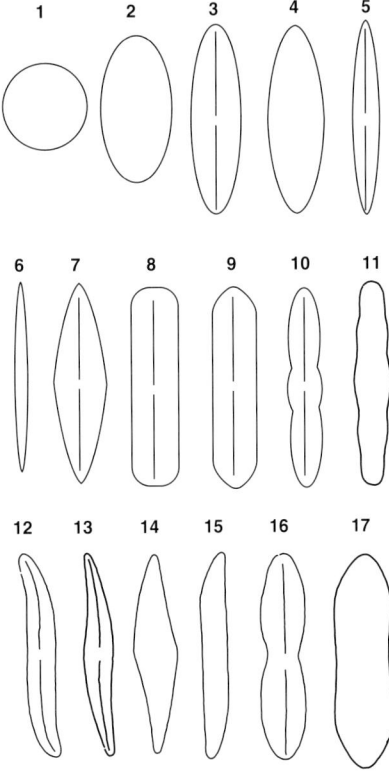

図15　左右相称で同極

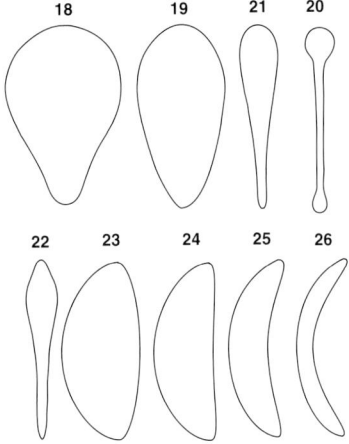

図16-1　左右相称で異極および左右不相称で同極

(27) 耳形 (auricular)
(28) じん臓形 (reniform)
(29) 台形 (trapezoidal)

c. 殻の先端の形（図17）
（1）広円，鈍形 (broadly rounded, obtuse)
（2）くさび形 (cuneate)
（3）広くちばし形 (broadly rostrate)
（4）くちばし形 (rostrate)
（5）へら形 (spatulate)
（6）頭状 (capitate)
（7）弱頭状 (subcapitate)
（8）微突頭で伸長 (apiculate and produced)
（9）微突頭 (apiculate)
(10) 鋭形 (acute)
(11) 鋭先形 (acuminate)
(12) S字状くさび形 (sigmoidly cuneate)
(13) 頭状 (capitate)
(14) 狭くちばし形 (narrowly rostrate)

2. 殻套 － valve mantle －（図18）
殻の縁にあって，傾斜しているか，場合によっては構造的に殻面と区別のつく部分をいう．

3. 殻面 － valve face －（図18）
殻套で囲まれた殻の部分をいう．

4. 殻肩 － valve shoulder －（図18）
殻面と殻套が明瞭に区別できる角度をもつ場合，この境界に相当する部分をいう．一般に，殻肩を言い表すのに殻面殻套接合部 (valve face and mantle junction もしくは valve face and mantle juncture) という長い用語を使っているが，殻肩が明瞭に区別できる種類も多い．また，この部分には針，縁辺稜，無紋域，唇状突起の開口など各種の構造物も見られ，この部分を端的に表現できる用語があるとたいへん便利と思われる．既述の用語委員会の提案にはないが，新たにこの用語を提案したい．

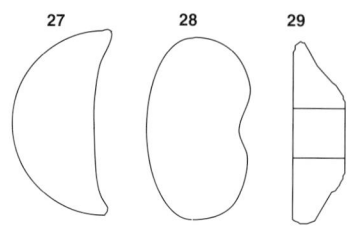

図16-2 左右不相称で異極

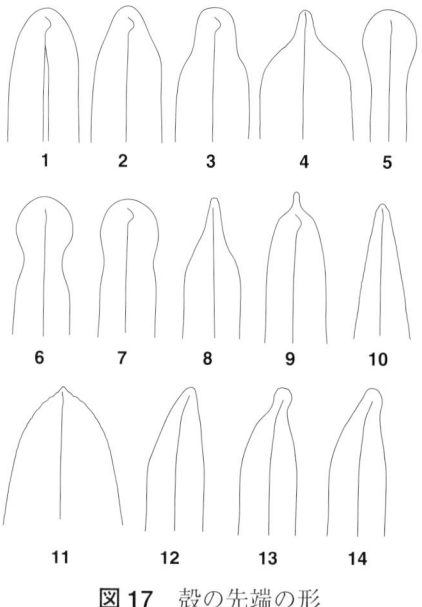

図17 殻の先端の形

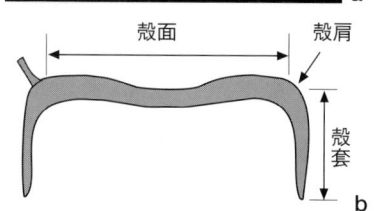

図18 殻面，殻套と殻肩
（bは模式図）

5. 隆起 — elevation —（図 19）

殻套より横にははみ出さない殻壁の隆起した部分．そこには特別の構造が見られる場合もあるが，大抵は殻と同じ構造である．

例：*Biddulphia biddulphiana* (Smith) Boyer

a. 角 — horn —（図 20）

細くて長い突起．隆起の特殊な例といえる．

例：*Hemiaulus polycystinorum* Ehrenb.

図 19　隆起

6. 剛毛 — seta —（図 21）

殻の縁から外側に突出する中空の突起．特に群体の端末細胞にある剛毛を端末剛毛 (terminal seta) という．群体を構成する細胞間の間隙は，開口 (aperture)，または窓 (window) と呼ばれてきた．

例：*Chaetoceros atlanticus* Cleve

7. 無紋域 — hyaline area —

胞紋によって貫通されていない珪酸質の区域をいう．羽状目珪藻では以下のようなタイプが見られる．

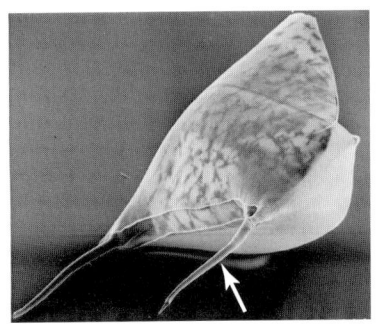

図 20　角

a. 軸域 — axial area —（図 22）

長軸に沿って存在する無紋域をいう．無縦溝珪藻の軸域を偽縦溝 (pseudoraphe) と呼ぶのは適当ではない．

b. 中心域 — central area —（図 22）

軸域の中間にある広がり，または目立った無紋域をいう．肥厚した中心節（"15. 縦溝" 参照）をもつ珪藻では，中心節を囲む肥厚しない無紋域が中心域である．横帯 (fascia) は，殻を横によぎる中心域をいう．横帯に似ているが構造的に異なる十字節 (stauros) については "15. 縦溝" 参照．

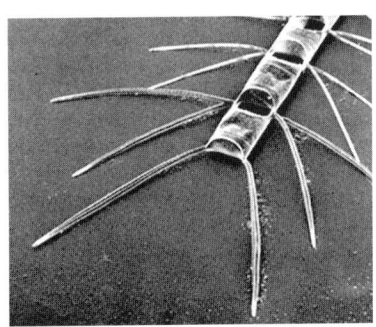

図 21　剛毛

c. 極域 — terminal area —（図 22）

軸域の殻頂での広がり，または極節を囲む無紋域をいう．

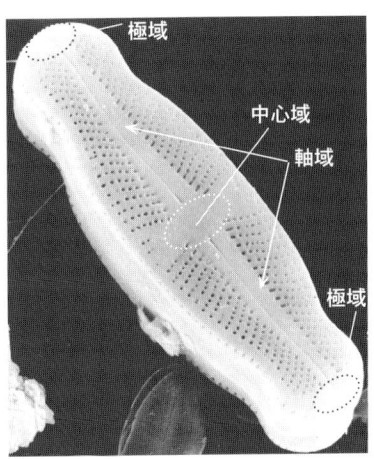

図 22　軸域，中心域と極域

(ⅰ) 洞 — cavity —(図 53 極節と洞参照)
極域の殻内面にある小さな丸いくぼみをいう．
例：*Sellaphora*

d. 側域 — lateral area —(図 23)
広がった中心域が長軸方向へ伸び，条線を分断している無紋域をいう．*Lyrella* 属は側域をもつことを1つの特徴としている．

e. 馬蹄域
— horseshoe-shaped area —(図 24)
中心域の片側または両側にあり，殻の内面に肥厚した周縁部をもつ無紋域 (hyaline area) をいう．ときにはずきん状 (hooded) の構造になる．
例：*Planothidium lanceolatum* (Bréb.) Bukhtiyarova & Round

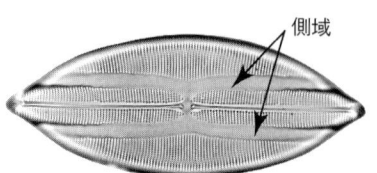
図 23 側域

8. 竜骨 — keel —(図 25)
縦溝をともなって突出した殻の一部分で，この縦溝を含む突出部全体をいう．
例：*Entomoneis, Nitzschia, Surirella, Plagiotropis*

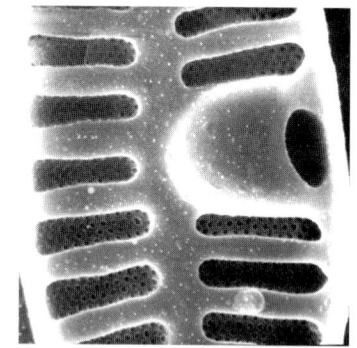
図 24 馬蹄域(ずきん状)

a. 翼 — wing —(図 26)
竜骨の一種で，*Surirella* 属や *Entomoneis* 属に見られるよく発達した竜骨をいう．翼の頂上には縦溝管と呼ばれる管が通り，この管と細胞内部は，翼管 (alar canal) と呼ばれる管で連絡している．翼管と翼管の間は翼窓 (fenestra) と呼ばれる窓のような構造があり，翼窓は単なる孔の場合もあれば，多数の平行に走る翼窓棒 (fenestral bar) がある場合もある．
例：*Surirella robusta* Ehrenb. (Paddock & Sims 1977)
注：翼の構造を言葉で表現するのは非常に難しいが，普通の珪藻の殻面と殻套が縦溝管の下でくっつき，そこが突出して翼ができたと考えるとわかりやすい．

図 25 竜骨

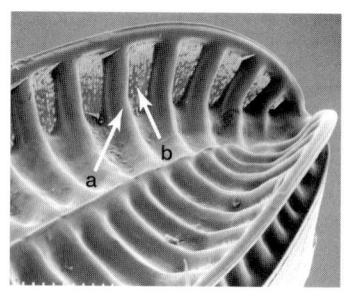
図 26 翼(a 翼管，b 翼窓)

9. 肋 — costa —（図 27）

殻の一部が伸長した中実の肥厚部をいう．
特殊なものとしては次のようなものがある．

a. 偽隔壁 — pseudoseptum —（図 28）

殻内面にある板状の肋をいう．特に殻の頂端から殻面に平行に伸びるものについて使われてきた．
例：*Biddulphia biddulphiana* (Sm.) C. S. Boyer, *Stauroneis anceps* W. Sm., *Eunotogramma weissei* Ehrenb.

b. 軸肋 — axial costa —（図 29）

殻内面で，縦溝に沿って存在する肋をいう．
例：*Frustulia rhomboides* (Ehrenb.) De Toni, *Gyrosigma attenuatum* (Kütz.) Rabenh.

従来，中心節の角 (horns of the central nodule) と呼ばれた構造は軸肋そのもので，軸肋の発達が弱い場合，それはわずかに軸域の細い線で分けられている (Sims & Paddock 1979)．

c. 縁辺稜 — marginal ridge —（図 30）

殻肩（殻面と殻套の間）にある縁取りをいう．縁辺稜には山の尾根のようにひと続きのものもあれば，図 30 のように突起が断続的に並んでできているものもある．
例：*Lithodesmium undulatum* Ehrenb., *Ditylum brightwellii* (T. West) Grunow, *Eunotogramma weissei* Ehrenb., *Rattrayella oamarensis* (Grunow) De Toni, *Biddulphia tuomeyi* Bailey

図 27　肋

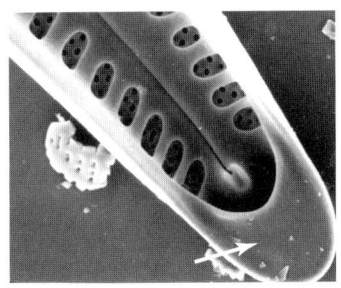

図 28　偽隔壁

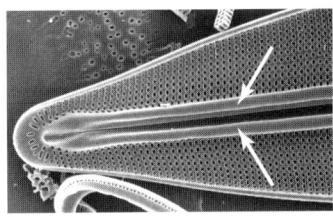

図 29　軸肋

図 30　縁辺稜

d. えり ― collar ―(図31)
殻の外側に突出した円形で膜状の肋をいう．
例： *Melosira nummuloides* (Dillwyn) C. Agardh

e. とさか ― crest ―
高くて薄い肋をいう．この肋はしばしば鋸歯状，または房状になる．
例： *Odontropis cristata* Grunow (Hustedt 1930, fig. 511)

f. 耳 ― otarium ―(図32)
Rhizosolenia 属に見られる構造で，1対の短い膜状の肋が唇状突起の基部，または基部近くに向き合って存在する．以前は翼 (wing) と呼ばれていた．
例： *Rhizosolenia styliformis* Brightw. (Hasle 1975, figs 3, 9)

g. 横輪 ― ring-costa ―(図33)
Aulacoseira 属に見られる殻套縁の内側に沿って張り出した肋で，ときに板状になる．

10. 天幕 ― canopy ―(図34)
軸域またはその付近から殻縁の方に向かってテントのように張り出した膜状体をいう．
例： *Sellaphora alastos* (M. H. Hohn & Hellerman) Lange-Bert. & Metzeltin

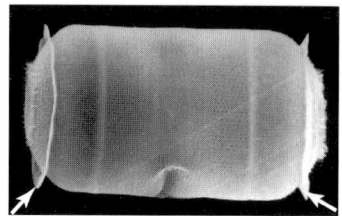

図31　えり

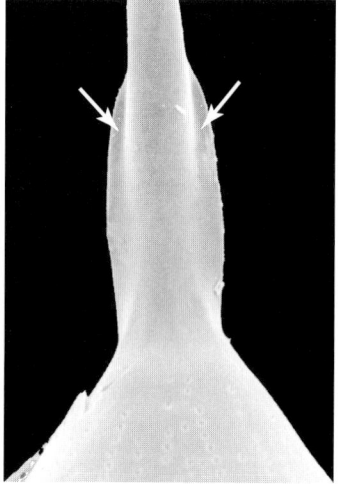

図32　耳

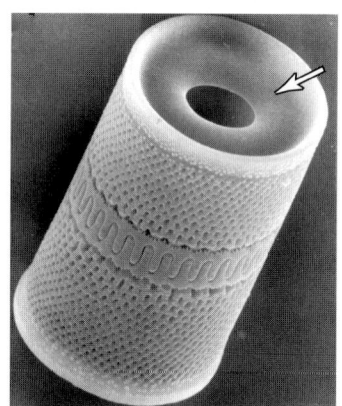

図33　横輪

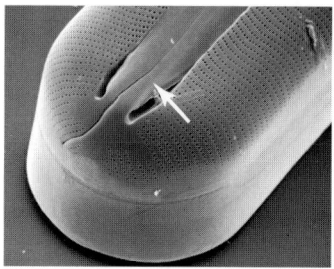

図34　天幕

11. 陥入 — fold —
殻壁の内側への陥入をいう.
例：*Hemiaulus capitatus* Grev.
注：用語統一委員会では，*Melosira ambigua* (Grunow) O. Müller (=*Aulacoseira ambigua* (Grunow) Simonsen) の輪溝 (sulcus) と呼ばれてきた構造を，陥入の特殊な例として挙げているが，それは偽隔壁の基部にできた中空の管状構造であり (小林・野沢 1981, figs 20, 21)，陥入には当たらない.

12. 遊離点 — stigma —（図35）
縦溝をもつ羽状類珪藻の中心域の珪酸基底層にある孔をいう．孔の外側の開口に覆いはなく，内側の開口はいろいろな構造物で閉塞されている.
例：ひび割れ状の膜状構造物で閉塞：*Cymbella cistula* (Ehrenb.) Kirchn., 弁状構造物で閉塞：*Luticola goeppertiana* (Bleisch) D. G. Mann (=*Navicula goeppertiana* (Bleisch) H. L. Sm.) (Mayama & H. Kobayasi 1986)

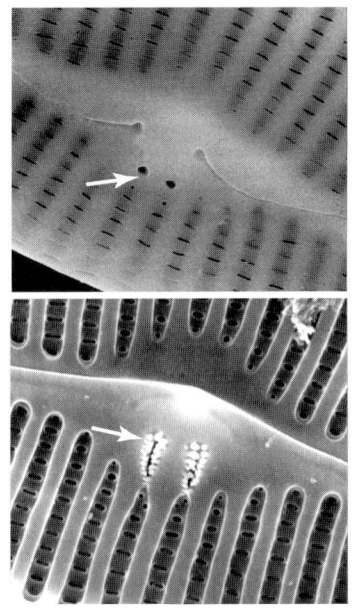

図35 遊離点（上；外面，下；内面）

13. 頂端と殻縁の区画域
 — apical and marginal field —
頂端と殻縁の部分にあって，構造的に殻の残りの部分と異なる区画をいう．次のようなタイプがある．

 a. 偽眼域 — pseudocellus —（図36）
殻の主たる部分の胞紋より小さい胞紋をもつことで区画された部分をいう．
例：*Biddulphia, Trigonium*

 b. 眼域 — ocellus —（図37）
一般に肥厚した模様のない縁 (rim) をもち，密につまった穴，すなわち小孔 (porellus) によって区画された部分をいう．
例：*Odontella edwardsii* (Febiger) Grunow, *Auliscus caelatus* Bailey, *Striatella unipunctata* (Lyngb.) C. Agardh
注：羽状類珪藻では，小孔は一般に縦に並ぶが，中心類珪藻では，それらは放射状，求心状または不規則に配列する．

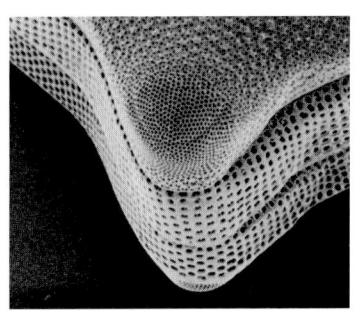

図36 偽眼域

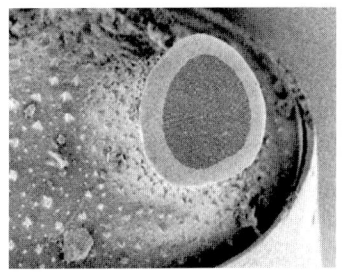

図37 眼域

c. 殻套眼域 — ocellulimbus —（図 38）
殻頂部の殻套にある網目状の小孔域で，全体にくぼんだ位置にある．
例：*Tabularia fasciculata* (C. Agardh) D. M. Williams & Round (=*Synedra tabulata* (C. Agardh) Kütz.)

d. 殻端小孔域 — apical pore field —（図 39）
殻端部の殻面から殻套にある網目状の小孔域で，粘液の分泌に関係している．
例：*Gomphonema truncatum* Ehrenb., *Cymbella tumida* (Brèb.) Van Heurck など

e. 偽節 — pseudonodulus —（図 40）
殻面の周辺部にあり，通常 1 殻に 1 個である．ときに胞紋状 (areolate) (*Actinocyclus normanii* (W. Greg.) Hust., 有蓋状 (operculate) (*A. octonarius* Ehrenb.), 窓開き状 (luminate) である (*Roperia tessellata* (Roper) Grunow).

14. 突起 — process —
均一に珪酸化した壁をもつ突出物をいう．次のようなタイプがある．

a. 唇状突起 — labiate process
またはrimoportula —（図 41）
殻を貫通する管または開口で，殻内面に偏平化した管，またはしばしば 2 枚の唇 (lips) に囲まれたスリットをもつ．
特殊なタイプとしては
（i）複唇状突起 — bilabiate process —
背中合わせの位置に 2 つの裂口 (slit-like opening) をもつもの．
例：*Streptotheca tamesis* Shrubsole (von Stosch 1977, text-fig. 2)
（ii）巻込突起 — periplekton —
殻の外側の部分が頂上で二叉になり，2 本の腕は隣の同様の突起の幹を留め金のようにつかんでいる．
例：*Rutilaria radiata* E. Grove & G. Sturt
注：縦溝 (raphe)（"15. 縦溝" 参照）は進化的には唇状突起 (labiate process) に由来すると考えられている．

b. 閉塞突起 — occluded process —（図 42）
殻の外側に突き出た，中空で先端の閉じた突起をいう．
例：*Thalassiosira angstii* (Gran) I. V. Makarova

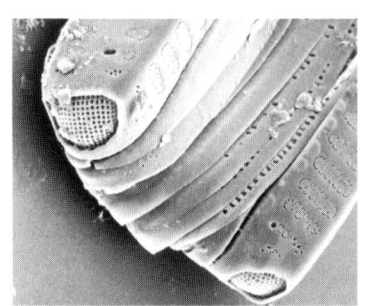

図 38　殻套眼域

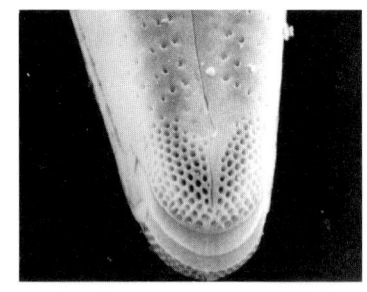

図 39　殻端小孔域

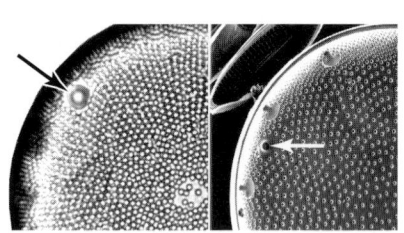

図 40　偽節

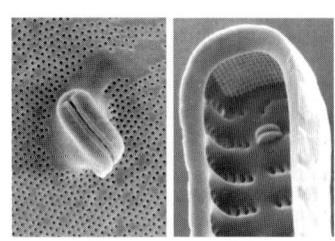

図 41　唇状突起

図 42　閉塞突起

c. 有基突起 — strutted process
　　　　　または fultoporutula —（図 43）
殻を貫通する 2〜5 個の箱 (chamber), または孔 (satellite pore = 付随孔) で囲まれ, 内部がアーチ状の支持物で分断された, 殻を貫通する管をいう. しばしば外部へ有機物質の糸を分泌する.
例：Thalassiosiraceae

d. 針 — spine —
殻の表面から突出する閉塞した中空, または中実の構造物.
例：*Thalassiosira eccentrica* (Ehrenb.) Cleve, *Corethron criophilum* Castrac.
　（i）　小針 — spinule —（図 44）
非常に小さい針.
例：*Diatoma mesodon* Kütz.
　（ii）　顆粒 — granule —（図 45）
殻表面の小さくて丸い突起.
例：*Melosira varians* C. Agardh.
　（iii）　連結針 — linking spine —（図 46）
被殻を鎖状に連結する互いに組み合った針.
例：*Aulacoseira italica* (Ehrenb.) R. Ross, *Staurosira construens* Ehrenb.

15. 縦溝 — raphe —（図 47）
殻壁を貫通する長く伸びた裂け目, または長軸方向に対になって存在する裂け目をいう. 対になるときはそれぞれを, 縦溝枝 (raphe branch) と呼ぶ. 縦溝が殻の中央を走る種類では, 2 つの縦溝枝の中間を中心節 (central nodule) と呼ぶ. また, 縦溝が殻の縁辺部を取り巻いて走る場合でも, 2 本の縦溝枝の中間は節 (nodule) である.
例：*Surirella*
中心節はいつも存在するとは限らない. たとえば Eunotiaceae にはない.
中心節が横に拡張し, ほとんどまたは完全に殻縁に達したものを十字節 (stauros) と呼ぶ. もし, 十字節が細くて厚く, また 1 本以上の細くて厚い部分が殻縁に達しているようなときは, 小梁十字節 (tigillate stauros) といい, その 1 本を小梁 (tigillum) と呼ぶ.
例：*Carpartogramma crucicula* (Grunow ex Cleve) R. Ross (Ross 1963, figs 8-11)

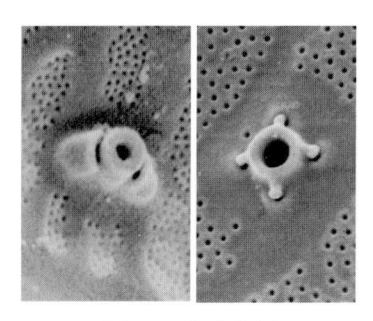

図 43　有基突起

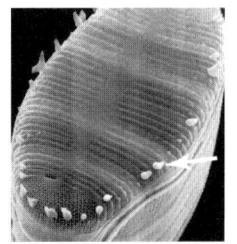

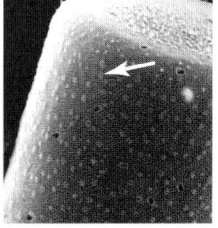

図 44　小針　　図 45　顆粒

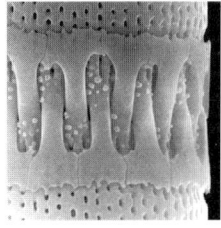

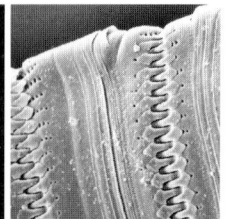

図 46　連結針

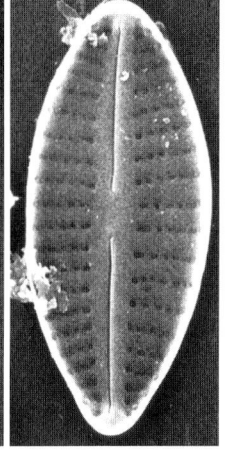

図 47　縦溝（右；内面観, 左；外面観）

a. 中心孔 — central pore —（図 48）
縦溝枝の中心側末端で孔状に拡大した部分.

b. 中心裂溝 — central fissure —
縦溝の裂け目が殻を貫通しないで，単なる裂け目として中心節の上を，またはそれをよぎって伸びるものをいう.
（ⅰ） 外中心裂溝
　　　　— external central fissure —（図 49）
殻外面にある中心裂溝.
（ⅱ） 内中心裂溝
　　　　— internal central fissure —（図 50）
殻内面にある中心裂溝.

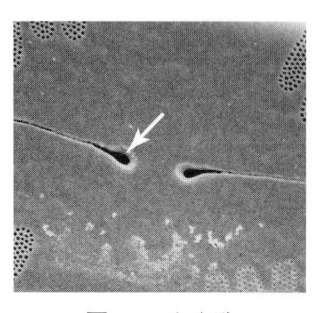

図 48　中心孔

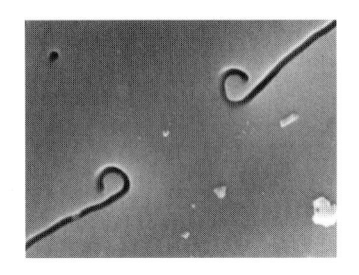

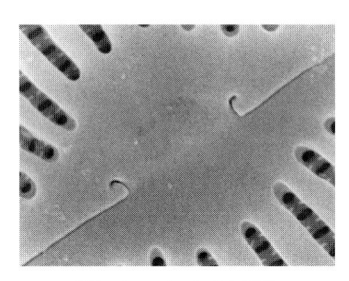

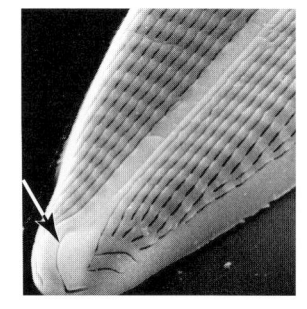

図 49　外中心裂溝　　図 50　内中心裂溝　　図 51　極裂

c. 極裂 — terminal fissure —（図 51）
殻端にある縦溝の裂け目で，殻を貫通していない部分をいう．一般に，極裂は殻の外側だけの溝である.

d. 蝸牛舌 — helictoglossa —（図 52）
殻内面の縦溝の終点部分にある突出した唇状構造をいう．
注：この用語はロート (infundibulum) に代わるものである．ロートは以前 *Surirella fastuosa* Ehrenb. やその類縁種のまったく異なる構造に使われていた．

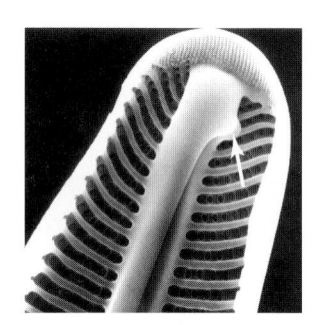

図 52　蝸牛舌（矢印）と極節

e. 極節 — terminal nodule —（図 53）
縦溝の極側末端にある珪酸基質層の肥厚した部分をいう．
例：*Sellaphora bacillum* (Ehrenb.) Mereschkowsky (Sims & Paddock 1979, fig. 43)

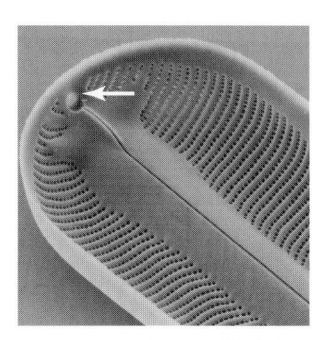

図 53　極節と洞（矢印）

f. 縦溝裂 － raphe fissure －(図 54)

縦溝の割れ目は斜行している (oblique) か，一部ないし全長にわたってひだ状 (plicate) になっており，切断面は横向きのV形 (<) をしている．割れ目が，斜行またはごくわずかにひだ状のときは，光顕での見え方から糸状縦溝 (filiform raphe) と呼ばれる．ひだ状縦溝 (plicate raphe) では，折れ目の上下半分ずつの割れ目が，それぞれ内裂溝 (inner fissure) および外裂溝 (outer fissure) と呼ばれる．しかし，これらの用語は単に殻の内外両表面での縦溝の割れ目にも使われる．したがって文章に書くときは，それらの語がどちらの場合を意味しているかがわかるように注意しなければならない．

Pinnularia 属 Complexae 節 の縦溝は複合 (complex) と呼ばれる．*Pinnularia* 属では，縦溝の2本の縦溝枝は中心節と極節の近くでは斜行するが，その中間部分ではひだ状になる．複合縦溝 (complex raphe) では，ひだ状部分の外裂溝は明瞭に殻表面に対して傾斜するが，複合でないものは，外裂溝は殻表面に対してほとんど垂直である．

g. 縦溝管 － raphe canal －(図 55)

縦溝の下にある管状の空間で，殻の内部とある程度隔離された部分をいう．このような構造を伴う縦溝を，特に管状縦溝 (canal raphe) という．

h. 小骨 － fibula －(図 56 f)

縦溝を挟んで両側の殻壁をつなぐ珪酸質の柱をいう．これは，1本の間条線 (interstria) が伸長したもの，あるいは2本またはそれ以上の間条線が融合して伸長したものである．

例：*Nitzschia, Bacillaria, Rhopalodia, Entomoneis, Surirella* など

i. 門口 － portula －(図 56 p)

小骨の間の開口，または翼管 (alar canal) の内側と外側の開口をいう．

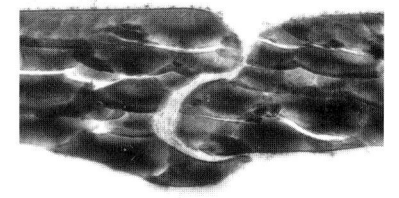

図 54 縦溝裂(TEM 横断面)

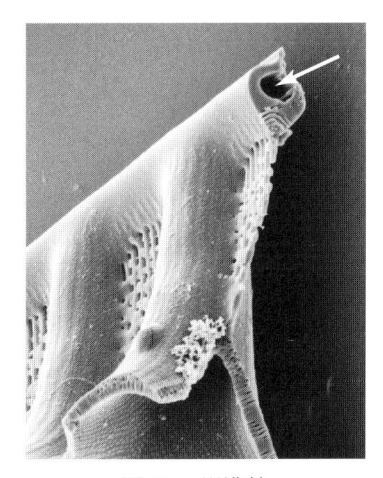

図 55 縦溝管

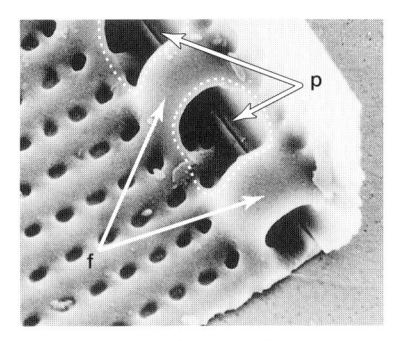

図 56 小骨 (f) と門口 (p)

16. 縦走管 — longitudinal canal —(図57)
しばしば中心部で中断されるが，ほぼ全殻長にわたって走る殻内部の管状の空間をいう．
例： *Diploneis, Neidium* (Sims & Paddock 1979, figs 3, 24 を参照)

17. 柵板 — craticula —(図58)
羽状類の特定の種類の殻の中につくられる頑丈なつくりの珪酸質の内生殻 (internal valve).
例： *Craticula cuspidata* (Kütz.) D. G. Mann

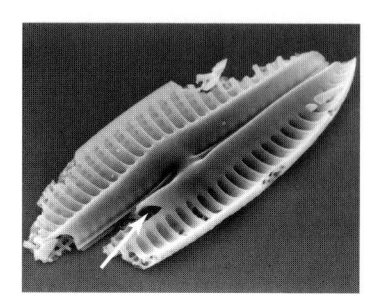

図57 縦走管

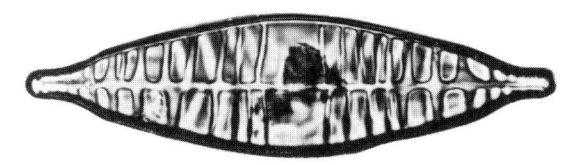

図58 柵板

3 殻帯 — girdle —(図12)
上殻と下殻の間の部分をいう．これは上半殻帯と下半殻帯に分けられる．

1. 半殻帯 — cingulum —(図13)
殻帯のうち，一方の殻と結合している1つまたは一連の帯片をいう．すなわち半被殻から殻を除いた部分．

2. 帯片 — band または segment —(図12)
殻帯の構成単位をいう．開放型 (open または sprit) と閉鎖型 (closed または complete) がある．

   a. 中間帯片 — copula —(図12)
   半殻帯のうちで，殻により近い方の構成単位をいい，より殻から離れた連結帯片とは構造的に異なっている．特に，殻に隣り合う帯片を接殻帯片 (valvocopula) という．

   b. 連結帯片 — pleura —(図12)
   半殻帯のうちで，中間帯片より先端部にある帯片をいう．中間帯片がない場合は，そこにあるすべての帯片を連結帯片と呼ぶ．
注：中間帯片と連結帯片は，紋様の有無や違いで明瞭に区別できる場合もあるが，それらが区別できない場合は，単に帯片として記載される．

c. 帯片中脈 － pars media －（図 59 a）
　帯片の長軸方向に走る中央部の肋．肋に沿って胞紋列を伴うこともある．

　d. 帯片内接部 － pars interior －（図 59 b）
　帯片の一部で，殻または他の帯片と重なる部分．

　e. 帯片表出部 － pars exterior －（図 59 c）
　帯片の一部で，殻または他の帯片と重ならず，露出した部分．

　f. 小舌 － ligula －（図 60）
　開放型帯片の帯片内接部側にある小さな突出部．帯片表出部側にあるものは副小舌 (antiligula) という．

　g. 隔壁 － septum －（図 61）
　帯片 (band) にある，殻面と平行な方向に伸長した珪酸質の薄板 (sheet)，ないしは稜 (ridge) をいう．
　例：*Tabellaria flocculosa* (Roth) Kütz., *Diatomella balfouriana* Grev.

3. 縫合 － suture －（図 62）
　殻と半殻帯の接合部位または半殻帯の各構成単位間の接合部位をいう．

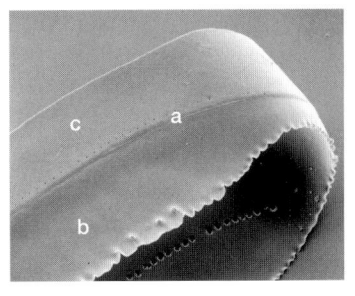

図 59　帯片の各部（a 帯片中脈，b 帯片内接部，c 帯片表出部）

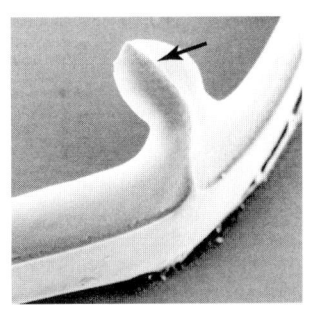

図 60　小舌

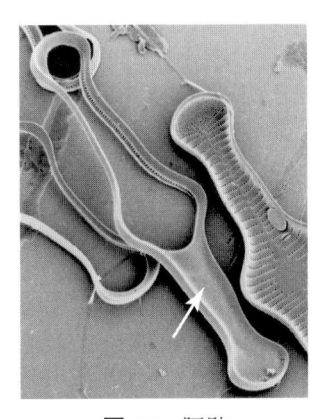

図 61　隔壁

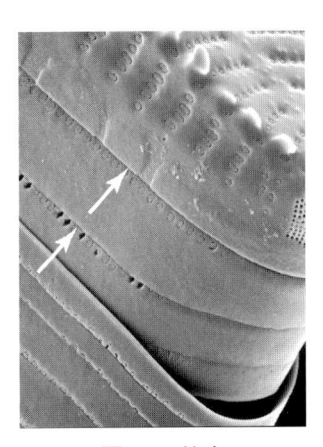

図 62　縫合

4. 区画環 ― partectal ring ―（図 63）

Mastogloia 属の特殊化した帯片をいう．2個またはそれ以上の区画 (partecta) をもつ．

a. 区画 ― partectum ―（図 63 矢印）

Mastogloia 属の特殊化した帯片にある1個の区画をいう．この区画は古くは箱 (chamber) または小箱 (loculus) と呼ばれてきたが，これらは胞紋構造に使われている．

b. 区画導管 ― partectal duct ―（図 64）

区画 (partectum) から殻外へつながる管をいう．

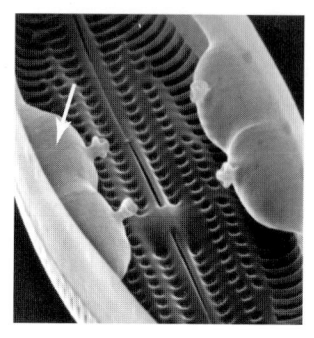

図 63　区画環と区画

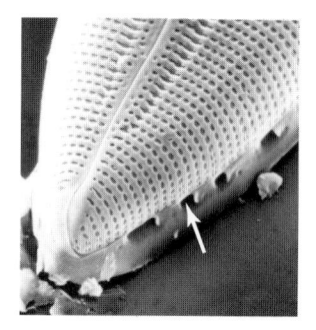

図 64　区画導管の外部開口

III. 珪酸質細胞壁の微細構造

1 珪酸基底層 — basal silliceous layer —
被殻のあらゆる構成要素の基礎的構造となる層をいう．

2 条線 — stria —
胞紋 (areola) または長胞 (alveolus) の列，または列の一部分ではない1個の長胞をいう．

1．中心珪藻 (centric diatom) の条線配列

a．射出 — radial —（図 65）
条線が中心から殻縁に向かって配列する状態をいう．
例：*Coscinodiscus nitidus* W. Greg., *C. oculusiridis* Ehrenb.
注：射出 (radial) と放射状 (radiate) は混同してはならない．

b．束出 — fasciculate —（図 66）
射出条線 (radial stria) に挟まれた部分で，束になった条線が互いに平行に走る場合，これを束出 (fasciculate) するといい，この束を条線束 (fascicle) という．
例：*Coscinodiscus rothii* (Ehrenb.) Grunow, *C. curvatulus* Grunow

c．正接 — tangential —（図 67）
条線が直線かまたは曲線で，射出しない状態をいう．
例：*Thalassiosira leptopus* (Grunow) Hasle & G. A. Fryxell (=*Coscinodiscus lineatus* Ehrenb., *Thalassiosira eccentrica* (Ehrenb.) Cleve)

2．羽状珪藻 (pennate diatom) の条線配列

a．平行 — parallel —（図 68）
条線が殻の中央線 (median line)，または縦溝に対して直角方向に走る状態をいう．
例：*Navicula plicata* Donkin, *Pleurosigma distinguendum* Hust.

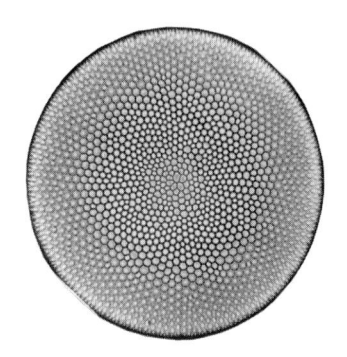

図 65 射出

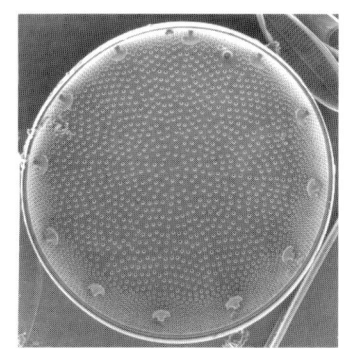

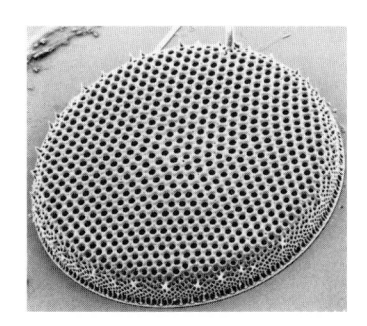

図 66 束出

図 67 正接

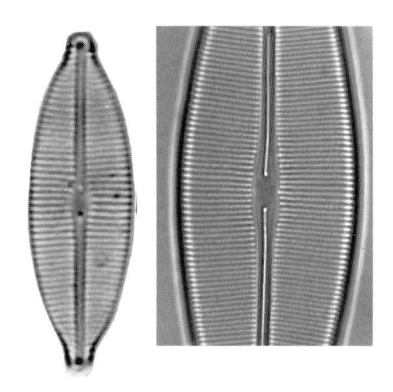

図 68 平行

b．放射状 － radiate －（図 69）
条線が殻縁 (valve margin) から殻の中心部に向かって走る状態をいう.
例：*Pinnularia major* (Kütz.) Rabenh. の中央の部分.

c．収れん － convergent －（図 70）
条線が殻縁から, 殻の先端に向かって走る状態をいう.
例：*Pinnularia major* (Kütz.) Rabenh. の殻端の部分.

3．ボアグの欠落 － Voigt fault －（図 71）
軸域に接する片側での条線配列の一部欠落をいう. 欠落は軸域に対し同じ側で, 殻の中心からほぼ同じ距離のところにできる (Voigt 1956 を見よ).

4．間条線 － interstria －（図 72）
隣り合う 2 本の条線の間の無紋の線をいう. 多くの羽状珪藻では, 条線は殻の内面で珪酸基底層の中に落ち込み, ときには外面も落ち込むこともある. このような場合, 間条線は肋 (costa：図 27 参照) と呼ばれるが, これらのうち特に珪酸基底層が肥厚している場合のみ肋と呼ぶべきである.

3　胞紋 － areola －
珪酸基底層を貫通し, 通常師板 (velum：" **4** 師板" 参照) または師皮 (rica：" **5** 師皮" 参照), またはその両方でふさがれている規則的に連続した穿孔をいう. 小形の胞紋はしばしば点紋 (puncta) と呼ばれることもある. その後 Cox & Ross (1981) は, 条線と条線の間の基底層に**横枝** (virga), 条線と構成する胞紋と胞紋の間の構造に**縦枝** (vimen) という用語を提唱している.

1．孔状胞紋
　　　－ poroid areola または poroid －（図 73）
殻の一方の表面で目立ったくびれのない胞紋.
例：*Diatoma*

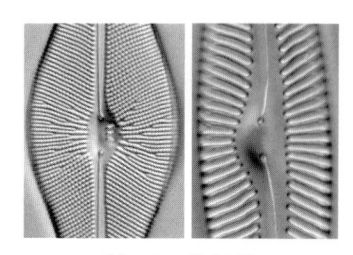

図 69　放射状

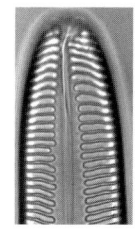

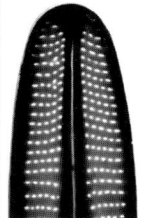

図 70　収れん

図 71　ボアグの欠落

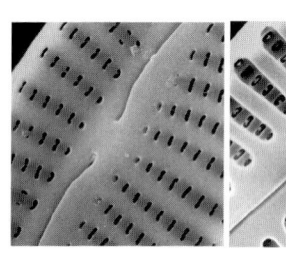

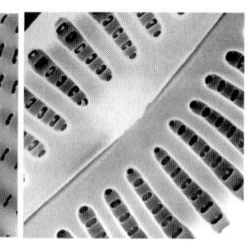

図 72　間条線
（左；外面観, 右；内面観）

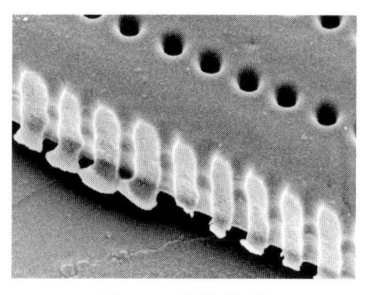

図 73　孔状胞紋

2. 小箱胞紋 — loculate areola
          または loculus —(図74)

殻の一方の表面で明瞭なくびれをもち，反対側が師板 (velum) または師皮 (rica)，またはその両方で閉ざされている胞紋．師板 (velum) または師皮 (rica) の反対側にあり，くびれによってつくられた開口を箱口 (foramen) と呼ぶ．

例：*Coscinodiscus oculusiridis* Ehrenb.

3. 偽小箱胞紋 — pseudoloculus —(図75)

交差ないしは網状に配列した肋が珪酸基底層の外側に形成されることによってつくられた箱をいう．

例：*Triceratium favus* Ehrenb.(Ross & Sims 1971, pl. 5, figs 4, 7 を見よ)，*Biddulphia reticulata* Roper (Ross & Sims 1972, fig. 7 を見よ)

4. 長胞 — alveolus —(図76)

殻の中軸，または中心から殻縁に向かって走る細長い1つの箱，またはいくつかの細長い箱の一連の並びをいう．殻内面では大きな開口をもち，その外側では胞紋層をもつ．

例：*Pinnularia viridis* (Nitzsch) Ehrenb. (Schrader 1973, pl. 6, figs 5-7, 10)，*Cyclotella* (Round 1970, figs 8E-H)

4 師板 — velum —

胞紋をふさぐ穿孔のある珪酸質の薄い層をいう．穿孔は通常直径 30 nm 以下で，円形とは限らない．師板には次のようなタイプがある．

1. 多孔師板 — cribrum —(図77)

網目状の師板，または規則的に配列した小孔 (pores) によって穿孔された師板をいう．

例：*Stephanodiscus minutullus* (Kütz.) Round

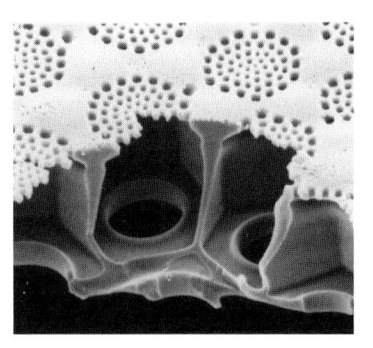

図74 小箱胞紋(外面観)

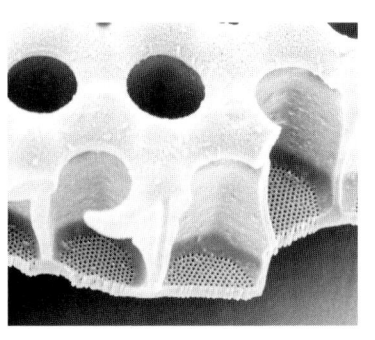

図75 偽小箱胞紋(外面観)

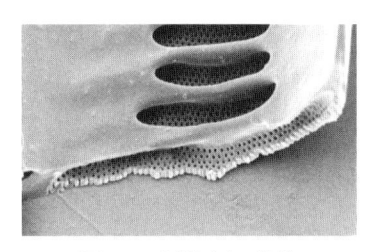

図76 長胞(内面観)

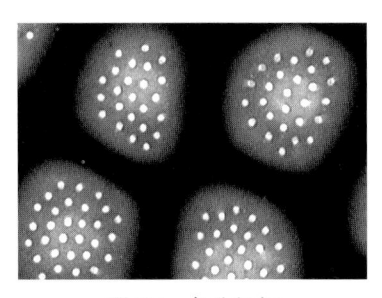

図77 多孔師板

2. 輪形師板 — rota —（図 78）
　胞紋をよぎる1本または数本の棒状体によって支えられた師板をいう．中心部に広い板状部をもつものと，もたないものがある．
例：*Triceratium shadboltianum* Grev.

3. 肉趾状師板 — vola —（図 79）
　胞紋の壁から突き出たいくつかの肉趾状の突起からできた師板をいう．
例：*Epithemia adonata* (Kütz.) Bréb.

5 師皮（薄皮）— rica (hymen) —（図 80）
　穿孔のある珪酸質の非常に薄い層をいう．穿孔は普通円形で直径 15 nm 以上にはならず，ほとんどの場合X字状 (decussately) に配列している．
例：*Navicula delognei* Van Heurck (Cox 1978, fig. 5)，*Frustulia rhomboides* (Ehrenb.) De Toni (Cox 1975, fig. 13)，*Nitzschia palea* (Kütz.) W. Sm. (Lange-Bertalot & Simonsen 1978, fig. 297).
注：Ross *et al.* (1979) は師皮に関し次のようなコメントを与えている．
　われわれの知るところでは，師皮は中心珪藻からは報告されていない．中心珪藻の師板が羽状珪藻の師皮と相同のものであるかどうか，*Cocconeis scutellum* Ehrenb. (Helmcke & Krieger 1953, pl. 48) や *Nitzschia amphibia* Grunow (Lange-Bertalot & Simonsen 1978, fig. 295) などの胞紋を閉塞している目の粗い網目構造と相同のものかどうかは定かではない．後者はわれわれの師板の定義に適合し，この師板の中に見られる穿孔 (perforation) は師皮で閉塞されたものである．
　その後 Mann (1981) は羽状珪藻の胞紋閉塞について，薄皮胞紋閉塞 (hymenate pore occulusion) および肉趾胞紋閉塞 (volate pore occulusion) と呼ぶ2つの基本的様式を提案している．前者は胞紋の閉塞が多数の小孔をもつ均一な厚さの薄皮 (hymen) により閉ざされるもので，後者は胞紋壁から伸長する肉趾の突起によって閉塞されるものである．肉趾は薄皮とは異なり，一般に胞紋壁付近は厚く，胞紋中央部では薄くなる．羽状珪藻の胞紋はこの2つのどちらか，あるいは両者の組み合わせ（例：*Diploneis* 属）により閉塞されている．薄皮には小孔が多数存在し，ある規則性をもって配列する．その配列様式には規則的散在 (reguler scatter)（図 80 a），六角整列 (hexagonal array)（図 80 b），中心整列 (central array)（図 80 c）な

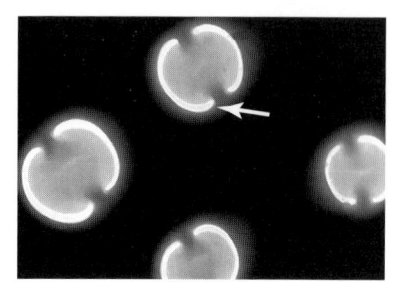

図 78　輪形師板

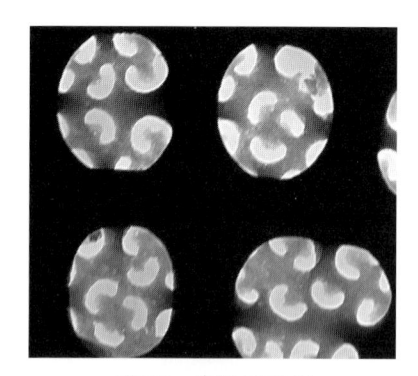

図 79　肉趾状師板

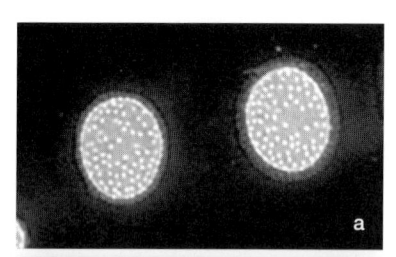

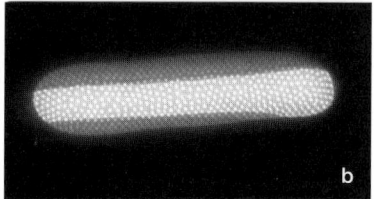

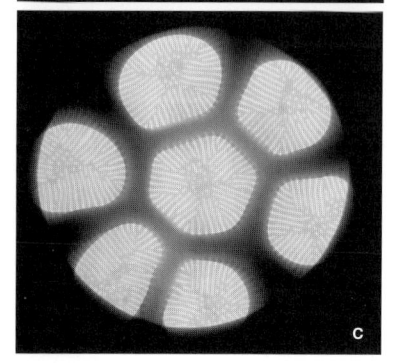

図 80　師皮（a 規則的散在, b 六角整列, c 中心整列）

どが知られている.

6　連絡口 － passage pore －（図 81）

1つの小箱胞紋と他の小箱胞紋をつなぐ開口をいう.

注：Sims & Paddock 1979, pl. 2, fig. 15 の *Diploneis* sp. では縦走管と小箱胞紋の連絡口に passage pore の語が使われている.

7　小泡 － bullula －（図 82）

分離された胞紋の間にある珪酸基底層内の泡状の空所の1つをいう.

例：*Aulacodiscus reticulatus* Pant.

8　内連絡 － hypocaust －（図 83）

薄い壁でよく分離された胞紋の間の珪酸基底層の中の連続した空間をいう (Ross & Sims 1972, fig. 13).

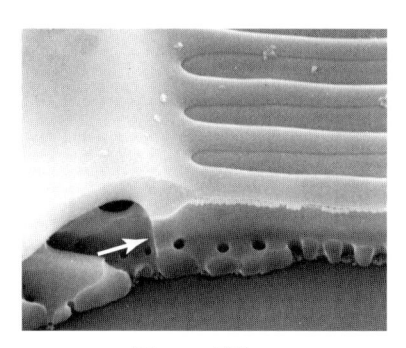

図 81　連絡口

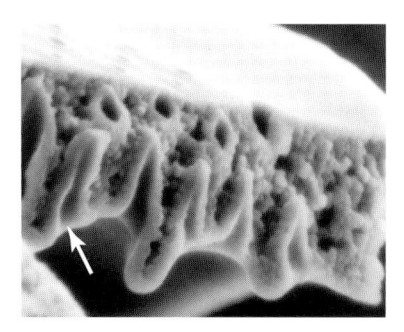

図 82　小泡

図 83　内連絡

珪藻用語対照表

ラテン語の( )内に主格複数の語尾変化，掲載ページに「構造と用語」中の用語解説ページを示す．

英語	日本語	日本語読み	ラテン語	掲載ページ
alar canal	翼管	よくかん	canalis (-es) alaris (-es)	30, 37
alveolus	長胞	ちょうほう	alveolus (-i)	43
annulus	中心環	ちゅうしんかん	annulus (-i)	
antiligula	副小舌	ふくしょうぜつ	antiligula (-ae)	39
aperture	開口	かいこう	apertura (-ae)	29
apical axis	長軸，頂軸	ちょうじく	axis (-es) apicalis (-es)	22
apical plane	縦断面	じゅうだんめん	planities (-es) apicalis (-es)	22
apical pore field	殻端小孔域	かくたんしょうこういき	area (-ae) porellis apicalibus	34
araphid valve	無縦溝殻	むじゅうこうかく	araphovalve (-a)	
areola	胞紋	ほうもん	areola (-ae)	42
auxospore	増大胞子	ぞうだいほうし	auxospora (-ae)	23
axial area	軸域	じくいき	area (-ae) axialis (-es)	29
axial costa	軸肋	じくろく	costa (ae) axialis (-es)	31
band	帯片	たいへん	taenia (-ae)	38, 39
basal silliceous layer	珪酸基底層	けいさんきていそう	stratum (-a) siliceum (-a) basale (-ia)	41
bilabiate process	複唇状突起	ふくしんじょうとっき	birimoportula (-ae)	34
bullula	小泡	しょうほう	bullula (-ae)	45
canal raphe	管状縦溝	かんじょうじゅうこう	raphe (-es) canalis (-es)	37
canopy	天幕	てんまく	conopeum (-ea)	32
cavity	洞	ほら	cavitas (-es)	30
central area	中心域	ちゅうしんいき	area (-ae) centralis (-es)	29
central array	中心整列	ちゅうしんせいれつ	ordinatio (-ones) centralis (-es)	44
central fissure	中心裂溝	ちゅうしんれっこう	centralifissura (-ae)	36
central nodule	中心節	ちゅうしんせつ	nodulus (-i) centralis (-es)	35
central pore	中心孔	ちゅうしんこう	porus (-i) centralis (-es)	36
cingulum	半殻帯	はんかくたい	cingulum (-a)	38
collar	えり	えり	collare (-ia)	32
complex raphe	複合縦溝	ふくごうじゅうこう	raphe (-es) comlexa (-ae)	37
connecting band	連結帯片	れんけつたいへん	pleura (-ae)	38
convergent	収れん	しゅうれん	convergens	42
copula	中間帯片	ちゅうかんたいへん	copula (-ae)	38
costa	肋	ろく	costa (-ae)	31
craticula	柵板	さくばん	craticula (-ae)	38
crest	とさか	とさか	crista (-ae)	32
cribrum	多孔師板	たこうしばん	criblum (-a)	43
cuniculus	暗渠	あんきょ	cuniculus (-i)	
elevation	隆起	りゅうき	elevatio (-ones)	29
endogenous resting spore	内生休眠胞子	ないせいきゅうみんほうし	hyponospora (-ae) endogena (-ae)	25

英語	日本語	日本語読み	ラテン語	掲載ページ
epicingulum	上半殻帯	じょう(うえ)はんかくたい	epicingulum (-a)	26
epitheca	上半被殻	じょう(うえ)はんひかく	epitheca (-ae)	26
epivalve	上殻	じょうかく	epivalva (-ae)	26
exogenous resting spore	外生休眠胞子	がいせいきゅうみんほうし	hyponospora (-ae) exogena (-ae)	25
external central fissure	外中心裂溝	がいちゅうしんれっこう	fissura (-ae) centralis (-es) externa (-ae)	36
fascia	横帯	おうたい	fascia (-ae)	29
fascicle	条線束	じょうせんそく	fasciculus (-i)	41
fasciculate	束出	そくしゅつ	fasciculatus	41
fenestra	翼窓	よくそう	fenestra (-ae)	30
fenestral bar	翼窓棒	よくそうぼう	barra (-ae) fenestralis (-es)	30
fibula	小骨	しょうこつ	fibula (-ae)	37
filiform raphe	糸状縦溝	しじょうじゅうこう	raphe (-es) filiformis (-es)	37
fissure	裂溝	れっこう	fissura (-ae)	36, 37
fold	陥入	かんにゅう	ruga (-ae)	33
footpole	足極	そくきょく	polus (-i) basalis (-es)	
foramen	箱口	はこぐち	foramen (-ina)	43
free auxospore	遊離増大胞子	ゆうりぞうだいほうし	auxospora (-ae) libera (-ae)	23
frustule	被殻	ひかく	frustulum (-a)	26
fultoportula	有基突起	ゆうきとっき	fultoportula (-ae)	35
girdle	殻帯	かくたい	cinctura (-ae)	38
girdle view	帯面観	たいめんかん	aspectus (-us) cincturalis (-es)	26
granule	顆粒	かりゅう	granulum (-a)	35
headpole	頭極	とうきょく	vertex (-tices)	
helictoglossa	蝸牛舌	かぎゅうぜつ	helictoglossa (-ae)	36
hexagonal array	六角整列	ろっかくせいれつ	ordinatio (-ones) hexagona (-ae)	44
horn	角	つの	cornu (-ua)	29
horseshoe-shaped area	馬蹄域	ばていいき	area (-ae) hippocrepica (-ae)	30
hyaline area	無紋域	むもんいき	area (-ae) hyalina (-ae)	29, 30
hymen	薄皮	はくひ	hymen (-es)	44
hypocaust	内連絡	ないれんらく	hypocaustum (-a)	45
hypocingulum	下半殻帯	か(した)はんかくたい	hypocingulum (-a)	26
hypotheca	下半被殻	か(した)はんひかく	hypotheca (-ae)	26
hypovalve	下殻	かかく	hypovalva (-ae)	26
initial cell	初生細胞	しょせいさいぼう	cellula (-ae) prima (-ae)	24
initial valve	初生殻	しょせいかく	valva (-ae) pirma (-ae)	25
inner fissure	内裂溝	ないれっこう	fissura (-ae) interior (-es)	37
intercalary auxospore	介在増大胞子	かいざいぞうだいほうし	auxospora (-ae) intercalaris (-es)	24
intercosta	間肋	かんろく	intercosta (-ae)	

英語	日本語	日本語読み	ラテン語	掲載ページ
internal centaral fissure	内中心裂溝	ないちゅうしんれっこう	fissura (-ae) centralis (-es) interna (-ae)	36
internal valve	内生殻	ないせいかく	valva (-ae) interna (-ae)	38
interstria	間条線	かんじょうせん	interstria (-ae)	37, 42
junction line	接合線	せつごうせん	linea (-ae) alae (-arum) basalis (-es)	
keel	竜骨	りゅうこつ	carina (-ae)	30
labiate process	唇状突起	しんじょうとっき	rimoportula (-ae)	34
lateral area	側域	そくいき	area (-ae) lateralis (-es)	30
lateral auxospore	側生増大胞子	そくせいぞうだいほうし	auxospora (-ae) lateralis (-es)	23
ligula	小舌	しょうぜつ	ligula (-ae)	39
linking spine	連結針	れんけつしん	spina (-ae) ligans (-ntes)	35
loculus (loculate areola)	小箱胞紋	こばこほうもん	aleola (-ae) loculata (-ae)	43
longitudinal canal	縦走管	じゅうそうかん	canalis (-es) longitudinalis (-es)	38
marginal ridge	縁辺稜	えんぺんりょう	crista (-ae) marginalis (-es)	31
nodule	節	せつ	nodulus (-i)	35
occluded process	閉塞突起	へいそくとっき	processus (-us) occlusus (-i)	34
ocellulimbus	殻套眼域	かくとうがんいき	ocellulimbus (-i)	34
ocellus	眼域	がんいき	ocellus (-i)	33
otarium	耳	みみ	otarium (-a)	32
outer fissure	外裂溝	がいれっこう	fissura (-ae) exterior (-es)	37
parallel	平行	へいこう	parallelus	41
pars exterior	帯片表出部	たいへんひょうしゅつぶ	pars (-tes) exterior (-es)	39
pars interior	帯片内接部	たいへんないせつぶ	pars (-tes) interior (-es)	39
pars media	帯片中脈	たいへんちゅうみゃく	pars (-tes) media (-ae)	39
partectal duct	区画導管	くかくどうかん	ductus (-us) partectalis (-es)	40
partectal ring	区画環	くかくかん	annulus (-i) partectalis (-es)	40
partectum	区画	くかく	partectum (-a)	40
passage pore	連絡口	れんらくこう	pervium (-ia)	45
periplekton	巻込突起	まきこみとっき	periplekton (-a)	34
perizonium	ペリゾニウム	ぺりぞにうむ	perizonium (-a)	23
pervalvar axis	貫殻軸	かんかくじく	axis (-es) parvalvaris (-es)	22
pleura	連結帯片	れんけつたいへん	pleura (-ae)	38
plicate raphe	ひだ状縦溝	ひだじょうじゅうこう	raphe (-es) plicata (-ae)	37
poroid areola (poroid)	孔状胞紋	こうじょうほうもん	arola (-ae) poroides (-es)	42
portula	門口	もんこう	portula (-ae)	37
primary valve	第1次殻	だいいちじかく	valva (-ae) prima (-ae)	25
process	突起	とっき	processus (-us)	34

英語	日本語	日本語読み	ラテン語	掲載ページ
pseudocellus	偽眼域	ぎがんいき	pseudocellus (-i)	33
pseudoloculus	偽小箱胞紋	ぎこばこほうもん	pseudoloculus (-i)	43
pseudonodulus	偽節	ぎせつ	pseudonodulus (-i)	34
pseudoseptum	偽隔壁	ぎかくへき	pseudoseptum (-a)	31
punctum (-a)	点紋	てんもん	punctum (-a)	42
radial	射出	しゃしゅつ	radialis	41
radiate	放射状	ほうしゃじょう	radiatus	42
raphe	縦溝	じゅうこう	raphe (-es)	35
rahhe (raphid) valve	縦溝殻	じゅうこうかく	raphovalva (-ae)	
raphe branch	縦溝枝	じゅうこうし	ramus (-i) raphis	35
raphe canal	縦溝管	じゅうこうかん	canalis (-es) raphis	37
raphe fissure	縦溝裂	じゅうこうれつ	fissura (-ae) raphis	37
raphe sternum	縦溝中肋	じゅうこうちゅうろく	sternum (-a) raphis	
reguler scatter	規則的散在	きそくてきさんざい	dispersio (-ones) regularis (-es)	44
resting spore	休眠胞子	きゅうみんほうし	hyponospora (-ae)	24
rica	師皮(薄皮)	しひ, はくひ	rica(-ae)	44
rimoportula	唇状突起	しんじょうとっき	rimoportula (-ae)	34
ring-costa	横輪	おうりん	annulicosta (-ae)	32
rota	輪形師板	りんけいしばん	rota (-ae)	44
satellite pore	付随孔	ふずいこう	porus(-i) satelliticus (-i)	35
scale	鱗片	りんぺん	squama (-ae)	23
secondary valve	第2次殻	だいにじかく	valva (-ae) secunda (-ae)	25
segment	分節帯片	ぶんせつたいへん	segmentum (-a)	
semi-endogenous resting spore	半内生休眠胞子	はんないせいきゅうみんほうし	hyponospora (-ae) semiendogena (-ae)	25
semi-intercalary auxospore	半介在増大胞子	はんかいざいぞうだいほうし	auxospora (-ae) semiintercalaris (-es)	24
septum	隔壁	かくへき	septum (-a)	39
seta	剛毛	ごうもう	seta (-ae)	29
spine	針	はり	spina (-ae)	35
spinule	小針	しょうしん	spinula (-ae)	35
stauros	十字節	じゅうじせつ	stauros (-i)	35
sternum	中肋	ちゅうろく	sternum (-a)	
stigma	遊離点	ゆうりてん	stigma (-ata)	33
stria	条線	じょうせん	stria (-ae)	41
strut	支柱	しちゅう	stignum (-a)	
strutted process	有基突起	ゆうきとっき	fultoportula (-ae)	35
stub	スタブ	すたぶ	truncus (-i)	
sulcus	輪溝	りんこう	sulcus (-i)	33
suture	縫合	ほうごう	sutura (-ae)	39
tangential	正接	せいせつ	tangentialis	41
terminal area	極域	きょくいき	area (-ae) terminalis (-es)	29
terminal auxospore	端生増大胞子	たんせいぞうだいほうし	auxospora (-ae) terminalis (-es)	23

英語	日本語	日本語読み	ラテン語	掲載ページ
terminal cap	極帽	きょくぼう	pileus (-i) telminalis (-es)	
terminal fissure	極裂	きょくれつ	fissura (-ae) terminalis (-es)	36
terminal nodule	極節	きょくせつ	nodulus (-i) terminalis (-es)	36
terminal seta	端末剛毛	たんまつごうもう	seta (-ae) terminalis (-es)	29
theca	半被殻	はんひかく	theca (-ae)	26
tigillate stauros	小梁十字節	しょうりょうじゅうじせつ	stauros (-i) tigillatus (-i)	35
tigillum	小梁	しょうりょう	tigillum(-a)	35
transapical axia	短軸, 切頂軸	たんじく, せっちょうじく	axis (-es) transapicalis (-es)	22
transapical costa	横走肋	おうそうろく	costa (-ae) transapicalis (-es)	
transapical plane	横断面	おうだんめん	planities (-es) transapicalis (-es)	22
transapical stria	横走条線	おうそうじょうせん	stria (-ae) transapicalis (-es)	
valvar plane	殻断面	かくだんめん	planities (-es) valvaris (-es)	22
valve	殻	から	valva (-ae)	26
valve face	殻面	かくめん	frons (-tes)	28
valve mantle	殻套	かくとう	limbus (-i)	28
valve margin	殻縁	かくえん	margo (-es) valvae	42
valve shoulder	殻肩	かくけん	humerus (-i) valvae	28
valve view	殻面観	かくめんかん	aspectus (-us) valvae	26
vegitative cell	栄養細胞	えいようさいぼう	cellula (-ae) vegetativa (-ae)	22
velum	師板	しばん	velum (-a)	43
vimen	縦枝	たてえだ	vimen (-ina)	42
virga	横枝	よこえだ	virga (-ae)	42
Voigt fault	ボアグの欠落	ほあぐのけつらく	inordinatio (-ones) Voigtii	42
vola	肉趾状師板	にくしじょうしばん	vola (-ae)	44
window	窓	まど	fenestra (-ae)	29
wing	翼	よく	ala (-ae)	30, 32

珪藻分類体系

珪藻植物門 Bacillariophyta

コアミケイソウ亜門 Coscinodiscophytina Medlin & Kaczmarska (2004)

コアミケイソウ綱 Coscinodiscophyceae Round & Crawford, emend, Medlin & Kaczmarska (2004)

キクノハナケイソウ目　Chrysanthemodiscales
　　キクノハナケイソウ科　Chrysanthemodiscaceae
　　　Chrysanthemodiscus
タルケイソウ目　Melosirales
　　タルケイソウ科　Melosiraceae
　　　Melosira (Pl. 1〜6),　*Druridgea*
　　クシダンゴケイソウ科　Stephanopyxidaceae
　　　Stephanopyxis
　　アミカゴケイソウ科　Endictyaceae
　　　Endictya
　　ドラヤキケイソウ科　Hyalodiscaceae
　　　Hyalodiscus,　*Podosira*
タルモドキケイソウ目　Paraliales
　　タルモドキケイソウ科　Paraliaceae
　　　Paralia,　*Ellerbeckia* (Pl. 7)
スジタルケイソウ目　Aulacoseirales
　　スジタルケイソウ科　Aulacoseiraceae
　　　Aulacoseira (Pl. 8〜24),　*Strangulonema*
ウスガサネケイソウ目　Orthoseirales
　　ウスガサネケイソウ科　Orthoseiraceae
　　　Orthoseira (Pl. 25, 26)
コアミケイソウ目　Coscinodiscales
　　コアミケイソウ科　Coscinodiscaceae
　　　Coscinodiscus,　*Palmeria*,　*Stellarima*,　*Brightwellia*,　*Craspedodiscus*
　　ハグルマケイソウ科　Rocellaceae
　　　Rocella
　　コウロケイソウ科　Aulacodiscaceae
　　　Aulacodiscus
　　ホネカサケイソウ科　Gossleriellaceae
　　　Gossleriella
　　ハンマルケイソウ科　Hemidiscaceae
　　　Hemidiscus,　*Actinocyclus*,　*Azpeitia*,　*Roperia*

タイヨウケイソウ科　Heliopeltaceae
 Actinoptychus, Lepidodiscus, Glorioptychus
オオコアミケイソウ目　Ethmodiscales
 オオコアミケイソウ科　Ethmodiscaceae
  Ethmodiscus
ニセヒトツメケイソウ目　Stictocyclales
 ニセヒトツメケイソウ科　Stictocyclaceae
  Stictocyclus
クンショウケイソウ目　Asterolamprales
 クンショウケイソウ科　Asterolampraceae
  Asterolampra, Asteromphalus
クモノスケイソウ目　Arachnoidiscales
 クモノスケイソウ科　Arachnoidiscaceae
  Arachnoidiscus
ハスノミケイソウ目　Stictodiscales
 ハスノミケイソウ科　Stictodiscaceae
  Stictodiscus
イガクリケイソウ目　Corethrales
 イガクリケイソウ科　Corethraceae
  Corethron
ツツガタケイソウ目　Rhizosoleniales
 ツツガタケイソウ科　Rhizosoleniaceae
  Rhizosolenia, Proboscia, Urosolenia, Guinardia, Dactyliosolen, Pseudosolenia
 トンガリボウシケイソウ科　Pyxillaceae
  Pyxilla, Gladius, Mastogonia, Pyrgupyxis, Gyrodiscus
ホソミドロケイソウ目　Leptocylindrales
 ホソミドロケイソウ科　Leptocylindraceae
  Leptocylindrus

クサリケイソウ亜門　Bacillariophytina Medlin & Kaczmarska (2004)

中間綱　Mediophyceae Medlin & Kaczmarska (2004)

ニセコアミケイソウ目　Thalassiosirales
 ニセコアミケイソウ科　Thalassiosiraceae
  Thalassiosira (Pl. 28〜36), *Planktoniella, Porosira, Minidiscus, Bacteriosira*
 ホネツギケイソウ科　Skeletonemataceae
  Skeletonema, Detonula
 トゲカサケイソウ科　Stephanodiscaceae
  Cyclotella (Pl. 41〜50), *Cyclostephanos* (Pl. 37〜40), *Discostella* (Pl. 51〜54), *Stephanodiscus* (Pl. 57〜61), *Mesodictyon, Pleurocyclus, Puncticulata* (Pl. 55, 56), *Stephanocostis*
 ヒメホネツギモドキケイソウ科　Lauderiaceae
  Lauderia

ミカドケイソウ目　Triceratiales
　　ミカドケイソウ科　Triceratiaceae
　　　　Triceratium, Sheshukovia, Odontella, Pseudoauliscus, Lampriscus, Auliscus, Eupodiscus, Pleurosira, Amphitetras, Cerataulus
　　ニセハネケイソウ科　Plagiogrammaceae
　　　　Plagiogramma, Glyphodesmis, Dimeregramma, Dimeregrammopsis
イトマキケイソウ目　Biddulphiales
　　イトマキケイソウ科　Biddulphiaceae
　　　　Biddulphia, Biddulphiopsis, Hydrosera, Isthmia, Trigonium, Pseudotriceratium
シマヒモケイソウ目　Hemiaulales
　　シマヒモケイソウ科　Hemiaulaceae
　　　　Hemiaulus, Eucampia (Pl. 62), *Climacodium, Cerataulina, Trinacria, Abas, Briggera, Pseudorutilaria, Keratophora, Kittonia, Strelnikovia, Riedelia, Baxteriopsis, Sphynctolethus, Ailuretta*
　　ヒモケイソウ科　Bellerocheaceae
　　　　Bellerochea, Subsilicea
　　ネジレオビケイソウ科　Streptothecaceae
　　　　Streptotheca, Neostreptotheca
ミズマクラケイソウ目　Anaulales
　　ミズマクラケイソウ科　Anaulaceae
　　　　Anaulus, Eunotogramma
サンカクチョウチンケイソウ目　Lithodesmiales
　　サンカクチョウチンケイソウ科　Lithodesmiaceae
　　　　Lithodesmium, Ditylum, Lithodesmioides
オビダマシケイソウ目　Cymatosirales
　　オビダマシケイソウ科　Cymatosiraceae
　　　　Cymatosira, Brockmanniella, Arcocellulus, Campylosira, Plagiogrammopsis, Minutocellus, Leyanella, Papiliocellulus, Extubocellulus
　　シンボウツナギケイソウ科　Rutilariaceae
　　　　Rutilaria, Syndetocystis
ツノケイソウ目　Chaetocerotales
　　ツノケイソウ科　Chaetocerotaceae
　　　　Chaetoceros, Bacteriastrum, Gonioceros
　　ジャバラケイソウ科　Acanthocerataceae
　　　　Acanthoceras
　　カクダコケイソウ科　Attheyaceae
　　　　Attheya
アミカケケイソウ目　Toxariales
　　アミカケケイソウ科　Toxariaceae
　　　　Toxarium
デカハリケイソウ目　Ardissoneales
　　デカハリケイソウ科　Ardissoneaceae
　　　　Ardissonea

クサリケイソウ綱 Bacillariophyceae Haeckel, emend. Medlin & Kaczmarska (2004)

＜無縦溝類＞

オビケイソウ目　Fragilariales
　　オビケイソウ科　Fragilariaceae
　　　　Fragilaria (Pl. 71〜78), *Centronella*, *Asterionella* (Pl. 63, 64), *Staurosirella* (Pl. 96〜98), *Staurosira* (Pl. 89〜95), *Pseudostaurosira* (Pl. 85〜87), *Punctastriata* (Pl. 88), *Fragilariforma* (Pl. 79), *Martyana* (Pl. 82, 83), *Diatoma* (Pl. 68〜70), *Hannaea* (Pl. 80, 81), *Meridion* (Pl. 84), *Ulnaria* (Pl. 101〜107), *Ctenophora* (Pl. 67), *Neosynedra*, *Tabularia* (Pl. 100), *Catacombas* (Pl. 66), *Hyalosynedra*, *Opephora*, *Trachysphenia*, *Thalassioneis*, *Falcula*, *Pteroncola*, *Asterionellopsis* (Pl. 65), *Bleakeleya*, *Podocystis*

ヌサガタケイソウ目　Tabellariales
　　ヌサガタケイソウ科　Tabellariaceae
　　　　Tabellaria (Pl. 108〜112), *Tetracyclus*, *Oxyneis*

オウギケイソウ目　Licmophorales
　　オウギケイソウ科　Licmophoraceae
　　　　Licmophora, *Licmosphenia*

オカメケイソウ目　Rhaphoneidales
　　オカメケイソウ科　Rhaphoneidaceae
　　　　Rhaphoneis, *Diplomenora*, *Delphineis*, *Neodelphineis*, *Perissonoë*, *Sceptroneis*
　　スナマルケイソウ科　Psammodiscaceae
　　　　Psammodiscus

ウミノイトケイソウ目　Thalassionematales
　　ウミノイトケイソウ科　Thalassionemataceae
　　　　Thalassionema, *Thalassiothrix*, *Trichotoxon*

ドウナガケイソウ目　Rhabdonematales
　　ドウナガケイソウ科　Rhabdonemataceae
　　　　Rhabdonema

ハラスジケイソウ目　Striatellales
　　ハラスジケイソウ科　Striatellaceae
　　　　Striatella, *Microtabella*, *Grammatophora*

シンツキケイソウ目　Cyclophorales
　　シンツキケイソウ科　Cyclophoraceae
　　　　Cyclophora
　　ミゾナシツメケイソウ科　Entophylaceae
　　　　Entophyla, *Gephyria*

オオヘラケイソウ目　Climacospheniales
　　オオヘラケイソウ科　Climacospheniaceae
　　　　Climacosphenia, *Synedrosphenia*

ハジメノミゾモドキケイソウ目　Protoraphidales
　　ハジメノミゾモドキケイソウ科　Protoraphidaceae
　　　　Protoraphis, *Pseudohimantidium*

<縱溝類>

イチモンジケイソウ目　Eunotiales
イチモンジケイソウ科　Eunotiaceae
Eunotia, *Actinella* (Pl. 113), *Semiorbis*, *Desmogonium*, *Eunophora*
ツマヨウジケイソウ科　Peroniaceae
Peronia (Pl. 114)

タテゴトモヨウケイソウ目　Lyrellales
タテゴトモヨウケイソウ科　Lyrellaceae
Lyrella, *Petroneis*

チクビレツケイソウ目　Mastogloiales
チクビレツケイソウ科　Mastogloiaceae
Aneumastus, *Mastogloia*

ニセチクビレツケイソウ目　Dictyoneidales
ニセチクビレツケイソウ科　Dictyoneidaceae
Dictyoneis

クチビルケイソウ目　Cymbellales
マガリクサビケイソウ科　Rhoicospheniaceae
Rhoicosphenia (Pl. 115〜117), *Campylopyxis*, *Cuneolus*, *Gomphoseptatum*, *Gomphonemopsis*
サミダレケイソウ科　Anomoeoneidaceae
Anomoeoneis, *Staurophora*
クチビルケイソウ科　Cymbellaceae
Brebissonia, *Cymbella*, *Cymbopleura*, *Delicata*, *Encynopsis*, *Encyonema*, *Gomphocymbella*, *Placoneis*
クサビケイソウ科　Gomphonemataceae
Gomphonema (Pl. 121〜135), *Didymosphenia*, *Gomphoneis* (Pl. 118〜120), *Reimeria*, *Gomphopleura*

ツメケイソウ目　Achnanthales
ツメケイソウ科　Achnanthaceae
Achnanthes (Pl. 136〜142)
コメツブケイソウ科　Cocconeidaceae
Cocconeis (Pl. 143〜151), *Campyloneis*, *Anorthoneis*, *Bennettella*, *Epipellis*
ツメワカレケイソウ科　Achnanthidaceae
Achnanthidium (Pl. 152〜163), *Eucocconeis*, *Lemnicola* (Pl. 164, 165), *Planothidium* (Pl. 166〜168), *Psammothidium* (Pl. 169〜173)

フナガタケイソウ目　Naviculales
ヒメクダズミケイソウ科　Berkeleyaceae
Parlibellus, *Berkeleya*, *Climaconeis*, *Stenoneis*
ニセコメツブケイソウ科　Cavinulaceae
Cavinula
フルイノメケイソウ科　Cosmioneidaceae
Cosmioneis
ネジレフネケイソウ科　Scolioneidaceae
Scolioneis
オビフネケイソウ科　Diadesmidaceae

   Diadesmis, Luticola
  アミバリケイソウ科　Amphipleuraceae
   Frickea, Amphipleura, Frustulia, Cistula
  サミダレモドキケイソウ科　Brachysiraceae
   Brachysira
  ハスフネケイソウ科　Neidiaceae
   Neidium
  シンネジモドキケイソウ科　Scoliotropidaceae
   Scoliopleura, Scoliotropis, Biremis, Progonoia, Diadema
  エリツキケイソウ科　Sellaphoraceae
   Sellaphora, Fallacia, Rossia, Caponea
  ハネケイソウ科　Pinnulariaceae
   Pinnularia, Diatomella (Pl. 175), *Östrupia, Dimidiata*
  デキソコナイケイソウ科　Phaeodactylaceae
   Phaeodactylum
  マユケイソウ科　Diploneidaceae
   Diploneis (Pl. 176〜178), *Raphidodiscus*
  フナガタケイソウ科　Naviculaceae
   Navicula, Trachyneis, Pseudogomphonema, Seminavis, Rhoikoneis, Haslea, Cymatoneis
  メガネケイソウ科　Pleurosigmataceae
   Pleurosigma, Donkinia, Rhoicosigma, Gyrosigma
  イカノフネケイソウ科　Plagiotropidaceae
   Plagiotropis, Stauropsis, Pachyneis
  ジュウジケイソウ科　Stauroneidaceae
   Stauroneis, Craticula
  ウミジュウジケイソウ科　Proschkiniaceae
   Proschkinia
ハンカケケイソウ目　Thalassiophysales
  ニセイチモンジケイソウ科　Catenulaceae
   Catenula, Amphora, Undatella
  ハンカケケイソウ科　Thalassiophysaceae
   Thalassiophysa
クサリケイソウ目　Bacillariales
  クサリケイソウ科　Bacillariaceae
   Bacillaria, Hantzschia, Psammodictyon, Tryblionella, Cymbellonitzschia, Gomphonitzschia, Gomphotheca, Nitzschia, Denticula, Denticulopsis, Fragilariopsis, Cylindrotheca, Simonsenia, Cymatonitzschia, Perrya, Ceratoneis
クシガタケイソウ目　Rhopalodiales
  クシガタケイソウ科　Rhopalodiaceae
   Epithemia, Rhopalodia, Protokeelia
コバンケイソウ目　Surirellales
  ヨジレケイソウ科　Entomoneidaceae
   Entomoneis (Pl. 179, 180)
  ミミタブケイソウ科　Auriculaceae

Auricula

コバンケイソウ科　Surirellaceae
Hydrosilicon, Petrodictyon, Plagiodiscus, Stenopterobia, Surirella, Campylodiscus, Cymatopleura

　珪藻では Simonsen（1979）の分類体系が，殻の特徴をよく捉えており，かつ単純明瞭なため，長年にわたり多くの人に使用されてきた．しかし Round *et al.*（1990）による分類体系が提示されると，これに取って替えられるようになってきた．それは，彼らの分類体系が詳細な殻の微細構造に加え，葉緑体の形状や性質までを含めた，総合的な形質により構築された詳細なものであったためである．彼らの体系には，それまで一般的であった中心目，羽状目という 2 つの分類群は存在せず，コアミケイソウ綱（Coscinodiscophyceae），オビケイソウ綱（Fragilariophyceae），クサリケイソウ綱（Bacillariophyceae）の 3 綱に珪藻は大別された．
　その後 Medlin & Kaczmarska（2004）は 18S rDNA の塩基配列の解析により，珪藻は最初に中心珪藻が 2 つのグループ，すなわちコアミケイソウ亜門（Coscinodiscophytina）とクサリケイソウ亜門（Bacillariophytina）に分かれることを示した．後者には中心珪藻のうち殻の中央部に突起をもつ属と 2 極性を示す属が，そしてすべての羽状珪藻属が分類された．これらは，それぞれ中間綱（Mediophyceae）およびクサリケイソウ綱（Bacillariophyceae）と呼ばれる．本書で提示した珪藻の分類体系は，Mayama & Kuriyama（2002）および Medlin & Kaczmarska（2004）による珪藻の分子系統解析と Round *et al.*（1990）の分類体系を組み合わせた後，若干の手を加えたものである．なお，Mediophyceae のタイプ属はツノケイソウ属（*Chaetoceros*）であるが，本書では綱のラテン語に従って中間綱の名称を提案した．
　近年 Cox（2006）は分岐分類に基づき，チクビレツケイソウ目（Mastogloiales）を改訂し，その中に新たに *Craspedostauros* 属と *Achnanthes* 属を追加した．また，筆者の一人（真山）の分子系統解析でも *Achnanthes* が他の単縦溝珪藻と異なる系統に所属することが示唆されている（未発表）．しかし，チクビレツケイソウ目に所属する他属の分子系統解析は，現時点では誰によってもなされておらず，*Achnanthes* 属の正しい系統学的位置の確定のためにはもうしばらく時間が必要である．このため，本書では *Achnanthes* 属の配属先として暫定的に Round *et al.*（1990）の体系を採用した．また，本書で取り上げる以下の分類群は，現時点では上述の分類体系のどこへ所属するか不明であるが，これらについても分子系統学的解析により，近い将来，所属が明らかになるものと思われる：*Brebisira, Fistulifera, Eolimna, Mayamaea, Nupela*.

和文解説

Plates 1, 2

Melosira moniliformis (O. Müller) C. Agardh

Agardh, C. A. 1824. Syst. Algarum. p. 8.
Basionym: *Conferva moniliformis* O. F. Müller 1783. Nov. Act. Holm. pl. 3. f. 6, 7.
Dimension: D. 25-70 μm,　H. 14-30 μm.

被殻・殻　被殻は茶筒のような円筒形で，殻面の中央部でゆるく結合し，長い糸状の群体を作る[1]．殻面は凸状で，丸い殻肩から垂直な殻套へと続く．そのため，帯面観では結合する姉妹殻の両サイドに明瞭な三角形の隙間が生じる[1]．生細胞の群体では2細胞が対となって見え，それぞれの細胞には殻帯がないように見えるが，実際には両細胞の下殻は多数の帯片からなる殻帯によって覆われている[1]．直径25～70 μm，殻套高 14～30 μm．

条線・胞紋　殻面にも殻套にも条線はなく，不規則で不揃いの胞紋が見られる．胞紋は中央より周辺部でやや大きく，さらに周辺部では不規則な仕切りが見られる[2]．電顕で見るとこの仕切りは，星状の突起が連続してできたもので，殻面から殻套まで殻一面に存在することがわかる[4,5]．また，殻表面には非常に細かな顆粒が密に分布する[5]．

唇状突起・突起　星状の突起が連結する部分に楕円形の開口が多数見られる[5]．これらの開口は唇状突起の外部開口であり，殻面にも殻套にも多数散在し，光顕でも殻面や殻套にこの唇状突起を点として捉えることができる．

半 殻 帯　開放型，数枚．

葉 緑 体　円盤状，多数．

有性生殖　増大胞子が天然でもしばしば見られ，両方または片方の母細胞殻を付けたまま膨潤する介在型増大胞子を作る．詳細は出井(1993)を参照．

生 活 形　付着性．

汚濁耐性　弱汚濁耐性種．

出 現 地　各地の汽水性の河川や湖沼に出現する．涸沼川（茨城県），藻琴湖（北海道）など．

Melosira nummuloides C. Agardh

Agardh, C. A. 1824. Syst. Algarum. p. 8.
Dimension: D. 9-42 μm,　H. 10-14 μm.

被殻・殻　被殻は俵形で，殻面中央部でゆるく結合し長い糸状の群体を作る[1,2,4]．殻は半球状で，殻面と殻套の区別が不明瞭[3,6,8,9]．殻の上部に膜状のえりと呼ばれる突出物があり，帯面観では殻の両脇に飛び出した角のように[1,4,6]，殻面観ではひとすじの輪のように見える．直径9～42 μm，殻套高10～14 μm．

条線・胞紋　殻面中央から放射状に伸びる細かな胞紋から成る条線があり，それらは殻套にまで連続している[2,6-8]．これらの胞紋は中心から放射状に伸びる肋（縦走肋）と，中心から同心円状に作られた肋（横走肋）によって規則的に区切られたものであり，外面では多孔師板によって閉塞されている[10-12]．また，2つの肋の交差点には半球状の顆粒がある[12]．

唇状突起・突起　殻面中央部にリング状に並ぶ不定形の突起があり，この突起のところで姉妹殻同士がゆるく結合する[1,9,11]．この突起の内面にはリング状に配列する円形の開口，および殻縁近くに1列に配置する開口は唇状突起の外部開口である[10-12]．

半　殻　帯　開放型，数枚．
葉　緑　体　円盤状，多数．
有 性 生 殖　増大胞子が天然でもしばしば見られ，両方または片方の母細胞殻を付けたまま膨潤する介在型増大胞子を作る[5]．
生　活　形　付着性．
汚 濁 耐 性　弱汚濁耐性種．
出　現　地　各地の汽水性の河川や湖沼によく出現する．青野川河口（静岡県）など．

Plates 5, 6

Melosira varians C. Agardh

Agardh, C. A. 1827. Flora Bot. Zeit. p. 628.
Dimension: D. 8-35 μm,　H. 9-13 μm.

被殻・殻　被殻は茶筒のような円筒形で，殻面でゆるく結合し長い糸状の群体を作る[1-9]．直径 8〜35 μm，殻套高 9〜13 μm.

条線・胞紋　殻面にも殻套にも条線はなく，散在する多数の点紋だけが見える[7-9]．殻面から殻套の外面は均質な多孔師板によって閉塞されている．

唇状突起・突起　殻面から殻套には多数の不定形または顆粒状の突起がある[10-13]．これらの突起に混じりほぼ円形の開口が見られるが，この開口は唇状突起の外部開口である[11, 13]．殻面にも殻套にも多数散在するが，特に殻套縁に沿って多数見られる[13]．殻内面は平坦で，多数の丸い突起が散在するが，これらはすべて唇状突起である[14]．光顕で殻面や殻套に特にはっきりと見える点[8]は，この唇状突起である．

半殻帯　開放型，数枚．

葉緑体　円盤状，多数．

有性生殖　増大胞子が天然でもしばしば見られ，両方または片方の母細胞殻を付けたまま膨潤する介在型増大胞子を作る[1]．詳細は von Stosch (1951) を参照．

生活形　付着性．

汚濁耐性　弱汚濁耐性種．

出現地　各地の河川や湖沼によく出現し，しばしば多量に産する．多摩川（東京都），相模川（神奈川県），霞ヶ浦（茨城県），琵琶湖（滋賀県），地蔵院沼（埼玉県），松本城堀（長野県）など．

ノート　本種の詳細な構造と分類については Crawford (1978) に述べられている．

Plate 7

Ellerbeckia arenaria f. *teres* (Brun) R. M. Crawford

Crawford, R. M. 1988. In: Round, F. E. (ed.) Algae and the aquatic environment. p. 419, 421.
Basionym: *Melosira teres* Brun in Schmidt, A. *et al*. 1892. Atlas Diat. pl. 179. f. 13, 14.
Dimension: D. 90-124 μm,　H. 15-23 μm.

被殻・殻　被殻は円筒形(1-3). 糸状群体を形成する(1,2). 殻面は円形(3), 直径90〜124 μm (38〜135 μm: Krammer & Lange-Bertalot 1991), 殻套高 15〜23 μm (10〜15 μm: l.c.). 殻面の周辺部には放射する刻み目があり, 隣接する細胞の殻面の刻み目と咬み合う. 殻面中央部は無紋で, わずかに凸あるいは凹に湾曲する(2,3). 殻套部は極めて厚く, 壁を貫通する小孔が多数開く(4,5). 殻套縁の端は他の多くの珪藻属のように薄くならず, 逆に内側に向かってわずかに張り出すためより厚みを増す(5,6). 接殻帯片と触れ合うこの部分には放射状の歯形が付いており(4-6), 帯片内接部がはずれにくいようになっている.

胞　紋　胞紋は通常の珪藻属と異なり, ぶ厚い殻套部を貫くため細長い管状をなす. 殻套外面および内面には胞紋の開口がびっしり並ぶ(4-6). 胞紋の閉塞は観察されていない.

突　起　殻套には貫通する管が多数存在する. 殻套外面ではこれらの管の開口は胞紋の開口より大きいため容易に識別することができる(4矢印). また, 殻套内側表面では独特の形態をした突起の頂点に開口をもつ(5,6). これらの突起はCrawford(1988)によって管状突起 (tube processes) と呼ばれたもので, 光学顕微鏡でも黒い影となって観察される(1矢印).

半殻帯　小孔を多数もつ帯片よりなる. 接殻帯片のみ閉鎖型で, 他の帯片は開放型とされる (Round *et al*. 1990).

葉緑体　円盤状, 多数 (Round *et al*. 1990).

有性生殖　Cholnoky(1933)が *Melosira arenaria* のオートガミーについて報告している.

生活形　浮遊性.

汚濁耐性　弱汚濁耐性種.

出現地　阿寒湖(北海道), 小野上層(群馬県).

[ノート]　本分類群およびf. *arenaria* の微細構造について, Crawford(1988)による詳細な研究がある.

Plates 8, 9

Aulacoseira ambigua (Grunow) Simonsen

Simonsen, R. 1979. Bacillaria **2**: 56.
Basionym: *Melosira crenulata* var. *ambigua* Grunow in Van Heurck 1882. Syn. Diat. Belg. pl. 88. f. 12-15.
Synonym: *Melosira ambigua* (Grunow) O. Müller 1903. Ber. Deut. Bot. Ges. **21**: 332.
Dimension: D. 4-18 μm, H. 4-16 μm, Str. 14-20 in 10 μm, Ar. 14-22 in 10 μm.

被殻・殻 被殻は円筒形(1-14). 連結針により糸状群体を形成. 直線状のもの(1-4, 7, 8, 11-13)と貫殻軸方向に湾曲するもの(9, 10)があるが, 微細構造上の相違はない. 湾曲するものは, らせん状の糸状群体を形成する. 殻面は円形(5, 6), 直径 4〜18 μm, 殻套高 4〜16 μm. 殻面外面は平滑(14). 殻肩は角ばり連結針がある(14, 18, 19). 殻套縁の内面に横輪がある(15, 16). この部分は光学切片像として観察すると, 殻の内側へくびれ込んだように見えるが(7), これは横輪が中空構造をしているためであって, 実際には殻套の外表面は平坦である(14, 17).

条線・胞紋 殻面周縁にやや角ばった丸い大きな胞紋が1列ある(5, 6, 14). より中心側の殻面には小さな胞紋が不規則に散在するが(14), これらは光学顕微鏡ではほとんど見えない. 殻套部では大きな胞紋が並んで条線を構成する. 条線は 10 μm あたり 14〜20 本(3, 8-13). 条線を構成する胞紋は 10 μm あたり 14〜22 個. 胞紋は内面で肉趾状篩板および薄い篩板により閉塞されている(14, 15, 18, 19).

針・突起 連結針は胞紋1個ほどの大きさで先広. 隣接する細胞の連結針と咬み合う(19). 分離殻では連結針は先細となる(14, 18). 唇状突起は殻套縁にある中空の横輪上に存在する(16). 唇状突起の外部開口は大きく目立つ(17).

半殻帯 開放型で多数の小孔をもつ帯片よりなる(17). 数枚.

葉緑体 円盤状, 多数(1, 2).

有性生殖 増大胞子は介在的(2, 4)に形成される. 形成過程の詳細な観察例はない.

生活形 浮遊性であるが, 付着基物上にもしばしば出現する.

汚濁耐性 中汚濁耐性種.

出現地 石狩川 (北海道), 尾瀬沼 (群馬県), 榛名湖 (群馬県), 近藤沼 (群馬県), 北浦 (茨城県), 仙女ヶ池 (埼玉県), 武蔵丘陵森林公園の沼 (埼玉県), 手賀沼 (千葉県), 三宝寺池 (東京都), 多摩川 (東京都, 神奈川県), 諏訪湖 (長野県), 琵琶湖 (滋賀県) など, 各地の湖沼や河川.

ノート *Aulacoseira* は, 属の設立当時は *Aulacosira* と綴られたため, 小林・野沢(1981, 1982)では後者の綴りが用いられた. ところがその後, これは正字法上の誤りであることがわかり Simonsen 自身によって国際植物命名法の規定に基づき正しい綴りに変更された.

Aulacoseira distans (Ehrenberg) Simonsen

Simonsen, R. 1979. Bacillaria **2**: 57.
Basionym: *Gallionella distans* Ehrenberg 1836. Ber. Bekanntm. Verh. Königl. Preuss. Akad. Wiss. Berlin **1**: 56.
Synonym: *Melosira distans* (Ehrenberg) Kützing 1844. Kies. Bacill. p. 54.
Dimension: D. 4.5-13.5 μm,　H. 16-20 μm,　Str. 14-16 in 10 μm,　Ar. 16-20 in 10 μm.

被殻・殻　被殻は円筒形[1-9]．連結針を有し糸状群体を形成[1,4,5]．殻面は円形[2,6,8,9]．直径4.5〜13.5 μm，殻套高16〜20 μm．殻套の外面には顆粒が散在する[8,9]．殻肩は角ばり連結針がある[8,9]．殻套縁は内側に柵板のように発達した偽隔壁様の横輪をもつが[3,5,6]，発達の程度はさまざまで，あまり発達しない個体もある[7]．

条線・胞紋　殻面全域にわたり胞紋が存在する[7-9]．これらの胞紋は光学顕微鏡観察でも明瞭に認められる[2]．殻套部では胞紋が貫殻軸方向に直線的に配列し条線を構成する[1,4]．条線は10 μmあたり14〜16本．条線を構成する胞紋は10 μmあたり16〜20個．胞紋は内面で肉趾状篩板および薄い篩板により閉塞される．

針・突起　連結針は小形で先細．このため，隣接する細胞と弱く連結していると思われるが，生細胞では細胞表面に分泌される粘液の粘着力も手伝って，糸状群体を形成しているものと思われる．細胞を硫酸処理して殻の観察を行った場合，糸状群体をほとんど見つけることができないのはこのためであろう．唇状突起は殻套縁にある横輪の上にあるため，殻を下側からSEM観察しても見ることはできないが[6,7]，光学顕微鏡を用い殻面観のピントを変えて観察すると，横輪の上部に黒い影として認めることができる[3]．

半殻帯　開放型で小孔を多数もつ帯片よりなる．数枚．
葉緑体　円盤状，多数（Cox 1996）．
有性生殖　不明．
生活形　浮遊性．
汚濁耐性　弱汚濁耐性種．
出現地　能登珪藻土（石川県）など．

ノート　通常，殻套長が殻の直径よりも短く，また酸処理など，被殻をクリーニングする過程で個々の殻が分離しやすいため，他の*Aulacoseira*種に比べ殻面を観察できる頻度が高い．

Plates 11, 12

Aulacoseira granulata (Ehrenberg) Simonsen

Simonsen, R. 1979. Bacillaria **2**: 59.
Basionym: *Gallionella granulata* Ehrenberg 1843. Abh. Königl. Akad. Wiss. Berlin **1841**: 415.
Synonym: *Melosira granulata* (Ehrenberg) Ralfs in Pritchard 1861. Hist. Infusoria. p. 820.
Dimension: D. 4.5-20 µm,　H. 4-16 µm,　Str. 8-12 in 10 µm,　Ar. 7-12 in 10 µm.

被殻・殻　被殻は円筒形(1,2,4,6-8)．連結針により糸状群体を形成．殻面は円形(3,9)，直径4.5〜20 µm，殻套高4〜16 µm．殻面外面は平滑(9)．殻肩は角ばり，連結針がある(9-11)．殻套縁の内側に横輪があるが(10)，光学顕微鏡では無紋域として観察される(1,2,4,7,8)．

条線・胞紋　殻面の周辺部には小さな胞紋が2列になって配列しているが，中心部は無紋である(3,9)．条線は殻套部で10 µmあたり8〜12本．条線を構成する胞紋は糸状群体の両端にある分離殻ではほとんど直線的に配列する(1,2,4,6-10)．しかし，それ以外の殻では胞紋はゆるいらせんを描いて配列する傾向がある(1,2,4,6-8)．条線を構成する胞紋は10 µmあたり7〜12個．1本の糸状群体中に，胞紋密度のまったく異なる殻を生じる場合がある(8)．胞紋は内面で肉趾状篩板により閉塞される．篩板は個体により，殻套外面近くまでせり出す場合もある(11)．

針・突起　結合殻の連結針は先広で，ほぼ胞紋1個分の大きさをもち，隣接する細胞の連結針と咬み合う(11)．分離殻では先細りの短い針と長い針をまばらに生じる．長い針は隣接する殻の殻套高にほぼ等しい長さにまでなる(4,6-10)．また，これらの長い針が差し込まれていた隣接殻の殻套部分には針の形の溝がある(6,9,10)．唇状突起は殻套縁の横輪上に存在する．

半殻帯　開放型で小孔を多数もつ帯片よりなる．8枚程度．

葉緑体　円盤状，多数(2)．

有性生殖　増大胞子は介在的に形成される(5)．形成過程の詳細な観察例はない．

生活形　浮遊性であるが，付着基物上にもしばしば出現する．

汚濁耐性　中汚濁耐性種．

出現地　阿寒湖（北海道），北浦（茨城県），三宝寺池（東京都），多摩川（東京都，神奈川県），津久井湖（神奈川県），諏訪湖（長野県），琵琶湖（滋賀県）など，各地の湖沼河川．

[ノート]　本種の最大の特徴は，分離殻に見られる長い針，およびその針が差し込まれていた殻套部の凹みである．これらは光学顕微鏡でも容易に観察できる．また，分離殻でのみ条線が直線になる傾向も特徴的である．

Plates 13, 14

Aulacoseira italica (Ehrenberg) Simonsen

Simonsen, R. 1979. Bacillaria **2**: 60.
Basionym: *Gaillonella italica* Ehrenberg 1838. Infusionsthier. p. 171.
Synonym: *Melosira italica* (Ehrenberg) Kützing 1844. Kies. p. 55; *Gallionella crenulata* Ehrenberg 1843. Abh. Königl. Akad. Wiss. Berlin **1841**: 376. pl. 4/1. f. 31; *Melosira crenulata* (Ehrenberg) Kützing 1844. Kies. p. 55; *Aulacoseira crenulata* (Ehrenberg) Thwaithes 1848. Ann. Mag. Nat. Hist. **1**: 167.
Dimension: D. 9-18 μm, H. 9-17.5 μm, Str. 16-22 in 10 μm, Ar. 12-16 in 10 μm.

被殻・殻 被殻は円筒形(2-5, 7, 9-11)．連結針により糸状群体を形成．殻面は円形(6)，直径9～18 μm，殻套高9～17.5 μm．殻面外面は平滑(9, 11)．殻肩はわずかに丸みを帯び(2, 5, 11)，そこから密に配列する頑丈な連結針が突出する(7, 9, 11, 12)．この部分の断面をSEM観察すると，隣り合う殻套の外面部分が連結針の橋渡しによって1枚の板のように見える(8矢印)．殻套縁の内側に中実の横輪がある(5矢印, 8, 14)．殻套外面の縁には，縦長の模様が1列存在する(7矢印, 10, 11)．

条線・胞紋 殻面に胞紋はないか，もしくは周辺に存在する．殻套部にはゆるく左巻きらせん（右下方から左上方へのぼる），もしくは貫殻軸とほぼ平行に胞紋が列をなし，条線を構成する(2-4, 7, 9, 11)．条線は直線的で10 μmあたり16～22本．条線を構成する胞紋は10 μmあたり12～16個．胞紋は内面で肉趾状師板および薄い師板により閉塞される(13, 14)．

針・突起 連結針は太く，間条線2本分ほどの幅がある(11, 12)．また，先端はへら状に平らに広がり，隣り合う殻の連結針と堅く咬み合う．連結針の基部に多数の顆粒が存在する(12)．殻套縁内側にある横輪の肋の上（胞紋にして2，3個分上）から唇状突起が斜め下の横輪の上へ伸び，スリット上の開口を開く(14矢印)．突起の柄の部分は殻套上に融合しており目立たない．

半殻帯 開放型で多数の胞紋もつ8枚ほどの帯片よりなる(10)．接殻帯片の胞紋は貫殻軸方向に列をなす．また，中間帯片の殻側の部分では胞紋は不規則に分布し，連結帯片側の部分では整列する．さらに連結帯片の胞紋は不規則に分布する．

葉緑体 円盤状，多数．
有性生殖 初生殻は半球形（Crawford *et al.* 2003）．
生活形 浮遊性であるが，付着基物上にも出現する．
汚濁耐性 弱汚濁耐性種．
出現地 支笏湖（北海道），竜ヶ窪（新潟県）など．

ノート 本種は*Aulacoseira ambigua*と似るが，*A. ambigua*では殻套縁の横輪が中空であるのに対し本種は中実であること，また連結針がより太く頑強なことから区別される．Krammer & Lange-Bertalot(1991)は，殻套部の条線が分離殻を除くすべての殻で明瞭にらせんを描くものに対し*A. italica*の名前を，また，殻套部の条線が貫殻軸とほぼ平行で直線的であり，かつ連結針上に5～10個の顆粒が認められるものには*A. crenulata*の名前をあてている．しかし，両種のタイプ標本に基づく研究を行ったCrawford *et al.*(2003)は，これらを同種と判定し，*A. crenulata*を新参の異名とした．彼らは殻套部における左巻きらせんの条線，および横輪の上4～5個上にある唇状突起を本種に特徴的な形質と報告した．本邦産個体では，殻套条線は左巻きらせんから直線状のものまで連続的に出現するが，右巻きらせんの個体は出現しておらず，彼らの提唱した分類形質は妥当なものであろう．唇状突起に関しては，本邦産の個体はやや下に位置している点で若干異なるが，この相違は分類群を分けるものでなく，種内変異の範囲にあるものと考えるべきである．小林・野沢(1982)による本邦産個体の微細構造の研究がある．

Plates 15, 16

Aulacoseira longispina (Hustedt) Simonsen

Simonsen, R. 1979. Bacillaria **2**: 61.
Basionym: *Melosira longispina* Hustedt in Huber-Pestalozzi, G. 1942. Phytoplank. Süsswass. p. 388. pl. 115. f. 469a.
Dimension: D. 4.5-16 µm, H. 11-16 µm, Str. 10-14 in 10 µm, Ar. 12-16 in 10 µm.

被殻・殻 被殻は円筒形(1-8)．連結針により糸状群体を形成．殻面は円形(9,10)．直径4.5〜16 µm，殻套高11〜16 µm．殻面外面は平滑(11)．殻肩は角ばり連結針がある(1-3,5-8,11-13)．殻套縁の内側に柵板のように発達した横輪をもつ(10,12,13)．殻套縁には胞紋はなく，殻套外面の端に沿って丸い顆粒と細長い顆粒がそれぞれ列をなす(6,11,12)．

条線・胞紋 殻面は胞紋をもたず無紋(9,11)．殻套では貫殻軸に対し若干斜めに胞紋が列をなして条線を構成する(1-3,5-8,11,12)．条線は10 µmあたり10〜14本．条線を構成する胞紋は10 µmあたり12〜16個．胞紋は内面で肉趾状師板により閉塞されている(15)．

針・突起 連結針は長く頑強，先端は尖り(1-3,5-8,11-13)，隣接する細胞の連結針としっかり咬み合う(16)．唇状突起は殻套縁内側の横輪の上に長く発達する．このため光学顕微鏡で横輪にピントを合わせると，その上にいくつかの大きな唇状突起の像が観察される(10)．SEM観察では，殻套から中心へ向かって伸長する蒲鉾形の構造が横輪の上に観察される(13矢印)．殻の断面の観察から，この唇状突起内部は管状で，殻套を貫通して外部に開口していること，内側の開口はスリット状であることがわかる(14矢じりおよび矢印)．

半 殻 帯 開放型で小孔を多数もつ帯片よりなる(7)．7枚ほど．
葉 緑 体 円盤状，数個(4)．
有性生殖 不明．
生 活 形 浮遊性であるが，付着基物上にもしばしば出現する．
汚濁耐性 弱汚濁耐性種．
出 現 地 中禅寺湖（栃木県）．

[ノート] 長い先細りの連結針，殻套縁の横輪の上に伸びた目立った唇状突起が本種の主な識別形質である．本種のタイプ産地は日光の中禅寺湖である．

Plates 17, 18

Aulacoseira nipponica (Skvortsov) Tuji

Tuji, A. 2002. Phycol. Res. **50**: 314.
Basionym: *Melosira solida* var. *nipponica* Skvortsov 1936. Philipp. J. Sci. **62**: 255. pl. 1. f. 1, 2, 21.
Dimension: D. 7-17 μm,　H. 8.5-12 μm,　Str. 12-14 in 10 μm,　Ar. 12-14 in 10 μm.

被殻・殻　被殻は円筒形[1-5, 10, 11]．連結針により糸状群体を形成する．殻面は円形[6-9]，直径7〜17 μm，殻套高8.5〜12 μm．殻面外面は平滑[13]．殻面周辺から殻肩にかけてわずかに丸みを帯びる[1]．殻肩には連結針がある[11]．殻套縁の内側に柵板のように発達した偽隔壁に似る大変肥厚した横輪をもつ[1, 3, 5, 6, 8, 12, 14]．横輪を生じる部分から殻套の端までに胞紋はないが，殻套外面には多数の顆粒が存在する[11, 12]．殻套は殻面と比べぶ厚い[14]．

条線・胞紋　殻面に胞紋はない[13]．殻套部で胞紋は貫殻軸と平行して直線的に，あるいは若干斜めになって並び，条線を構成する．条線は10 μmあたり12〜14本．条線を構成する胞紋は10 μmあたり12〜14個．胞紋は内面で肉趾状師板により閉塞される[14]．

針・突起　連結針は頑強，基部は中空で間条線2本分の幅をもち，先端は尖る[1, 2, 11, 13]．連結針は光学顕微鏡でもたやすく見ることができる[1, 2, 10]．唇状突起は比較的長く突出し，殻套縁にある横輪の上に[14] 存在し，4〜7個ほど観察される[6, 8, 12]．

半殻帯　開放型の帯片数枚からなる．
葉緑体　不明．
有性生殖　不明．
生活形　浮遊性であるが，付着基物上にもしばしば出現する．
汚濁耐性　弱汚濁耐性種．
出現地　Britton湖（カリフォルニア），琵琶湖（滋賀県）．

[ノート]　隣接する殻はこの立派な連結針が咬み合うことで糸状群体を作るのであるが，針の先端が先広ではないためはずれやすい．殻套縁の内側にある横輪が著しく厚く，殻面の厚みの3, 4倍ほどある．本種は従来 *Aulacoseira solida* (Eulenstein) Krammer と呼ばれることの多かった珪藻で，本邦では琵琶湖から出現報告（辻・伯耆 2001, Tuji 2002）がある．

Plates 19, 20

Aulacoseira subarctica (O. Müller) Haworth

Haworth, E. Y. 1988. In: Round, F. E. (ed.) Algae and the aquatic environment. p. 143.
Basionym: *Melosira italica* subsp. *subarctica* O. Müller 1906. Jahrb. Wiss. Bot. **43**: 78. pl. 2. f. 7, 8.
Dimension: D. 3.5-16.5 µm,　H. 6-14 µm,　Str. 15-20 in 10 µm,　Ar. 14-18 in 10 µm.

被殻・殻　被殻は円筒形(1-3, 5-7, 10-12, 15)．連結針により糸状群体を形成．殻面は円形(4, 8, 9)，直径3.5〜16.5 µm，殻套高6〜14 µm．殻面外面は平滑(15)．殻肩は角ばり連結針がある(15, 17, 18)．殻套縁の内側に発達した横輪をもつ(2, 7, 9, 11, 14, 17, 18, 21)．殻套縁に胞紋はなく，この部分の殻套外面には縦長の顆粒が列をなす(15)．初生殻は半球形(13, 14, 20, 21)．

条線・胞紋　殻面には胞紋が多数散在する(4, 8, 15)．殻肩には細長い胞紋が連結針の間に分布している(15)．殻套では若干らせんを描く胞紋列が条線を構成する(1, 3, 5, 6, 10, 12, 15)．条線は10 µmあたり15〜20本．条線を構成する胞紋は10 µmあたり14〜18個．胞紋は内面で肉趾状師板により閉塞される(17-19)．初生殻では胞紋は半球形をした殻全体に広がるが，それらが条線状に列をなすのは半球の縁付近のみで，中央を含む大部分では不規則に散在する(20, 21)．

針・突起　連結針は頑強で基部は2本の間条線につながっている．先端は先広ではなく先細りしている(1, 3, 5, 6, 10, 12, 15-18)．このため隣接する被殻は糸状群体のどの場所でも比較的分離しやすいようである．唇状突起は殻套縁にある横輪の上(18 矢印)，あるいは横輪と殻套部の境目付近にある(17, 19 矢印)．比較的小形であり，光学顕微鏡で帯面観を観察しても目立たない(9 矢印)．

半 殻 帯　開放型で無紋の帯片からなる．6, 7枚のものが観察されている．
葉 緑 体　円盤状，数個(Cox 1996)．
有性生殖　不明．
生 活 形　浮遊性であるが，付着基物上にもしばしば出現する．
汚濁耐性　弱汚濁耐性種．
出 現 地　阿寒湖（北海道），パンケ沼（北海道），木崎湖（長野県），津久井湖（神奈川県），中之条（群馬県），余呉湖（滋賀県），琵琶湖（滋賀県）など．

ノート　殻面全域に胞紋が存在すること，連結針が大きくかつ先細りであること，殻套縁内側の横輪がよく発達すること，唇状突起が小形であることが本種の主な識別形質である．なお本珪藻の名称は*Aulacoseira subarctica*にすべきでないとの見解がある（辻・伯耆 2001）．

Plates 21, 22

Aulacoseira tenuis (Hustedt) H. Kobayasi

Kobayasi, H. in Mayama, S., Idei, M., Osada, K. & Nagumo, T. 2002. Diatom **18**: 89.
Basionym: *Melosira longipes* var. *tenuis* Hustedt in Huber-Pestalozzi, G. 1942. Phytoplank. Süsswass. p. 389. pl. 115. f. 469b.
Synonym: *Aulacoseira longispina* var. *tenuis* (Hustedt) Simonsen 1979. Bacillaria **2**: 61.
Dimension: D. 3.5-13 µm, H. 10-18 µm, Str. 11-16 in 10 µm, Ar. 15-20 in 10 µm.

被殻・殻 被殻は円筒形[1-5, 9-13]．連結針により糸状群体を形成．糸状群体は直線状のものと湾曲するものがある．湾曲した糸状群体を作るものでは，個々の殻が湾曲している[10-12]．殻面は円形[6-8]．直径3.5〜13 µm，殻套高10〜18 µm．殻肩は角ばり連結針がある[9-13]．殻套縁の内側には柵板状の横輪が発達する[4, 8, 14]．殻套縁には胞紋はなく，この部分の殻套外面には不定形の顆粒が縁に沿って1, 2列存在する[9, 10, 12, 13]．

条線・胞紋 殻面で胞紋は周辺部にまばらにあるのみで，中央部にはない[6]．殻套では比較的小形の胞紋がらせん状に列をなし，条線を構成する．条線は10 µmあたり11〜16本．条線を構成する胞紋列はしばしば2重になる[1, 3, 9-13]．条線を構成する胞紋列は10 µmあたり15〜20個．胞紋は内面で肉趾状師板によって閉塞される[15, 16]．

針・突起 連結針は頑強，基部では間条線2本分の幅をもち，先端は尖る[11-13]．連結針は光学顕微鏡でもたやすく観察できる[1, 3]．唇状突起は横輪のすぐ上部にあり，渦巻き形で[14]，その先端には唇形の開口がある[16]．唇状突起の殻套外面にある開口は胞紋列の中にあるが，直径が大きいため区別できる[13矢印]．光学顕微鏡で殻套縁にある横輪にピントを合わせても唇状突起は観察されないが[8]，わずかに殻面よりにピントを合わせると，殻套に張り付いたような数個の黒い影として認めることができる[7]．

半殻帯 開放型の小孔を多数もつ帯片．数枚．
葉緑体 円盤状，数個[5]．
有性生殖 不明．
生活形 浮遊性であるが，付着基物上にもしばしば出現する．
汚濁耐性 弱汚濁耐性種．
出現地 湯ノ湖（栃木県）．

ノート　条線を構成する胞紋が，他の*Aulaoseira*種と比べて整然とせず，よろよろして配列していること，また部分的に2重列になること，殻面の周辺部にのみ胞紋が存在すること，顕著に大きい先細りの連結針，渦巻き状になる唇状突起が本種の特徴．なお，本種のタイプ産地は日光の湯ノ湖である．

Plates 23, 24

Aulacoseira valida (Grunow) Krammer

Krammer, K. 1991. Nova Hedwigia **52**: 98.
Basionym: *Melosira crenulata* var. *valida* Grunow in Van Heurck 1882. Syn. Diat. Belg. pl. 88. f. 8.
Dimension: D. 7-18 μm,　H. 9-18 μm,　Str. 14-16 in 10 μm,　Ar. 8-12 in 10 μm.

被殻・殻　被殻は円筒形(1-9, 13, 14). 連結針により糸状群体を形成. 殻面は円形(10-12), 直径7〜18 μm, 殻套高9〜18 μm. 殻面外面は平滑(13). 殻面の周辺から殻肩にかけ, わずかに丸みを帯びる(5, 9). 殻套縁内側に柵板のように発達した中実の横輪をもつ(2, 5, 14, 16, 17). 光学顕微鏡で帯面を観察すると, 横輪から殻套縁の端までは胞紋がないため, 透明に見える(3, 4, 8, 9). しかしSEM観察をすると, この部分の殻套外面には多数の顆粒が存在することがわかる(14).

条線・胞紋　殻面中央部には胞紋はなく, 殻面の周囲から殻肩にかけて, 細長い大きな胞紋が1列に配列する(13). 光学顕微鏡で殻面を観察すると, 一見 *Cyclotella* 属の種のように見えるのはこのためである(10, 11). 殻套部には胞紋がらせん状に配列した条線をもつ. 条線は10 μm あたり14〜16本. 条線を構成する胞紋は10 μm あたり8〜12個. 殻套壁は二重の層からできており, 胞紋は殻外面から見ると1つ1つが網目状の胞紋壁によって区画されているが, その下にはスポンジ状の厚い層が殻内面まで続いている(13, 16, 17). 唇状突起は殻套縁にある横輪の上部に存在する(16, 18 矢印).

針・突起　連結針は殻肩にあり, 胞紋と胞紋の間から突出する(13). 基部は棒状で, 先端はへらのように平らに広がる(15). SEMによる帯面の観察では, 両殻から伸びる連結針により, 隣り合う細胞の殻面と殻面が接する部分が, 完全に隠されてしまうため, 若干丸みを帯びた殻面周辺部を見ることはできない(15).

半 殻 帯　開放型で小孔を多数もつ帯片よりなる. 7枚ほど.
葉 緑 体　不明.
有性生殖　不明.
生 活 形　浮遊性であるが, 付着基物上にもしばしば出現する.
汚濁耐性　弱汚濁耐性種.
出 現 地　武蔵丘陵森林公園の沼 (埼玉県), 三宝寺池 (東京都), 下末吉層 (神奈川県), 大堤 (福島県), 尾瀬沼 (群馬県), 宮床湿原 (福島県) など, 弱酸性の止水域.

ノート　殻套部でらせんに配列する条線と, 連結針の形状が本種の特徴である. 連結針は基部がたいへん太いため(13), 光学顕微鏡による糸状群体の観察においても, その存在を簡単に認めることができる(1, 3, 4, 6, 8). 佐竹・小林(1991)による被殻微細構造の詳細な報告がある.

Orthoseira asiatica (Skvortsov) H. Kobayasi

Kobayasi, H. in Mayama, S., Idei, M., Osada, K. & Nagumo, T. 2002. Diatom **18**: 89.
Basionym: *Melosira roseana* var. *asiatica* Skvortsov 1938. Philipp. Journ. Sci. **65** (3): 265. pl. 3. f. 1, 3.
Dimension: D. 12-47 μm, H. 5.5-16 μm, Str. 11-18 in 10 μm, Ar. 16-18 in 10 μm.

被殻・殻 被殻は円筒形(7,8,9). 殻肩上方に伸びる発達した三角形の板状の刺(8,9)により糸状群体を形成する. 殻面は円形(1-4,6), 直径12〜47 μm. 殻面中央部の外面には不定形の皺や瘤が多数存在する(5,9). 殻肩はやや丸みを帯びる. 殻套高 5.5〜16 μm. 殻套縁の付近の外面は平滑な無紋域である(5,9).

条線・胞紋 殻面中心部には胞紋はないが, その少し外側から殻面縁辺へ胞紋列が伸びる. 中央近くでは条線は単列の胞紋からなるが, 縁近くになるに従い胞紋列の数が2列, 3列に増え, 殻外面から観察すると刺によって仕切られた条線束のように見える(1-4,6). 条線は10 μm あたり 11〜18 本. 胞紋は 10 μm あたり 16〜18 個, 孔状で, 内面でかさぶた状のものにより閉塞されている個体をいくつか確認したが, 詳細は不明.

突　起 板状の3角形の刺は頑強で, 光学顕微鏡で殻面を観察すると, 強コントラストの像となって現れる(2,4,6). 刺の殻套部での基部はしばしば二叉になる(5,7).

半殻帯 開放型の小孔をもつ帯片よりなる(7,8). 小孔は多くの個体の接殻帯片と2枚目の帯片で少ないことが観察された. 5〜8枚程度.

葉緑体 円盤状, 多数.

有性生殖 不明.

生活形 付着性の群体のみを観察している.

汚濁耐性 弱汚濁耐性種.

出現地 橋立鍾乳洞（埼玉県），日原川（東京都），神流川（群馬県）など，蘚苔類の表面に付着.

ノート 三角形の刺の先端は, 多くの殻では殻面より上方まで伸びており(8,9), 隣接する細胞の刺と組み合い, さらに粘液で接合部を覆うことにより, 糸状群体を形成しているものと思われる. しかし, ときおり, 殻面と同じ程度のレベルまでしか刺が発達していない殻が見受けられる(5,7). これらはおそらく分離殻であろう. 殻面中央部には *O. epidendron* 同様の, 胞紋よりはるかに大きな穴が3〜4個存在するが, その機能は不明である.

Orthoseira epidendron (Ehrenberg) H. Kobayasi

Kobayasi, H. in Mayama, S., Idei, M., Osada, K. & Nagumo, T. 2002. Diatom **18**: 89.
Basionym: *Stephanosira epidendron* Ehrenberg 1848. Ber. Bek. Verh. Köngl. Preuss. Akad. Wiss. Berlin **1848**: 219.
Synonym: *Melosira roseana* var. *epidendron* (Ehrenberg) Grunow in Van Heurck 1882. Syn. Diato. Belg. pl. 89. f. 17, 18; *Aulacoseira epidendron* (Ehrenberg) R. M. Crawford, 1981. Phycologia **20**: 190. f. 1, 2, 8, 9, 32-43.
Dimension: D. 9-32 μm, H. 6-13 μm, Str. 10-16 in 10 μm, Ar. 15-20 in 10 μm.

被殻・殻 被殻は円筒形[5,7]，殻面は円形[1-4,6]．直径9〜32 μm．殻套高6〜13 μm．殻面，殻套ともに外面に不定形の小さな瘤を多数もつ[5,8]．やや丸みを帯びた殻肩からは先端の尖った三角形の板状の刺が伸長する[5]．刺の基部はしばしば2本の肋となる[8]．

条線・胞紋 殻面では中心部を除き胞紋が周囲まで放射配列し条線を形成する[1-4]．条線は10 μmあたり10〜16本．条線を構成する胞紋は10 μmあたり15〜20個．殻套では条線は貫殻軸に平行に走り，その密度は殻面の条線の約2倍となる[5,6,8]．胞紋は孔状で，内面で肉趾状師板によって閉塞される．

突　起 殻面中心部に2,3個の穴がある[1-4]．これらはSEMでは殻面の外面で短い管状の突起として，内面では目立った穴として観察される[5,8]．また，突起の周囲には浅い溝がある[8]．

半殻帯 開放型で胞紋のある帯片からなる．4〜6数程度．

葉緑体 円盤状，多数（Crawford 1981）．

有性生殖 増大胞子と初生殻のSEMによる詳細な微細構造がRoemer & Rosowski (1980) により報告されている．

生活形 付着性の群体のみを観察している．

汚濁耐性 弱汚濁耐性種．

出現地 橋立鍾乳洞（埼玉県），中津峡（埼玉県），大正洞（山口県），屋久島湊川（鹿児島県）など，蘚苔類の表面に付着．

ノート　殻面中央部の突起についてCrawford (1981) は竜骨突起（carinoportulae）の名前を与えている．その機能については，おそらく粘液を出して隣接する細胞とつなぎあっているのだろうとの類推がされているにすぎない（Round *et al.* 1990）．なお，Round *et al.*（1990, p.96）に記された *O. epidendron* の名は，基礎異名の引用を伴っておらず，正式なものではない．本属の2種の殻の形態変異について，Spaulding & Kociolek (1998) が詳細な報告を行っている．

Brebisira arentii (Kolbe) Krammer

Krammer, K. 2001. In: Jahn *et al*. (eds) Lange-Bertalot-Festscherift. A.R.G. Ganter Verlag. p. 19.
Basionym: *Cyclotella arentii* Kolbe 1948. Sven. Bot. Tidsk. **42**: 464. f. 9.
Synonym: *Coscinodiscus arentii* (Kolbe) Cleve-Euler 1951; *Melosira arentii* (Kolbe) Nagumo et Kobayasi 1977.
Dimension: D. 11-17.5 μm,　H. 6-11 μm,　Str. 21-24 in 10 μm.

被殻・殻　被殻は偏平な太鼓形で，殻は円形[1-3]．殻套にある刺列でゆるく結合し数個の細胞が連なった糸状の群体を作る[4,5]．直径 11〜17.5 μm，殻套高 6〜11 μm．

条線・胞紋　殻面縁辺部の条線は放射状で半径の約 1/2 を占め，10 μm あたり 21〜24 本．中心部は正接配列となる．縁辺の条線は殻套にも連続して見られる[2-5]．

唇状突起・突起　殻面，殻套にも顆粒が散在し，特に殻套縁に多数見られる[6]．殻套には小針列が見られるが，殻内面は平坦で唇状突起や有基突起などの突起物はない[7]．

半殻帯　数枚．

葉緑体　円盤状，多数．

有性生殖　不明．

生活形　不明．

汚濁耐性　腐植酸性，貧腐水性種．

出現地　各地のミズゴケ湿原．仙女ヶ池（埼玉県），大峰沼（群馬県），藺牟田池（鹿児島県）など．

[ノート] 本種は南雲・小林(1977)によって *Melosira* 属に組み替えられたが，Krammer(2001) によって記載された新属 *Brebisira* に移された．

Thalassiosira allenii Takano

Takano, H. 1965. Bull. Tokai Reg. Fish. Res. Lab. **1965** (42): 4. text-f. 2a-f. pl. 1. f. 9-11.
Dimension: D. 8-18 μm,　Str. 20 in 10 μm.

被殻・殻　殻は円形⁽¹⁻⁵⁾．細胞は単体から4細胞程度の短い群体を形成する．殻套部は狭いが斜行するため，帯面観では被殻は4つの角が欠けた八角形に見える⁽⁶⁾．直径8〜18 μm．

条線・胞紋　条線は正接であるが放射状にも見え，10 μmあたり20本．殻套部ではさらに細かい．

針・突起　殻のほぼ中央には中心有基突起が1個あり，これから放出される粘液によって群体を形成する⁽⁷⁾．殻肩部に殻縁有基突起の外管が環状に配列し，10 μmあたり5〜7本．その列の中に1個の唇状突起の開口である外管があり，縁辺有基突起のものより長い⁽⁷,⁸ 矢印⁾．

半殻帯　開放型．

葉緑体　円盤状，多数．

有性生殖　不明．

生活形　浮遊性．

汚濁耐性　不明．

出現地　汽水域に広く分布する．多摩川河口（東京都），東京湾（東京都），渥美湾（愛知県）など．

Thalassiosira eccentrica (Ehrenberg) Cleve

Cleve, P. T. 1904. Bull. Cons. Expor. Mar. 1903-1904. 216.
Basionym : *Coscinodiscus eccentricus* Ehrenberg. 1841. Abh. Königl. Akad. Wiss. Berlin. **1839**: 146.
Dimension: D. 12-101 μm, Str. 5-9 in 10 μm.

- **被殻・殻** 殻は円形[1,2]．殻面から放出される粘液糸によってつながる群体を形成する．直径 12〜101 μm．
- **条線・胞紋** 条線は正接に配列し，中心部では 10 μm あたり 5〜8 個，殻縁では 7〜9 個．胞紋は外面に開口をもち，内面が閉塞される小室構造となる[4]．
- **針・突起** 殻外面の殻肩部には先端が尖った小針で不規則な間隔で配列する．その中に 1 本の長い突起が認められるが，これは唇状突起の外管である[3,5矢印]．殻外面には胞紋と胞紋に挟まれて多数の小孔が認められるが[4矢印]，これは殻面有基突起の開口であり，殻内面には 4 脚をもつ短い内管がある[6矢印]．
- **半殻帯** 開放型．
- **葉緑体** 円盤状，多数．
- **有性生殖** 不明．
- **生活形** 浮遊性．
- **汚濁耐性** 不明．
- **出現地** 沿岸から汽水域まで広く分布する．多摩川河口（東京都），八郎潟（秋田県）など．

Thalassiosira faurii (Gasse) Hasle

Hasle, G. R. 1978. Phycologia **17**: 282.
Basionym: *Coscinodiscus faurii* Gasse 1975. Doct. Sci. Nat. l'Univ. Paris, vol. 2-3. p. 24. pl. 32. f. 1, 2.
Dimension: D. 19-55 µm, Str. 22 in 10 µm, Ar. 22-25 in 10 µm.

被殻・殻 殻は円形(1-5)．殻面は平坦で殻高は直径よりも短い．単体で浮遊したり，粘質で連なった群体を形成する場合もある．直径 19〜55 µm．

条線・胞紋 条線は放射状で 10 µm あたり 22 本(1-5)，条線は明瞭な点紋で構成され 10 µm あたり 22〜25 個．光顕で条線が黒く"抜けた"ように見える箇所が数カ所観察されるが，それは殻面有基突起である(2,3 矢印)．

針・突起 殻面有基突起は殻面の中間辺りに数個ずつまとまって存在する(7)．殻肩部にはほぼ等間隔で 1 列の突起が認められるが，これらは殻套部有基突起の外管である(6)．殻縁有基突起は殻内面では短い内管と 4 脚をもつ(8)．唇状突起は複数あり，殻縁有基突起の 7〜9 個ごとに見られる．

半 殻 帯 開放型．
葉 緑 体 不明．
有性生殖 不明．
生 活 形 浮遊性．
汚濁耐性 中汚濁耐性種．
出 現 地 淡水域から汽水域の河川に広く分布．富栄養化した池沼にも見られる．垳（がけ）川（東京都），多摩川河口（東京都），舟田池（千葉県），上田城跡の堀（長野県）など．

Thalassiosira guillardii Hasle

Hasle, G. R. 1978. Phycologia **17**: 274. f. 28-50.
Dimension: D. 4-35 µm.

被殻・殻 殻は円形(1-6)．群体は知られていない．殻面は平坦．直径4〜35 µm．
条線・胞紋 条線は光顕では観察できない．透過電顕観察から，細かな胞紋がやや不規則に配列した放射条線であることがわかる(9,10)．
針・突起 殻面に有基突起が見られる個体(3,9,10)と，見られない個体(2,8)があり，その数は0〜9と多様である．この殻面有基突起は3脚(9,10矢印)．殻套部有基突起は1列で4脚，短い外管をもつ(7,8)．唇状突起は殻套部有基突起より殻縁にあり(9矢印)，その外管は有基突起の外管に比べ少し長い(7矢印,11)．
半殻帯 不明．
葉緑体 不明．
有性生殖 不明．
生活形 浮遊性．
汚濁耐性 不明．
出現地 淡水域から汽水域まで広く分布し，多少汚濁した池沼にも見られる．涸沼川（茨城県），松本城堀（長野県）など．

Thalassiosira lacustris (Grunow) Hasle

Hasle, G. R. in Hasle, G. R. & Fryxell, G. A. 1977. Nova Hedwigia Beih. **54**: 40.
Basionym: *Coscinodiscus lacustris* Grunow in P. T. Cleve & Grunow 1880. Kongl. Svensk. Vet.-Akad. Handl. **17** (2): 114.
Dimension: D. 20-75 μm, Str. 12 in 10 μm, Ar. 10-14 in 10 μm.

被殻・殻 殻は円形[1-3]．殻面は大きく波打ち，光顕でも殻の半分にピントが合うが，一方に合わないことで確認できる．SEM観察では中心部がS字状に波打つのが認められる[5]．直径20〜75 μm．

条線・胞紋 条線は放射状で10 μmあたり12本[1-3]，条線は明瞭な点紋で構成され10 μmあたり10〜14個．胞紋は外面に開口をもち，内面では薄皮によって閉塞される[5,6]．

針・突起 殻外面には小顆粒が多数，不規則にある[5]．殻肩から殻套部には2種類の突起列がある．殻肩の長く先端が尖った突起は中空の針状突起で，殻套の短い突起は縁辺縁辺有基突起の外管である[5,7]．殻内面の凹んだ側に，通常数個の中心域有基突起が見られ，4脚をもつ[4]．また，縁辺有基突起列に挟まって唇状突起が1個見られるが，内面の唇状部分はよく発達し，縦裂溝は間条線に対して斜めの方向に位置する[6矢印]．唇状突起は内面観で凸を左にした場合，ほぼ時計の9時の位置にある．

半殻帯 開放型．
葉緑体 円盤状，多数．
有性生殖 不明．
生活形 浮遊性．
汚濁耐性 不明．
出現地 淡水域から汽水域まで広く分布し，多少汚濁された池沼によく見られる．多摩川河口（東京都），山中湖（山梨県），パンケ沼（北海道），長面浦（宮城県）など．

ノート 本種は一時 *Thalassiosira bramaputrae* (Ehrenberg) Håkansson & Locker とされたが，Hasle & Lange (1989) によって両種の区別が明確にされた．

Thalassiosira nordenskioeldii Cleve

Cleve, P. T. 1873. Bih. Kongl. Svenska Vet.-Akad. Handl. **1** (13): 7. pl. 1. f. 1.
Dimension: D. 12-43 μm, Str. 16-18 in 10 μm, Ar. ca. 15 in 10 μm.

被殻・殻 殻は円形(2,3). 中心部有基突起から放出される粘液によって, 長い群体を形成する(1,4). 殻中央部がくぼむため, 光顕では殻面が大きく波打って見える. 殻套部が幅広いため, 帯面観では被殻は八角形に見える. 直径 12〜43 μm.

条線・胞紋 条線は正接であるが, 放射状にも見え, 中心から殻縁まで連続し, 10 μm あたり 16〜18 本. 条線は明瞭な点紋で構成され 10 μm あたり約 15 個.

針・突起 殻中心部に外管が発達した中心有基突起の開口が 1 個あり, そこから粘液を放出する(5,6矢印). 殻套部には先端が広がった突起が全周に 8〜20 個ほどあり, これは縁辺有基突起の外側への開口である(5,6).

半 殻 帯 開放型.

葉 緑 体 円盤状, 多数.

有性生殖 不明.

生 活 形 浮遊性.

汚濁耐性 不明.

出 現 地 冷水域に多い. 東京湾(東京都), 渥美湾(愛知県)など.

Thalassiosira pseudonana Hasle & Heimdal

Hasle, G. R. & Heimdal, B. R. 1970. Nova Hedwigia Beih. **31**: 565. f. 27-38.
Dimension: D. 4-9 μm, Str. 12 in 10 μm.

被殻・殻　殻は円形 (1-9)．群体は知られていない．直径 4〜9 μm．

条線・胞紋　条線は微細であるため，光顕では認められない．透過電顕観察から，条線は放射状に配列していることがわかる (11)．10 μm あたり 12 本．胞紋は外面ではスリット状の開口をもつが，内面では薄皮により閉塞される．

針・突起　殻面殻肩部には短い中空の突起が 1 列に配置するが，これらは縁辺有基突起であり，内面では 2 脚である (10-13)．縁辺有基突起列に挟まれて 1 個の唇状突起がある (13 矢印)．

半 殻 帯　開放型．

葉 緑 体　不明．

有性生殖　不明．

生 活 形　浮遊性．

汚濁耐性　不明．

出 現 地　汽水域に広く分布する．坩（がけ）川（東京都），多摩川河口（東京都），涸沼川（茨城県）など．

Thalassiosira tenera Proshkina-Lavrenko

Proshkina-Lavrenko, A. I. 1961. Notul. Syst. Inst. Crypt. Acad. Sci. URSS. **14**: 33. pl. 1. f. 1-4. pl. 2. f. 5-7.

Dimension: D. 10-29 μm, Str. 9-16 in 10 μm.

被殻・殻 殻は円形(1-8),殻面は平坦であるが,個体によっては殻外面に多数の突起をもつ.突起が著しく肥厚した個体では,帯面観で殻縁が鋸歯状となる(9).生細胞は殻中心から放出された粘質糸によって,2細胞程度が連なることがあるが,長い群体は観察されていない.直径10〜29 μm.

条線・胞紋 条線は正接配列が多いが,なかには中心から放射状の束列になる場合もある.殻面の胞紋は明瞭で10 μmあたり9〜16個,殻肩部だけは小さな胞紋で構成される.

針・突起 殻肩部には2種類の突起がある.殻面側のものは先端が尖った閉塞突起であり,殻縁側のものは縁辺有基突起の外管である(5-7).殻中央には1個の中心有基突起の開口がある(6, 7, 10 矢印).

半 殻 帯 開放型.

葉 緑 体 不明.

有性生殖 不明.

生 活 形 浮遊性.

汚濁耐性 不明.

出 現 地 温帯沿岸汽水域から湾に広く分布する.多摩川河口(東京都),松川浦(福島県),三河湾(愛知県)など.

Thalassiosira weissflogii (Grunow) G. A. Fryxell & Hasle

Fryxell, G. A. & Hasle, G. R. 1977. Nova Hedwigia Beih. **54**: 68.
Basionym: *Micropodiscus weissflogii* Grunow in Van Heurck, H. 1880-1885. Types du Synopsis. No. 11, 416.
Dimension: D. 5-35 µm, Str. 30-40 in 10 µm.

被殻・殻 殻は円形[1-4]．殻外面はほぼ平坦で，放射状の細い肋が殻全面を覆っている．細胞は単体か，粘質で多数の細胞が絡まった群体を作り浮遊する．直径5～35 µm．

条線・胞紋 条線は細かく，放射状で10 µmあたり30～40本[1-5]．透過電顕観察から，放射肋が網目状の模様を形作っていることがわかる[8]．

針・突起 殻外面の殻肩部には多数の突起が縁取っているが，これらは縁辺有基突起の外管である[5,8]．殻中央部には4～9個の有基突起があり，光顕観察でも明瞭に存在を確認できる[3,4,6,7]．縁辺有基突起の内管は4脚で，それらの列に挟まって唇状突起が1個見られる[9]．この突起の唇状部分はよく発達し，裂溝は中心方向に向いている．

半 殻 帯 開放型．

葉 緑 体 円盤状，多数．

有性生殖 不明．

生 活 形 浮遊性．

汚濁耐性 不明．

出 現 地 淡水域から汽水域まで広く分布し，富栄養の地域で大量発生することがある．八郎潟（秋田県），多摩川河口（東京都）など．

Cyclostephanos dubius (Fricke) Round

Round, F. E. 1982. Arch. Protistenk. **125**: 326.
Basionym: *Cyclotella dubius* Fricke 1900. A. S. Atlas. pl. 222. f. 23, 24.
Synonym: *Stephanodiscus dubius* (Fricke) Hustedt 1928. Kies. p. 367.
Dimension: D. 8-20 μm,　F. 9-11 in 10 μm.

- **被殻・殻**　殻は円形で中央部が凹凸している(1-19)．中央部で条線束は明瞭な1列胞紋であるが周辺部では不明瞭となる(1-8)．被殻は凹殻と凸殻の組み合わせが多いが(10)，両方が凸殻の場合もある(9)．中心環は不明瞭である．殻肩の束間肋（条線束と条線束の間）上に刺がある(11,12)．殻面周辺部では束間肋が発達し，その間が溝状に見えることもある(11)．直径8〜20 μm．
- **条線・胞紋**　条線束は中央で粗い1列の胞紋，殻縁に向かって広がり，より細かな2列から4列の胞紋をもつ(13,17,21)．胞紋は外面では円形に開口し，内面では円形のドーム状の多孔篩板によって閉塞される(13,14,17,20)．束間肋は外面では中心近くから発達するが，内面では周縁部のみで発達し，溝を形成する(15,19-21)．条線束は10 μmあたり9〜11本．
- **針・唇状突起・有基突起**　殻肩の束間肋上に刺がある(11)．1個から数個の中心有基突起が中心から少しはずれた位置にあり，2個の付随孔をもつ(14,15,19)．この中心有基突起の外面の開口は，胞紋の開口と区別が難しい．縁辺有基突起は殻套にあり，3,4本の条線束ごとに刺のすぐ下に開口し，内面では2個の付随孔をもつ(11,19-21)．これらの突起からは粘液が分泌される(11)．唇状突起は1個で，有基突起よりわずかに中央寄りにあり，いずれかの有基突起の隣の束間肋上にある(20,21)．
- **半　殻　帯**　開放型で，幅広の接殻帯片と細い4枚の帯片．
- **葉　緑　体**　多数．
- **有性生殖**　不明．
- **生　活　形**　浮遊性．
- **汚濁耐性**　弱汚濁耐性種．
- **出　現　地**　諏訪湖（長野県），多摩川河口（東京都），北浦（茨城県），山中湖（山梨県），池田湖（鹿児島県）など．

Cyclostephanos invisitatus (Hohn & Hellerman) Theriot *et al.*

Theriot, E., Stoermer, E. & Håkansson, H. 1987. Diatom Research **2**: 256.
Basionym: *Stephanodiscus invisitatus* Hohn & Hellerman 1963. Trans. Amer. Microsc. Soc. **87**: 325. pl. 1. f. 7.
Dimension: D. 5-14 μm, F. 14-20 in 10 μm.

被殻・殻 殻は円形で殻面は平たく，中央に中心環がある[1-5,9]．殻面と殻套の境となる殻肩には刺があり，すべて束間肋（条線束と条線束の間）上に規則的に配置する[8]．各々の束間肋は殻面から殻套には達するとV字形に分岐する[5,11]．殻の珪化の程度により，殻外面の形状が大きく変わる．厚く珪化した殻では，胞紋は小箱状になり外面に小円形の開口をもつが[8]，珪化の弱い殻では小箱状にならず，多孔師板のみによってふさがれる[9,11]．直径5～14 μm．

条線・胞紋 中央に無紋の中心環があり，その内側に数個の胞紋がある[9]．条線束は中央側で1列，殻縁側で2列の胞紋から成る[7-11]．束間肋は殻套に達するとV字形に分岐する．殻套の胞紋は殻面の胞紋に比べ細かい．胞紋は珪化の強い殻では，内面にドーム状に張り出した多孔師板によって閉塞され[10]，外面では小円形の開口をもつ[8]．珪化の弱い殻では，束間肋の中段に師板が存在する．条線束は10 μmあたり14～20本．

唇状突起・有基突起 先の尖った刺が，すべての束間肋が殻肩に達するところに配置されている[6,8,11]．縁辺有基突起は4～7条線束ごとに1個存在し，小針の直下の殻套に短い管状突起として開口し[8,11]，殻内面ではより長い内管と2個の付随孔をもつ[7]．中心有基突起は中心環のすぐ外側にあり，2個の付随孔をもち[7,9]，外面には丸く開口する[8,11矢印]．唇状突起は1個で殻套にあり[7]，外面に有基突起の開口より小さな開口をもつ．

半殻帯 開放型，数枚[6]．
葉緑体 多数．
有性生殖 不明．
生活形 浮遊性．
汚濁耐性 中汚濁耐性種．
出現地 八郎潟（秋田県），中川（東京都），舟田池（千葉県），涌池（長野県），北浦（茨城県）など，淡水から汽水の湖沼や河川．

ノート 本種は，小形の *Stephanodiscus hantzschii* f. *tenuis* (Hustedt) Håkansson et Stoermer と似るが，中心有基突起があることで区別できる．本種の詳しい分類と形態は，小林・井上(1985)によって報告されている．

Cyclotella atomus Hustedt

Hustedt, F. 1937. Arch. Hydrobiol. Suppl. **15**: 143. pl. 9. f. 2-4.
Dimension: D. 4-7 μm, Str. 18-20 in 10 μm.

被殻・殻　殻は円形(1-10)．中央部には広い無紋域があり，縁辺条線は短い．殻が小形であるため，殻面の凹凸は光顕ではほとんど見分けがつかないが，SEM 観察ではわずかに波打つのが認められる(12)．直径 4～7 μm．

条線・胞紋　縁辺条線は 10 μm あたり 18～20 本(1-10)，外面は多孔師板で覆われる(13)．間条線は，殻内面で肥厚し(1-11)，この発達した肋の 1～3 本おきに殻肩部の内側に有基突起がある．

針・突起　殻面には通常 1 個の中心有基突起が見られるが(12)，まれにない場合もある(4)．有基突起は，内面では明瞭な 3 個の付随孔をもち(11)，その外部開口は丸く外管は認められない(12,13)．また，縁辺有基突起は発達した肋上にあり，2 個の付随孔をもつ(11)．唇状突起は発達した肋上の殻肩部に 1 個見られ(11 矢印)，この外部開口は有基突起の開口より小さく楕円形で，外管は認められない(13 矢印)．唇状突起の裂口は間条線に対して斜めの方向に位置する．なお，この唇状突起の位置は，中心有基突起と殻の中心点を結ぶ線のほぼ延長線上にある．

半殻帯　開放型，数枚．
葉緑体　円盤状，数個．
有性生殖　不明．
生活形　浮遊性．
汚濁耐性　中汚濁耐性種．
出現地　淡水域から汽水域まで広く分布．中川（東京都），多摩川河口（東京都）など．

Cyclotella atomus Hustedt var. *gracilis* Genkal & Kiss

Genkal, S. I. & Kiss, K. T. 1993. Hydrobiologia Suppl. **269/270**: 43.
Dimension: D. 3.5-12.5 µm, Str. 20-26 in 10 µm.

被殻・殻 殻は円形(1-5,7). 縁辺条線域の幅は殻半径の1/5～1/3を占める. 殻が小形であるため殻面の凹凸は殻面観ではほとんど見分けがつかないが, 帯面観ではS字状にわずかに波打つのが認められる(6). 殻中央部には不規則な紋様が見られるが(2-4), これは殻外面にある細かな凸凹であることがわかる(8). このような凸凹は小山状 (colliculate; Håkansson 1982) と呼ばれることもある. それに対し殻内面は平滑である(9). 直径3.5～12.5 µm.

条線・胞紋 縁辺条線は10 µmあたり20～26本(1-5,7). 条線は殻縁では4, 5列の胞紋列より成り, それらの外面は薄皮で閉ざされ(7), 内面では楕円形の開口をもつ明瞭な長胞構造となる(10).

針・突起 殻面には通常1個の中心域有基突起が見られるが(9), 個体によっては2個の場合もある(7). この殻外面への開口は丸い孔として開き, 外管はほとんど認められない(8). しかし, 内面には明瞭な内管が見られ, そのほとんどは3脚をもつ(9). 縁辺有基突起は間条線上にあり, 間条線の3～4本ごとに存在する. これらの縁辺有基突起は殻内面に長い内管をもち, それぞれ貫殻軸方向に2脚である(9,10). また, 縁辺有基突起列に挟まって唇状突起が1個見られるが(9 矢印), 外側への開口は有基突起の開口より小さく楕円形で, 外管は認められない(8 矢印). 内側の唇状部分は小さいがよく発達し, 縦裂溝は間条線に対して斜めの方向に位置する.

半 殻 帯 開放型.
葉 緑 体 不明.
有性生殖 不明.
生 活 形 不明.
汚濁耐性 中汚濁耐性種.
出 現 地 淡水域から汽水域まで広く分布. 中川 (東京都), 多摩川 (東京都), 八郎潟 (秋田県), 猪鼻湖 (愛知県) など.

[ノート] 本変種は *Cyclotella caspia* Grunow として報告されていたが (南雲・小林 1985), その後タイプ標本が精査された結果, *C. caspia* には殻面に多数の有基突起があることなどが確認された(Krammer & Lange-Bertalot 1991). そのため, 本邦にしばしば出現する種類は, *C. caspia* ではなく, Genkal & Kiss (1993) の見解に従って *C. atomus* var. *gracilis* とした.

Plate 43

Cyclotella criptica Reimann *et al.*

Reimann, B. E. F., Lewin, J. M. C. & Guillard, R. R. L. 1963. Phycologia **3**: 82.
Dimension: D. 6.5-9.0 µm,　Str. 8-10 in 10 µm.

被殻・殻　殻は円形(1-5). 縁辺条線域の幅は殻半径のほぼ3/4を占める. 殻面はほとんど平坦であるが, 条線部がややうね状に盛り上がる. 直径6.5～9.0 µm.

条線・胞紋　縁辺条線は10 µmあたり8～10本(1-5), 条線は殻縁では7～8列の胞紋列より成るが, 殻の中心に向かうに従って減少し, 最後には1～2列になる(5-7). 間条線は内面で肥厚し肋状になる. 間条線の有基突起が, 光顕では殻縁に黒い明瞭な肥厚肋として写る.

針・突起　殻面には通常1個の中心域有基突起が見られるが(2-8), 個体によってはない場合もある. この外面への開口は丸い孔として開き, 外管はほとんど認められない(6矢印). しかし内面には明瞭な内管が見られ, そのほとんどは3脚をもつ(7,8矢印). また, 縁辺有基突起列に唇状突起が1個見られるが(8矢じり), これの外面への開口は有基突起の開口より小さく楕円形で, 外管は短い(6矢印). 内面の唇状部分は小さいがよく発達し, 縦裂溝は間条線に対して斜めの方向に位置する(8矢じり). なお, この唇状突起の位置は, 中心域有基突起と殻の中心点を結ぶ線のほぼ延長線上にある.

半 殻 帯　開放型.
葉 緑 体　円盤状, 多数.
有性生殖　不明.
生 活 形　浮遊性.
汚濁耐性　中汚濁耐性種.
出 現 地　淡水域から汽水域まで広く分布. 涌池（長野県）, 中川（東京都）など.

Cyclotella litoralis Lange & Syvertsen

Lange, C. B. & Syvertsen, E. E. 1989. Nova Hedwigia **48** (3-4): 343, 344. f. 1-30.
Dimension: D. 20-80 μm,　Str. 7-9 in 10 μm.

被殻・殻　殻は円形(1-4,6)．縁辺条線域の幅は殻半径のほぼ1/2を占める．殻面中心域には全面に渡って不規則な凹凸があり(6)，中心部がS字状に大きく波打つ(6,8)．直径20～80 μm．

条線・胞紋　縁辺条線は10 μmあたり7～9本(1-4)，条線は殻縁では5～7列の胞紋より成るが，殻の中心に向かうに従って減少し，最後には2，3列になる(6,7)．間条線は無紋で，わずかに肥厚する．殻内面は中心部が大きく凸凹となる以外は平滑で，縁辺部に長胞の殻内面への開口が見られる(8,9)．殻中心部の隆起した側（内面観では凹む側）の無紋域には，数個の有基突起が円弧状に配列する中心域有基突起列が見られる(5,8)．間条線部は殻内面では短い肋となって肥厚し，縁辺部に有基突起をもつ(8,9)．

針・突起　中心域有基突起列は光顕でも容易に観察できる(2,3)．これらの有基突起は2～3脚をもつ(5)．外面への開口は丸い小孔として開き，外管は認められない．殻縁有基突起はほぼ2～3本おきの肋上にあり，内面には明瞭な内管が見られ，2脚をもつ(9)．また，縁辺有基突起列に挟まって唇状突起が1個見られるが(8矢印,9)，内面の唇状部分はよく発達し，縦裂溝は殻面に対して垂直に位置する．なお，この唇状突起の位置は，中心域有基突起と殻の中心点を結ぶ線のほぼ延長線上にある(8)．

半 殻 帯　開放型．
葉 緑 体　円盤状，多数．
有性生殖　不明．
生 活 形　浮遊性．
汚濁耐性　α-中腐水，貧塩–好塩性，好アルカリ性種．
出 現 地　主に内湾の底泥試料に見られるが，汽水域からも出現する．

Cyclotella meduanae Germain

Germain, H. 1981. Flore Diat. **36**. pl. 8. f. 28; pl. 154. f. 4, 4a.
Dimension: D. 5-7 μm, Str. 13-16 in 10 μm.

被殻・殻 殻は円形(1-7)．縁辺条線は粗く，殻半径の約2/3まで達する．殻が小形であるため殻面の凹凸は，光顕ではほとんど見分けがつかないが，殻外面では縁辺条線部がうね状に大きく盛り上がり(9)，殻内面では間条線が肋状に隆起している(8)．直径5～7 μm．

条線・胞紋 縁辺条線は10 μmあたり13～16本，条線は殻縁では7～10列の胞紋列より成るが，殻の中心に向かうに従って減少する．殻内面の肥厚肋の殻套部には2～3本おきに縁辺有基突起がある(8,9)．TEMによって観察すると，条線を構成する胞紋はさらに小さな約27 nmの小孔を有する師皮で閉塞されていることがわかる(10)．

針・突起 殻套部外面には多数の小顆粒が見られ，間条線の2～3本おきに外管のない縁辺有基突起の開口が見られる(9)．縁辺有基突起は3脚で明瞭な内管が見られる(8)．殻套部には唇状突起が1個存在する(8矢印)．この唇状突起の外側への開口は，有基突起の開口より小さな円形で外管は認められない(9矢印)．

半 殻 帯 開放型，数枚．
葉 緑 体 不明．
有性生殖 不明．
生 活 形 浮遊性．
汚濁耐性 中汚濁耐性種．
出 現 地 汽水域に広く分布．中川（東京都），多摩川（神奈川県），涸沼（茨城県），八郎潟（秋田県），長良川河口（三重県）など．

Cyclotella meneghiniana Kützing

Kützing, F. T. 1844. Bacill. p. 50. pl. 30. f. 68.
Dimension: D. 6-35 μm, Str. 7-9 in 10 μm.

- **被殻・殻** 殻面は円形[1-5,10]．縁辺条線は明瞭で，殻半径の1/2〜1/3を占める．中心域はほとんど無紋で，大きく波打つ．SEM観察から中心部の半分ずつで凹凸していることがわかる[6]．直径6〜35 μm．
- **条線・胞紋** 縁辺条線は10 μmあたり7〜9本[1-5,10]，条線部は盛り上がり，殻縁では7〜8列の胞紋列より成るが，殻の中心に向かうに従ってやや減少する[10,11]．殻中心部の隆起した側の無紋域の条線域寄りに通常2, 3個の中心域有基突起が見られる[1-5,7,10]．条線部は小孔で穿孔された1層の珪酸壁でできている[10,11]．間条線部は殻内面では放射肋となって肥厚し，各肋の殻縁辺部に有基突起をもつ[7,8]．
- **針・突起** 殻面には通常1〜3個の中心域有基突起が見られる[7,10]．この外面への開口は丸い小孔で，外管は認められない[6,11]．また，内面には明瞭な内管が見られ3脚をもつ[8]．また，縁辺有基突起列の中に唇状突起が1個見られるが[7矢印]，唇状部分はよく発達し，縦裂溝は間条線に対してわずかに斜めの方向に位置する．なお，この唇状突起の位置は，中心域有基突起と殻の中心点を結ぶ線のほぼ延長線上にある．
- **半 殻 帯** 開放型．
- **葉 緑 体** 円盤状，多数．
- **有性生殖** 不明．
- **生 活 形** 浮遊性．
- **汚濁耐性** α-中腐水，貧塩-好塩性，好アルカリ性種．
- **出 現 地** 淡水域から汽水域まで広く分布し，多少汚濁された池沼によく見られる．諏訪湖（長野県），小千谷の水田（新潟県），東京湾（東京都），多摩川（東京都），宝蔵寺沼（埼玉県），松本城堀（長野県），市ヶ谷堀（東京都）など．

Cyclotella ocellata Pantocsek

Pantocsek, J. 1901. Kies. Balaton. p. 104. pl. 15. f. 318.
Dimension: D. 6-35 μm,　Str. 7-9 in 10 μm.

被殻・殻　殻は円形[1-5, 10]．縁辺条線域の幅は殻半径の 1/2〜1/3 を占める．条線の長さが不揃いであるため，中心域の形が丸くならない[1-5]．殻面中心域は平滑であるが，内部まで貫通していない大きなくぼみが数個存在する[7]．直径 6〜35 μm．

条線・胞紋　縁辺条線は 10 μm あたり 7〜9 本[1-5]，条線は 4〜5 列の胞紋から成り，少し大きな胞紋が両側に配列し，その間に小さな胞紋が配列する[8]．間条線は殻内面の殻縁部から殻肩部では肥厚し，肋を形成する[9]．殻縁部では肋の 2〜4 本ごとに有基突起をもつ[9, 10]．

針・突起　殻面には通常 1〜3 個の中心域有基突起が見られる[9, 10]．この外面への開口は丸い小孔として開き，外管は認められない[7]．内面には明瞭な内管が見られ，2 脚をもつ[9, 10]．殻套に近い殻面縁辺部に唇状突起が 1 個見られる[9, 10 矢印]．

半 殻 帯　開放型．
葉 緑 体　円盤状，多数．
有性生殖　不明．
生 活 形　浮遊性．
汚濁耐性　α-中腐水，貧塩-好塩性，好アルカリ性種．
出 現 地　淡水域から汽水域まで広く分布し，多少汚濁された池沼によく見られる．

Cyclotella pantaneliana Castracane

Castracane, C. A. F. 1886. Proc. Verb. Soc. Toscana Sci. Nat. p. 171.
Dimension: D. 17.5-30 μm,　Str. 12-16 in 10 μm.

- **被殻・殻**　殻は円形(1-3)．縁辺条線域の幅は狭く，殻半径の約 1/3．殻中央部が広く，不規則に分布する点紋が見られ，わずかに盛り上がる．直径17.5〜30 μm．
- **条線・胞紋**　縁辺条線は 10 μm あたり 12〜16 本(1-3)，条線は殻套では 3〜5 列の胞紋列より成るが，殻面では 2 列になる(5)．殻内面の肥厚肋は条線の 3〜5 本おきにある(1-3)．殻面のやや短い条線域寄りに遊離点が見られる．殻外面では間条線の肥厚の程度はいずれも同様であるが，有基突起をもつ間条線は光顕ではより明瞭な肥厚肋として写る．
- **針・突起**　殻内面中央部には数個の中心有基突起が見られ，外面へは丸い孔として開き，外管は認められない(4 矢印, 5 黒矢印)．また，縁辺有基突起の外面への開口は中心有基突起の開口より小さく楕円形で，外管は認められない(5 白矢印)．
- **半 殻 帯**　開放型．
- **葉 緑 体**　不明．
- **有性生殖**　不明．
- **生 活 形**　浮遊性．
- **汚濁耐性**　中汚濁耐性種．
- **出 現 地**　淡水域から汽水域まで広く分布．中川（東京都），多摩川河口（東京都）など．

Cyclotella striata (Kützing) Grunow

Grunow, A. in Cleve, P. T. & Grunow, A. 1880. Kongl. Svensk. Vet.-Akad. Handl. **17** (2): 119.
Basionym: *Coscinodiscus striata* Kützing 1844. Bacill. p. 131. pl. 1. f. 8.
Dimension: D. 10-50 µm,　Str. 8-10 in 10 µm.

被殻・殻　殻は円形[1-3]．縁辺条線域の幅は殻半径のほぼ1/3を占める．殻面の中央部の凸凹は顕著で，光顕でも十分にS字状に波打つのが認められる[2,3]．直径10〜50 µm．

条線・胞紋　縁辺条線は10 µmあたり8〜10本[2,3]，条線は殻縁では4〜6列の胞紋より成るが，中央側の端では少なくなる[4]．有基突起をもつ間条線は，殻内面においては短く落ち込んだ肋である．

針・突起　殻面中央部のうちで外側に隆起した側，すなわち内面では凹んだ側には，1〜7個の中心域有基突起が見られる[2,3,6]．この外面への開口は丸い孔として開き，外管は認められない[5]．しかし，内面には明瞭な内管が見られ，そのほとんどは3脚をもつ[6]．縁辺有基突起は間条線2, 3本ごとに1個存在し，この外面への開口は丸い孔で，外管は認められない[4]．また，縁辺有基突起列に挟まって唇状突起が1個見られる[6 矢印]．唇状突起は小さいがよく発達し，縦裂溝は間条線に対して同一方向に位置する．なお，この唇状突起の位置は，中心域有基突起をもつ隆起する部分と殻の中心点を結ぶ線のほぼ延長線上にある[6]．

半殻帯　開放型．
葉緑体　円盤状，多数．
有性生殖　不明．
生活形　浮遊性．
汚濁耐性　中汚濁耐性種．
出現地　汽水性で内湾の河口域に多く出現する．涸沼川（茨城県）など．

Discostella asterocostata (Lin *et al.*) Houk & Klee

Houk, V. & Klee, R. 2004. Diatom Research **19**: 220.
Basionym: *Cyclotella asterocostata* Lin *et al.* in Xie, S., Lin, B. & Cai, S. 1985. Acta. phytotax. Sinica. **23** : 473. pl. 1. f. 1-6.
Dimension: D. 11.5-29.0 μm,　Str. 12-16 in 10 μm.

被殻・殻　殻は円形[1-5]．殻中心域と縁辺条線域は無紋域によって分断される．この無紋域は殻の内面に隆起したドーナツ状の構造である[6-8]．中心部は広い無紋域で，その外側に縁辺条線より少し粗い条線が放射状に配列する．縁辺条線域の幅は殻半径の1/2～1/3を占める．また，大きな個体ではこの縁辺条線部が2層に分かれて見える[1,2]．殻面の中央部が大きく突出する殻とくぼむ殻がある．この凸凹が大きいため，光顕観察では殻縁辺と中央部を同時にピントを合わせることは難しい．直径11.5～29.0 μm．

条線・胞紋　縁辺条線は10 μmあたり12～16本[1-4]．条線はふつう2列の胞紋より成るが，ときどきその間に非常に小さな胞紋が見られることがある．縁辺有基突起は肋と肋の間（条線上）にあるが，そこから殻縁側に肋が生じ，条線が2本になる．

針・突起　縁辺有基突起の外管は明瞭で先端がわずかに広がる[6]．内側には明瞭な内管が見られ，2脚をもつ[7]．また，縁辺有基突起に挟まって唇状突起が1個見られる[7矢印]．内側の唇状部分は大きくよく発達し，縦裂溝は間条線に対して直角方向に位置する[7]．

半殻帯　開放型．
葉緑体　不明．
有性生殖　不明．
生活形　浮遊性．
汚濁耐性　不明．
出現地　淡水域から汽水域まで広く分布．北浦（茨城県），都川（千葉県），道灌堀（東京都）など．

Discostella pseudostelligera (Hustedt) Houk & Klee

Houk, V. & Klee, R. 2004. Diatom Research **19**: 223.
Basionym: *Cyclotella pseudostelligera* Hustedt 1939. Abh. Naturw. Ver. Bremen **31**: 581. f. 1, 2.
Dimension: D. 4.5-12.0 μm,　Str. 18-24 in 10 μm.

被殻・殻　殻は円形(1-18). 縁辺条線域が広く, 殻半径の 2/3 以上を占める. 殻が小形であるため殻面の凹凸は殻面観ではほとんど見分けがつかないが, 帯面観では S 字状にわずかに波打つのが認められる(21). 直径 4.5〜12.0 μm.

条線・胞紋　縁辺条線は 10 μm あたり 18〜24 本(1-18) である. 殻中央部で条線が短い放射状となる個体と無紋域になる個体が認められる(4, 15, 21). 有基突起は, 光顕でも明瞭な陰影として写る(6-12). 条線を構成する胞紋は両側に大きな胞紋が配列し, 間に細かい胞紋が存在する(14, 15). 間条線の肥厚は強くないが, 分枝が著しい(13-15, 18).

針・突起　縁辺有基突起の外管は短いが先端が T 字形となる(21). 殻縁有基突起, 唇状突起とも殻肩に位置し条線上にある. 唇状突起は 1 個で(15 矢印), 外面への開口は小さく楕円形, 外管は認められない(21 矢印).

半 殻 帯　開放型.
葉 緑 体　円盤状, 多数.
有性生殖　不明.
生 活 形　浮遊性.
汚濁耐性　中汚濁耐性種.
出 現 地　淡水域から汽水域まで広く分布. 涌池（長野県）, 中川（東京都）, 井の頭公園池（東京都）, 狩野川三日月湖（静岡県） など.

ノート　Houk & Klee (2004) は本種を始めとする 15 分類群を *Cyclotella* 属から *Discostella* 属に移し換えている.

Plates 53, 54

Discostella stelligera (Ehrenberg) Houk & Klee

Basionym: *Discoplea graeca* var. *stelligera* Ehrenberg 1854. Microg. pl. 2, 6. f. 3a-e.
Synonym: *Cyclotella stelligera* Cleve & Grunow in Van Heurck, H. 1882. Synop. Diat. Belg. pl. 94.
　　f. 22-26.
Dimension: D. 6.0-15.5 μm,　Str. 12-18 in 10 μm.

被殻・殻　殻は円形(1-12)．縁辺条線域の幅は殻半径の 1/2～1/3 を占める．殻中心部には 1 個の胞紋と短い放射状の条線があり，ドーム状に盛り上がった殻(11,13)とくぼんだ殻がある．しかし，小形であるため殻面の凸凹は殻面観ではほとんど見分けがつかない．直径 6.0～15.5 μm．

条線・胞紋　縁辺条線は 10 μm あたり 12～18 本で(1-10)，中心部は特徴的な放射条線となる．1 本の条線は殻縁では 4, 5 列の胞紋より成り(12,13)，胞紋列外側の胞紋は大きく，内側は小さい．

針・突起　縁辺有基突起は縁辺条線の延長上で 2, 3 本おきにある(11-14)．それらの外面への開口は丸い孔であり，外管はほとんど認められないが肥厚している(11,13)．内面には明瞭な内管があり，そのほとんどは 2 脚をもつ(14,15)．唇状突起が 1 個あり，縁辺有基突起の 1 個に近接して見られる(14 矢印,16)．唇状突起の外面への開口は有基突起の開口より小さく楕円形で，外管は認められない．

半殻帯　開放型．
葉緑体　円盤状，多数．
有性生殖　不明．
生活形　不明．
汚濁耐性　中汚濁耐性種．
出現地　淡水域から汽水域まで広く分布．印旛沼（千葉県），中川（東京都），多摩川河口（東京都），東京学芸大学中庭の池（東京都）など．

ノート　Houk & Klee(2004)は，本種の基礎異名として，*Cyclotella meneghiniana* var. *stelligera* P. T. Cleve & Grunow in Cleve, P. T. (1881)をあげているが，これは著者引用の間違いである．しかし，国際植物命名規約（セントルイス規約）33 条によれば，これによって，この組み合わせが正式発表とみなされなくなることはない．

Puncticulata praetermissa (Lund) Håkansson

Håkansson, H. 2000. Diatom Research **17**: 116.
Basionym: *Cyclotella praetermissa* Lund 1951. Hydrobiologica **3**: 98. f. 1 A-H, 2 A-L.
Dimension: D. 9.5-29 μm, Str. 16-18 in 10 μm.

被殻・殻 殻は円形(1-3,5)．殻中央部と縁辺条線の区別は明瞭で，縁辺条線域の幅は殻半径の1/2〜2/3を占め，中央部は放射状の点紋列をもつ．殻面はわずかに凹凸となるが，光顕では見分けがつきにくい．直径9.5〜29μm．

条線・胞紋 殻中央部は明瞭な放射状点紋で粗い．縁辺条線は10μmあたり16〜18本(2-4)．条線は多列胞紋であるが，ほとんどの場合両側に大きい胞紋が配列し，その間に細かい胞紋が配列している(9)．殻内面の肥厚肋は条線の3〜6本おきにある(2-4)．条線部は長胞構造となり，殻外面は胞紋で覆われ，殻内部への開口は殻套にある(7,8)．間条線は内側では肥厚して肋を形成するが，肥厚の程度は2通りで，より太い肋に有基突起が存在する(7,8)．有基突起をもつ肋は，光顕でも明瞭な黒線として確認できる(2-4)．

針・突起 殻面中央部には多数の中心域有基突起が見られる(7,8)．この外面への開口は丸い小孔で，胞紋より明らかに小さいため区別できる(6,9矢印)．中心域有基突起に外管は認められないが，内側には明瞭な内管が見られ，そのほとんどは3脚をもつ(7,8)．また，殻套部の太い肋上には縁辺有基突起があり，短い内管は2脚をもつ(8)．唇状突起は殻面に見られ，その数は1個のことが多いが2個の場合もある(7)．この外側への開口は有基突起の開口より小さく楕円形で，外管は認められない(9矢印)．内側の唇状部分は小さいがよく発達し，縦裂溝は間条線に対して斜めの方向に位置する(7,8)．

半殻帯 開放型．
葉緑体 円盤状，多数．
有性生殖 不明．
生活形 浮遊性．
汚濁耐性 中汚濁耐性種．
出現地 淡水域から汽水域まで広く分布．中川（東京都），多摩川河口（東京都）など．

ノート 本種は従来しばしば *Cyclotella radiosa* と誤同定されていたが，本来の *C. radiosa* は殻中央部の放射状点紋は，より粗いものである(4)．Håkansson(2002)は，これらの種の微細構造の違いを示すとともに，両種を *Puncticulata* 属に移属した．Tanaka & Nagumo (2005)はこれらの種と類似する *Puncticulata ozensis* を新種記載すると共に，類似分類群の形質比較表を示している．

Puncticulata shanxiensis (S. Q. Xie & Y. Z. Qi) Nagumo comb. nov.

Basionym: *Cyclotella shanxiensis* S. Q. Xie & Y. Z. Qi 1984. In: Mann, D. G. (ed.), Proc. 7th Intern. Diat. Symp. p. 188. pl. 1-4.
Dimension: D. 10-17 μm, Str. ca. 15 in 10 μm.

被殻・殻 殻は円形(1-5)．縁辺条線域の幅は狭く殻半径の 1/5～1/6．殻面の大部分は放射状の点紋条線に占められている．殻外面はほとんど平坦である(7)．直径 10～17 μm．

条線・胞紋 縁辺条線は 10 μm あたり約 15 本(1-5)．殻縁は 7～8 列の小さな胞紋列から構成される(7,9)．*Cyclotella* 属に特徴的な長胞条線は非常に短く，殻内部への開口もほとんど殻套部のみで殻面まで発達しない(6,8)．殻内面では薄皮が胞紋をドーム状に閉塞する(8,10)．

針・突起 殻面の間条線上には小さな顆粒状突起が多数見られる(7)．有基突起は殻縁条線の 4～7 列おきに 1 個ある(6)．この有基突起は 2 脚で長く，殻内面に突出し(8,10)，外面では丸く開口し外管はない(9 矢印)．唇状突起は殻面の縁に通常 1～2 個見られるが(10 矢印)，この位置の規則性は不明瞭．

半 殻 帯 開放型．
葉 緑 体 不明．
有性生殖 不明．
生 活 形 不明．
汚濁耐性 β-中汚濁耐性種．
出 現 地 員弁川（三重県）．

ノート 本種は，中国山西省南部，中條山脈，雪花峰の谷底を流れる小川から記載された（Xie & Qi 1984）．現在のところ日本では員弁川から報告されている．

Stephanodiscus hantzshii f. *tenuis* (Hustedt) Håkansson & Stoermer

Håkansson, H. & Stoermer, E. F. 1984. Nova Hedwigia **39**: 486.
Basionym: *Stephanodiscus tenuis* Hustedt 1939. Abh. Nat. Ver. Bremen **31**: 583. f. 3.
Dimension: D. 7-16 μm,　F. 8-16 in 10 μm.

被殻・殻　円形で殻面は平たく，中央に中心環がある[1-9,13,14]．殻面と殻套の境となる殻肩には刺があり，すべて束間肋上に規則的に配置する[9]．束間肋は殻面のみで，殻套には達しない[8,9]．殻の珪化の程度により，殻外面の形状が大きく変わる[9,10]．通常の殻は胞紋や束間肋が明瞭であるが[9]，厚く珪化した殻では，殻全体の模様が不明瞭になる[10]．直径7〜16 μm．

条線・胞紋　中央に無紋の中心環があり，その内側に10個前後の胞紋がある[9,16]．条線束は中央側で1列，殻縁側で3列の胞紋からなる[9,16]．殻套の胞紋は殻面の胞紋に比べ細かく，均一である．胞紋は内面では円形のドーム状に張り出した多孔師板によって閉塞され[11]，外面では五角形または六角形の開口をもつ[17]．条線束は10 μmあたり8〜16本．

唇状突起・有基突起　殻肩に1周する先の尖った刺があるが[9]，ときに先が平らなこともある[17]．縁辺有基突起は条線束ごとに3〜5本，短い管を伴って殻套外面に開口し[9]，殻内面では3個の付随孔をもつ[12]．唇状突起は1個で殻肩にあり，刺の1本と置き換わる．

半 殻 帯　開放型，数枚[15]．
葉 緑 体　多数．
有性生殖　不明．
生 活 形　浮遊性．
汚濁耐性　弱汚濁耐性種．
出 現 地　利尻島姫沼（北海道），琵琶湖（滋賀県），湯ノ湖（栃木県），涌池（長野県），涸沼（茨城県）など，淡水から汽水の湖沼に出現．

[ノート]　小形の *Stephanodiscus minutulus* や *S. invisitatus* と似るが，中心付近に有基突起がないことで区別できる．本種の詳しい分類と形態については，小林ら（1985）によって報告されている．

Stephanodiscus minutulus (Kützing) Round

Round, F. E. 1981. Arch. Protistenk. **124**: 462.
Basionym: *Cyclotella minutula* Kützing 1844. Bacill. p. 50. pl. 2. f. 3.
Dimension: D. 6-12 μm,　F. 10-13 in 10 μm.

被殻・殻　殻は円形である．殻面は全体に平坦なものから(1-11)，中央部で凹凸するものまである(12-24)．中心環は不明瞭．殻肩に目立つ刺がある(8,9,21,22)．束間肋は殻によって目立つものと目立たないものがあるが(8,9,21,22)，殻面のみで殻套には達しない(9,21)．珪化の程度によって殻外面の形状が大きく変わる(8,9,21,22)．直径6〜12 μm．

条線・胞紋　条線束は，中央側で1列，殻縁側で3列の胞紋をもつ(11,24)．殻套の胞紋は殻面の胞紋に比べやや細かい．胞紋は内面では円形でドーム状に張り出した多孔師板によって閉塞される(10,23)．外面の開口は，基本的には五角形または六角形であるが(9,11)，珪化の程度が強い殻では不明瞭で，ほとんどふさがったように見える(8,22)．条線束は10 μmあたり10〜13本．

唇状突起・有基突起　殻肩上を1周する刺があり(8)，その先端はときに平らなこともある(10)．中心有基突起は中心から少しはずれた位置に1個あり，内面では2個の付随孔をもち，外面では丸く開口する(8,10,23)．縁辺有基突起は3〜6本の条線束ごとにあり，刺のすぐ下に開口し，内面では3個の付随孔をもつ(10,23)．唇状突起は1個で，殻肩上にあり，刺の1本と置き換わる．

半殻帯　開放型，数枚．
葉緑体　多数．
有性生殖　不明．
生活形　浮遊性．
汚濁耐性　弱汚濁耐性種．
出現地　琵琶湖（滋賀県），利尻島姫沼（北海道），八郎潟（秋田県），井の頭公園池（東京都）など，淡水から汽水の湖沼に出現．

ノート　*Stephanodiscus hantzschii* f. *tenuis* と似ているが，殻中央の凹凸と，中心有基突起の存在で識別できる．本種の分類と形態については，小林ら（1985）によって詳細に報告されている．

Stephanodiscus rotula (Kützing) Hendey

Hendey, N. I. 1964. An introductory account of the smaller algae of British Coastal Waters. Part 5. Bacillariophyceae (Diatom). p. 75.
Basionym: *Cyclotella rotula* Kützing 1844. Bacill. p. 50. pl. 2. f. 4.
Dimension: D. 26-50 µm,　F. 6-7 in 10 µm.

被殻・殻　殻は円形で比較的大形，中央部が凹凸する[1-4]．中心環は不明瞭．殻肩に先の尖った細い刺がある[4]．直径26〜50 µm（Håkansson 2002）．

条線・胞紋　条線束は殻面中央部では1列，周辺部では明瞭な2列の胞紋から成るが[1-3]，殻套では束間肋がなくなり3，4列の胞紋列となる[4,5]．殻套部の胞紋は殻面の胞紋に比べやや細かい．胞紋は外面では円形の開口をもち，内面ではドーム状に張り出した多孔師板によって閉塞される[5,6]．条線束は10 µmあたり6〜7本．

唇状突起・有基突起　刺，有基突起，および唇状突起が見られる．殻面中央の凹凸の境界に沿って2個の付随孔をもつ有基突起が，リング状かつ不均一に偏在している．細長い刺は殻肩にあり，ほぼ1本おきの束間肋上にある[10]．縁辺有基突起も刺と同じように束間肋上にあるが，やや殻縁側に寄った殻套部にある．縁辺有基突起は内面では3個の付随孔をもち，その中心管は長い[5,6]．唇状突起は1〜数個で，殻肩上にあり，刺と置き換わる[5,6]．

半 殻 帯　開放型，3枚(?)．
葉 緑 体　不明．
有性生殖　不明．
生 活 形　浮遊性．
汚濁耐性　不明．
出 現 地　中之条・湖成層（群馬県）．

ノート　本種についてはRound(1981)がKützingのタイプ試料を電顕で観察している．また，Håkansson & Locker(1981)によって本種の命名上の検討がなされている．*Stephanodiscus, Cyclostephanos, Cyclotella*に関する比較研究はHåkansson(2002)を参照．

Eucampia zodiacus Ehrenberg

Ehrenberg, C. G. 1840. Abh. Akad. Wiss. Berlin **1839** (1841): 151. vol. 4. f. 8.
Dimension: L. 15-100 μm, W. 15-25 μm, Str. 16-20 in 10 μm.

被殻・殻 細胞は偏平で，突出した殻の両端部で連結し，らせん状群体を形成する[1,4]．殻は細長い楕円形であるが，両端の突出の程度が異なるため異形となる[2]．帯面観ではその突出が明瞭で，らせんの外側に当たる殻端の突出が長く，内側が短いことがわかる[3]．殻長 15〜100 μm，殻幅 15〜25 μm．

条線・胞紋 条線は殻中央から放射状に配列し，明瞭な点紋となる．10 μm あたり 16〜20 本．

針・突起 殻面中央の少し凹んだ部分に唇状突起が 1 個存在し[7,8]，外面にはスリット状の開口がある[5]．この唇状突起は光顕でも明瞭に識別できる[2]．

半 殻 帯 開放型，多数．

葉 緑 体 円盤状，多数．

有性生殖 不明．

生 活 形 浮遊性．

汚濁耐性 不明．

出 現 地 沿岸海域に多く出現する．

Asterionella formosa Hassall

Hassall, A. H. 1850. Micros. Exam. Water. p. 10. pl. 2 (2). f. 5.
Dimension: L. 40-130 μm,　W. 1.5-3 μm (center), 3-12 μm (footpole), 1.5-4 μm (headpole),
　Str. 24-28 in 10 μm.

被殻・殻　殻は狭線形，両殻端は頭状で異極[1 a-d]．殻長 40〜130 μm，中央部の殻幅 1.5〜3 μm．細胞は幅の広い方の殻端で殻同士が粘液によって接着し，星形の群体を作る[1 e]．群体の基部となる幅広の殻端（足極；footpole）は幅 3〜12 μm，もう一方の殻端（頭極；headpole）は幅 1.5〜4 μm．殻端小孔域は両極にある．殻面は平坦で，角ばった殻肩には 10 μm あたり 10〜12 本の密度で配列する小針をもつ（南雲 1982）．殻面観における被殻は中央部より両極部の方が厚くなる[1 e]．

軸　　域　軸域は直線的で極めて狭く，殻の中央を縦走する．縦溝はない[1 a-d]．軸域の両末端に横向きの唇状突起をもつ（南雲 1982）．

条線・胞紋　条線の配列は平行で[1 a-d]，条線密度は 10 μm あたり 24〜28 本．各条線は 1 列の孔状胞紋から成る．胞紋は外面で薄皮によって閉塞される．

半 殻 帯　少なくとも 3 枚の開放型帯片で構成される．各帯片は 1 列または 2 列の胞紋列もつ（南雲 1982）．

葉 緑 体　板状で小さく，多数．

有性生殖　不明．

生 活 形　浮遊性．

汚濁耐性　弱汚濁耐性種．

出 現 地　湯ノ湖（栃木県），諏訪湖（長野県），琵琶湖（滋賀県）など，本邦および世界各地の淡水湖沼に広く分布する．

ノート　本種は，しばしば春と秋に多量発生し水道の濾過池を詰まらせる．群体を構成する細胞数は，増殖が盛んなときには 16 個にも達することがある．形態的には後出の *Asterionella gracillima* と酷似するが，群体の形状，異極の殻およびより粗い条線数などによって区別される．また，クローン培養で *formosa* 型の殻が *gracillima* 型に変わることも確かめられてはいるが（Körner 1970），生態的に重要な種であるため，詳細が明確にされるまでこれらの 2 種を区別して扱うことにした．

Plate 63 (Figs 2a-g)

Asterionella gracillima (Hantzsch) Heiberg

Heiberg, P. A. C. 1863. Consp. Crit. Diat. Danicarum. p. 68. pl. 6. f. 19.
Basionym: *Diatoma gracillimum* Hantzsch, in Rabenhorst, G. L. 1861. Agl. Eur. No. 1104.
Dimension: L. 40-130 μm,　W. 1.5-3 μm,　Str. ca. 30 in 10 μm.

- **被殻・殻**　細胞は星形群体またはジグザグ形の群体を作る(2 f, g)．殻は線形，両殻端は頭状で異極(2 a-e)．殻長 40～130 μm，中央部の殻幅 1.5～3 μm．殻端小孔域は両殻端にある．角ばった殻肩には不規則に配列する小針をもつ（河島・小林 1995）．殻面観では被殻の両極部は厚くなる(2 a)．
- **軸　　域**　軸域は直線的で極めて狭く，殻全体にわたって走る．縦溝はない(2 b-e)．殻端に横向きの唇状突起をもつことが知られている（河島・小林 1995）．
- **条線・胞紋**　条線の配列は平行で(2 b-e)，条線密度は 10 μm あたり約 30 本．各条線は 1 列の孔状胞紋から成る．
- **半　殻　帯**　3 枚以上の開放型帯片で構成される．
- **葉 緑 体**　板状で小さく，多数(2 f)．
- **有性生殖**　不明．
- **生 活 形**　浮遊性．
- **汚濁耐性**　弱汚濁耐性種．
- **出　現　地**　阿寒湖（北海道），琵琶湖（滋賀県）など，本邦の湖沼や貯水池に広く分布し，しばしば *Asterionella formosa* に混じって出現する．

ノート　悪臭を発する代表的な種類として知られる．数が少ないときは芳香のゼラニウム臭だが，多くなると生臭い魚臭となる．Körner (1970) や Krammer & Lange-Bertalot (1991) などは，本種を *A. formosa* の異名として扱っている．

Asterionella ralfsii W. Smith

Smith, W. 1856. Syn. Brit. Diat. vol. 2. p. 81.
Dimension: L. 20-50 μm, W. ca. 3 μm, Str. ca. 32 in 10 μm.

被殻・殻　細胞は星形群体を作る[6]．殻は線形で異極，*A. formosa* などに比べて短く太い．殻の一端は頭状に大きく膨らむが，他方は丸く小さい[1,2,4,5]．殻端小孔域をもつ．殻長20～50 μm，中央部の殻幅約3 μm．殻面観では被殻は中央部より両極部の方が厚くなる[3]．

軸　　域　軸域は直線的で極めて狭く，両極に達する[1,2,4,5]．両極に横向きの唇状突起をもつ（Krammer & Lange-Bertalot 1991）．

条線・胞紋　条線は平行で細かく，10 μm あたり約32本ある[4,5]．

半 殻 帯　知られていない．

葉 緑 体　小さく板状，多数．

有性生殖　不明．

生 活 形　浮遊性．

汚濁耐性　弱汚濁耐性種．

出 現 地　苗場山湿地（新潟県）など腐植酸を含む湿原にしばしば出現する．

ノート　Körner(1970)は，条線密度や殻端部の形状によって var. *ralfsii*, var. *hustedtiana* Körner および var. *americana* Körner の3変種を区別している．

Asterionellopsis glacialis (Castracane) Round

Round, F. E. in Round, F. E., Crawford, R. M. & Mann, D. G. 1990. The diatoms. p. 664.
Basionym: *Asterionella glacialis* Castracane 1886. Rep. Sci. Res. H. M. S. Challenger. Bot. **2**: 50. pl. 14. f. 1.
Dimension: L. 29-98 μm, W. 8-11 μm, Str. 24-32 in 10 μm.

被殻・殻 被殻は異極，帯面観では足極（footpole）が三角形に膨らみ，足極の上部から頭極（headpole）までが非常に狭い線形となる(1-3)．殻は大部分が細長い線形で，足極は楕円形またはひし形(6)．殻長29〜98 μm，殻幅8〜11 μm．両殻端に細長い開口の列を伴った楕円形の眼域をもつ．殻中央部から頭極までの殻縁部には先端が鋭く尖った多くの針があり，その先端は頭極側を向く(4)．細胞は向かい合った殻同士が足極で接着し(3,5矢じり)，らせん状にねじれた糸状の群体を形成する(1,2)．

軸　　域 軸域は殻の中央を縦走するが，極めて狭い(5矢印,6)．頭極の眼域付近に1個の唇状突起をもつことが知られている（Round *et al.* 1990）．

条線・胞紋 条線は10 μmあたり24〜32本で，それぞれ1列の胞紋で構成される(5,6)．胞紋は小円形から長方形で(5)，その外面は裂孔師板によって閉塞される（Round *et al.* 1990）．

半 殻 帯 多数の帯片で構成される．帯片は小舌をもち，帯片中脈の両側に胞紋列を伴う（Round *et al.* 1990）．

葉 緑 体 細胞の足極部に2枚の葉緑体をもつ(2)．
有性生殖 不明．
生 活 形 浮遊性．
汚濁耐性 不明．
出 現 地 東京湾（東京都），三崎（神奈川県），三河湾（愛知県），志摩（三重県）など本邦および世界各地の沿岸域に広く分布する．

ノート 本属は殻形，胞紋構造，両極にある眼域，頭極の唇状突起および殻縁に配列する針などの特徴により *Asterionella* 属とは明瞭に区別されている．本種は古くから *Asterionella japonica* Cleve（Cleve & Müller 1878）と呼ばれていたものであるが，Körner(1970)によりこの名前が正式発表の条件を欠く裸名であることが指摘され，それに代わる合法名として *Asterionella glacialis* の名が用いられてきた．

Catacombas obtusa (Pantocsek) Snoeijs

Snoeijs, D. J. M. in Snoeijs, D. J. M., Hällfors, G. & Leskinen, E. 1991. Diatom Research **6**: 156.
Basionym: *Synedra fasciculata* var. *obtusa* Pantocsek 1889. Beitr. Kenntn. Foss. Bacill. Ungarns 2: 64. pl. 26. f. 377, 380.
Dimension: L. 175-340 μm, W. 7-9 μm, Str. 10-11 in 10 μm.

被殻・殻　殻は線形．殻端近くで細くなり，くちばし状の殻端をもつ[1-5]．軸域は非常に広い．数細胞が殻面で結合し，その先端から分泌する粘液質によって基物に付着し，叢状の群体を形成する．殻長 175〜340 μm，殻幅 7〜9 μm（Snoeijs *et al.* 1991）．

条線・胞紋　条線は殻縁にのみあり，全体に平行に配列し，10 μm あたり 10〜11 本．条線は外面観では殻肩で分離され，それぞれ殻面と殻套に大きな 1 個の胞紋をもつ[4,5]．胞紋は卵形または楕円形で，外表面に多孔篩板をもつが[5]，内面では完全に開口する[6]．

唇状突起・突起　両殻端にそれぞれ 1 個の唇状突起があり，最後の条線の先に水平かやや斜めに位置する[4]．その外部開口は長方形から楕円形で，軸域の中心からややそれた位置にある．両殻端には殻套眼域があり[4]，その上にはいくつかの小さな突起が見られる．しかし，殻によってはないこともあり，処理の途中で欠けたものと考えられる．

半殻帯　開放型，数枚．接殻帯片は無紋で，その内接部は鋸歯状となる．

葉緑体　円盤状，多数．

有性生殖　不明．

生活形　付着性．

汚濁耐性　弱汚濁耐性種．

出現地　荒川（東京都），墨田川（東京都）など，河川の河口域の汽水域によく出現する．

ノート　本種はその殻形や広い軸域をもつ点で，*Tabularia fasciculata* に類似するが，より線形で，より大きく，条線も粗く，両端の唇状突起が明瞭である．また，胞紋の微細構造が異なり，*T. fasciculata* では明瞭な縦小肋（cross bar）が見られるが，本種の胞紋にはない．

Ctenophora pulchella (Rakfs ex Kützing) D. M. Williams & Round

Williams, D. M. & Round, F. E. 1986. Diatom Research **1**: 330.
Basionym: *Synedra pulchella* Rakfs ex Kützing 1844. Bacil. p. 68. pl. 29. f. 87.
Dimension: L. 30-130 μm, W. 5-7 μm, Str. 13-16 in 10 μm, Ar. ca. 20 in 10 μm.

被殻・殻 殻は線状皮針形．殻端はくちばし状からやや頭状[1-5]．軸域は狭い．中心域は横帯状となり，ほぼ正方形から縦長の長方形．ただし，殻内面が丸くくぼむため[6,9]，楕円形に見える．また，ここには条線の痕跡が見られるものもある．細胞はその先端から分泌する粘液によって基物に付着し，叢状の群体を形成する．殻長 30~130 μm，殻幅 5~7 μm．

条線・胞紋 条線は明瞭な点紋からなり，全体に平行に配列し，10 μm あたり 13~16 本．条線は1列の胞紋から成り，殻面から殻套まで連続する．胞紋は四角形または楕円形で，外表面に肉趾状師板をもち[7]，内面に開口する[6,8]．10 μm あたり約 20 個．

唇状突起・突起 両殻端にそれぞれ1個の唇状突起があり，軸域に接し，水平かやや斜めに位置する[8]．唇状突起の外部開口は長方形から楕円形で，軸域からややそれた位置にある[7]．両殻端には殻套眼域（ocellulimbus）がある[7]．

半殻帯 開放型，数枚．接殻帯片は無紋で，その内接部は鋸歯状となる．
葉緑体 板状，2枚．
有性生殖 不明．
生活形 付着性．
汚濁耐性 弱汚濁耐性種．
出現地 天竜川（静岡県），涸沼川（茨城県）など，河川や湖沼の汽水域によく出現する．淡水にもまれに出現する．

ノート 本属は横帯状の中心域と複雑な師板によって閉塞される胞紋をもつことを特徴としている．現在のところ1属1種である．

Diatoma mesodon (Ehrenberg) Kützing

Kützing, F. T. 1844. Bacill. p. 47.
Basionym: *Fragilaria mesodon* Ehrenberg 1839. Abh. Köngl. Akad. Wiss. Berlin **1838**: 57. pl. 2. f. 9.
Synonym: *Diatoma hiemalis* var. *mesodon* (Ehrenberg) Grunow in Van Heurck 1881. Syn. Diat. Berg. pl. 51. f. 3, 4.
Dimension: L. 8.5-21 µm, W. 6-10.5 µm, Str. 20-24 in 10 µm, Ar. ca. 20 in 10 µm, T. C. 6-8 in 10 µm.

被殻・殻 殻は広皮針形，殻端部は広円または先端くちばし形となる(1-5)．帯面は長方形(6-8)．殻端部もしくは殻面全体より分泌した粘液により帯状群体を形成する(6,7)．殻長8.5〜21 µm（10〜40 µm: Krammer & Lange-Bertalot 1991），殻幅6〜10.5 µm（6〜14 µm: l.c.）．殻外面は平滑(10,11,13)．殻肩は角ばる．殻肩には小針をもつ個体もある(10)．環境の変化に応じ，内生殻を作ることがある(8 矢印)．

軸　　域 線形．殻中央部を縦走する(1,10,11)．

条線・胞紋 条線は孔状胞紋よりなり軸域から殻套部まで伸びる(9,12)．条線は平行で10 µmあたり約20〜24本．また，条線を構成する胞紋の密度は10 µmあたり約20個．殻内面では殻套部より伸びる横走肋が発達し，ほとんどが軸域を越え殻面を横断する．横走肋は条線とほぼ並行に走る．10 µmあたり6〜8本．

殻端小孔域・唇状突起 両殻端には殻端小孔域がある(10,11)．また，一方の殻端部に唇状突起が1個ある(9)．唇状突起の外部開口は単純なスリットで，条線と平行に並ぶ(10,11)．

半 殻 帯 胞紋をもつ開放型の多数の帯片よりなる．10枚程度(12)．各帯片は両殻端部で特徴的に著しく湾曲する．

葉 緑 体 円盤状，数個(6)．

有性生殖 不明．

生 活 形 付着性．

汚濁耐性 弱汚濁耐性種．

出 現 地 藻興部（もおこっぺ）川（北海道），鬼怒川（栃木県），浅川（東京都），荒川（埼玉県），神無川（群馬県），千曲川（長野県），児野沢（長野県木曽郡）など，各地の河川上流の冷水域や地蔵院沼（埼玉県），琵琶湖（滋賀県）などの湖沼．

ノート 本種は，かつて *Diatoma hiemalis* として，しばしば同定されてきた．しかし，*D. hiemalis* はより条線が粗く，横走肋もより頑強なこと，また各帯片が殻端部でもあまり湾曲せず，殻の長さを通して平行であることから，本種と区別することができる．

Diatoma tenuis C. Agardh

Agardh, C. A. 1812. Svensk Botanik vol. 7. pl. 491. f. 4,5.
Dimension: L. 21-83 μm,　W. 2.5-4.5 μm,　Str. ca. 50 in 10 μm,　Ar. ca. 70 in 10 μm,
　　T. C. 5-9 in 10 μm.

被殻・殻　殻は線形またはわずかに紡錘状皮針形，殻端部は頭状～弱頭状となる(1-4)．帯面は長方形．殻長 21～83 μm（22～120 μm: Krammer & Lange-Bertalot 1991），殻幅 2.5～4.5 μm（2～5 μm: l.c.）．殻外面は平滑(6,7)．殻肩は丸みを帯びる．細胞の1端で粘液により結合し，ジグザグ状の群体を形成する（Williams 1985）．

軸　　域　線形．殻中央部を縦走する(3,5)．

条線・胞紋　孔状胞紋よりなる条線が，軸域から殻套部まで伸びる(5)．条線は平行で 10 μm あたり約 50 本と細かく，光学顕微鏡観察は困難である．また，孔状胞紋の密度は 10 μm あたり約 70 個あり，殻外面で師板により閉塞される．殻内面の殻套部より伸びる横走肋には幅の広いものと狭いものがあり，幅広のものは軸域を越え殻面を横断するが，狭いものは軸域まで達せずに消失してしまうものが多い(5,8,9)．これら横走肋は条線にほぼ並行に配置し，殻面を横断するものは 10 μm あたり 5～9 本．

殻端小孔域・唇状突起　両殻端部には殻内面，外側表面ともに胞紋より大きめの均一な直径の孔より構成される殻端小孔域がある(5-9)．また，一方の殻端部で横走肋の上に唇状突起が1個存在する(5,9 矢印)．唇状突起の外部開口は単純なスリットである(6,7 矢印)．

半殻帯　胞紋をもつ開放型の帯片よりなる．4枚（Williams 1985）．

葉緑体　長円盤状，8枚（Tschermak-Wess 1973）．

有性生殖　Tschermak-Wess(1973)が *Diatoma elongatum* の名の下に報告をしたものがある．*D. elongatum* は *D. tenuis* の異名である．

生 活 形　付着性．

汚濁耐性　弱汚濁耐性種．

出 現 地　藻興部(もおこっぺ)川（北海道），パンケ沼（北海道），支笏湖（北海道），中禅寺湖（栃木県），湯ノ湖（栃木県），北浦（茨城県），霞ヶ浦（茨城県），千曲川（長野県）など，各地の湖沼，河川．

ノート　本種の種小名（種形容語）は，長い間 *tenue* と綴られていた．しかし，これは植物国際命名規約における正字法上の誤りであり，近年になり正しい語尾(-is)をつけて綴られるようになった．Potapova & Snoeijs(1997)は，*Diatoma moniliformis* Kützing（*D. tenuis* の変種とみなす研究者もいる）の野外個体群における生活環中のサイズの変化や初生細胞の観察を行っている．

Diatoma vulgaris Bory

Bory de Saint Vincent, J. B. M. 1824. In: Dict. Class. Hist. Natur. **5**: 461.
Dimension: L. 22-59 μm, W. 11-13 μm, Str. ca. 40-50 in 10 μm, Ar. ca. 70 in 10 μm,
 T. C. 6-8 in 10 μm.

被殻・殻　殻は狭楕円形〜線形．殻端部は広円もしくはわずかに広くちばし形となる(1-4)．帯面は長方形．殻端部より分泌した粘液によりジグザグの群体を形成する(5)．殻長 22〜59 μm (8〜75 μm: Krammer & Lange-Bertalot 1991)，殻幅 11〜13 μm (7〜18 μm: l.c.)．殻外面は平滑(6)．殻肩は角ばる．

軸　　域　線形．殻中央部を縦走する(1-4)．

条線・胞紋　条線は平行に配列し，軸域から殻套部まで伸びる(7)．10 μm あたり約 40〜50 本．条線を構成する孔状胞紋の密度は 10 μm あたり約 70 個．条線は光学顕微鏡ではほとんど見ることができない．殻内面では殻套部より伸びる横走肋が発達し，その多くは軸域を越え殻面を横断する．間条線が単純に肥厚した構造ではないが，条線に平行な横走肋が多い．しかし，しばしば斜行し条線を横切るものもある(2,7)．10 μm あたり 6〜8 本．

殻端小孔域・唇状突起　両殻端には殻端小孔域があり(7)，群体形成のための粘液を分泌する．また，一方の殻端部にわずかに突出する唇状突起が 1 個ある(7)．唇状突起の外部開口は単純なスリットである．

半 殻 帯　胞紋をもつ開放型の帯片よりなる．6 枚 (Williams 1985)．

葉 緑 体　円盤状，多数(5)．

有性生殖　Geitler (1958) が若干の報告を行っているが，図は添えられていない．

生 活 形　付着性．

汚濁耐性　弱汚濁耐性種．

出 現 地　鬼怒川（栃木県），荒川（埼玉県），多摩川（東京都），河口湖（山梨県），諏訪湖（長野県）など，各地の河川，湖沼．

ノート　本種の種小名（種形容語）は長らく *vulgare* と綴られていたが，これは正字法上の誤りであった．*Diatoma* は女性であるのに種小名（種形容語）の語尾(-e)は中性だからである．近年では国際植物命名規約に基づき正しい女性の語尾(-ris)に直されて使用されるようになっている．*Pleurosira* などある種の中心珪藻では光刺激や接触刺激により葉緑体が凝集・拡散運動をすることが知られているが (Makita & Shihira-Ishikawa 1997)，羽状珪藻の本種も同様の運動を行う．季節による殻形の変異 (seasonal form) が報告されている (Hartmann 1967)．本種のバンドの微細構造が Williams (1985) によって報告されている．

Fragilaria capitellata (Grunow) J. B. Petersen

Petersen, J. B. 1946. Dan. Biol. Medd. **20** (1): 54.
Basionym: *Synedra capitellata* Grun. in V. H. 1881. Syn. Diat. Belg. pl. 40. f. 26.
Dimension: L. 15-38 μm, W. 4.5-7 μm, Str. 16-18 in 10 μm, Ar. ca. 60 in 10 μm.

被殻・殻 殻は線形〜広皮針形[1-8]．大きいものでは殻側がほぼ並行な線形であるが，小さくなるにつれ中央が膨れた皮針形となる．殻端近で急に細くなり，頭状の殻端をもつ[1-8]．時々中央部でS字状に曲がった個体も見られる[13-21]．軸域は非常に狭い．中心域は左右が不揃いで，片側の膨らんだ側だけが広く馬蹄形(紋)になり，その反対側は短い条線を伴う．この馬蹄紋には光学顕微鏡では痕跡状の条線が見られるが，走査電顕観察から，外面に弱い線状の凹凸を伴い，内面では全体がややくぼんでいることがわかる[11, 19]．殻面は条線部がややくぼみ，間条線部がやや隆起した波板状となる．殻長15〜38 μm，殻幅4.5〜7 μm．

条線・胞紋 条線は左右が交互に配列し，10 μmあたり16〜18本．条線は1列の胞紋からなり，殻面から殻套まで連続する[10, 12]．胞紋は小円形で外面に輪形師板をもつ[10, 20, 21]．胞紋は非常に密で，10 μmあたり約60個．

唇状突起・突起 片側の殻端に1個の唇状突起があり[11]，外面に丸い開口をもつ[12, 17-18]．連結針などの突起はない．殻端には小孔が規則的に配列した広い殻套眼域がある[10, 12, 18]．

半殻帯 開放型で4枚の帯片．
葉緑体 板状，2枚．
有性生殖 不明．
生活形 付着性または浮遊性．
汚濁耐性 弱汚濁耐性種．
出現地 鬼怒川（栃木県），荒川（埼玉県），奥多摩湖（東京都），十和田湖（青森県），阿寒川（北海道）など，河川や湖沼に出現する．

ノート *Fragilaria vaucheriae* (Kützing) J. B. Petersen と似るが，全体により広皮針形で，条線がより細く密である．

Plates 73, 74

Fragilaria crotonensis Kitton

Kitton, F. 1869. Sci.-Gossip **5**: 110. f. 81.
Dimension: L. 38-140 μm, W. 2-4 μm, Str. 15-18 in 10 μm.

被殻・殻・軸域 被殻は帯面観では中央で幅広く，両端に向かって細くなるが，殻端で再び太くなるタイプと[1]，そのまま徐々に細くなって終わるタイプがある[10,17]．しかし，両タイプとも殻面の中央部で結合し，帯状の群体を作る．殻は線状皮針形で，中央部で膨らむかやや波打ち，両端に向かって急に細くなり，弱い頭状となる[1-10,12-16]．軸域は非常に狭い．中心域は縦長の長方形で，無紋または条線の痕跡が見られる．殻面の条線部はややくぼみ，間条線部が肥厚するため，殻面は波板状になる．殻長38～140 μm，殻幅2～4 μm．

条線・胞紋 条線は軸域を挟んで左右が交互に配列し，10 μmあたり15～18本．条線は1列の胞紋から成り，殻面から殻套まで連続する[11,20]．胞紋は小円形で外面に輪形師板をもつ．胞紋は光顕では識別できない．

唇状突起・突起 唇状突起は片側の殻端に1個あり，殻面のほぼ中央に開口する[11]．連結針は間条線部の殻肩にあり，殻中央付近の連結針は先がへら状で，これによって隣り合う殻同士が強く結合する[18,20]．中央以外の殻肩部にも先の尖った小さな突起がある．殻端の2本の角状の突起が特に目立つ[11,19]．殻端には広い殻套眼域がある[11]．

半殻帯 開放型で3，4枚の帯片からなり，それぞれに1列の胞紋がある．

葉緑体 板状で2枚．

有性生殖 Nipkow(1953)による報告がある．

生活形 浮遊性．

汚濁耐性 弱汚濁耐性種．

出現地 湯ノ湖（栃木県），木崎湖（長野県），津久井湖（神奈川県），山中湖（山梨県），一碧湖（静岡県），多摩川（東京都），荒川（埼玉県）など，各地の湖沼や河川に出現する．

ノート 河川に出現する種は上流のダム湖などからの流下と考えられる．好アルカリ性で，富栄養性または中栄養性種．浄水場の濾過池を詰まらせる珪藻として有名．

Fragilaria mesolepta Rabenhorst

Rabenhorst, L. 1861. Alg. Sachsens resp. Mitteleuropas. No. 1041.
Dimension: L. 26-50 μm, W. 3-4 μm, Str. 13-16 in 10 μm, Ar. ca. 60 in 10 μm.

被殻・殻 殻は線状または線状皮針形．殻は中央でややくびれ，殻端に向かってわずかに細くなり，くちばし状かやや頭状の殻端をもつ[1-5]．中心域は縦長の長方形でほとんど無紋であるが，痕跡的な条線が見られることもある．軸域は非常に狭い．被殻は殻肩の突起によって結合して，帯状群体を作る[6]．殻長26〜50 μm，殻幅3〜4 μm．

条線・胞紋 左右の条線が交互に配列し，10 μm あたり13〜16本．条線は1列の胞紋からなり，殻面から殻套まで連続する[7,8]．胞紋は小円形で外面に輪形篩板をもつ．胞紋は非常に密で，10 μm あたり約60個．

唇状突起・突起 片側の殻端の殻套に1個の唇状突起がある[8矢印, 9矢印]．唇状突起が殻面ではなく殻套部に見られる．先端がへら状に広がった連結針をもち，これらによって隣り合う殻と結合する[8,10]．殻端には小孔が規則的に配列した広い殻套眼域がある[8,10]．

半殻帯 開放型で4枚の帯片があり，すべてに1列の胞紋がある．

葉緑体 板状，2枚．

有性生殖 不明．

生活形 付着性または浮遊性．

汚濁耐性 弱汚濁耐性種．

出現地 琵琶湖（滋賀県），阿寒湖（北海道）など．

[ノート] *Fragilaria capucina* の1変種として扱われていることが多いが，唇状突起が殻套にある点で *F. capucina* とは大きく異なると考え，別種とした．

Fragilaria neoproducta Lange-Bertalot

Lange-Bertalot, H. 1993. Biblioth. Diatomol. **27**: 48.
Dimension: L. 18-47 μm,　W. 5-7 μm,　Str. 14-16 in 10 μm,　Ar. ca. 50 in 10 μm.

被殻・殻　殻は線状または線状皮針形．殻側はほぼ並行で，くちばし状の殻端もつ(1-5)．中心域はなく，軸域は線状で狭い．被殻は殻肩の連結針によって結合して，帯状群体を作る(6)．殻長 18～47 μm，殻幅 5～7 μm．

条線・胞紋　条線はほぼ平行で，左右が交互に配列し，殻端近くでやや放射状となる(7,8,10)．条線は 1 列の胞紋からなり，殻面から殻套まで連続し，10 μm あたり 14～16 本(9,10)．胞紋は小円形で外面に開口し，10 μm あたり約 50 個．条線は内面で全体に閉塞されている(8)．

唇状突起・突起　唇状突起はない(8)．連結針は殻肩の間条線上にあり，その先端はへら状に広がる(7,9,10)．殻套眼域は比較的広く，多数の小孔をもつ(11)．

半 殻 帯　開放型で 4 枚の帯片．
葉 緑 体　不明．
有性生殖　不明．
生 活 形　付着性．
汚濁耐性　弱汚濁耐性種．
出 現 地　手賀沼（千葉県），三宝寺池（東京都），阿寒湖（北海道）など．

ノート　*Fragilariforma virescens* (Ralfs) D. M. Williams & Round と類似するが，唇状突起が見られないことや軸域がやや広いことで区別できる．

Fragilaria perminuta (Grunow) Lange-Bertalot

Lange-Bertalot, H. 2000. In: Krammer & Lange-Bertalot (eds) Bacill. **2** (3): 2. p. 581.
Basionym: *Synedra perminuta* Grunow in Van Heurck 1881. Syn. Diatom. Belg. pl. 40. f. 23.
Dimension: L. 22-43 µm, W. 3-4 µm, Str. 16-18 in 10 µm, Ar. ca. 50 in 10 µm.

被殻・殻 殻は線状皮針形，中央から殻端に向かい徐々に細くなり，頭状の殻端をもつ(1-8)．中心域は片側が広く，馬蹄紋状になる．軸域は狭い．殻面は波板状になる．帯状群体を作らない．殻長22～43 µm，殻幅3～4 µm．

条線・胞紋 条線は左右が互い違いに配列し，殻端までほぼ平行．10 µmあたり16～18本．条線は1列の胞紋からなり，殻面から殻套まで連続する(10)．胞紋は小円形で外面に輪形師板をもつ(9)．胞紋は非常に密で，10 µmあたり約50個．

唇状突起・突起 唇状突起は片方の殻端に1個．殻肩に連結針は見られない(9-11)．殻套眼域は広く明瞭．

半 殻 帯 開放型で4枚の帯片(?)．

葉 緑 体 不明．

有性生殖 不明．

生 活 形 付着性．

汚濁耐性 弱汚濁耐性種．

出 現 地 支笏湖（北海道），琵琶湖（滋賀県）など．

Fragilaria vaucheriae (Kützing) J. B. Petersen

Petersen, J. B. 1938. Bot. Not. **1938**: 167. f. 1c-g.
Basionym: *Exilaria vaucheriae* Kützing 1833. Linnaea **8**: 560. pl. 15. f. 38.
Dimension: L. 15-40 µm,　W. 2-4 µm,　Str. 12-16 in 10 µm,　Ar. ca. 60 in 10 µm.

被殻・殻・軸域　殻は線状皮針形，中央の片側がやや膨らみ，殻端はくちばし状または弱い頭状[1-9]．軸域は狭い．中心域は片側に広く，馬蹄紋となるが，大きさはいろいろ．馬蹄紋は内面でくぼみ[10]，外面は平坦で模様をもたない[11]．殻長15～40 µm，殻幅2～4 µm．細胞は殻面で結合し，帯状の群体を作る[9]．

条線・胞紋　条線はやや放射状で，向かい合う条線は交互に配列する．条線は1列のやや縦長の胞紋から成り，10 µmあたり12～16本．条線は外面観では殻肩部で分断しているように見えるが[10,12]，内面観では殻面から殻套まで1本の溝となる．胞紋は輪形篩板によって閉塞され，10 µmあたり約60個．

唇状突起・突起　片方の殻端に1個の唇状突起がある．殻肩の条線上にへら状の連結針があり，これによって結合するが，連結はそれほど強固ではない．殻端の殻套部には，ほぼ長方形の広い殻套眼域がある[12]．

半殻帯　開放型で4枚の帯片からなり，各々に1列の胞紋をもつ．

葉緑体　板状，2枚(?)．

有性生殖　自家生殖との報告がある（Geitler 1958）．

生活形　付着性．

汚濁耐性　中汚濁耐性種．

出現地　河川，湖沼，水田などに広く出現する汎布種．荒川（埼玉県），鬼怒川（栃木県），天竜川（長野県），阿寒湖（北海道），川越の水田（埼玉県），宝蔵寺沼（埼玉県）など．

Fragilariforma bicapitata (A. Mayer) D. M. Williams & Round

Williams, D. M. & Round, F. E. 1988. Diatom Research **3**: 265.
Basionym: *Fragilaria bicapitata* A. Mayer 1916. Beit. Diat. Bayerns **21**. pl. 1. f. 26, 27.
Dimension: L.15-34 μm, W. 4-5 μm, Str. 16-18 in 10 μm.

被殻・殻 殻は線状皮針形で，殻端は頭状(1-5)．大きな殻では殻側が平行になるが，小さな殻では中央部が膨らむ傾向がある．軸域は非常に狭い．細胞は殻端で粘液によって結合し，ジグザグ群体を作る(6)．殻長15〜34 μm，殻幅4〜5 μm．

条線・胞紋 条線は全体にほぼ平行で，10 μmあたり16〜18本．条線は単列の円形胞紋からなり，殻面から殻套まで連続する．

連結針・唇状突起・眼域 殻肩の間条線上に小さな刺があるが，連結針の働きはしない．唇状突起は片側の殻端近くに1個ある．殻端小孔域は両端にあり，ここから分泌する粘液によって細胞同士が結合し群体を作る．プラークが殻套縁に沿って点在する．

半 殻 帯 開放型．幅広の接殻帯片と4枚の帯片．いずれも1列の胞紋をもつ．
葉 緑 体 不明．
有性生殖 不明．
生 活 形 付着性．
汚濁耐性 弱汚濁耐性種．
出 現 地 三宝寺池（東京都），勇払川（北海道），阿寒湖（北海道），朝日池（新潟県）など．

Hannaea arcus (Ehrenberg) R. M. Patrick

Patrick, R. M. in Patrick, R. & Reimer, C. W. 1966. Monogr. Acad. Nat. Sci. Philad. **13** (1): 132. pl. 4. f. 20.
Basionym: *Navicula arcus* Ehrenberg 1838. Infusionsthier. p. 182. pl. 21. f. 10.
Dimension: L. 15-150 μm,　W. 4-7 μm,　Str. 15-18 in 10 μm.

被殻・殻　殻は中央でくの字に曲がる．両端に向かい細くなり，頭状の殻端をもつ．折れ曲がった中央部の腹側に馬蹄形の中心域がある[1-4]．中心域には条線の痕跡が見られる．軸域は狭くほぼ一定．被殻は背側がやや厚く，そのため殻面で結合した群体は，バナナの房状になる．殻長 15〜150 μm，殻幅 4〜7 μm．

条線・胞紋　条線は平行で，10 μm あたり 15〜18 本．条線は長胞状で，殻面から殻套まで連続する[6]．胞紋は外面に縦長のスリット状の開口をもつ[5,7]．胞紋は光顕では識別できない．中心域には胞紋の開口はないが，弱い横走肋が見られる[5]．

唇状突起・突起　一方の殻端に 1 個の唇状突起があり[6]，外側に細長い小さな開口をもつ[7]．条線の殻肩上に連結針がある[7]．殻端に横長の殻套眼域がある[7]．

半 殻 帯　開放型，4 枚．接殻帯片およびその他の帯片にも胞紋列がある．
葉 緑 体　板状，2 枚．
有性生殖　不明．
生 活 形　付着性．
汚濁耐性　弱汚濁耐性種．
出 現 地　忠類川（北海道）など，寒冷地，主に北海道の流水に出現する．高酸素要求種として知られている．

[ノート]　本種についてのタイプ標本に基づく詳細な研究は Bixby & Jahn (2005) に記されている．なお，本属は *Ceratoneis* と呼ばれていたことがあるが，そのタイプ標本の観察から *Cylindrotheca* であることが示されている（Jahn & Kusber 2005）．

Plate 81

Hannaea arcus var. *recta* (Cleve) M. Idei comb. nov.

Basionym: *Fragilaria arcus* var. *recta* Cleve 1898. Bih Kongl. Svenska Vet.-Akad. Handl. **24**: 9.
Dimension: L. 29-71 μm, W. 6-7 μm, Str. 12-14 in 10 μm.

被殻・殻 殻は線状皮針形．中央の片側が膨らみ，馬蹄形の中心域をつくる(1-7)．中心域には条線の痕跡が見られる．両端に向かい細くなり，頭状の殻端をもつ．軸域は直線的で，狭くほぼ一定．被殻に背腹性なく，殻面で結合した板状の群体をつくる(4)．殻長 29～71 μm，殻幅 6～7 μm．

条線・胞紋 条線は平行で，10 μm あたり 12～14 本．条線は長胞状で，殻面から殻套まで連続する．胞紋は外面に縦長のスリット状の開口をもつ(8)．胞紋は光顕では識別できない．中心域には胞紋の開口はないが，弱い横走肋が見られる(6)．

唇状突起・突起 一方の殻端に 1 個の唇状突起があり，小さな細長い外部開口をもつ(8)．条線の殻肩上に先の二分した連結針があり，殻面全体で結合する(9)．殻端の先端部に横長の殻套眼域がある(8)．

半殻帯 開放型，4 枚．接殻帯片およびその他の帯片にも胞紋列がある．
葉緑体 板状，2 枚．
有性生殖 不明．
生活形 付着性．
汚濁耐性 弱汚濁耐性種．
出現地 荒川（埼玉県），鬼怒川（栃木県），木崎湖（長野県）など．

ノート　*Hannaea arcus* var. *arcus* と異なり，殻がくの字に曲がらず，真っ直ぐである．

Plates 82, 83

Martyana martyi (Héribaud) Round

Round, F. E. in Round, F. E., Crawford, R. M. & Mann, D. G. 1990. The diatom. p. 673.
Basionym: *Opephora martyi* Héribaud 1902. Diat. Foss. Auvergne **1**: 43. pl. 8. f. 20.
Dimension: L. 9-58 µm,　　W. 5-8.5 µm,　　Str. 6-8 in 10 µm.

被殻・殻　殻はこん棒形の上下異極（幅の広い頭部と狭い足部）で，丸い殻端をもつ(2-9)．軸域は狭い(1-8)．帯面観では縦長の長方形で，一方（頭部）の角がわずかに凹む(4)．細胞は一方の端（足部）で基物に付着する．初生殻は中央で膨れ，両極に向かって細くなり，異極性が不明瞭となる(1,14)．殻長9～58 µm，殻幅5～8.5 µm．

条線・胞紋　条線は太く明瞭でほぼ並行．10 µmあたり6～8本．条線は単列のスリット状胞紋で，殻面から殻套まで連続し(10-13)，内側では複雑な薄皮で閉塞される(9)．初生殻は通常の栄養細胞殻とは大きく異なり，殻面が湾曲した凸状となる(14-17)．また，条線は栄養殻同様に単列のスリットであるが，幅が狭い(14-17)．

連結針・唇状突起・眼域　殻端小孔域は両端にあるが，頭部側は目立たない．頭部殻端の殻面がやや凹む(10,13)．連結針，唇状突起，プラークはない．

半　殻　帯　開放型で2～5枚．接殻帯片は無紋で幅が広いが，両端で急に幅が狭くなる(13)．

葉 緑 体　2枚(?)．

有性生殖　不明．

生 活 形　付着性．

汚濁耐性　弱汚濁耐性種．

出 現 地　阿寒湖（北海道）など．

ノート　*Opephora martyi* として長く知られていた種である．しかし，*Opephora* の属名は元々海産種に与えられたものであることや，本種のような淡水産種は海産種とは構造的にも異なるため，*Martyana* という新しい属名が与えられた（Round *et al*. 1990）．

Meridion circulare (Greville) C. Agardh

Agardh, C. 1831. Consp. Cirt. Diat. part 3. p. 40.
Basionym: *Echinella circularis* Greville 1822. Mem. Wernerian Nat. Hist. Soc. **4**: 213. pl. 8. f. 2.
Dimension: L. 19.5-68 μm,　W. 5-7 μm,　Str. ca. 16-18 in 10 μm,　Ar. ca. 60-65 in 10,
　　　　　 T. C. 3-4 in 10 μm.

被殻・殻　殻はこん棒形(1-4,6). 頭部側の殻端は広円～弱頭状～頭状で変異に富む. 足部側の殻端は広円～弱頭状である. 帯面はくさび形. 殻長 19.5～68 μm, 殻幅 5～7 μm. 殻外面は平滑(7-10). 殻肩は若干丸みを帯びる. 殻套にはプラークが見られる(8, 10 矢印). 細胞から分泌される粘液により殻面が密着して扇形の群体を形成する(5).

軸　　域　線形. 殻中央部を縦走する(1-4,6,8).

条線・胞紋　条線は孔状胞紋の列より構成され, 軸域から殻套部まで途切れることなく伸びている(8,10). 殻面全域において平行で 10 μm あたり約 16～18 本. また, 条線を構成する胞紋の密度は 10 μm あたり約 60～65 個で, 外面では師板によって閉塞される. 殻内面では横走肋が殻套から伸張する. 大半の横走肋は殻面の両側間を横断するが, 何本かの横走肋は軸域付近で消失する(1-4,6). これらはいずれも条線にほぼ並行に配置する. 10 μm あたり 3～4 本.

殻端小孔域・唇状突起　足部側の殻端には, 胞紋より若干大きめの孔が碁盤の目状に並んだ殻端小孔域がある(6). 頭部側の殻端には殻端小孔域は存在せず, 通常の殻套部と同様の胞紋が存在する(8). 唇状突起の外部開口はスリット状で, 頭部側の殻端付近の条線上に 1 個存在する(7 矢印, 8).

半 殻 帯　胞紋をもつ開放型の帯片よりなる(8,10). 4 枚 (Williams 1985).

葉 緑 体　円盤状, 多数.

有性生殖　Geitler (1940, 1966) による報告がある.

生 活 形　付着性.

汚濁耐性　弱汚濁耐性種.

出 現 地　多摩川 (東京都), 清里高原の渓流 (山梨県), 大戸川 (滋賀県).

(ノート) 本種では内生殻を作ることが知られている (Geitler 1971). また, Geitler (1966) は増大胞子形成後奇形の殻ができてしまっても, その後, 細胞分裂を繰り返すことにより, 正常な殻を作る過程を示した. 本属や *Diatoma* の師板の構造の詳細は不明であるが, おそらく *Eunotia* に見られる肉趾状師板の特殊化したもの (Mayama & Kobayasi 1991) に類似すると考えられる.

Pseudostaurosira brevistriata (Grunow) D. M. Williams & Round

Williams, D. M. & Round, F. E. 1987. Diatom Research **2**: 276. f. 28-31.
Basionym: *Fragilaria brevistriata* Grunow in Van Heurck, H. 1881. Syn. Diat. Belg. pl. 45. f. 32.
Dimension: L. 8-38 μm,　W. 4-6 μm,　Str. 13-14 in 10 μm.

被殻・殻　殻は披針形で，殻端はくちばし状[1-6]．軸域は広い．被殻は帯面観では長方形．細胞は殻面の連結針で結合し，糸状の群体を作る．殻長 8〜38 μm，殻幅 4〜6 μm．殻面と殻套が明瞭に区別できる[9, 10]．

条線・胞紋　条線は短く殻縁に沿って見られる．ほぼ並行で殻端近くでわずかに放射状となる[1-9]．10 μm あたり 13〜14 本．条線は殻外面では，殻肩の連結針を挟んで 2 つに分断されるが[9, 10]，内面では，殻面から殻套までひとつながりの長胞状で，全体が繊細な薄皮のようなもので閉塞されている[8, 11]．殻面の胞紋は 1 個（まれに 2 個）の横長で，外表面から少し奥に分枝する師板がある[9]．殻套の胞紋は丸く 1 個で，2, 3 個でへら状のフラップによって閉塞される[9]．

連結針・唇状突起・眼域　連結針は条線上の殻肩にあり，その先端は平たく，ややへら状になる[10]．唇状突起はない．殻端小孔域は両端にあるが，小さく目立たない[12]．プラークが殻套縁に沿って点在する[9 矢印]．

半殻帯　開放型．胞紋のない幅広の接殻帯片と細い 5 枚の帯片[12]．
葉緑体　板状．
有性生殖　不明．
生活形　付着性．
汚濁耐性　中汚濁耐性種．
出現地　阿寒湖（北海道），榛名湖（群馬県），山中湖（山梨県），和村珪藻土（長野県），大峰沼（群馬県），木崎湖（長野県），野尻湖（長野県），荒川（埼玉県），千曲川（長野県）など．

Pseudostaurosira brevistriata var. *nipponica* (Skvortsov) H. Kobayasi

Kobayasi, H. in Mayama, S., Idei, M., Osada, K. & Nagumo, T. 2002. Diatom **18**: 90.
Basionym: *Fragilaria brevistriata* var. *nipponica* Skvortsov 1936. Philipp. Journ. Sci. **61** (1): 17. pl. 16. f. 7.
Dimension: L. 16-25 μm, W. 4.5-5.5 μm, Str. ca. 14 in 10 μm.

被殻・殻 殻は殻縁が2回波打ち, 殻端は細くくちばし状となる(1-6). 軸域は非常に広い. 条線は短く, 殻縁にのみ偏在する. 被殻は帯面観では長方形. 殻面の連結針で結合し, 糸状の群体を作る. 殻長16～25 μm, 殻幅4.5～5.5 μm. 殻面と殻套が明瞭に区別できる(10).

条線・胞紋 条線は短く殻縁に沿って見られる. ほぼ並行か殻端近くでわずかに放射状となり(1-8), 10 μmあたり約14本. 条線は外面では殻肩を挟んで2個の胞紋に分かれ, 殻面の胞紋は横長の楕円形で, 殻套の胞紋は丸い(7, 10). しかし, 内面では胞紋は分離せずひとつながりの長胞状で, 胞紋壁から張り出した師板によって閉塞される(8, 9, 11-13).

連結針・唇状突起・眼域 連結針は間条線上の殻肩にあり, 根元は楕円形であるが, 先端は平らなへら状となり, 殻端から殻端まで強く結合する(9, 11). 唇状突起はない. 殻端小孔域は両端にあり, 整列した小孔をもつ(12).

半 殻 帯 不明.
葉 緑 体 不明.
有性生殖 不明.
生 活 形 付着性.
汚濁耐性 弱汚濁耐性種.
出 現 地 阿寒湖（北海道）.

ノート Skvortsovによって木崎湖から記載されたものである. 殻縁が波打つ点で基本種とは異なり, さらに電顕的には連結針が間条線上にある点でも異なる.

Pseudostaurosira robusta (Fusey) D. M. Williams & Round

Williams, D. M. & Round, F. E. 1987. Diatom Research **2**: 278.
Basionym: *Fragilaria construens* var. *binodis* f. *robusta* Fusey 1951. Bull. Microsc. Appliq. ser. 2 (1/2). 34. pl. 1. f. 2.
Synonym: *Fragilaria robusta* (Fusey) Manguin 1954. 78. pl. 1. f. 6a, b.
Dimension: L. 14-21 μm, W. 5-6.5 μm, Str. 14-16 in 10 μm.

被殻・殻 殻は中央で強くくびれた双瘤形で，殻端はくちばし状となる(1-4)．軸域は非常に広い．被殻は帯面観では長方形で，瘤の部分が上下2本の黒い縞として見える．殻肩の連結針で結合し，糸状の群体を作る．殻長14～21 μm，殻幅5～6.5 μm．殻面と殻套が明瞭に区別できる(5,6)．

条線・胞紋 条線は光顕でも明瞭な点紋として観察でき，1～3個が殻縁に偏って存在する(1-4)．条線はほぼ並行か殻端近くでわずかに放射する．条線は10 μmあたり14～16本．胞紋は円形またはやや横長の楕円形で，殻面に1～3個，殻套に1個ある(5-9)．このように殻面に数個，殻套に1個の胞紋をもつことが本属の大きな特徴である．胞紋の内面は師皮で閉塞される(7,8,10)．

連結針・唇状突起・眼域 連結針は間条線の殻肩にあり，その先端は扇状に広がり，殻端から殻端まで強く結合する(9,10)．唇状突起はない．殻端小孔域は両殻端にあり，小さいが明瞭な小孔をもつ(10,11)．プラークが殻套縁に沿って点在する(11)．

半殻帯 開放型．胞紋のない幅広の接殻帯片と細い4,5枚の帯片(11)．各帯片の縁は殻の湾曲に合わせるように波打つ(9)．

葉 緑 体 不明．
有性生殖 不明．
生 活 形 付着性．
汚濁耐性 弱汚濁耐性種．
出 現 地 阿寒湖（北海道），支笏湖（北海道），青木湖（長野県）など．

[ノート] 本種は光顕ではその外形が，*Staurosira construens* var. *binodis* と類似するが，条線が点紋であることで容易に識別できる．

Punctastriata linearis D. M. Williams & Round

Williams, D. M. & Round, F. E. 1987. Diatom Research **2**: 278. f. 38-42.
Dimension: L. 5-32 μm,　W. 3-6 μm,　Str. 8-10 in 10 μm.

被殻・殻　殻は皮針形から楕円形で，わずかに上下不相称(1-8)．殻端はくちばし状で，しばしば一方の殻端に小さな凹みが見られる．軸域は狭い．被殻は帯面観では長方形．細胞は殻面の連結針で結合し，短い糸状の群体を作る．殻長5～32 μm，殻幅3～6 μm．

条線・胞紋　条線は太く明瞭で，ほぼ並行．向かい合う条線は交互に配列．10 μmあたり8～10本．条線は縦横の小肋によって区画された網目状の師板によって閉塞され，殻套まで連続する(10-14)．条線の幅は間条線より広い(11)．間条線が肥厚して肋となるため，殻面は波板状になる(10)．

連結針・唇状突起・眼域　連結針は間条線上の殻肩にあり，先細で先端はT字状になる(12)．唇状突起はない．殻端小孔域は片側にのみで，それも小さく目立たない(10)．

半 殻 帯　開放型．胞紋のない接殻帯片と細い数枚の帯片(12)．
葉 緑 体　不明．
有性生殖　不明．
生 活 形　付着性．
汚濁耐性　中汚濁耐性種．
出 現 地　阿寒湖（北海道），鬼怒川（栃木県），東京学芸大学構内万葉池（東京都），三宝寺池（東京都），中禅寺湖（栃木県）など．

ノート　本種は*Punctastriata*属のタイプ種であり，従来*Fragilaria pinnata*として分類されていたものの一部を含む．本属は，従来*Fragilaria*属や*Opephora*属の種として扱われていたものの中から，その実体となる種の正式名称をさかのぼりタイプを指定することが困難であるため，新属名が与えられた．条線が縦横の小肋によって区画された網目状胞紋である点が本属の最大の特徴である．*Staurosirella pinnata*と区別するためには電顕観察が必須となる．

Staurosira construens Ehrenberg var. *construens*

Ehrenberg, C. G. 1843. Abh. Akad. Wiss. Berlin **1**: 424.
Synonym: *Fragilaria construens* (Ehrenberg) Grunow 1862. Ver. Kaiserl.-Königl. Zool.-bot. Ges. Wien. **12**: 371.
Dimension: L. 10-17 μm, W. 6-9 μm, Str. 12-14 in 10 μm.

被殻・殻 殻は中央で殻側が大きく膨らみ十字形となり，殻端は弱く頭状となる[1-4]．軸域は狭く線状で，中央部でやや広がる．細胞は殻面同士が連結針で結合し，糸状の群体を作る[5]．被殻の帯面観は縦長の長方形であるが，殻側が中央で突出するため，全体にピントが合うことはない[5]．殻長 10〜17 μm，殻幅 6〜9 μm．

条線・胞紋 条線はほぼ並行で，中央でわずかに放射状．10 μm あたり 12〜14 本．条線は単列の胞紋からなり，円形かやや縦長の楕円形の開口をもつ[6,8]．胞紋は殻面から殻套まで連続する[6-8]．

連結針・唇状突起・眼域 連結針は殻肩の間条線上にあり，特に中央部の連結針は大きく，その先端はへら状に広がるが[8]，両端寄りの連結針は小さく先が尖る[9]．唇状突起はない．殻端小孔域は両方にあるが，小孔の数も少ない．プラークは目立たない．

半殻帯 開放型．胞紋のない幅広の接殻帯片と細い 6〜8 枚の帯片．

葉緑体 2 枚．

有性生殖 不明．

生活形 付着性．

汚濁耐性 中汚濁耐性種．

出現地 阿寒湖（北海道），弁天沼（北海道），八幡平大沼（青森県），つつじ池（長野県），和村珪藻土（長野県），三宝寺池（東京都），一碧湖（静岡県），手賀沼（千葉県）など．

ノート 本種が *Staurosira* 属のタイプ種．殻帯の枚数が多く，それらが両端近くで大きく湾曲すること，殻端小孔域が小さいこと，単純な胞紋をもつことが本属の特徴である．

Staurosira construens var. *binodis* (Ehrenberg) P. B. Hamilton

Hamilton, P. B. in Hamilton, P. B., Poulin, M., Charles, D. F. & Angell, M. 1992. Diatom Research 7: 29.
Basionym: *Fragilaria binodis* Ehrenberg 1854. Microgeol. pl. 5. f. 26 ; pl. 6. f. 43 ; pl. 11. f. 15.
Dimension: L. 15-26 μm, W. 6-9 μm, Str. 11-14 in 10 μm.

被殻・殻 殻は小形で殻側が大きく2つに波打ち，殻端は細く突出する[1-4,6,7]．軸域は狭く線状で中央部はやや広がる．被殻の帯面観は縦長の長方形で，上下それぞれ1/4のところに濃い横縞が見える[5]．細胞は殻面の連結針で結合し，糸状の群体を作る．殻長15〜26 μm，殻幅6〜9 μm．

条線・胞紋 条線はほぼ並行で殻端近くでわずかに放射状．10 μmあたり11〜14本．条線は単列の胞紋からなり，殻面から殻套まで連続する[8-10]．胞紋は外面で縦長の楕円形から円形に開口し，内面では薄皮によって閉塞される[8-10]．

連結針・唇状突起・眼域 連結針は殻肩の間条線上にあり，その先端は三葉状となる[8,9,11]．唇状突起はない．殻端小孔域は両端にあり，小孔列が明瞭で目立つ．

半 殻 帯 開放型．胞紋のない幅広の接殻帯片と細い6〜8枚の帯片[11]．

葉 緑 体 2枚．

有性生殖 不明．

生 活 形 付着性．

汚濁耐性 中汚濁耐性種．

出 現 地 阿寒湖（北海道），多々良沼（群馬県），尾瀬沼（群馬県），手賀沼（千葉県），木崎湖（長野県），朝日池（新潟県），一碧湖（静岡県）など．

Staurosira construens var. *exigua* (W. Smith) H. Kobayasi

Kobayasi, H. in Mayama, S., Idei, M., Osada, K. & Nagumo, T. 2002. Diatom **18**: 90.
Basionym: *Triceratium exigua* W. Smith 1856. Brit. Diat. vol. 2. p. 87.
Dimension: L. 12-15 μm,　W. 4-5 μm,　Str. 16-18 in 10 μm.

被殻・殻　殻は三角形で，それぞれの殻端が突出し弱い頭状となり，中央部は弓形かやや膨らむ．軸域は線状で細く，中央部でやや広がる(1-6)．細胞は殻肩の連結針で結合し群体を作る．殻長（本種では隣り合う殻頂から殻頂までの長さ）12～15 μm，殻幅4～5 μm．

条線・胞紋　条線はやや放射状で，殻端近くで平行．10 μm あたり16～18本．条線は1列の胞紋からなり，殻面から殻套まで連続する．また，条線は内面では1つの溝状となる(7-11)．胞紋は，外面では縦長の楕円形から円形に開口し，内面では薄皮によって閉塞される(10, 11)．

連結針・唇状突起・眼域　連結針は殻肩の間条線上にあり，先端は尖る(8, 11)．唇状突起はどの殻端にもない．殻端小孔域はそれぞれの殻端にあり，小さく数個の小孔をもつ(9, 10)．

半殻帯　開放型(?)．
葉緑体　不明．
有性生殖　不明．
生活形　付着性．
汚濁耐性　弱汚濁耐性種．
出現地　八郎潟（秋田県）など．

[ノート]　非常に特徴的な殻形をもつため，分類は容易であるが，出現地は少なく，出現頻度も低い．

Staurosira construens var. *triundulata* (H. Reichelt) H. Kobayasi

Kobayasi, H. in Mayama, S., Idei, M., Osada, K. & Nagumo, T. 2002. Diatom **18**: 90.
Basionym: *Fragilaria construens* var. *triundulata* H. Reichelt in Hartz, N. et Østrup, E. 1899. Denm. geol. Unders. **2** (9): 57. pl. 2. f. 5. 1899.
Dimension: L. 17-28 µm,　W. 5-5.5 µm,　Str. ca. 15 in 10 µm.

被殻・殻　殻は殻縁が3回波打ち，両端はやや頭状からくちばし状となる[1-7]．軸域は狭く線状で，中央でやや広がり縦長の中心域を作る[2,4,7,9]．殻面の連結針で結合し，短い帯状の群体を作る．殻長17～28 µm，殻幅5～5.5 µm．

条線・胞紋　条線は全体にわたってほぼ並行で，10 µmあたり約15本．条線は縦長の楕円形の開口をもった単列胞紋からなり，殻面から殻套まで連続する[8]．胞紋は内面で閉塞される[12]．

連結針・唇状突起・眼域　連結針は殻肩の間条線上にあり，その先端は三角形に広がり，さらに細かな突起が見られる[10,11,13]．唇状突起はない．殻端小孔域は両殻端にあり，縦に並ぶ小孔列をもつ[10]．

半殻帯　開放型．胞紋のない幅広の接殻帯片[3,4]と数枚の帯片[13]．
葉 緑 体　不明．
有性生殖　不明．
生 活 形　付着性．
汚濁耐性　弱汚濁耐性種．
出 現 地　手賀沼（千葉県），琵琶湖（滋賀県）．

ノート　殻縁が3回波打つのが大きな特徴．胞紋や連結針の構造は，*Staurosira construens* の仲間に共通する特徴が見られる．

Staurosira elliptica (Schumann) D. M. Williams & Round

Williams, D. M. & Round, F. E. 1987. Diatom Research **2**: 272. f. 18-20.
Basionym: *Fragilaria elliptica* Schumann 1867. Schr. Königl. Phys. Ges. Königsb. **8**: 52.
Dimension: L. 7-9 μm,　W. 3.5-5 μm,　Str. 12-14 in 10 μm,　Ar. 35-40 in 10 μm.

被殻・殻　殻は小形で楕円形[1-4]．軸域は狭く線状．被殻の帯面観は横長の長方形[5-7]．細胞は殻面の連結針で結合し，糸状の群体を作る．殻長7〜9μm，殻幅3.5〜5μm．

条線・胞紋　条線はほぼ並行で殻端近くでわずかに放射状．10μmあたり12〜14本．条線は丸い単列の胞紋からなり，殻面から殻套まで連続し，10μmあたり35〜40個[8,9]．

連結針・唇状突起・眼域　連結針は殻肩の間条線上にあり，その先端は平たくへら状になっている[8-10]．唇状突起はない．殻端小孔域は両殻端にあるが，小さく目立たない．プラークが殻套縁に沿って点在する．

半 殻 帯　開放型．胞紋のない幅広の接殻帯片と細い6〜8枚の帯片．

葉 緑 体　板状，2枚．

有性生殖　不明．

生 活 形　付着性．

汚濁耐性　中汚濁耐性種．

出 現 地　阿寒湖（北海道），三宝寺池（東京都），多摩川（東京都），朝日池（新潟県），仙女ヶ池（埼玉県），湯ノ湖（栃木県）など．

Staurosira venter (Ehrenberg) H. Kobayasi

Kobayasi, H. in Mayama, S., Idei, M., Osada, K. & Nagumo, T. 2002. Diatom **18**: 90.
Basionym: *Fragilaria venter* Ehrenberg 1854. Mikrogeol. pl. 14. f. 50 ; pl. 9 (1). f. 6, 7.
Dimension: L. 11-22 μm,　W. 5-7 μm,　Str. 13-14 in 10 μm,　Ar. 50-60 in 10 μm.

被殻・殻　殻は広皮針形のものから，中央が膨れ両端がくちばし状から頭状に突出するものもある(1-9)．軸域は線状で，中央部でやや広がる．被殻の帯面観は縦長の長方形．細胞は殻面の連結針で結合し，糸状の群体を作る(9)．殻長 11～22 μm，殻幅 5～7 μm．

条線・胞紋　条線はほぼ並行で殻端近くでわずかに放射状．10 μm あたり 13～14 本．条線は 1 列の胞紋からなり，殻面から殻套まで連続し，10 μm あたり 50～60 個(10,15)．胞紋は外面が縦長の楕円形に開口し，内面が薄皮によって閉塞される(12)．

連結針・唇状突起・眼域　連結針は殻肩の間条線上にあり，先端はへら状に広がり，先端は分岐する(10,11,14)．唇状突起はない．殻端小孔域は両殻端にあり，小さく数個の小孔をもつものもある(13-15)．プラークが殻套縁に沿って点在する．

半 殻 帯　開放型．胞紋のない幅広で粗い鋸歯縁をもつ接殻帯片と細い 4 枚の帯片(14-16)．
葉 緑 体　2 枚．
有性生殖　不明．
生 活 形　付着性．
汚濁耐性　弱汚濁耐性種．
出 現 地　阿寒湖（北海道），湯ノ湖（栃木県），中禅寺湖（栃木県），一碧湖（静岡県）など．

Staurosira venter (Ehrenberg) var. *binodis* H. Kobayasi

Kobayasi, H. in Mayama, S., Idei, M., Osada, K. & Nagumo, T. 2002. Diatom **18**: 90.
Basionym: *Fragilaria venter* Ehrenberg 1854. Mikrogeol. pl. 14. f. 50 ; pl. 9 (1). f. 6, 7.
Dimension: L. 8-20 μm,　W. 4.5-5.5 μm,　Str. 12-14 in 10 μm.

被殻・殻　殻は中央でくびれ，殻縁が2回波打ち，両端はくちばし状となる[1-6]．波打ちの程度はさまざまで，左右で異なることも多い．軸域は狭く線状．被殻の帯面観は縦長の長方形で両端近くに黒いすじが入る[8]．殻面の連結針で結合し，帯状の群体を作る．殻長8〜20 μm，殻幅4.5〜5.5 μm．

条線・胞紋　条線はほぼ並行で，殻端近くでわずかに放射状となり，10 μmあたり12〜14本．条線は縦長の楕円形の開口をもった単列胞紋からなり，殻面から殻套まで連続する[9-11]．胞紋は内面で閉塞される[12,13]．

連結針・唇状突起・眼域　連結針は殻肩の間条線上にあり，その先端はへら状に広がり，さらにいくつかに分岐する[13-15]．唇状突起はない．殻端小孔域は両殻端にあり，縦に並ぶ小孔列をもつ[11,14]．プラークが殻套縁に沿って点在する．

半殻帯　開放型．胞紋のない幅広の接殻帯片と細い数枚の帯片[15]．
葉緑体　2枚．
有性生殖　不明．
生活形　付着性．
汚濁耐性　弱汚濁耐性種．
出現地　阿寒湖（北海道），鬼怒川（栃木県），和村珪藻土（長野県），琵琶湖（滋賀県）など．

Staurosirella lapponica (Grunow) D. M. Williams & Round

Williams, D. M. & Round, F. E. 1987. Diatom Research **2**: 274.
Basionym: *Fragilaria lapponica* Grunow in Van Heurck 1881. Syn. Diat. Belg. pl. 45. f. 35.
Dimension: L. 10-38 μm,　W. 5-7 μm,　Str. ca. 6 in 10 μm.

被殻・殻	殻は線状かわずかに皮針形で，丸い殻端をもつ[1-5]．軸域は広く，皮針形．被殻は帯面観では長方形．殻面の連結針で結合し，糸状の群体を作る[6]．殻長10～38 μm，殻幅5～7 μm．
条線・胞紋	条線は太く明瞭で，並行かやや放射状．10 μmあたり約6本．条線は殻面から殻套まで連続し，それぞれの胞紋は縦長のスリット状となる[7-9]．間条線が太く明瞭．
連結針・唇状突起・眼域	先端が分岐した連結針をもつ．唇状突起はない[7]．
半殻帯	開放型．幅広の接殻帯片と細い数枚の帯片．
葉緑体	板状，2枚．
有性生殖	不明．
生活形	付着性．
汚濁耐性	弱汚濁耐性種．
出現地	青木湖（長野県），山中湖（山梨県），中禅寺湖（栃木県），尾瀬沼（群馬県），阿寒湖（北海道）など．

ノート　本種はこの属のタイプ種である．この属は間条線が肥厚すること，条線がスリット状の胞紋であることが特徴．

Staurosirella leptostauron (Ehrenberg) D. M. Williams & Round

Williams, D. M. & Round, F. E. 1987. Diatom Research **2**: 276.
Basionym: *Biblarium lapponica* Ehrenberg 1854. Microgeol. pl. 12. f. 35, 36.
Dimension: L. 14-25 μm,　W. 8.5-14 μm,　Str. 8-9 in 10 μm,　Ar. ca. 50 in 10 μm.

被殻・殻　殻は中央で大きく膨れた十字形で，丸い殻端をもつ[1-6]．左右の膨らみが不揃いの場合も多く，また上下も非対称で，両殻端の太さが異なることも多い．軸域は狭いが，中央でやや広がり皮針形の中心域となる．被殻は帯面観では長方形．殻面の連結針で結合し，糸状の群体を作る．殻長 14〜25 μm，殻幅 8.5〜14 μm．

条線・胞紋　条線は太く明瞭で，並行かやや放射状．10 μm あたり 8〜9 本．条線は殻面から殻套まで連続し，それぞれの胞紋は外面では縦長のスリット状となり，10 μm あたり約 50 個見られる[5,9]．胞紋は内面で薄皮によって閉塞される[6,10]．間条線は太く明瞭．

連結針・唇状突起・眼域　連結針は間条線上にあり，平板状で，やや先端が尖り，ときに根元で 2, 3 に分かれていることもある[7,8,12]．殻端小孔域は円形の小孔ではなく，縦のスリット状となる[7,8]．唇状突起はない．

半殻帯　開放型．幅広で鋸歯状の内接部をもった接殻帯片と 3, 4 枚の帯片からなる[11,12]．

葉緑体　不明．

有性生殖　不明．

生活形　付着性．

汚濁耐性　弱汚濁耐性種．

出現地　三宝寺池（東京都），苗場山湿原（新潟県），地獄沢（栃木県），阿寒湖（北海道）など．

ノート　本種は同じ十字形をした *Staurosira construens* に一見似るが，明らかに条線が粗く，光顕でも容易に識別できる．本種の 10 μm あたりの胞紋密度は，Patrick & Reimer (1966) では 25〜30 個，Krammer & Lange-Bertalot (1991) では 25〜35 個となっているが，本邦のものは細かく約 50 個であった．

Staurosirella pinnata (Ehrenberg) D. M. Williams & Round

Williams, D. M. & Round, F. E. 1987. Diatom Research **2**: 274.
Basionym: *Fragilaria pinnata* Ehrenberg 1843. Phys. Abh. Akad. Wiss. Berlin. **1841**: 415. pl. 3 (6), f. 8.
Dimension: L. 5-17 μm, W. 3-6 μm, Str. 10-12 in 10 μm.

被殻・殻 殻は中央でやや膨らみのある皮針形から楕円形で,丸い殻端をもつ(1-7).殻端の大きさは上下でわずかに異なる.軸域は狭く,全体にほぼ一定の幅をもつ.被殻は帯面観では長方形.殻面の連結針で結合し,糸状の群体を作る.殻長5〜17 μm,殻幅3〜6 μm.

条線・胞紋 条線はやや太く明瞭で,並行か殻端近くでわずかに放射状.向かい合う条線は交互に配列する.10 μmあたり10〜12本.条線はスリット状の胞紋からなり,殻面から殻套まで連続する(8-14).条線は殻肩部で最も幅広となり,両側に向かって狭くなる.間条線は肥厚するが,条線の幅よりやや狭い.胞紋は薄皮のようなもので閉塞されている(13).

連結針・唇状突起・眼域 連結針は間条線上の殻面殻套境界部にあり,その付け根は丸く,先端は2回二叉分岐している(8, 12, 14).唇状突起はない(10).殻端小孔域は両端で異なり,幅の狭い側の殻端には多数の小孔が見られるが,幅の広い側の殻端には小孔がほとんど見られない(11, 12).

半殻帯 開放型.幅広で接殻帯片と数枚の帯片(8, 14).
葉緑体 板状.
有性生殖 不明.
生活形 付着性.
汚濁耐性 弱汚濁耐性種.
出現地 八郎潟(秋田県),中禅寺湖(栃木県),山中湖(山梨県),阿寒湖(北海道)など.

(ノート) 従来*Fragilaria pinnata* Ehrenbergとして同定されていたもので,条線がスリット状の胞紋をもつのが大きな特徴である.本種と類似するため混同され*F. pinnata* Ehrenbergとして同定されていたもので,条線がやや粗く,多列胞紋をもつ種類は,*Punctastriata linearis* D. M. Williams & Roundである.しかし,この2種を光顕だけで正確に識別するのは困難である.

Synedrella parasitica (W. Smith) Round & Maidana

Round, F. E. & Maidana, N. I. 2001. Diatom **17**: 24.
Basionym: *Odontidium parasiticum* W. Smith 1856. Brit. Diat. **2**: 19. pl. 60. f. 375.
Dimension: L. 10-30 μm,　W. 3-5 μm,　Str. 16-20 in 10 μm.

被殻・殻　殻は皮針形で中央で膨らみ，両端で急に細くなる(1-4)．軸域は広く，全体として皮針形となる．殻面は平坦で，明瞭な殻肩を境に垂直的な殻套へと続く．細胞は他の大形珪藻（たとえば *Nitzschia sigmoidea*）や藻類の表面に付着し(6,7)，叢状群体を作る．殻長10〜30 μm，殻幅 3〜5 μm．

条線・胞紋　条線はほぼ並行で左右交互に配置し，殻端近くでわずかに放射状．10 μm あたり 16〜20本．条線は殻肩で2つに分離され，殻面には長い胞紋が1個，殻套には短い胞紋が1個あり，胞紋の内面の縁から不規則に張り出した師板によって閉塞される(8,9)．

連結針・唇状突起・眼域　連結針も唇状突起もない．殻端小孔域は両端にあり，明瞭で，ややくぼんだ殻套眼域状にもなる(8)．プラークが殻套縁に沿って点在する．

半殻帯　開放型．5〜8枚の殻帯．
葉緑体　板状．
有性生殖　不明．
生活形　付着性．
汚濁耐性　弱汚濁耐性種．
出現地　琵琶湖（滋賀県），川内川（熊本県），多々良沼（群馬県），山中湖（山梨県），阿寒湖（北海道）など．

ノート　本種は，*Synedra parasitica* (W. Smith) Hustedt や *Fragilaria parasitica* (W. Smith) Grunow とされてきたが，Round & Maidana (2001) によって設立された新属 *Synedrella* に組み替えられた．本種は元々 *Nitzschia sigmoidea* や他の大形の管状縦溝珪藻に特異的に付着して生育することが知られている．この生態的特徴に加え，まず，唇状突起をもたない点で広義の *Synedra* 属ではないこと，さらに，結合針をもたない点で広義の *Fragilaria* 属でもないこと，そして特徴的な胞紋閉塞をもつ点から新属とされた．

Tabularia affinis (Kützing) Snoeijs

Snoeijs, P. 1992. Diatom Research **7**: 343.
Basionym: *Synedra affinis* Kützing 1844. Kies. Bacill. p. 68. pl. 15. f. 6, 11.
Dimension: L. 20-80 μm, W. 4-5 μm, Str. 14-18 in 10 μm.

被殻・殻 殻は線形または狭皮針形で，丸い殻端をもつ．殻幅はほとんど変わらないが殻長の変異が大きい(1-9)．軸域は，タイプ標本の個体では幅が狭いが(1,2)，本邦の個体は概して広い．細胞は一方の端で基物に付着して叢状の群体を作る．殻長 20～80 μm，殻幅 4～5 μm．

条線・胞紋 条線はほぼ並行で短く殻縁にあり，10 μm あたり 14～18 本．条線は長胞状で，外面に 2 列の胞紋が並び，殻面から殻套まで連続し(10,11)，内面は開口する(12,13)．胞紋の外面の開口は円形か三日月形となる．

連結針・唇状突起・眼域 連結針はない．唇状突起は片側の端に 1 個(11,12)で，最端の条線の中心寄りに開口する(10)．殻套眼域は両殻端にあり，広い(11-13)．

半 殻 帯 開放型．
葉 緑 体 板状，2 枚．
有性生殖 不明．
生 活 形 付着性．
汚濁耐性 弱汚濁耐性種．
出 現 地 明石川河口（兵庫県），涸沼川（茨城県）など，汽水域によく見られる．

ノート 本邦産の個体はタイプ標本に比べ軸域が広い傾向があるが，条線が 2 列の胞紋である点から同種と考えた．

Ulnaria acus (Kützing) M. Aboal

Aboal, M. 2003. In: Witkowski, A.(ed) Diat. Monogr. **4**: 105.
Basionym: *Synedra acus* Kützing 1844. Bacill. p. 66. pl. 15. f. 7.
Dimension: L. 100-250 μm,　W. 4-6 μm,　Str. 12-15 in 10 μm.

被殻・殻　殻は線状皮針形．中央部から両端に向かって徐々に細くなり，殻端は細く弱い頭状となる(1-4)．軸域は狭く，直線的でほぼ一定．中心域は横長の長方形で，条線の痕跡が見られる．殻長100～250 μm，殻幅4～6 μm．細胞は一方の端で基物に付着し，叢状の群体をつくる(5)．

条線・胞紋　条線はほぼ並行で，10 μmあたり12～15本．条線は1列の胞紋から成り，殻面から殻套まで連続する(9)．胞紋は小円形で外面に輪形師板をもつ．胞紋は細かく光顕では識別できない．

唇状突起・突起　両殻端にそれぞれ1個の唇状突起がある(6-8矢印小)．殻肩や殻端に突起はない．殻端の殻套にはほぼ全面に広い殻套眼域（ocellulimbus；Williams 1986）がある(8矢印大)．

半　殻　帯　閉鎖型，数枚の帯片をもつと思われる．
葉　緑　体　板状，2枚．
有性生殖　不明．
生　活　形　付着性または浮遊性．
汚濁耐性　弱汚濁耐性種．
出　現　地　塩沢の水田（新潟県），川内川（鹿児島県），宝蔵寺沼（埼玉県），多々良沼（群馬県）など，河川や湖沼に広く出現する．プランクトンとしても出現することがある．

Ulnaria biceps (Kützing) Compère

Compère, P. 2001. In: Jahn, R. *et al.* (eds) Lange-Bertalot-Festschrift. p. 100.
Basionym: *Synedra biceps* Kützing 1844. Bacill. p. 66. pl. 14. f. 18.
Dimension: L. 160-600 μm, W. 6-8 μm, Str. 8-9 in 10 μm, Ar. 33 in 10 μm.

被殻・殻　殻は線形で，全体としてやや弓状に曲がる．殻側はほぼ並行で，両端近くでややくびれ，殻端はくちばし状か頭状となる[1-4]．軸域は狭く，直線的でほぼ一定．中心域はあったりなかったりする．殻面と殻套がほぼ垂直．殻長 160～600 μm，殻幅 6～8 μm．

条線・胞紋　条線は平行で，左右でほぼ向かい合って配列し，10 μm あたり 8～9 本．条線は 1 列の胞紋から成る[4-6]．胞紋は円形または楕円形で，10 μm あたり 33 個，外面に輪形篩板をもつ．胞紋は光顕では識別できない．

唇状突起・突起　両殻端にはそれぞれ 1 個の唇状突起がある．外面には小さな丸い開口をもつ（4 矢印）．殻肩は明瞭で，殻端に 2 本の小さな突起がある[6]．殻端には長方形の殻套眼域がある[6]．

半　殻　帯　閉鎖型，数枚の帯片をもつと思われる．
葉　緑　体　板状，2 枚と思われる．
有 性 生 殖　不明．
生　活　形　付着性．
汚 濁 耐 性　弱汚濁耐性種．
出　現　地　湯ノ湖（栃木県），榛名湖（群馬県），阿寒湖（北海道）．

Ulnaria capitata (Ehrenberg) Compère

Compère, P. 2001. In: Jahn, R. *et al.* (eds) Lange-Bertalot-Festschrift. p. 100.
Basionym: *Synedra capitata* Ehrenberg 1836. Ber. Bek. Verh. Königl. Preuss. Acad. Wiss. Berlin **1836**: 53.
Dimension: L. 250-360 μm, W. 10-11 μm, Str. 9-10 in 10 μm.

被殻・殻 殻は線形．殻側はほぼ並行で，両端で膨らみ，殻端はくさび形[1-3]．軸域は狭く，直線的でほぼ一定．中心域はない．殻長250〜360 μm，殻幅10〜11 μm．

条線・胞紋 条線は平行に配列し，10 μmあたり9〜10本．条線は1列の胞紋から成る[4-6]．胞紋は円形または楕円形で，外面に輪形師板をもつ．胞紋は光顕では識別できない．

唇状突起・突起 両殻端にはそれぞれ1個の唇状突起がある．外面には小さな丸い開口をもつ[4,6]．殻肩は明瞭で，くさび形の殻端の先端部に小さな殻套眼域がある[6]．

半 殻 帯 閉鎖型，数枚の帯片をもつと思われる．
葉 緑 体 板状，2枚と思われる．
有性生殖 不明．
生 活 形 付着性．
汚濁耐性 弱汚濁耐性種．
出 現 地 阿寒湖（北海道）．

Ulnaria inaequalis (H. Kobayasi) M. Idei comb. nov.

Basionym: *Synedra inaequalis* H. Kobayasi in Kobayasi, H., Idei, M., Kobori, S. & Tanaka, H. 1987. Diatom **3**: 9 ; Kobayasi, H. 1965. Journ. Jap. Bot. **40**: 347. f. 1a-d.
Dimension: L. 30-90 µm, W. 7.5-9 µm, Str. 11-14 in 10 µm, Ar. ca. 40 in 10 µm.

- **被殻・殻** 殻は皮針形であるが，中央部でややくびれる．両端に向かって急に細くなり，殻端は頭状となる(1-4)．殻の左右がやや非対称．軸域は狭く，ほぼ一定であるが中央からややはずれ，また直線的でないものも多い．中心域は正方形から縦長の長方形であるが，多少とも条線痕が見られることが多い(1-3)．殻長 30〜90 µm，殻幅 7.5〜9 µm．
- **条線・胞紋** 条線は直線的ではなく不規則に曲がり，配列にむらがあり，10 µm あたり 11〜14 本．条線は 1 列の胞紋から成り，殻面から殻套まで連続する(6)．胞紋は楕円形で，外面に輪形師板をもつ(8)．胞紋は 10 µm あたり約 40 個と細かく，光顕ではほとんど識別できない．
- **唇状突起・突起** 両殻端にそれぞれ 1 個の唇状突起がある(7)．殻肩や殻端に突起はない．殻端にはほぼ長方形の広い殻套眼域がある(5)．
- **半 殻 帯** 閉鎖型，3 枚．接殻帯片には 1 列の胞紋．
- **葉 緑 体** 板状，2 枚．
- **有性生殖** 不明．
- **生 活 形** 付着性．
- **汚濁耐性** 弱汚濁耐性種．
- **出 現 地** 荒川（埼玉県），石狩川（北海道），神流川（群馬県），三面川（新潟県）など，河川に出現する．

> ノート　*Ulnaria ulna* と似ているが，殻の左右が非対称であること，条線がゆれ，密度にむらがある点などで区別できる．本種は Kobayasi(1965) によって最初に記載されたが，国際命名規約に規定されているタイプの指定を欠いていたため，後に Kobayasi *et al.* (1987) によって正式出版に必要な条件が付加された．

Ulnaria lanceolata (Kützing) Compère

Compère, P. 2001. In: Jahn, R. *et al*. (eds) Lange-Bartalot-Festschrift. p. 100.
Basionym: *Synedra lanceolata* Kützing 1844. Bacill. p. 66. pl. 30. f. 31.
Dimension: L. 57-110 μm, W. 7-9 μm, Str. 9-10 in 10 μm.

被殻・殻 殻は線形で殻側はほぼ並行．両端近くで細くなり，殻端はくちばし状(1-4)．軸域は狭く，直線的でほぼ一定であるが，やや片側に寄っている．中心域は楕円形でどちらか一方に片寄り，その反対側に短い条線が見られる(1-4,9)．これが本種の特徴の1つである．殻長57～110 μm，殻幅7～9 μm．

条線・胞紋 条線は左右が向かい合って配列し，10 μm あたり9～10本．条線は2列の胞紋から成り(8)，殻面から殻套まで連続する(10)．胞紋は小円形で外面に輪形師板をもつ(8)．胞紋は細かく光顕では識別できない．

唇状突起・突起 両殻端にそれぞれ1個の唇状突起がある(6,7)．殻肩や殻端に突起はない．殻端の殻套にはほぼ長方形の広い殻套眼域がある(5)．

半殻帯 閉鎖型，3枚（まれに2枚）．接殻帯片の縁は鋸歯状で1列の胞紋がある．他の帯片にも1列の胞紋がある．

葉緑体 板状，2枚．

有性生殖 不明．

生活形 付着性．

汚濁耐性 弱汚濁耐性種．

出現地 酒匂川（神奈川県），紀ノ川（和歌山県），筑後川（福岡県）など，関東以南の河川に出現する．

ノート *Ulnaria ulna* と非常によく似ており，混同されてきた可能性がある．中心域の形がやや異なるが，光顕観察だけでは正確な同定は困難である．電顕観察により条線の胞紋が1列か2列かを確認する必要がある．Kobayasi *et al*. (1987) は本種の詳細な観察を報告している．

Ulnaria pseudogaillonii (H. Kobayasi & M. Idei) M. Idei comb. nov.

Basionym: *Fragilaria pseudogaillonii* H. Kobayasi & M. Idei 1979. Jap. J. Phycol. **27**: 196. f. 1-3.
Dimension: L. 220-410 μm,　W. 8-10 μm,　Str. 7-9 in 10 μm,　Ar. 24-28 in 10 μm.

被殻・殻　殻は長線形．殻側はほぼ並行で，殻端は丸いくちばし状[1-3]．8～20個程度の細胞が殻面同士で結合し，板状の群体を作る[4,7]．特に多数の細胞からなる大きな群体は肉眼で確認することもできる．軸域は狭くほぼ一定．中心域はない．殻長220～410 μm，殻幅8～10 μm．

条線・胞紋　条線は平行で左右が向かい合い，10 μmあたり7～9本．条線は1列の胞紋から成り，殻面から殻套まで連続する[5,6]．胞紋は円形または楕円形で，10 μmあたり24～28個，外面に輪形篩板をもつ[5]．

唇状突起・突起　両殻端にそれぞれ1個の唇状突起があり，それらは対角線上に位置することが多い[2]．間条線の殻肩には先の丸い連結針があり，これらにより互いの殻面が端から端まで強く結合する[6]．殻端には広い殻套眼域があり，その上方に小さな角状の2本の突起が見られる[5]．

半殻帯　閉鎖型，4枚．連結帯片およびその他の帯片にも1列の胞紋．
葉緑体　板状，2枚．
有性生殖　不明．
生活形　付着性．
汚濁耐性　弱汚濁耐性種．
出現地　筑後川（福岡県），浅川（東京都），狩野川（静岡県）など，河川に出現する．

(ノート)　本種は，連結針によって結合して群体を形成することから，*Fragilaria*属の1種として記載したが，新たな分類基準に基づいて，*Ulnaria*属に組み替えた．本種のような明瞭な連結針をもつ種類は本属ではまれである．Williams (1986) は，本種を *Synedra ungeriana* (Grunow) D. M. Williams (= *Ulnaria ungeriana* (Grunow) Compère) の異名としているが，*U. ungeriana* は本種に比べ明らかに小さく（殻長85～230 μm），中心域をもつなどの点で異なる．また，*U. ulna* と類似しているが，本種は条線も点紋も粗く，殻端が細くびれないことで識別できる．

Ulnaria ulna (Nitzsch) Compère

Compère, P. 2001 In: Jahn, R. *et al.* (eds) Lange-Bartalot-Festschrift. p. 100.
Basionym: *Bacillaria ulna* Nitzsch 1817. Schr. Nat. Ges. Halle **3** (1): 99. pl. 5.
Dimension: L. 50-350 µm, W. 5-9 µm, Str. 8-12 in 10 µm, Ar. ca. 40 in 10 µm.

被殻・殻 殻は長線形，殻側はほぼ並行．殻端はくちばし状から頭状[1-4]．殻外面は条線部がややくぼみ，弱い波板状[6]．軸域は狭く，中心域は正方形から長方形で，条線の痕跡が見られるものもある．細胞はその先端から分泌する粘液質によって基物に付着し，叢状の群体を形成する．殻長50〜350 µm，殻幅5〜9 µm．

条線・胞紋 両側の条線は平行で向かい合い，10 µm あたり8〜12本である．条線は1列の胞紋から成り，殻面から殻套まで連続する[5]．胞紋は円形または楕円形で，外表面に輪形篩板をもち[5-7]，10 µm あたり約40個．胞紋列は殻肩で分断しているように見えるが，殻内面では明らかに1本の溝の中にある[4]．

唇状突起・突起 両殻端にそれぞれ1個の唇状突起があり，軸域に接し，やや斜めに位置している[4]．両端の殻套には殻套眼域（ocellulimbus; Williams 1986）があり，その上方に小さな角状の突起がしばしば見られる[6]．

半殻帯 閉鎖型，3枚．接殻帯片およびその他の帯片にも1列の胞紋がある．

葉緑体 板状，2枚．

有性生殖 対合した各母細胞には2個の異型配偶子が生じ，2個の接合子が作られる（Geitler 1939a, b）．

生活形 付着性．

汚濁耐性 中汚濁耐性種．

出現地 多摩川（東京都），相模川（神奈川県），川越の湧泉（埼玉県），鬼怒川（栃木県），筑後川（福岡県），池田湖（鹿児島県）など，河川や湖沼によく出現する普通種．

Plates 108, 109

Tabellaria fenestrata (Lyngbye) Kützing

Kützing, F. T. 1844. Bacill. p. 127.
Basionym: *Diatoma fenestratum* Lyngbye 1819. Tent. Hydrophytol. Danic. p. 180. pl. 61. f. E3.
Dimension: L. 38-84 μm,　W. 6-8 μm,　Str. 17-19 in 10 μm,　Ar. ca. 50 in 10 μm.

被殻・殻　殻は中央で膨らんだ線形，もしくは中央で膨らんだ紡錘状皮針形，殻端部は頭状である[3,5]．帯面は長方形[1,2,6]．殻端部より分泌した粘液により直線状の群体を作るが[1]ときにY字に群体を形成することもある[9]．殻長38〜84 μm（33〜116 μm: Krammer & Lange-Bertalot 1991）．殻幅6〜8 μm（4〜10 μm: l.c.）．殻外面は平滑[10,11]．殻肩は角ばる．殻肩で間条線の上に小針を生じる個体もあるが，すべての個体が生じるわけではない．

軸　　域　殻面中央部を縦走する．線形で中央部でもほとんど広がらない[3,5]．

条線・胞紋　条線は平行に配列し，殻面から殻套部まで断続せずに続く，10 μmあたり17〜19本．条線を構成する胞紋は孔状胞紋で，10 μmあたり約50個，殻外面で師板により閉塞される．個々の胞紋は光学顕微鏡では認識できない．

殻端小孔域・唇状突起　両方の頂端には殻端小孔域があり[10,13,14]，この部分から分泌する粘液によって隣接細胞と結合する．殻面中央の膨れた部分に1個の唇状突起があるが，これは条線の延長上に存在する[5,11,16]．唇状突起の外部開口は横長のスリットで，光学顕微鏡で認めることができる[3]．

半 殻 帯　開放型の帯片のみからなる[8,13-15]．3〜6枚．帯片はすべて中脈上に1列の胞紋をもつが[13,14]，閉鎖末端部では数列の胞紋をもつ帯片もある．隔壁[7,16,17]は1細胞あたり4枚[6]．副隔壁をもつ帯片はない．

葉 緑 体　円盤状，多数．黄緑色を呈す．

有性生殖　不明．

生 活 形　浮遊性であるが，しばしば付着基物上にも出現する．

汚濁耐性　弱汚濁耐性種．富栄養の湖沼や河川に出現するが，腐植質の水域には出現しない．

出 現 地　勇払川（北海道），上駒月のため池（滋賀県）など．

[ノート]　直線状の群体を作ること，軸域が中央部で膨らまないこと，帯片がすべて開放型であることが本種の特徴である．

Plates 110, 111

Tabellaria flocculosa (Roth) Kützing

Kützing, F. T. 1844. Bacill. p. 127.
Basionym: *Conferva flocculosa* Roth 1797. Catalecta Bot. Fasc. **1**. p. 192. pl. 4. f. 4 ; pl. 5. f. 6.
Dimension: L. 13-47 μm,　W. 5.5-8.5 μm,　Str. 15-20 in 10 μm,　Ar. ca. 50 in 10 μm.

被殻・殻　殻は中央が膨らんだ線形，もしくは紡錘状皮針形である．殻端部は著しく頭状となる(1-9, 11, 14)．帯面は長方形〜正方形．殻端部より分泌した粘液によりジグザグの群体を作る(10, 12, 13)．この形が神事に使う幣帛（へいはく，ぬさ）に似ているため，*Tabellaria* にはヌサガタケイソウの和名がついた．殻長 13〜47 μm（6〜130 μm: Krammer & Lange-Bertalot 1991）．殻幅 5.5〜8.5 μm（3.8〜8.5 μm: l.c.）．殻外面は平滑(15, 17)．殻肩は角ばり，間条線の上に小針を生じる(15-17)．

軸　　域　殻面中央部を縦走し線形，中央部ではわずかに皮針形に広がる(1, 3-6, 11)．

条線・胞紋　条線は平行に配置し，殻面から殻套まで断続せずに続く．10 μm あたり 15〜20 本，条線を構成する胞紋は孔状である．胞紋は 10 μm あたり約 50 個あり，光学顕微鏡では個々の胞紋を認識できない．

殻端小孔域・唇状突起　両方の頂端には殻端小孔域があり(15, 16)，この部分から分泌する粘液によって隣接細胞と結合する(17)．殻面中央の膨れた部分に 1 個の唇状突起があるが，これは条線の延長上に存在する(11, 15)．唇状突起の外部開口は横長のスリットで，光学顕微鏡で認めることができる(4-6)．

半殻帯　帯片はすべて中脈上に 1 列の胞紋をもっているが，帯片の閉鎖末端部では数列の胞紋をもつ帯片もある(16, 17)．上半殻帯の帯片は 10 枚まである．殻側から数えて 2〜7 枚目までは閉鎖型であるが，半殻帯の端側 4, 5 枚は開放型の帯片である(15-17)．隔壁は閉鎖型帯片にあり(2, 7-9, 14)，1 細胞あたり 3〜7 枚の隔壁をもつが，これらは，帯面観で不規則に曲がる傾向にある(13)．通常の隔壁と向き合って位置する，小さな副隔壁をもつ帯片も存在する(2, 8, 9, 14, 18)．

葉 緑 体　円盤状，多数(13)．珪藻の中では珍しく黄緑色を呈す．

有性生殖　Schütt（1896, p. 104）に *Tabellaria* の有性生殖の報告がある．Geitler（1932）はこれを IVa 型（アポミクシスにより 1 個の母細胞から 2 個の増大胞子を形成）に分類している．

生 活 形　浮遊性であるが，しばしば付着基物上にも出現する．

汚濁耐性　弱汚濁耐性種．淡水の富栄養から腐植栄養までの水域に出現する．

出 現 地　小千谷の水田（新潟県），野々海湿原（長野県），菅平の野池（長野県），川内川（鹿児島県）など．

(ノート) 閉鎖型と開放型との帯片の枚数は個体により変動し一定しない．副隔壁をもつ帯片の存在はこの種の特徴である．歴史上初めて学術雑誌に報告された珪藻は 1703 年に英国の地方のとある紳士が稚拙な顕微鏡で観察した本種であると考えられている．ただし，当時学名はおろか，珪藻（diatom）の名前さえ存在していなかった．

Tabellaria pseudoflocculosa H. Kobayasi ex Mayama sp. nov.

Dimension: L. 35-45 μm, W. 5-6.5 μm, Str. ca. 16 in 10 μm, Ar. ca. 50 in 10 μm.

被殻・殻 殻は中央で膨らんだ紡錘状皮針形で，殻端部は弱頭状である(1,4,5)．帯面は長方形(3,7-9)．殻端部より分泌した粘液によりジグザグの群体を作る(7-9)．ときに細胞同士が約120度まで広がり，星形の群体を作る場合もある．殻長 35〜45 μm．殻幅 5〜6.5 μm．殻外面は平滑(10-12)．殻肩は角ばり，間条線の上に小針を生じる(10-12)．

軸　　域 線形で殻面中央部を縦走する(1,4,5)．中央部ではわずかに皮針形に広がる．

条線・胞紋 条線が平行で，殻面から殻套部まで断続せずに続く(11)．条線は 10 μm あたり約 16 本．条線を構成する胞紋は孔状胞紋で，10 μm あたり約 50 個ある．個々の胞紋は光学顕微鏡では認識できない．

殻端小孔域・唇状突起 両方の頂端には殻端小孔域があり(12)，この部分から分泌する粘液によって隣接細胞と結合する．殻面中央の膨れた部分に 1 個の唇状突起があるが，これは条線の延長上に存在する(1,4,5)．唇状突起の外部開口は横長のスリットで，光学顕微鏡で認めることができる．

半 殻 帯 帯片はすべて中脈上に 1 列の胞紋をもつが(10,11)，閉鎖末端部では数列の胞紋をもつ帯片もある(12)．殻側から数えて 2〜3 枚までは閉鎖型であり，それに続く 2〜4 枚の帯片は開放型である．隔壁は閉鎖型帯片に生じ，1 細胞あたり 3〜6 枚ある．これらの隔壁は帯面観では直線的である(8)．

葉 緑 体 円盤状，多数．黄緑色を呈す．

有性生殖 不明．

生 活 形 浮遊性であるが，しばしば付着基物上にも出現する．

汚濁耐性 弱汚濁耐性種．

出 現 地 勇払川（北海道），上駒月のため池（滋賀県）．

ノート　閉鎖型と開放型との帯片の枚数は個体により変動し一定しないが，成熟した細胞では 4 枚の隔壁をもつものが多い．*Tabellaria flocculosa* に似るが，本種は隔壁が直線的なこと，隔壁の枚数が少ないこと，副隔壁をもつ帯片が存在しないこと，帯面観がより細長いこと，また殻面中央部と殻端部の間の部分があまり細くならないことから識別できる．また *T. quadriseptata* も本種に類似するが，殻面の両側が平行であること，また副隔壁を形成することで本種とは異なる．また，本種は *T. fenestrata*（腐植質の水域には出現しない）と一緒に富栄養の水域（中性〜アルカリ性）に出現していたが，*T. quadriseptata* は pH 6 以上になると出現しない(Flower & Battarbee 1985)点でも異なっている．

Actinella brasiliensis Grunow

Grunow, A. in Van Heurck, H. 1881. Syn. Diat. Belg. pl. 35. f. 19.
Dimension: L. 38-238 μm, W. 9-13 (head), 6-9 (mid) μm, Str. 13-15 in 10 μm,
Ar. 32-36 in 10 μm.

被殻・殻 殻はゆるく弓状に曲がったほこ形，幅広の極（頭極）側では微突頭で伸長し，幅狭の極（足極）側はくさび形(1-4)．帯面は頭極側が幅広になるくさび形(5)．足極側で基物に付着し叢状群体を形成する．殻長38～238 μm，殻幅は頭部で9～13 μm，中央部で6～9 μm．殻外面(5)および殻内面(6,7)は共に平滑である．殻肩に小針が配列する(6,7)．

軸域・縦溝 縦溝は短く，両極付近の腹側に近い殻面から，腹側殻套部へ向かって伸長しているが(2)，殻套部の方がより長い(6,7)．縦溝は極末端では殻套に接する殻面上に斜めに配置される蝸牛舌を形成する(6,7)．蝸牛舌およびそれに接する極域は肥厚のため，光学顕微鏡では高コントラスト部位として観察され(1,3,4)，従来は極節と呼ばれていた．縦溝の中心末端の殻内面は縦溝周辺部の弱い肥厚を伴って直線的に終わる(6)．狭い軸域が縦溝の両極末端間に存在するが，頭局側の部分を除き，殻面の腹側縁に非常に近い所を走るため(2)，光学顕微鏡では観察しにくい．

条線・胞紋・唇状突起 条線は平行で10 μmあたり13～15本．頂端部では求心的に配列し，足極側では密に配列し小孔域状になる(7)．また，縦溝と接する殻套部側の条線は密に配列している．殻面の胞紋は10 μmあたり32～36個．胞紋は殻外面で篩板によって閉塞される(5)．1殻に1個殻端部に唇状突起をもつが，上殻と下殻とでは対角線の位置に配置するため，片方の殻では頭極に，もう一方の殻では足極に観察される．(6)および(7)の写真は共に唇状突起をもたない殻端を示す．

半 殻 帯 開放型で胞紋をもつ帯片よりなる．帯片の開放末端と閉鎖末端は被殻の両極において交互に配列する(5)．成熟した半殻帯は4枚．

葉 緑 体 板状．2枚．

有性生殖 配偶子母細胞にそれぞれ1個の配偶子が形成され，双方の配偶子が両母細胞の間で接合し，1個の増大胞子を形成する（真山1993）．

生 活 形 付着性．

汚濁耐性 弱汚濁耐性種．

出 現 地 郡殿ノ池（新潟県），宝蔵寺沼（埼玉県），地蔵院沼（埼玉県）など，腐植質に富む湖沼に出現する．

ノート 本邦の*Actinella*属には本種以外に，頭極の形が異なる*Actinella punctata* Lewisもまれに出現する．

Peronia fibula (Brébisson ex Kützing) R. Ross

Ross, R. 1956. Ann. Mag. Natur. Hist. ser. 12. **9**: 78.
Basionym: *Gomphonema fibula* Brébisson ex Kützing 1849. Spec. Alg. p. 65.
Dimension: L. 25-56 μm,　W. 3.5-4.5 μm,　Str. 14-18 in 10 μm,　Ar. ca. 70 in 10 μm.

被殻・殻　縦溝殻と不完全縦溝殻より構成される．殻は狭いこん棒形で，幅の広い極（頭極）側の殻端は弱頭状であるが，短い個体ではやや広円となる[1,2,5-10]．帯面はくさび形[3,4]．殻長25～56μm，殻幅3.5～4.5μm，縦溝殻，不完全縦溝殻共に殻外面は平滑で，殻肩には太く短い針をもつ[13]．この針は比較的大きめの被殻では光学顕微鏡による帯面観察において認めることが容易である[3]．また光学顕微鏡による殻面観察においても，ピントの操作によりそれらを黒ずんだ点として認めることも可能である[5,8]．

軸域・縦溝　縦溝殻，不完全縦溝殻共に軸域は線形[1,2,5-10]．縦溝殻のそれぞれの縦溝枝は長い殻では殻長の約1/4，小さな殻では約1/3程度の長さとなる[1,5-7]．縦溝枝の中心末端はゆるく曲がり丸く膨れて終わる[11,12]．極末端は殻頂から少し殻の中心に寄った位置で直線的に終わる．不完全縦溝殻では縦溝は頭極側には存在せず，足極側の殻端部にのみ存在する[13]．不完全縦溝殻の縦溝はたいへん短く，光学顕微鏡観察では注意深く観察しないと見落としてしまう[8-10]．

条線・胞紋・唇状突起　条線は縦溝殻，不完全縦溝殻ともに平行で，頂端部では短い条線が円弧を描く殻肩に対し直角になるように放射配列する[1,2,11-13]．縦溝殻，不完全縦溝殻ともに10μmあたり14～18本．条線を構成する胞紋は孔状で殻内面の開口も外面の開口も丸い[11-13]．条線を構成する胞紋は縦溝殻，不完全縦溝殻ともに10μmあたり約70個．胞紋の閉塞については不明である．両殻端に唇状突起をもつ[13]，外部の開口は胞紋の開口より大きいので走査型電子顕微鏡でよく観察できる[11,12]．

半 殻 帯　開放型の帯片よりなる．
葉 緑 体　板状のもの2枚，それぞれ上下殻の殻面の下に広がる（Cox 1996）．
有性生殖　不明．
生 活 形　付着性．
汚濁耐性　弱汚濁耐性種．
出 現 地　郡殿ノ池（新潟県），男池（新潟県），八島ヶ池（長野県）など，ミズゴケが生える腐植性の弱酸性湿地にしばしば分布．

ノート　*Peronia* 属は国際植物命名規約上の保存名となっている．

Rhoicosphenia abbreviata (C. Agardh) Lange-Bertalot

Lange-Bertalot, H. 1980. Bot. Notiser **133**: 586.
Basionym: *Gomphonema abbreviatum* C. Agardh 1831. Conspectus Criticus Diat. Part 2. p. 34.
Synonym: *Gomphonema curvatum* Kützing 1833. Linnaea **8**: 567. pl. 16. f. 51 ; *Rhoicosphenia curvata* (Kützing) Grunow ex Rabenhorst 1864. Fl. Europaea Alg. p. 112, 342.
Dimension: L. 12-46 (max. 82) μm,　W. 4-7 (max. 9) μm,　Str. 11-16 in 10 μm,
　Ar. ca. 30-40 in 10 μm.

被殻・殻　縦溝殻と不完全縦溝殻より構成される．殻はこん棒状の皮針形で，殻端は広円(1-13)．帯面はくさび形で，中央ではなくわずかに"幅の広い極"(頭極)側に寄った部位で屈曲する(14-17)，縦溝殻が凹状，不完全縦溝殻が凸状に屈曲する(17)．殻長12～46(最大82) μm，殻幅4～7(最大9) μm，縦溝殻，不完全縦溝殻共に殻外面は平滑で，殻肩は丸みを帯びる(17,18)．両殻共に殻套の縁部に偽隔壁をもつ(25)．この偽隔壁は光学顕微鏡による帯面観の観察においても認めることができる(14-16)．

軸域・縦溝　縦溝殻，不完全縦溝殻共に軸域は線形(1-13,18)．縦溝殻の長い殻では小さな中心域を形成するが(1-3)，短い殻では中心域は不明瞭となる(4-6)．殻外面では縦溝枝の中心末端は中心孔を形成する(20)．極末端ではどちらの極においてもゆるくS字形に曲がる極裂を形成する(19,22)．極裂の曲がる方向は両極で同じである．縦溝殻の内表面で縦溝枝の中心末端は互いに同じ方向に釣針状に曲がる(25)．極末端の形状は小さな蝸牛舌を形成して終わる．不完全縦溝殻の縦溝枝は双方ともたいへん短く殻端部にのみ存在する(7-13)．電子顕微鏡観察では殻外面で中心末端はわずかに丸く膨れる程度で，縦溝殻に見られるような顕著な中心孔ではない(18)．極末端は"幅の狭い極"(足極)では縦溝殻の縦溝に見られるようなゆるいS字形の極裂を形成するが(18)，頭極側では頂端まで伸びず，手前で直線的に終わる(18,25)．殻内面で縦溝枝の両末端は直線的に終わる(23)．

条線・胞紋・殻端小孔域　条線は縦溝殻，不完全縦溝殻ともに平行，殻端部ではゆるく放射状になるが短い個体ではほとんど平行のままである(1-13)．縦溝殻，不完全縦溝殻ともに11～16本．観察されたボアグの欠落は，縦溝殻ではかなり極裂に近い部位であり(19)，不完全縦溝殻においても殻端小孔域近くで観察された(18)．条線を構成する胞紋列は縦溝殻，不完全縦溝殻共に殻内面では発達した間条線に挟まれた溝の中に存在する(21,23,24)．条線を構成する胞紋は縦溝殻，不完全縦溝殻ともに10 μmあたり約30～40個．胞紋の開口は殻内面では丸く，外面では長軸方向に伸びたスリット状であるが，殻中央部では若干丸みを帯びるものもある(18-20,22)．両殻とも内面で閉塞されているようだが詳細は不明．胞紋列は殻面から殻套部まで途切れることなく続く(17)．殻端小孔域は縦溝殻，不完全縦溝殻共に足極側にのみ存在する(17,18,22)．殻端小孔域より粘質の柄が分泌される．

半殻帯　開放型の帯片で帯片中脈には1列の胞紋がある(25)．接殻帯片は，殻の偽隔壁に密着する部分で，殻断面方向に直角に折れ込んでいる(21,23-25)．成熟した半殻帯は3枚(25)．
葉緑体　帯面観はH形で，殻面観では片側に寄る．ピレノイドを中心にもつ(Mann 1982a)．
有性生殖　Mann(1982b, 1984)の観察が知られている．
生活形　付着性．
汚濁耐性　弱汚濁耐性種．
出現地　青池(青森県)，鬼怒川(栃木県)，霞ヶ浦(茨城県)，荒川(埼玉県)，多摩川(東京都)，河口湖(山梨県)，千曲川(長野県)，竜返しの滝(長野県)，琵琶湖(滋賀県)など．

ノート 本種のサイズおよび胞紋密度の変異幅は大きく，今後，より詳細な研究が必要であろう．

Gomphoneis heterominuta Mayama & Kawashima

Mayama, S. & Kawashima, A. in Mayama, S., Idei, M., Osada, K. & Nagumo, T. 2002. Diatom **18**: 89.
Basionym: *Licmophora minuta* C. Agardh 1827. Flora oder Bot. Zeit. **10**: 629.
Synonym: *Gomphonema minutum* (C. Agardh) C. Agardh 1831. Consp. Crit. Diat. part 2. p. 34.
Dimension: L. 15-38 μm,　W. 4-6.5 μm,　Str. 11-15 in 10 μm,　Ar. ca. 40 in 10 μm.

被殻・殻　殻は細い卵形から異極の皮針形で，幅の狭い殻足部をもち[1-4]，殻長15～38 μm，殻幅4～6.5 μm．頭極は鈍形，足極は鋭形となり，双方とも内面に偽隔壁をもつ[5,7]．殻端小孔域は足極にある[9]．

縦溝・軸域　軸域は狭く，ほぼ線形．中心域は小さく，殻の片側にある[1-4,8]．中心節は片側に1個の遊離点をもち，内面では遊離点側に隆起する[5,7,8]．遊離点は殻外面で円形の明瞭な開口をもち[8]，殻内面で楕円形のくぼみの中に開口する[13]．両縦溝枝は糸状である．外裂溝は弱く波打つが[6]，内裂溝は真っ直ぐである[7]．中心末端は外面では丸く膨らんだ中心孔で終わり，内面では遊離点側に傾いて釣針状に曲がる[7]．両方の極裂は遊離点側に湾曲し[6,9]，殻内面の極末端は小さな蝸牛舌で終わる[5,7]．

条線・胞紋　条線は比較的太く，殻の全域で放射状に配列し[1-4]，10 μm あたり 11～15本．条線の点紋は不明瞭．両側の中央条線は共に短い．各条線は長胞構造を示し，外面が二重胞紋，内面が条線のほぼ全域に広がった大きな開口となる[5-9]．胞紋は 10 μm あたり約40個あり，それぞれ外面ではわずかに張り出した肉趾状師板を伴ってじん臓形に開口するか，またはまれに丸く単純に開口する[6,8,9]．殻の中央部の条線は，殻套側の先端に1個の大きな胞紋をもつ[6]．

半殻帯　不明．
葉緑体　不明．
有性生殖　不明．
生活形　付着性．
汚濁耐性　弱汚濁耐性種．
出現地　阿寒湖（北海道），湯ノ湖（栃木県），多摩川（東京都），山中湖（山梨県），狩野川（静岡県），員弁川（三重県），川辺川（球磨川支流，熊本県）など，淡水域に見られる．

ノート　本種は，条線が二列胞紋を伴った長胞であることから *Gomphoneis* 属に分類されたが，その際，本属には *Gomphoneis herculeana* var. *minuta* Stone を基礎異名とする *G. minuta* (Stone) Kociolek & Stoermer がすでに存在していたため，本種には，種形容語 *heterominuta* が新たに与えられた（Mayama *et al.* 2002）．*G. olivacea* (Hornemann) Ross & Sims（Dawson 1974, 河島・真山 2001）に類似するが，中心域の形状，遊離点の有無，殻端部の条線配列および極裂の形状によって区別できる．

Gomphoneis rhombica (Fricke) V. Merino *et al.*

Merino, V., García, J., Hernández-Mariné, M. & Fernández, M. 1994. Diatom Research **9** (2): 343. f. 1-46.

Basionym: *Gomphonema rhombicum* Fricke in A. Schmidt 1904. pl. 248. f. 1.

Synonym: *Gomphonema sumatrense* Fricke sensu H. Kobayasi 1964. Chichibu Museum Nat. Hist. **12**: 74. pl. 10. f. 63.

Dimension: L. 16.5-44 µm, W. 4-7 µm, Str. 8-10 in 10 µm, Ar. 30-40 in 10 µm.

被殻・殻 被殻は帯面観でくさび形[9]である．殻はこん棒状の線状皮針形から皮針形となり[1-7,10,11]，殻長 16.5～44 µm，殻幅 4～7 µm．頭極は鈍形，足極はやや鋭形となる．両殻端の内面に発達した偽隔壁をもつ[15,17]．殻端小孔域は足極にあり，胞紋と同様な大きさの小孔で構成される[14]．

軸域・縦溝 軸域はやや広く，中央部でひし形に開き，両殻端部で狭まる[1-7,9-11]．軸域は内面も外面も共に滑らかである[12-17]．遊離点は中心節の片側に 1 個あり，外面では顕著な丸い開口をもち[13]，内面では横に伸びたスリットとなる[16]．中心節は内面で片側に肥厚し，肥厚部の外形が縦に長い楕円形となる[16]．縦溝は糸状で，ほぼ真っ直ぐに伸びる．中心末端は外面では丸く膨らんで終わり[13]，内面では遊離点側に直角に屈曲した後，先端部が鋭角に折り返る[16]．双方の極裂は遊離点側に小さく湾曲する[12,14]．内面の両極末端は，共に遊離点の反対側にわずかに偏った蝸牛舌で終わる[15,17]．

条線・胞紋 条線は太く，二重点紋から成り，平行またはわずかに放射状に配列し[1-7,9-11]，10 µm あたり 8～10 本．各条線は 1 個の長胞で，外面に互い違いに並ぶ 2 列（足極の条線では 1 列）の胞紋，内面に楕円形または丸みを帯びた長方形の開口をもつ[15-17]．長胞の内壁は殻の内表面で内側に突出して縁辺部をわずかに覆う[20]．そのため長胞の開口は広くなり，光顕下で縦走線は見られない．胞紋は殻の内面と外面で円形または楕円形に開口する[12-17]．

半殻帯 4 枚の開放型帯片から成っている．すべての帯片は縫合部に沿って 1 列の胞紋をもつ[8,18,19]．

葉緑体 不明．

有性生殖 不明．

生活形 付着性．

汚濁耐性 弱汚濁耐性種．

出現地 荒川（埼玉県），日原川（東京都），児野沢（長野県），おいらん淵（柳沢川，山梨県），小菅川（山梨県），狩野川下流の三日月湖（静岡県）など，淡水域に出現する．

Gomphonema clevei Fricke

Fricke, F. in Schmidt, A. 1902. Atlas, pl. 234. f. 44-46.
Dimension: L. 11.5-34 µm, W. 3.5-7 µm, Str. 10-12 (15) in 10 µm, Ar. 30-40 in 10 µm.

被殻・殻 被殻は帯面観でくさび形．殻は楕円形またはひし形に近いこん棒形で，殻中心から頭極側に多少離れた位置で殻幅が最も広くなる(1-6)．殻長11.5〜34 µm，殻幅3.5〜7 µm．両殻端は共に鈍形であり，頭極は足極よりわずかに幅広となる．両殻端の内面に偽隔壁をもつ(8矢印)．足極に殻端小孔域をもつ．

縦溝・軸域 軸域は皮針形に広く開き，中心節の片側に1個の遊離点をもつ(1-6)．中心節は内面で遊離点側に肥厚し，肥厚部の外形が両中心末端と遊離点を取り囲む縦に長い楕円形となる(8,11)．遊離点は殻外面に縁辺が肥厚した円形の開口をもち(10)，殻内面では楕円形のくぼみの中に開口する(11)．両方の縦溝枝は糸状．外裂溝は大きく波打つが(7,10)，内裂溝は真っ直ぐである(8)．中心末端は殻外面では横に膨らんだ中心孔で終わり(10)，殻内面では遊離点側に屈曲した後，先端部が釣針形となる(11)．極裂は遊離点側に湾曲し(10)，内面の極末端は小さな蝸牛舌で終わる(8)．

条線・胞紋 条線は短く，わずかに放射状に配列し(1-6)，10 µmあたり10〜12本（小形の殻ではまれに15本）．各条線は殻面から殻縁部に達する1個の長胞で，外面に1列の胞紋(7,10)をもち，内面が大きな開口となる(8,11)．胞紋は縦に長い楕円形で(11)，外面では盛り上がった肉趾状師板（vola）のフラップとC字形のスリットをもつ(10)．軸域に沿った2,3列のフラップは各胞紋壁から軸域側に，その他のフラップでは殻縁側に突出する(10)．

半殻帯 4枚の開放型帯片から成る．閉鎖末端（closed end）は第1帯片（接殻帯片）と第3帯片では頭極に(9)，第2および第4帯片では足極にある．第4帯片は短く，その開放末端（open end）は殻端部に達しない．第1，第2および第3帯片は，縫合部に沿って1列の胞紋をもつ．接殻帯片の胞紋は長軸方向に伸びた楕円形で，他の2帯片の胞紋に比べて粗い．

葉緑体 不明．
有性生殖 不明．
生活形 付着性．
汚濁耐性 弱汚濁耐性種．
出現地 カリマベンゲ川（タンガニーカ湖流入河川）．

ノート 本種の原試料が残っていないこともあって，これまで研究者によってかなり異なる分類群が*Gomphonema clevei*に同定されてきた（Krammer & Lage-Bertalot 1986）．本図鑑では，原産地の旧ドイツ領東アフリカ（現タンザニア）に近いカリマベンゲ川から採集された個体群に基づいて記載したが，その条線は明らかに1列の点紋から構成されていた．しかし，Cholnoky(1953)は*G. clevei*の条線は2列点紋であるとして，1列点紋のものを*G. schweickerdii* Cholnokyとして区別している．

Gomphonema curvipedatum H. Kobayasi ex K. Osada sp. nov.

Dimension: L. 21.5-44.5 μm, W. 4.5-8 μm, Str. 13-17 in 10 μm, Ar. 30-40 in 10 μm.

被殻・殻 殻はこん棒状の細長い皮針形で[1-7]，殻長21.5〜44.5 μm，殻幅64.5〜8 μm. 両殻端は遊離点と反対側に曲がり，大形の個体では足極が頭極に比べて強く湾曲する傾向がある[1-4]．頭極と足極は共に内側に偽隔壁をもち，足極には殻端小孔域がある[10,11]．

縦溝・軸域 軸域は皮針形で広く，外表面には胞紋列の延長線上に1列に並んだ丸いくぼみをもつ[8,12]．遊離点は中心節の片側に1個あり，殻の内外面に小さな丸い開口をもつ[12,13]．中心節は内面で遊離点側に肥厚し，肥厚部の外形が遊離点のところでくびれた形となる[12]．縦溝の外裂溝は大きく波打つが[8]，内裂溝は直線的である[11]．中心末端は，殻外面ではほんのわずかに膨らんで終わり[12]，殻内面では遊離点側に屈曲し，末端部が釣針形に湾曲する[13]．極裂は遊離点側に曲がる[8]．内面の末端は遊離点側に傾いた小さな蝸牛舌で終わる[11]．

条線・胞紋 条線は短く，わずかに放射状に配列し[1-7]，10 μmあたり13〜17本ある．各条線は外面に1列の胞紋をもった長胞から成り，殻面から殻套部まで伸長する[11,12]．胞紋は縦に長い楕円形．胞紋の外面は1個または2個のフラップで閉ざされ，C字形，I字形または3字形などのスリットをもつ[12]．

半殻帯 4枚の開放型帯片で構成される．殻帯片の閉鎖端（closed end）は，第1帯片（接殻帯片）と第3帯片では頭極にあり[9]，第2帯片および第4帯片では足極にある[10]．それぞれの帯片は縫合部に沿って1列の胞紋（第1帯片で10 μmあたり30〜40個，他の帯片では10 μmあたり70〜80個）をもつ．

葉緑体 H形，1枚．
有性生殖 不明．
生活形 付着性．
汚濁耐性 弱汚濁耐性種．
出現地 東京学芸大学構内のため池（東京都），一碧湖（静岡県），霧ヶ峰（長野県），小川原湖（青森県）など，淡水域に見られる．

[ノート] 本種は条線が短く軸域が広く開く点で*Gomphonema clevei*と類似するが，殻端部が遊離点と反対側に曲がることや胞紋のスリットがC字形，I字形およびS字形であることなどによって区別できる．

Gomphonema gracile Ehrenberg

Ehrenberg, C. G. 1838. Infusionsthier. p. 217. pl. 18. f. 3.
Dimension: L. 31-77 μm, W. 7.5-9.5 μm, Str. 10-15 in 10 μm, Ar. 24-30 in 10 μm.

被殻・殻 殻は幅の広い線状皮針形または細いひし形で，足極がわずかに長くなる[1-4]．殻長31〜77 μm，殻幅 7.5〜9.5 μm．両殻端は鋭形で，足極には，多数の丸い小孔が配列した殻端小孔域がある[9]．

縦溝・軸域 軸域は狭く，中心域は殻の片側に偏る[1-4]．遊離点は片側の中央条線の先端に1個あり，殻外面では比較的大きな丸い開口をもち[6]，殻内面では横に伸長したくぼみの中に開口する[7]．両縦溝枝は，ひだ状縦溝[8]．外裂溝は比較的大きく波打つが[5,9]，内裂溝は真っ直ぐである[7]．中心末端は外面ではわずかに拡張した中心孔を伴って真っ直ぐに終わり，内面では遊離点側に曲がって釣針形となる[7]．両方の極裂は殻面で遊離点側に湾曲したのち，反対側の殻縁部に達する[5,9]．

条線・胞紋 条線はほんのわずかに放射配列し[1-4]，10 μm あたり10〜15本．条線の点紋は比較的明瞭．中央条線に隣接する間条線は他の間条線よりも幅が広く，遊離点と反対側の中央条線は短い[1-4]．各条線は1個の長胞であり，外面では 10 μm あたり 10〜15 個の胞紋が1列に配列し[5,6]，内面では隣接する胞紋の間にスタブ（stub）または支柱（strut）を伴った細長い開口となる[7]．胞紋は 10 μm あたり 24〜30 個で，外面にフラップを伴ったC字形または三日月形の開口をもつ[5,6,8,9]．

半殻帯 不明．

葉緑体 1枚でH形となる．殻帯の片側中央に1個のピレノイドをもつ．

有性生殖 各母細胞に2個の同型配偶子の形成，生理学的異形配偶子接合，および対合体に2個の増大胞子の形成が報告されている（Geitler 1951, Geitler 1973b）．

生活形 付着性．

汚濁耐性 弱汚濁耐性種．

出現地 小千谷の水田（新潟県），荒川（埼玉県），宝蔵寺沼（埼玉県），川越の湧泉池（埼玉県），東京学芸大学構内のため池（東京都），芦ノ湖（神奈川県）など，湖沼にも河川にも見られるが，淡水の止水域により多く出現する．

Gomphonema inaequilongum (H. Kobayasi) H. Kobayasi

Kobayasi, H. in Mayama, S., Idei, M., Osada, K. & Nagumo, T. 2002. Diatom **18**: 89.
Basionym: *Gomphonema clevei* var. *inaequilongum* H. Kobayasi 1965. Journ. Jap. Bot. **40**: 350. f. 12a, b.
Dimension: L. 16.5-44 μm,　W. 4-7 μm,　Str. 13-17 in 10 μm,　Ar. 30-40 in 10 μm.

被殻・殻　被殻の帯面観はくさび形．殻はこん棒状の皮針形で，殻長16.5～44 μm，殻幅4～7 μm．殻頭部は幅広で広円形の頭極をもち，殻足部はやや急に細くなり鈍形の足極で終わる(1-6)．足極に殻端小孔域をもつ(11)．両殻端の内面に発達した偽隔壁がある．

縦溝・軸域　軸域は皮針形に広く開き，外面に細かい凹凸がある(11)．中心節は内面の片側で楕円形に肥厚するが(10)，しばしばその中央部は陥没したものとなる．遊離点は中心節の片側に1個あり，外面に小さな丸い開口(9)，内面には横に長い楕円形のくぼみをもつ(10)．縦溝はほぼ真っ直ぐ伸びた糸状縦溝である．縦溝枝の外裂溝も内裂溝も直線的である．中心末端は殻外面では遊離点と反対側にわずかにそれてやや膨らんで終わる(9)．殻内面の中心末端は遊離点側に屈曲した後，先端部が鋭く曲がった釣針形となり，その湾曲部に刺状の突出部を形成する(10)．極裂は遊離点側に湾曲するが，頭極では湾曲した後，反対側の殻套部に向かう(7)．内面の極末端は小さな蝸牛舌で終わる．

条線・胞紋　条線は短く，長短交互して不揃いとなる．殻の中央部で平行，殻端部でわずかに放射状に配列し(1-6)，10 μmあたり13～17本．頭極では極節の周りに放射する(7)．各条線は1個の長胞で，外面には1列の胞紋(7,9,11)，内面には一部が内壁によって閉ざされた楕円形の開口をもつ(10)．そのため，光顕観察において縦走線が見られる．胞紋は外面に肉趾状篩板（vola）を伴ってC字形，S字形または3字形のスリットをもつ(7-9)．

半殻帯　4枚の開放型帯片から成る．閉鎖端（closed end）は第1帯片（接殻帯片）と第3帯片では頭極に，第2帯片および第4帯片では足極にある(8)．第1帯片（接殻帯片）は縫合部に沿って1列の胞紋をもつ(8)．

葉緑体　不明．
有性生殖　不明．
生活形　付着性．粘質物の柄をもった樹状の群体を形成する．
汚濁耐性　弱汚濁耐性種．
出現地　蔦沼（八甲田，青森県），波久礼川（荒川支流，埼玉県），井の頭公園池（東京都），東京学芸大学構内のため池（東京都），一碧湖（静岡県），諏訪湖（長野県）など，淡水域に見られる．

ノート　種形容語の*inaequilongum*は，本種に特有な不揃いの条線の形状に由来する．

Gomphonema kinokawaensis H. Kobayasi ex K. Osada sp. nov.

Dimension: L. 17-46 μm, W. 6.5-9 μm, Str. 14-16 in 10 μm, Ar. 35-40 in 10 μm.

被殻・殻 殻はこん棒状のひし形で(1-4)，殻長 17～46 μm，殻幅 6.5～9 μm．頭極はくさび形で，足極は鈍形．両殻端の内側に偽隔壁をもつ．足極には殻端小孔域がある(7)．

縦溝・軸域 軸域はひし形で広い(1-4)．軸域の外表面は，胞紋列の延長線上に丸いくぼみの列または細長いくぼみをもつ(5)．中心節は内面で片側に肥厚し，肥厚部の条線側の縁に 1 個の遊離点をもつ(9)．遊離点は外面に小さな円形の開口をもち，殻内面では横にわずかに伸長した細長いくぼみの中に開口する(8)．遊離点と中心末端がそれぞれ丸く肥厚する．外裂溝は大きく波打ち(5)，内裂溝は真っ直ぐとなる．中心末端は外面では丸い孔で終わり(5, 8)，内面では遊離点側に湾曲した釣針形となる(9)．両方の極裂は遊離点側に曲がり(5)，内面の極末端は小さな蝸牛舌で終わる．

条線・胞紋 条線は短く，放射状に配列し，軸域の両側に 1 本の縦走線（longitudinal line）をもち(1-4)，10 μm あたり 14～16 本．各条線は，殻外面に 1 列の胞紋を伴った 1 個の長胞となる(5, 9)．長胞の内面は，軸域に近い部分が内壁によって閉ざされた細長い開口となる(9)．そのため，光学顕微鏡では条線領域を縦に走る縦走線が観察される．殻外面にある胞紋の開口はそのほとんどが長軸方向に伸びた直線的なスリットであるが，軸域に隣接する胞紋では C 字形となる(8)．

半 殻 帯 4 枚の開放型帯片で構成される．第 1 帯片（接殻帯片）と第 3 帯片は足極で開き(7)，第 2 および第 4 帯片は頭極で開く(6)．第 1 帯片では，縫合部に沿って 10 μm あたり 30～40 個の胞紋が 1 列に配列する．

葉 緑 体 不明．
有性生殖 不明．
生 活 形 付着性．
汚濁耐性 弱汚濁耐性種．
出 現 地 紀ノ川（和歌山県），一碧湖（静岡県）など，淡水域に出現する．

> ノート 本種は短い条線と広い軸域をもつことで *Gomphonema clevei*，*G. curvipedatum* および *G. inaequilongum* などと類似するが，*G. clevei* と *G. curvipedatum* とは内面の一部が閉ざされた長胞をもつ点で，*G. inaequilongum* とは殻形，条線の形状および胞紋の開口の形状によって区別できる．

Gomphonema micropus Kützing

Kützing, F. T. 1844. Bacill. p. 84. pl. 8. f. 12.
Synonym: *Gomphonema parvulum* var. *micropus* (Kützing) Cleve 1894. Kongl. Sven. Vet. Akad. Handl. **26**: 180.
Dimension: L. 16.5-43 μm, W. 6-8.5 μm, Str. 10-14 in 10 μm, Ar. 40-50 in 10 μm.

被殻・殻 殻はわずかにこん棒形をした広皮針形または狭皮針形で[1-8]，殻長 16.5〜43 μm，殻幅 6〜8.5 μm．両殻端は鈍く突出するか，もしくはくちばし形となる．頭極は足極に比べてわずかに幅が広い．足極には，多数の丸い小孔が放射配列する殻端小孔域をもつ[10]．

縦溝・軸域 軸域は狭く，線形[1-8]．中心域は殻の片側に偏る．遊離点は片側の中央条線の先端に 1 個あり，外面では横に伸長した楕円形の明瞭な開口[12]をもち，内面では中心節から中央条線の長胞まで伸びる細長いスリットを形成する (Reichardt 1999)．縦溝は糸状で，直線的に伸びる．両縦溝枝の中心末端は外面では丸い中心孔で終わり[9,10,12]，内面では単純な先端部を伴って遊離点側に折れ曲がる (Reichardt 1999)．双方の極裂は遊離点と反対側の方向にそれる[9,10]．

条線・胞紋 条線は殻面全域でわずかに放射配列し，10 μm あたり 10〜14 本であるが，中心部では条線間隔が著しく広がる[1-8]．頭極には極節の周りに放射する短い条線がある[9]．条線の点紋は光顕では不明瞭である．遊離点と反対側の中央条線は極端に短い．各条線は殻套部まで達する 1 個の長胞で，外面に 1 列の胞紋[9,10,12]，内面では隣接する胞紋の間に支柱 (strut) をもつ (Reichardt 1999)．胞紋は 10 μm に 40〜50 個の密度で配列する．胞紋の殻外面の開口は孔状で，縁に小さなフラップをもつ[11]．

半殻帯 不明．
葉緑体 不明．
有性生殖 各母細胞に 2 個の同型配偶子を形成し，生理学的異形配偶子接合の後，2 個の増大胞子が形成される (Locker 1950, Geitler 1973b)．
生活形 付着性．
汚濁耐性 弱汚濁耐性種．
出現地 南浅川 (東京都)，多摩川 (東京都)，大栗川 (東京都)，丹波川 (山梨県) など．

Gomphonema nipponicum Skvortsov

Skvortsov, B. W. 1936. Philip. Journ. Sci. **61** (1): 54. pl. 12. f. 3 ; pl. 13. f. 24.
Dimension: L. 36-83.5 μm,　W. 7-10.5 μm,　Str. 7-9 in 10 μm,　Ar. 30-35 in 10 μm.

被殻・殻　殻はこん棒状の線状皮針形で，中央部の両側がわずかに膨らむ(1-4)．殻長36～83.5 μm，殻幅7～10.5 μm．頭極は弱頭状または頭状に膨らみ，足極は細く，鈍形となる．殻端小孔域は足極にあり，多数の丸い小孔で構成される(6)．両殻端の内面には発達した偽隔壁をもつ．

縦溝・軸域　軸域は狭いが，殻の中央に向かって徐々に広がる(1-4)．中心域は殻の片側に広く開く．遊離点は片側の長い中央条線の先端にあり，外面では周囲がわずかに肥厚した丸い開口(7)，内面では横に長いスリット状となる．縦溝はほぼ真っ直ぐで，糸状．両方の縦溝枝の中心末端は殻外面では互いに接近して真っ直ぐに終わるが(7)，殻内面では互いに少し離れて遊離点側に折れ曲がる．両方の極裂は遊離点側に湾曲する(5, 6)．

条線・胞紋　条線は太く，10 μm あたり7～9本．殻の中央部では粗く，殻端部では放射状に密に配列し（10 μm あたり12～16本）(1-4)，頭極では極節の周りに強く放射する(5)．遊離点をもたない中央条線は極めて短い．各条線は1個の長胞から成り，外面には1列の胞紋，内面には多くのスタブ（stub）をもつ．各胞紋は，外面にある三日月形またはC字形の凹みの中にスリット状の開口をもつ(5-7)．

半 殻 帯　不明．
葉 緑 体　不明．
有性生殖　不明．
生 活 形　付着性．
汚濁耐性　弱汚濁耐性種．
出 現 地　長沼（宮城県），多々良沼（群馬県），近藤沼（群馬県），宝蔵寺沼（埼玉県），木崎湖（長野県），庄内川（愛知県），池田湖（鹿児島県）など．

ノート　本種は，縦溝の特徴，条線構造および条線配列などの点で *Gomphonema vibrio* Ehrenberg（Reichardt & Lange-Betalot 1991）に類似するが，殻がより小形であることや頭極が頭状に膨れることによって区別できる．

Gomphonema parvulum var. *lagenula* (Kützing) Frenguelli

Frenguelli, J. 1923. Bol. Acad. Nac. Cienc. Córdoba **27**: 68. pl. 6. f. 16.
Basionym: *Gomphonema lagenula* Kützing 1844. Bacill. p. 85. pl. 30. f. 60.
Dimension: L. 15-27 µm, W. 6.5-8 µm, Str. 11-16 in 10 µm, Ar. 35-40 in 10 µm.

被殻・殻 殻は楕円形で，殻長15～27 µm，殻幅6.5～8 µm．頭極と足極は共に狭く，明瞭な頭状またはくちばし状となる(1-5)．両殻端の内面には狭い偽隔壁をもつ(7)．殻端小孔域は足極にあり，放射配列した多数の丸い小孔で構成される(7,10)．

縦溝・軸域 軸域は線形で，中心域は狭い(1-5)．遊離点は中心節の片側に1個あり，外面では中央条線の先端の胞紋から少し離れた位置に丸い開口をもち(23,25)，内面では横に長いスリットとなる(8矢印)．縦溝は糸状で，真っ直ぐに伸びる．縦溝枝の外裂溝はわずかに波打ち(6)，内裂溝は真っ直ぐに走る(8)．中心末端は外面では殻の中央線上で丸く膨らみ(6)，内面では遊離点側に直角に折れ曲がり先端部が釣針状となる(8)．双方の極裂は共に遊離点側に湾曲した後，反対側の殻套部に達する(9,10)．内面の両極末端は小さな蝸牛舌で終わる(7)．

条線・胞紋 条線はわずかに放射もしくは平行に配列し(1-5)，10 µmあたり11～16本で，頭端では短条線が極節の周りを放射する(9)．条線の点紋は不明瞭．殻の片側の中央条線は長く，先端に遊離点をもつが，反対側の中央条線は非常に短いかまたは欠如する(1-5,6,8)．各条線は，縦溝中肋の縁辺から殻縁部まで1列の胞紋（10 µmあたり35～40個）をもち(6-8)，殻面では長胞構造，殻套部では胞紋列のみとなる(7,8)．長胞の部分は外面に胞紋，内面に多数のスタブ(stub)または支柱(strut)を伴った細長い開口をもつ(8)．胞紋の外面の開口はC字形，S字形または3の字形のスリットとなる(9,10)．

半殻帯 4枚の開放型帯片から成る．各帯片は1列の胞紋をもつ．
葉緑体 基本種と同様と考えられるが，詳細は不明．
有性生殖 不明．
生活形 付着性．
汚濁耐性 弱汚濁耐性種．
出現地 多摩川（東京都），大栗川（東京都），野川（東京都），地蔵院沼（埼玉県），宝蔵寺沼（埼玉県），狩野川下流の三日月湖（静岡県）．

[ノート] 本変種は，全体的に丸みを帯びた幅広の殻，明らかに頭状またはくちばし状に膨らんだ殻端，および殻面に限定された長胞をもつことなどによって基本種と区別できる．なお，*lagenula*の形容語は主格単数の女性名詞"lagena"（酒がめ，瓶を意味する）に縮小辞が付いた名詞であるため，その語尾は属名の性とは無関係に維持される．

Gomphonema parvulum var. *neosaprophilum* H. Kobayasi ex K. Osada var. nov.

Dimension: L. 12-36 μm, W. 4.5-8 μm, Str. 10-16 in 10 μm, Ar. 35-40 in 10 μm.

被殻・殻 殻はこん棒状皮針形，または小形のものでは卵形となり，殻長12～36 μm，殻幅4.5～8 μm．頭極は短い広くちばし状で，足極は鈍形または弱頭状となる(1-8)．両殻端の内面には非常に狭い偽隔壁をもつ(13, 16, 18)．殻端小孔域は足極にあり，放射配列した小孔列で構成される(12, 14)．

縦溝・軸域 軸域は線形で，中心域は狭い(1-8)．遊離点は片側の中央条線の先端に1個あり，外面では丸い開口(9, 11)，内面では横に細長い開口をもつ(17)．縦溝は糸状で真っ直ぐに伸びる．縦溝枝の外裂溝はほんのわずか波打つが(8)，内裂溝はほぼ真っ直ぐに伸びる(13)．中心末端は外面ではわずかに膨らんで真っ直ぐに終わり(9, 11)，内面は遊離点側に釣針状に曲がる(13, 17)．両方の極裂は共に遊離点側に湾曲した後，反対側の殻頭部まで伸びる(8, 10, 12)．内面の極末端は，反遊離点側にわずかに偏った蝸牛舌で終わる(13, 16, 18)．

条線・胞紋 条線はわずかに放射または平行に配列し，10 μm あたり10～16本(1-7)．頭極では短条線が極節の周りを放射する(10, 16)．殻の片側の中央条線は先端に遊離点を伴って長く伸びるが，他方の中央条線は非常に短い(1-9, 13)．各条線は殻縁部まで続く1個の長胞で(13, 16-18)，外面に1列の胞紋をもち(8-12)，内面では細長い開口となる(13, 16, 18)．長胞は，隣接する胞紋間の内表面に支柱（まれにスタブ）をもつ(15, 17)．各胞紋は外面にフラップを伴ったC, S, X, N字形または3の字形などのスリット状の開口をもつ(9-12, 14)．

半殻帯 1列の胞紋をもつ4枚の開放型帯片から成る．
葉緑体 不明．
有性生殖 不明．
生活形 付着性．
汚濁耐性 汚濁に強い耐性をもつと思われるが，詳細は不明．
出現地 多摩川（東京都），境川（東京都）．

[ノート] 本変種は殻が皮針形で，殻端があまり突出しないこと，条線がより粗いこと，および長胞内の胞紋が支柱によって区画されることなどの点で他の種内分類群と区別できる．

Plates 131, 132

Gomphonema parvulum (Kützing) Kützing var. *parvulum*

Kützing, F. T. 1849. Species algarum. p. 65.
Basionym: *Sphenella parvula* Kützing 1844. Bacill. p. 83. pl. 30. f. 63.
Dimension: L. 15-40 μm,　W. 4.5-6.5 μm,　Str. 12-20 in 10 μm,　Ar. 32-38 in 10 μm.

被殻・殻　殻はこん棒形，線状皮針形，皮針形または楕円形で殻形の変異が大きく(1-21)，殻長15〜40 μm，殻幅4.5〜6.5 μm．両殻端の形も変異し，頭極は弱頭状または短いくちばし状で，足極はくさび形，弱頭状，くちばし状または微突頭に伸長したものとなる．両殻端の内面には狭い偽隔壁をもつ(27,29)．殻端小孔域は足極にあり，放射配列した多数の丸い小孔で構成される(26,29)．

縦溝・軸域　軸域は狭く，線形(1-21)．中心域は小さく，殻の片側による．遊離点は片側の中心条線の先端に1個あり，外面では中央条線の先端の胞紋から少し離れた位置に円形の開口をもち(23,25)，内面ではスリット状の溝の中に開口する(28)．縦溝は糸状で直線的に伸びる．両縦溝枝の外裂溝はわずかに波打ち(22,23)，内裂溝は直線的に走行する(27,29)．中心末端は外面ではわずかに丸く膨らんで真っ直ぐに終わり(23,25)，内面では遊離点側に直角に折れ曲がり先端部が小さな釣針形となる(28)．両方の極裂は共に遊離点側に湾曲した後反対側の殻套部に達する(24,26)．内面の両極末端は小さな蝸牛舌で終わる(27,29)．

条線・胞紋　条線はわずかに放射もしくは平行に配列し(1-21)，10 μmあたり12〜20本で，頭端では短条線が極節の周りを放射する(27)．条線の点紋は不明瞭．向かい合う2本の中央条線のうち，一方は長く伸びて先端付近に遊離点をもつが，他方は非常に短い．各条線は1個の長胞で，外面では10 μmあたり32〜38個の胞紋が1列に配列し，内面では多数のスタブを伴った細長い開口となる(22,23)．スタブは各縦枝(vimen)の上に1対ずつ存在する(28)．胞紋は円形または楕円形で，外面にフラップとC字形または3字形などのスリット状の開口もつ(23-26)．

半殻帯　4枚の開放型帯片から成る．各帯片は1列の胞紋をもつ．
葉緑体　H形で1枚．殻帯の片側中央に1個のピレノイドをもつ．
有性生殖　各母細胞に同型配偶子が2個ずつ形成され，生理学的異形配偶子接合の後，2個の増大胞子が形成される（Geitler 1958, 1973b）．
生活形　付着性．
汚濁耐性　強汚濁耐性種．
出現地　大沼（北海道），霞ヶ浦（茨城県），近藤沼（群馬県），道塚池（大網白里町，千葉県），東京学芸大学構内のため池（東京都），大栗川（東京都），荒川（埼玉県），地蔵院沼（埼玉県），狩野川（静岡県），一碧湖（静岡県），児野沢（長野県），川辺川（熊本県）など，湖沼や河川，および止水・流水を問わずあらゆる淡水域に出現する．

ノート　従来は多くの変種に分類されていたが，いろいろな個体群でそれらの中間的な個体が多く出現することや，同一株でも形態変異が大きく現れること（Dawson 1972）などから，近年では変種分けをしないでひとまとめにして扱う傾向にある．ただし，Krammer & Lange-Bertalot (1986) によって本種の異名とされている *Gomphonema micropus*（*G. parvulum* var. *micropus* の基礎異名）および *G. parvulum* var. *lagenula* は殻形の特徴や殻微細構造などの点で本種と明らかに異なるため，それぞれ別の分類群とした．

Gomphonema pseudoaugur Lange-Bertalot

Lange-Bertalot, H. 1979. Arch. Hydrobiol. Suppl. **56**: 202. f. 11-16, 80, 81.
Dimension: L. 16.5-50.5 μm, W. 8-10 μm, Str. 11-14 in 10 μm, Ar. 28-30 in 10 μm.

被殻・殻 殻は卵形から皮針形に近いこん棒形で，頭極に比べて幅の狭い足極をもち(1-5)，殻長 16.5〜50.5 μm，殻幅 8〜10 μm．両殻端は鈍くわずかに突出する．足極は殻端小孔域をもつ(6)．

縦溝・軸域 軸域は狭く，直線的．中心域は非常に小さく，殻の片側に偏る(1-5)．遊離点は片側の中央条線の先端に密接して 1 個あり，殻外面では横にわずかに伸びた楕円形の開口をもち(8)，殻内面では横に伸長したスリット状となる(9)．両縦溝枝はひだ状縦溝で(8)，わずかに殻の片側に寄る．外裂溝は比較的大きく波打っているが(6,8)，内裂溝は比較的真っ直ぐである(7)．中心末端は外面では明瞭な中心孔を伴って真っ直ぐに終わり，内面では遊離点側に曲がり，さらに先端部が釣針形となる(9)．両方の極裂は殻面で遊離点側に湾曲する(6)．

条線・胞紋 条線はわずかに放射配列し，10 μm あたり 11〜14 本．条線の点紋は不明瞭(1-5)．中心域側の中央条線は，遊離点をもつ中央条線に比べて多少短い．各条線は，外面に胞紋と内面に細長い開口をもつ 1 個の長胞から成る(6-9)．長胞の内面に明瞭な支柱やスタブの構造を欠く(7,9)．胞紋は 10 μm あたり 28〜30 個の密度で 1 列に配列し，外面にフラップを伴った C 字形，I 字形またはまれに不規則な形のスリット状の開口をもつ(8)．

半 殻 帯 詳細不明．
葉 緑 体 不明．
有性生殖 不明．
生 活 形 付着性．
汚濁耐性 中汚濁耐性種．
出 現 地 荒川（埼玉県），多摩川（東京都），隅田川（東京都），境川（東京都）など，淡水の流水域に多く出現する．

Gomphonema truncatum Ehrenberg

Ehrenberg, C. G. 1832. Abh. Königl. Akad. Wiss. Berlin **1831**: 88; 1838. Infusionsthier. p. 216. pl. 18. f. 1.
Synonym: *Gomphonema constrictum* Ehrenberg (nom. nud.), *G. capitatum* Ehrenberg 1838. Infusionsthier. p. 217. pl. 18. f. 2., *G. turgidum* Ehrenberg 1854. Mikrogeol. pl. 2 (2). f. 40.
Dimension: L. 26.5-53 µm, W. 9.5-13 µm, Str. 10-13 in 10 µm, Ar. 25-30 in 10 µm.

被殻・殻 被殻は明瞭なくさび形．殻はくさび形からこん棒形で，殻頭部と殻中央部の間で強くくびれるものからそれほどくびれないものまで多様に変異する[1-3,7]．殻長 26.5〜53 µm，殻幅 9.5〜13 µm．頭極は幅が広く，切頭形もしくは広円形で，足極は狭い鈍形となる．両殻端の内面に偽隔壁がある．殻端小孔域は足極にあり，多数の丸い小孔で構成される[8]．

縦溝・軸域 軸域は明瞭で狭く，長軸に沿ってほぼ真っ直ぐに伸びる[1-3,7]．遊離点は中心節と片側の中央条線の間に 1 個あり，外面に小さな丸い開口[5]，内面に孔状またはスリット状の開口をもつ[6]．縦溝は糸状で，大きく波打つ．両縦溝枝の中心末端は殻外面ではわずかに丸く広がり[5]，殻内面では遊離点側に折れ曲がって釣針状となる．双方の極裂は殻の遊離点側に湾曲する[4,8]．内面の極末端は小さな蝸牛舌で終わる[6]．

条線・胞紋 条線は太く，弱く放射し[1-3,6,7]，10 µm あたり 10〜13 本．各条線は 1 個の長胞から成り，外面では通常 1 列の胞紋をもつが[4,5,8]，殻套部では 2 列になることもある（河島・真山 2001）．長胞の内面は全体が細長い開口となる．各胞紋は外面に肉趾状篩板と C 字形のスリットをもつ[4,5,8]．

半殻帯 胞紋の列をもった開放型帯片をもつ．

葉緑体 H 形で，1 個．殻帯の片側中央に 1 個のピレノイドをもつ．

有性生殖 粘質層内で 2 個の母細胞が互いに殻頭部を反対側に向けて帯面で対合し，各母細胞内に同型配偶子が 2 個ずつ形成される．配偶子の生理学的異形配偶子接合によって対合体あたり 2 個の増大胞子が形成される（Geitler 1951, 1973a）．殻頭部にくびれをもつ個体とそれをもたない個体で有性生殖を行うこともある（Geitler 1969）．ペドガミー（1 個の配偶子嚢にできた 2 個の配偶子同士による接合）も行う（Geitler 1952, 1973a，小林 1993）．

生活形 付着性．分枝した粘質柄をもつ．

汚濁耐性 弱汚濁耐性種．

出現地 支笏湖（北海道），蔦沼（八甲田，青森県），湯ノ湖（栃木県），牛久沼（茨城），多々良沼（群馬県），養老川（千葉県），東京学芸大学構内のため池（東京都），三宝寺池（東京都），荒川（埼玉県），宝蔵寺沼（埼玉県），川越の湧泉（埼玉県），郡殿ノ池（新潟県），青木湖（長野県）など．

ノート 従来，殻頭部にくびれをもたないものを *Gomphonema constrictum* var. *capitatum*，くびれをもつものを *G. constrictum* var. *constrictum* としてきたが，*G. constrictum* は判別文も図も伴わないで発表されていること，異なる変種間で対合すること（Geitler 1952, 1973a），および殻頭部の形は中間形も多く連続して変異することなどによって，変種分けをせずに *G. truncatum* の種に統合されている．なお，Reichardt（2001）は Ehrenberg の試料から *G. truncatum* および *G. capitatum* のレクトタイプ（選定基準標本）を選定している．

Gomphonema yamatoensis H. Kobayasi ex K. Osada sp. nov.

Dimension: L. 15-33 μm, W. 4-5 μm, Str. 11-12 in 10 μm, Ar. 40-50 in 10 μm.

被殻・殻 殻は線形または線状皮針形で(1-8)，殻長 15〜33 μm，殻幅 4〜5 μm．両殻端は鈍形で共に同様の幅をもつか，もしくは足極がほんのわずかに狭くなる．足極に殻端小孔域をもつ(10)．両殻端の内面には偽隔壁(11)がある．

縦溝・軸域 軸域は狭く，線形で，中心域は殻の片側に偏る(1-8)．中心節は片側に 1 個の遊離点をもち，内面では遊離点の開口と縦溝中心末端を伴って縦に長い楕円形に隆起する(12)．遊離点は，殻外面では縁辺が肥厚した円形の開口(13)，殻内面ではわずかに横に伸長した楕円形の開口(12)をもつ．両縦溝枝はひだ状縦溝．外裂溝はわずかに波打つが(10, 12)，内裂溝は真っ直ぐである(11, 12)．中心末端は外面では中心孔を形成して真っ直ぐに終わり，内面では遊離点側に直角に屈曲し，さらに先端部が鋭角に折り返す(12)．殻内面の極末端は明瞭な蝸牛舌で終わる(11)．両方の極裂は遊離点側にわずかに湾曲する(10, 14)．

条線・胞紋 条線はわずかに放射配列し，10 μm あたり 11〜12 本．条線の点紋は不明瞭．両側の中央条線は共に短い．遊離点と反対側の中央条線は短い．各条線は 1 個の長胞から成り，外面に 10 μm あたり 40〜50 個の胞紋が 1 列に配列し(10, 13, 14)，内面では丸みを帯びた細長い開口となる(11, 12)．各胞紋は，外面に肉趾状篩板と C 字形（まれに S 字形）のスリットをもち(10, 13, 14)，内面に円形または縦に長い楕円形の開口をもつ(11, 12)．C 字形スリットの湾曲は，頭極以外の領域では殻肩付近を境にして向きが逆転し，殻面の胞紋では軸域側に，殻套部のものでは殻縁側に凸面を向ける(10, 13, 14)．

半 殻 帯 4 枚の開放型帯片で構成される．
葉 緑 体 不明．
有性生殖 不明．
生 活 形 付着性．
汚濁耐性 弱汚濁耐性種．
出 現 地 員弁川（三重県），橋野川（上栗林付近，岩手県）に出現する．

[ノート] 本種は，*Gomphonema pumilum* (Grunow) Reichardt & Lange-Bertalot (1991) および *G. procerum* Reichardt & Lange-Bertalot (1991) に類似するが，前者とは中心域が広く開かないこと，後者とは遊離点の殻内面の開口がスリット状にならないこと，などの点で異なる．

Achnanthes coarctata (Brébisson) Grunow

Grunow, A. in Cleve, P. T. & Grunow, A. 1880. Kongl. Svenska Vet.-Akad. Handl. **17** (2): 20.
Basionym: *Achnanthidium coarctatum* Brébisson in W. Smith 1855. Ann. Mag. Nat. Hist. **15** (2): 8. pl. 1. f. 10.
Dimension: L. 7-25 µm,　W. 2.5-4 µm,　Str. 12-13 in 10 µm,　Ar. 11-15 in 10 µm.

被殻・殻　縦溝殻と無縦溝殻より構成される．殻は中央のくびれた皮針形で，殻端は広円である[1,2,4-9]．帯面は中央部が屈曲し，殻端部が反対側に反り返った長方形で[3,18]，縦溝殻が凹状[10]，無縦溝殻が凸状（殻内面では凹状[20]）に屈曲する．殻長7〜25 µm，殻幅2.5〜4 µm，縦溝殻，無縦溝殻共に殻外面は平滑．縦溝殻の殻肩は丸みを帯びるが[10,13]，無縦溝殻の殻肩は反り返って角ばり，無紋域を生じる[15]．

軸域・縦溝　軸域は縦溝殻では長軸に沿って存在し線形[1,5,7,8]，中心域は横帯を形成する．無縦溝殻では偏心的に殻肩に沿って存在し線形[2,4,6,9]．縦溝はひだ状縦溝[10,11]．殻外面では縦溝枝の中心末端は広く膨れて終わる[10]．両縦溝枝は極末端で同方向に曲がって終わる[5,13]．殻内面では両方の中心末端は釣針状に同じ向きに曲がって終わる[14]．この曲がりは，光学顕微鏡でもピントの合わせ方によって観察することが可能である[7]．極末端は小さな蝸牛舌を生じ，直線的に終わる．

条線・胞紋　条線は縦溝殻では殻全体にわたりゆるく放射状に配列し[1,5,7,8]，10 µmあたり12〜13本．無縦溝殻の条線はほとんど平行で[2,4,6,9]，10 µmあたり12〜13本．ボアグの欠落はあまり観察されない．条線を構成する胞紋列は縦溝殻，無縦溝殻共に殻内面では発達した間条線に挟まれて存在する[14,20]．条線を構成する胞紋は縦溝殻，無縦溝殻共に10 µmあたり11〜15個．両殻とも胞紋は外面で肉趾状師板によって閉塞される[11-13,15-17,19]．胞紋の殻内面の開口はほぼ円形だが[14]，殻外面の開口は丸みを帯びた四角形である[12,19]．殻套部には両殻とも胞紋が存在する[15,18]．

半殻帯　開放型の胞紋をもつ帯片よりなる．3枚[16]．
葉緑体　2枚（Van der Werff & Huls 1960）．
有性生殖　不明．
生活形　付着性．
汚濁耐性　弱汚濁耐性種．
出現地　阿寒湖（北海道），十和田湖（青森県），神流川（群馬県），荒川（埼玉県）．

[ノート] 本種の殻の形態形成についてBoyle *et al*. (1984) は詳細な観察を行っている．

Plates 138, 139

Achnanthes crenulata Grunow

Grunow, A. in Cleve, P. T. & Grunow, A. 1880. Kongl. Svenska Vet.-Akad. Handl. **17** (2): 20.
Dimension: L. 26-61 μm, W. 11-17 μm, Str. 8-9 in 10 μm (RV), 7-8 in 10 μm (ARV),
Ar. ca. 8-10 in 10 μm.

被殻・殻 縦溝殻と無縦溝殻より構成される．殻は縁が多数波打った楕円形〜狭楕円形，殻端は広円[1-4]．帯面は中央部が屈曲した長方形で[5]，縦溝殻が凹状[6]，無縦溝殻が凸状に屈曲する[11]．殻長26〜61 μm，殻幅11〜17 μm，縦溝殻，無縦溝殻共に殻外面は平滑で，角ばった殻肩には無紋域を生じる[6,10,11]が，無縦溝殻のそれは縁取りのようにわずかに隆起する[10,11]．

軸域・縦溝 軸域は縦溝殻では長軸に沿って存在し，やや紡錘状皮針形〜線形[1,3,6]，無縦溝殻では殻肩と完全に重なってしまっており，殻肩上の無紋域としてのみ認められる[10,11]．縦溝はひだ状縦溝[7]．殻外面では縦溝枝の中心末端は直線的に膨れて終わる[6]．極末端は両縦溝枝ともほぼ直線的に終わる[3,6]．殻内面では中心末端は釣針形に曲がる[7]．極末端はわずかに殻套部へ入り込み，小さな蝸牛舌で終わる．

条線・胞紋 条線は縦溝殻では殻全体にわたりゆるく放射状に配列し[1,3]，10 μmあたり8〜9本ある．無縦溝殻の条線はほとんど平行で，10 μmあたり7〜8本[2,4]．明瞭なボアグの欠落は観察されない．条線を構成する胞紋列は縦溝殻，無縦溝殻共に殻内面ではよく発達した間条線に挟まれた浅い溝の中に存在する[7,12,14,15]．無縦溝殻の殻内面の観察では，殻面の片方の縁沿いにある軸域から肋状の間条線が殻面側と殻套側に伸張していることがよくわかる[12]．条線を構成する胞紋は縦溝殻，無縦溝殻ともに10 μmあたり約8〜10個．胞紋の開口は殻内面では円形であるが[14,15]，外面の開口は丸みを帯びた四角形[9,10,13]で，両殻とも外面が肉趾状篩板によって閉塞される[6,8-11,13-15]．殻套部には両殻とも胞紋列が存在する[5,11]．

半殻帯 開放型の帯片で帯片中脈には1列の胞紋がある．3枚[10,11]．
葉緑体 2枚．
有性生殖 不明．
生活形 付着性．
汚濁耐性 弱汚濁耐性種．
出現地 越辺川（埼玉県），入間川（埼玉県），秋川（東京都），案内川（高尾山，東京都），狩野川（静岡県）など．

ノート あまり多産する種ではないが，たまに清水域に出現する．

Plates 140, 141

Achnanthes inflata (Kützing) Grunow

Grunow, A. in Cleve, P. T. & Grunow, A. 1880. Kongl. Svenska Vet.-Akad. Handl. **17** (2): 19.
Basionym: *Stauroneis inflata* Kützing 1844. Bacill. p. 105. pl. 30. f. 22.
Dimension: L. 38-59 μm,　W. 14-16 μm,　Str. 10-12 in 10 μm (RV), 9-11 in 10 μm (ARV),
　Ar. 10-14 in 10 μm.

被殻・殻　縦溝殻と無縦溝殻より構成される．殻は中心の膨れた線形で，殻端は幅の広い弱頭状(1-4,6,7)．帯面は中央部が屈曲し，殻端側で逆方向に反り返った長方形で(5)，縦溝殻の中央部が凹状(8)，無縦溝殻の中央部が凸状に屈曲する．殻長38～59 μm，殻幅14～16 μm，縦溝殻，無縦溝殻共に殻外面は平滑．縦溝殻の殻肩は丸みを帯び(8)，無縦溝殻の殻肩は若干反り上がって角ばり無紋域を生じる(16)．

軸域・縦溝　軸域は縦溝殻では長軸に沿って存在し線形(1,3,6)，無縦溝殻では偏心的に片側の殻肩に沿って存在し線形(2,4,7)．縦溝殻では中心域は横帯を形成する(1,3,6,12)が，無縦溝殻に横帯は形成されない．縦溝はひだ状縦溝(13)．殻外面では縦溝枝の中心末端は大きく膨れて終わる(9,11)．極末端は両縦溝枝とも同方向へ曲がる(6,10)．殻内面では両方の中心末端は同方向へ釣針形に曲がる(12)．極末端は直線状に終わり小さな蝸牛舌を作る．

条線・胞紋　条線は縦溝殻では殻全体にわたりゆるく放射状に配列し(1,3,6)，10 μmあたり10～12本．無縦溝殻の条線はほとんど平行で(2,4,7)，10 μmあたり9～11本．一般にボーグの欠落は観察されない．条線を構成する胞紋列は縦溝殻，無縦溝殻共に殻内面では発達した間条線に挟まれて存在する(12,13,15,19)．無縦溝殻の間条線は殻内面で殻面の片方の縁沿いに存在する軸域により断続するが(15)，この部分の観察から，無縦溝殻の間条線は軸肋より肥厚していることがわかる．胞紋は縦溝殻，無縦溝殻ともに10 μmあたり10～14個，外面では肉趾状篩板によって閉塞される(13,14,17-19)．胞紋の開口は殻外面，内面共にほぼ円形である(9,10,12,14,16,19)．殻套部には両殻とも胞紋が存在する(5,8,10,15,16)．

半 殻 帯　胞紋列のある開放型の帯片よりなる．3枚．
葉 緑 体　4枚．
有性生殖　Geitler(1932)はジャワ産の個体群で，増大胞子形成後に生じた大きな細胞と，それ以外の小さな細胞との間でサイズの比較を行った．彼は増大胞子形成後の細胞では96～90 μmの殻長を記録し，個体群全体として殻長36～96 μm，殻幅14～21 μmを記載している．
生 活 形　付着性．
汚濁耐性　弱汚濁耐性種．
出 現 地　神流川（群馬県），高須賀沼（埼玉県幸手市），堀切菖蒲園の池（東京都），野川公園の湧水（東京都三鷹市），滄浪泉園の湧水（東京都小金井市），狩野川（静岡県），明神池（静岡県戸田村井田）など．

ノート　本種は無縦溝殻の殻肩が反り上がり，丸い殻肩をもつ無縦溝殻の姉妹殻（縦溝殻）がその中にわずかにくい込むようになるため，細胞分裂後も姉妹細胞は分離しにくくなり，生細胞の観察ではしばしば数個の細胞が連なる偽群体が認められる．

Achnanthes kuwaitensis Hendey

Hendey, N. I. 1958. Journ. Microscop. Soc. **77** (3): 55. pl. 6. f. 8-10.
Dimension: L. 32-70 µm, W. 8-9.5 µm, Str. 14-16 in 10 µm, Ar. 14-16 in 10 µm.

被殻・殻 縦溝殻と無縦溝殻より構成される．殻は線形～やや狭楕円形で，殻端は広円(1-6)．帯面は中央部が屈曲し，殻端部で逆方向に反り返った長方形で(7)，縦溝殻中央部が凹状，無縦溝殻中央部が凸状に屈曲する．殻長 32～70 µm，殻幅 8～9.5 µm．縦溝殻の殻外面は平滑(10)で，殻肩は丸みを帯びる．

軸域・縦溝 軸域は縦溝殻では長軸に沿って存在し線形(1,3,5)，中央部に長軸方向が短い横帯を形成する．無縦溝殻では偏心的に殻肩に沿って存在し線形であるが(2,4)，小さな殻では軸域は長軸よりに位置することもある．殻外面で，両縦溝枝の中心末端はわずかに同方向にそれ，大きく膨れて終わる(5,8)．極末端は両縦溝枝とも同じ方向へ曲がって終わる(1,5)．殻内面で縦溝の両中心末端は同方向へ曲がって終わり(9)，極側末端では蝸牛舌を作って直線的に終わる．

条線・胞紋 条線は縦溝殻では殻全体にわたりほぼ平行に配列し(1,3,5)，10 µm あたり 14～16 本ある．無縦溝殻の条線もほとんど平行で，10 µm あたり 14～16 本．ボアグの欠落が認められる個体のその位置は極裂が曲がる側である(1)．条線を構成する胞紋列は縦溝殻，無縦溝殻共に発達した間条線に挟まれて存在する(9,10,11)．また，条線を構成する胞紋列における，お互いの胞紋間の壁もよく発達している(11)．条線を構成する胞紋は縦溝殻，無縦溝殻ともに 10 µm あたり 14～16 個．外面では両殻とも胞紋が肉趾状篩板によって閉塞される(8-11)．胞紋の内面および外面の開口は丸みを帯びた四角形である(8-11)．無縦溝殻の両殻端部では，間条線や胞紋壁のない広範囲を大きな 1 枚の板が閉塞しているが(6,11)，これは光学顕微鏡観察では無紋域のように観察される(2,4)．殻套部には両殻とも胞紋が存在する(7)．

半 殻 帯 胞紋をもつ帯片よりなる(7)．数枚．
葉 緑 体 2枚．
有性生殖 不明．
生 活 形 付着性．
汚濁耐性 弱汚濁耐性種．
出 現 地 鯨波海岸（新潟県），森戸海岸（神奈川県），青野川河口域（静岡県）など河口域から沿岸に出現する．

ノート 本図鑑では細分化した分類体系を採用したため，従来 *Achnanthes* 属に分類されていた多くの種は，*Achnanthidium* など他属の種として扱われている．これらの種の数と比べ，狭義の *Achnanthes* に分類される従来の種数は圧倒的に少ない．本図鑑に含められた *Achnanthes* 種は淡水産種が多いが，実際は多くのものが海産種である．

Cocconeis diminuta Pantocsek

Pantocsek, J. 1902. Kies. Balaton. p. 67. pl. 7. f. 181; pl. 17. f. 374.
Dimension: L. 11-13 µm,　W. 7-8 µm,　Str. 30-32 in 10 µm (RV), 14-16 in 10 µm (ARV),
　Ar. 35-45 in 10 µm (RV), 8-12 in 10 µm (ARV).

被殻・殻　殻は小形で，広楕円形(1-7)．殻長 11〜13 µm，殻幅 7〜8 µm．縦溝殻の外面は殻面が平坦で殻縁がわずかにせり上がった浅い皿状，内面は肥厚したリング状の縁辺隆起を伴う(8 矢印)．一方，無縦溝殻の殻面は軸域部でくぼみ，その両側でやや膨らむ(10)．

縦溝・軸域　縦溝は直線的で，外裂溝の中心末端および極末端は丸く広がる(9)．また，内裂溝の中心末端は互いに逆方向を向いて終わり，極末端はリング状の縁辺隆起まで達する(8)．無縦溝殻の軸域は直線状または被針形で，外表面はわずかに沈下した浅い溝をもつ(10)．

縦溝殻の条線・胞紋　条線は細かく，10 µm あたり 30〜32 本．条線を構成する胞紋は 1 列で，10 µm あたり 35〜45 個．各胞紋は円形で，殻外面に開口をもち，殻の内面で薄皮によって閉塞される(8,9)．薄皮の穿孔は細長く，放射状に配列する．

無縦溝殻の条線・胞紋　条線は 10 µm あたり 14〜16 本．各条線は，1 列に並んだ 1〜4 個の細長い長方形の胞紋で(10)構成される．胞紋密度は 10 µm あたり 8〜12 個．各胞紋は殻外面の近くにある薄皮によって閉塞される．薄皮の周辺部には，縁から中央に向かって垂直に伸びる細長い穿孔が配列する．間条線に相当する横走肋は，殻の内側に強く肥厚する．

半殻帯　縦溝殻に連結する接殻帯片は閉鎖型で，横走肋の縁辺部と重なる舌状の突出部をもつ．この突出部は短く，殻のリング状縁辺隆起にようやく届くものである．一方，無縦溝殻に連結する接殻帯片は開放型であり，その突出部は縦溝殻のものより長い．

葉緑体　不明．
有性生殖　不明．
生活形　付着性．
汚濁耐性　弱汚濁耐性種．
出現地　乙女沼（福島県），霞ヶ浦（茨城県），青木湖（長野県），木崎湖（長野県），琵琶湖（滋賀県）など，湖沼にしばしば見られる．

ノート　わが国における本種の報告は少ない．これは，殻が小形で，縦溝殻の構造が光顕では識別しにくく，また無縦溝殻の変異が激しいために，しばしば *Cocconeis placentula* var. *euglypta* や *C. pediculus* と誤同定されてきたことによると思われる．

Cocconeis pediculus Ehrenberg

Ehrenberg, C. G. 1838. Infusionsthier. p. 194. pl. 21. f. 11.
Dimension: L. 11-56 μm, W. 6-37 μm, Str. 16-24 in 10 μm (RV and ARV),
　Ar. 18-23 in 10 μm (RV), 10-13 in 10 μm (ARV).

被殻・殻 被殻は横軸方向に強く湾曲する(5,10,11)．殻はひし形状楕円形もしくは広楕円形(1-4)．殻長 11〜56 μm，殻幅 6〜37 μm．縦溝殻は殻外面が大きく凹み，殻縁部に無紋域をもつ(5)．この無紋域は殻の内面で隆起しない．無縦溝殻の殻面は外側に強く湾曲する(11)．

縦溝・軸域 縦溝は直線的で，外裂溝の中心および極末端はわずかに丸く広がる(5)．また，内裂溝の中心末端は互いに逆方向に曲がり，極末端は蝸牛舌で終わる．無縦溝殻の軸域は直線状で狭い(4,6)．

縦溝殻の条線・胞紋 条線は 10 μm あたり 16〜24 本で，各条線は 1 列の胞紋で構成される(5)．胞紋は円形で殻外面に開口をもち，殻内面では放射状配列の穿孔を伴った薄皮によって閉塞される．

無縦溝殻の条線・胞紋 条線は 10 μm あたり 16〜24 本．各条線は，1 列の短い長胞で構成される(6,11)．各長胞は外面に胞紋列，内面に楕円形の開口をもつ(6,8)．光学顕微鏡で見られる殻面の顕著な点紋はこの開口部に相当する．胞紋は外面に半月形，じん臓形または不定形の開口をもち，縁辺に穿孔を伴った薄皮によって閉塞される．

半 殻 帯 3 枚の帯片で構成される (Holmes *et al*. 1982)．接殻帯片は閉鎖型．縦溝殻の接殻帯片の帯片内接部は，中央部に多数の樹枝状の突出部をもつ(6,7)．一方，無縦溝殻の接殻帯片の突出部は短い舌状となる(9)．

葉 緑 体 板状でC字形，1枚．
有性生殖 不明．
生 活 形 付着性．
汚濁耐性 弱汚濁耐性種．
出 現 地 小櫃川（千葉県），多摩川（東京都），河口湖（山梨県），琵琶湖（滋賀県）など，本邦各地の淡水または汽水域に見られる．

Plates 146, 147

Cocconeis placentula Ehrenberg var. *placentula*

Ehrenberg, C. G. 1838. Infusionsthier. p. 194.
Dimension: L. 10-70 µm, W. 8-40 µm, Str. 20-23 in 10 µm (RV), 24-26 in 10 µm (ARV),
 Ar. 18-22 in 10 µm (RV), 10-12 in 10 µm (ARV).

被殻・殻 殻は楕円形から狭楕円形(1-5, 7, 8). 殻長 10～70 µm, 殻幅 8～40 µm. 縦溝殻は殻内面に無紋のリング状縁辺隆起(11矢印)をもつが, 無縦溝殻には見られない.

縦溝・軸域 縦溝は直線的で, 殻外面の中心末端と極末端はわずかに丸く広がる(9). 殻内面の中心末端は互いに逆向きに終わり, 極末端はリング状の縁辺隆起の上に達し細長い蝸牛舌で終わる(11). 両殻(縦溝殻と無縦溝殻)とも軸域は狭い. 縦溝殻の中心域は, 小さいが明瞭で楕円形(1, 2, 9).

縦溝殻の条線・胞紋 条線は放射状に配列し, 10 µm あたり 20～23 本で, 縁辺部でリング状の無紋域によって分断される. 各条線は, 1列の胞紋から成る(9). 胞紋は円形または楕円形で, 10 µm あたり 18～22 個の密度で配列し(10), 殻内面で薄皮によって閉塞される(15).

無縦溝殻の条線・胞紋 条線は 10 µm あたり 24～26 本で, 1列の胞紋で構成される. 各胞紋は横に細長く, 殻外面で薄皮によって閉塞され(10), 殻内面では円形もしくは楕円形の開口をもつ. 胞紋密度は 10 µm あたり 10～12 個. 縦走条線(長軸方向の胞紋列)は不規則で, 目立たない.

半殻帯 縦溝殻の接殻帯片は開放型で, 内接部に大小2種類の突出部をもつ(6, 12, 13). 大形の突出部はところどころで肥厚し, 殻のリング状縁辺隆起の上まで伸張する(15). 無縦溝殻の接殻帯片は閉鎖型で, 内接部に鋸歯状の突出部をもつ(14).

葉 緑 体 板状でC字形, 1枚.
有性生殖 不明.
生 活 形 付着性.
汚濁耐性 弱汚濁耐性種.
出 現 地 大血川(埼玉県), 多摩川(東京都), 南浅川(東京都), 柿田川(静岡県), 三面川(新潟県)など, 本邦各地の河川によく見られる.

[ノート] 本変種と他の変種は無縦溝殻の構造によって区別されている. 本変種は, 胞紋の密度が高いこと, および縦走条線が目立たないことが主な特徴である.

Plates 148, 149

Cocconeis placentula var. *lineata* (Ehrenberg) Van Heurck

Van Heurck, H. 1885. Syn. Diat. p. 133.
Basionym: *Cocconeis lineata* Ehrenberg 1843. Abh. Königl. Akad. Wiss. Berin **1841**: 81.
Dimension: L. 20-80 μm, W. 8-40 μm, Str. 16-23 in 10 μm (RV), 14-18 in 10 μm (ARV),
  Ar. 18-22 in 10 μm (RV), 6-12 in 10 μm (ARV).

被殻・殻 殻は楕円形から線状楕円形で[1-5]，殻長20〜80 μm，殻幅8〜40 μm．縦溝殻は浅い皿状で[6]，殻内面にはリング状に肥厚した縁辺隆起をもつ[10 矢印]．無縦溝殻の殻面はドーム状に膨らむ[7,10]．

縦溝・軸域 縦溝は直線的で，外裂溝の中心および極末端はわずかに丸く膨らむ[6,13]．また，内裂溝の中心末端は互いに逆向きで[14]，極末端はリング状の縁辺隆起の上に達し蝸牛舌で終わる．両殻（縦溝殻と無縦溝殻）とも軸域は狭いが[1-5,10]，無縦溝殻の軸域では，しばしば長軸に沿って浅い溝をもつ[10,12 矢印]．縦溝殻の中心域は明瞭で多少とも楕円形[1,6,13]．

縦溝殻の条線・胞紋 条線は10 μmあたり16〜23本で，1列の胞紋から成るが[6,13]，殻縁部を1周するリング状の無紋域（殻内面の縁辺隆起に相当する）によって分断される[1,6]．胞紋は円形または楕円形で小さく，10 μmあたり18〜22個あり，殻内面にある薄皮によって閉塞される[14]．薄皮の穿孔は，縁辺部では放射状に配列し，中央部では平行に配列する[8]．

無縦溝殻の条線・胞紋 条線は10 μmあたり14〜18本で，1列の胞紋で構成される．また，これらの胞紋が殻の中央付近では縦方向に並んで列を成すため，両側にそれぞれ4〜5本の縦走条線が見られる．胞紋密度は10 μmあたり6〜12個．各胞紋は横長の丸みを帯びた長方形で，外面で薄皮によって閉塞され[9,11,12]，内面では円形または楕円形の小さな開口をもつ[11 矢印,15]．薄皮は，縁辺部に平行配列する細長い穿孔と中央部に散在する小さな丸い穿孔をもつ[9]．薄皮の一部は，横走肋の外面から張り出したフラップによって覆われる[11 矢印]．

半殻帯 縦溝殻の接殻帯片は開放型で[12]，内接部に突出部をもつ．突出部は所々で肥厚し，殻内面のリング状縁辺隆起の上まで伸張する（Krammer & Lange-Bertalot 1991）．無縦溝殻の接殻帯片は閉鎖型で，帯片内接部には一様に伸張した鋸歯状の突出部をもつ[10]．

葉緑体 1枚，板状でC字形．

有性生殖 単為生殖を行い，厚い粘質物に包まれた増大胞子を形成することが報告されている（Geitler 1927, 1932）．

生活形 付着性．

汚濁耐性 弱汚濁耐性種．

出現地 大血川（埼玉県），多摩川（東京都）など，本邦の河川や湖沼から頻繁に出現する．

[ノート] 本邦では，*Cocconeis placentula* var. *placentula* よりむしろ本変種の方が多く出現するように思われる．本変種同様，無縦溝殻の縦走条線が明瞭な var. *euglypta* が知られているが，無縦溝殻の胞紋がより長いことによって区別できる．また，var. *euglypta* は流水域によく出現することが知られている．

Cocconeis scutellum Ehrenberg

Ehrenberg, C. G. 1838. Infusionsthier. p. 194. pl. 14. f. 8.
Dimension: L. 20-60 μm,　W. 10-40 μm,　Str. 10-15 in 10 μm (RV), ca. 7 in 10 μm (ARV),
　Ar. 16-20 in 10 μm (RV-middle portion), ca. 10 in 10 μm (ARV-middle portion).

被殻・殻　殻は楕円形で，殻長20〜60 μm，殻幅10〜40 μm[1-5]．縦溝殻は殻面が平坦で殻縁がわずかにせり上がり，全体的に浅い皿状となる[6]．縦溝殻内面の殻縁にはリング状に肥厚した縁辺隆起がある[7矢印]．無縦溝殻では，殻面は平坦で殻套部はゆるやかに傾斜する[7]．

縦溝・軸域　縦溝は直線的で，外裂溝の中心および極末端はわずかに丸く広がる[6]．また，内裂溝の極末端は，リング状の縁辺隆起まで達し，蝸牛舌で終わる[7]．

縦溝殻の条線・胞紋　条線は10 μmあたり10〜15本．条線を構成する胞紋は，殻面では1列で（10 μmあたり16〜20個），殻縁部では2列となる[6]．胞紋は円形で，殻内面で薄皮によって閉塞される．薄皮では細長い穿孔が放射状に配列する．

無縦溝殻の条線・胞紋　条線は10 μmあたり約7本．殻面では条線は粗く大きな1列の胞紋（10 μmあたり約10個）で構成されるが，殻套部では2列または3列になり胞紋も小さくなる[3-5,7]．各胞紋は円形または楕円形で，殻外面では縁辺に穿孔をもつ薄皮によって閉ざされる．

半殻帯　Holmes *et al*.(1982)が記述しているように，縦溝殻の接殻帯片は開放型で，条線の2〜4本間隔に突出部をもつ[6]．この突出部は殻のリング状縁辺隆起まで達し，先端でわずかに肥厚する．光学顕微鏡で見られる殻縁部の区画された構造は，この突出部と縁辺隆起によって囲まれた部分に相当する．一方，無縦溝殻の接殻帯片は閉鎖型であり，その突出部は殻套部の各間条線を裏打ちするように長く伸張する[7]．

葉緑体　C字形で殻面を向く，1枚．

有性生殖　1個の母細胞から1個の配偶子を形成し，雌雄同株の同形配偶を行う．増大胞子は周囲に粘質物を伴わず，各配偶子に由来する2枚の葉緑体をもち，初生細胞の長さは43.3〜59.5 μmであることなどが報告されている（Mizuno 1987）．

生活形　付着性．

汚濁耐性　不明．

出現地　三河湾竹島（愛知県），泊（秋田県），伊勢湾（愛知県），加茂湖（新潟県）など，本邦各地の海水・汽水域でふつうに見られ，しばしば淡水域にも出現する．また，新潟県村上産のテングサ（Takano 1961）および東京都式根島産ユカリ（鈴木ら 1999）などの海藻では優占的に着生することが報告されている．

ノート　*Cocconeis placentula* とよく似るが，無縦溝殻の胞紋がより粗く大きいこと，さらに殻縁で2〜3列に分岐し，胞紋も極端に小さくなることで区別できる．

Cocconeis stauroneiformis (Rabenhorst) Okuno

Okuno, H. 1957. Bot. Mag. Tokyo **70**: 217. pl. 6. f. 2a-c.
Basionym: *Cocconeis scutellum* var. *stauroneiformis* Rabenhorst 1864. Flora Eur. Alg. p. 101.
Dimension: L. 12-20 μm,　W. 10-15 μm,　Str. ca. 10 in 10 μm (RV), 9-10 in 10 μm (ARV),
　　Ar. 11-12 in 10 μm (RV and ARV).

被殻・殻　殻は長楕円形(1-4)．殻長 12〜20 μm，殻幅 10〜15 μm．縦溝殻の内面には，リング状に肥厚した縁辺隆起がある(5 矢印)．

縦溝・軸域　縦溝は直線的で，外裂溝の中心および極末端はわずかに丸く広がる(5)．縦溝殻も無縦溝殻も軸域は共に狭い(5,8)．縦溝殻の中心域は横に伸張して顕著な横帯（fascia）となる(1,3,5)．

縦溝殻の条線・胞紋　条線は 10 μm あたり約 10 本で，それぞれ 1 列の胞紋から成る．胞紋は大きく，丸みを帯びた長方形で 10 μm あたり 11〜12 個の密度で配列し，薄皮を伴う網目状の師板によって内表面が閉塞される(5)．薄皮の穿孔は細長いスリット状で，放射状に配列する（Okuno 1957）．

無縦溝殻の条線・胞紋　条線は 10 μm あたり 9〜10 本．各条線は 1 列の胞紋で構成される．胞紋の配列密度は縦溝殻と同様であるが，薄皮を伴った網目状の師板は殻外面にある(6-8)．

半殻帯　接殻帯片は，間条線の縁辺部に重なる突出部をもつ(5,6 矢じり)．縦溝殻の接殻帯片では，この突出部が殻のリング状縁辺隆起を越えて伸張することはない(5)．

葉緑体　不明．

有性生殖　不明．

生活形　付着性．

汚濁耐性　不明．

出現地　三河湾竹島（愛知県）など本邦各地の海水および汽水域に出現する．

Achnanthidium convergens (H. Kobayasi) H. Kobayasi

Kobayasi, H. 1997. Nova Hedwigia **65**: 159.
Basionym: *Achnanthes convergens* H. Kobayasi in Kobayasi, H. *et al*. 1986. Diatom **2**: 84. f. 1-7, 11-18, 37-43, 51-54.
Dimension: L. 7-25 μm, W. 3-5.5 μm, Str. ca. 18 in 10 μm (RV, center), 36-40 in 10 μm (RV, ends), ca. 24 in 10 μm (ARV), Ar. 40-50 in 10 μm.

被殻・殻 縦溝殻と無縦溝殻より構成される．殻は共に狭楕円形〜線状皮針形で，殻端は広円(1-10)．帯面は中央部が屈曲した長方形で，縦溝殻が凹側，無縦溝殻が凸側になる．殻長7〜25 μm，殻幅3〜5.5 μm．縦溝殻，無縦溝殻共に殻外面は平滑．殻肩は両殻共に角ばり無紋域を生じる(11, 13, 14)．

軸域・縦溝 軸域は縦溝殻では紡錘状皮針形(1,3,5,9)，無縦溝殻では線形(2,4,6,8,10)．縦溝は糸状縦溝．殻外面では縦溝枝の中心末端は直線的に膨れて終わる(11)．極末端では両縦溝枝とも同じ側に曲がる(11, 13)．殻内面では両方の中心末端は反対方向へ曲がる(12)．極末端は蝸牛舌に終わる(12)．

条線・胞紋 条線は縦溝殻では中央部ではほぼ平行，殻端部では強く収斂する(1,3,5,9)．中央部で10 μmあたり約18本，殻端部では密になり36〜40本．無縦溝殻の条線はほぼ平行だが殻端部ではわずかに放射状で(2,4,6,8,10)，10 μmあたり約24本．ボアグの欠落は両殻とも電顕使用で観察できる(9,10,共に殻の右側)．条線を構成する胞紋列は殻内面では浅い溝の中に配置する(12, 16)．胞紋は縦溝殻，無縦溝殻とも10 μmあたり40〜50個．両殻とも胞紋は内側が薄皮によって閉塞され(12, 16, 17)，殻外面では単軸方向にわずかに細長い開口をもつが，この傾向は無縦溝殻でより顕著である(11, 13-15)．薄皮の穿孔は丸でなく短い線形で，薄皮の周縁にのみ放射配列する中心整列である(17, 18)．隣り合った薄皮同士は珪酸基質層によって明瞭に区切られず，くびれをもって連続するものも多い(17)．縦溝殻，無縦溝殻共に殻套部では殻面の胞紋列から少し距離をおいて細長い1列の胞紋が存在する(11, 13, 14)．

半殻帯 開放型の無紋の帯片よりなる．3枚．
葉緑体 板状，1枚．
有性生殖 不明．
生活形 付着性．
汚濁耐性 弱汚濁耐性種．
出現地 奥入瀬川（青森県），荒川（埼玉県），多摩川（東京都），平井川（東京都），由良川（京都府），児野沢（長野県木曽郡），狩野川（静岡県），宇陀川（奈良県），銅山川（愛媛県），四万十川（高知県），遠賀川（福岡県），三川川（長崎県），川内川（鹿児島県）など．

ノート 本種は普通に見られる珪藻であるが，しばしば *A. japonicum* と誤認されやすい．これは，縦溝殻の殻端部での収斂状に配列する条線が見えにくいことによる．被殻が屈曲しているため，上手にピントを合わせないと縦溝殻の殻中央部と無縦溝殻の殻端部を同一光学切片上に観察してしまうことがある．これは別の種のように見えてしまうので(7)，注意が必要である．また本種の縦溝殻の軸域は *A. japonicum* と比べより狭く，条線配列が観察しづらい場合でも，慣れるとこれを指標として2種を識別できる．さらに，無縦溝殻の条線が *A. japonicum* のものより若干細かく，やや観察にしくくなることも識別点である．Kobayasi (1997) は，本種および類似する3種（*A. japonica, A. pirenaicum, A. latecephalum*）の詳細な微細構造の報告を行っている．Potapova & Ponader (2004) は，本種に類似する *A. rivalare* Potapova & Ponader と *A. crassum* (Hustedt) Potapova & Ponader の微細構造を示し，比較を行っている．

Achnanthidium exiguum (Grunow) Czarnecki

Czarnecki, D. B. 1994. In: Kocioleck, J.P. (ed.) 11th Int. Diat. Symp. p. 157.
Basionym: *Stauroneis exilis* Kützing 1844. Bacill. p. 105. pl. 30. f. 21.
Synonym: *Achnanthes exigua* Grunow in Cleve & Grunow 1880. Kongl. Svenska Vet.-Akad. Handl. **17** (2): 21.
Dimension: L. 11.5-17 μm, W. 5-6 μm, Str. 20-24 in 10 μm (center), ca. 26 in 10 μm (ends), Ar. ca. 60 in 10 μm.

被殻・殻 縦溝殻と無縦溝殻より構成される．殻は円形〜線形，中央部が若干くびれるものもある．殻端近くで強くすぼまり，先端は広くちばし形，まれに弱頭状となる[1-8]．殻面は中央部が屈曲した長方形で[11]，縦溝殻が凹側に[10,11]，無縦溝殻が凸側に[12]なって屈曲する．殻長11.5〜17μm，殻幅5〜6μm．縦溝殻，無縦溝殻共に殻外面は平滑で，殻肩は若干丸みを帯びる[10,12]．

軸域・縦溝 軸域は縦溝殻，無縦溝殻共に線形[1-3,5,7,8,10]であるが，無縦溝殻では中央部がわずかに広がる個体もある[4,6,12]．縦溝は糸状縦溝．殻外面では縦溝枝の中心末端は直線的に終わる[9]．縦溝の外裂溝は中心末端付近では殻表面の浅い溝の中に存在する[9,10]．極末端は両縦溝枝で反対側に曲がって終わる[10]．内裂溝の中心末端付近は直線的であるが，末端そのものは互いにかすかに反対方向へ曲がる．また極末端はかすかに側方へそれて終わる．無縦溝殻の殻端付近の観察では，殻内側の軸域内に殻形成過程で二次的に埋められ消失した縦溝の痕跡が認められる個体もある[14矢印]．

条線・胞紋 条線は縦溝殻，無縦溝殻共に弱く放射状に配列するが，殻端部ではほぼ並行となる[1-8,10,13]．縦溝殻では中央部に短い横帯が形成されるが[1,3,5,7,10,11]，無縦溝殻では相当する部分の片側に条線が存在し（長い条線の場合も，短い条線の場合もある），その反対側も殻套側から短い条線が1本挿入されるため横帯と呼べるものは存在しない[2,4,6,8,12]．条線は中央部では10μmあたり20〜24本，殻端部では約26本ある．ボアグの欠落は電顕的にのみ観察が可能である．観察例は少ないが，ボアグの欠落は両殻端で極裂が終わる側に，すなわち殻の横断面によって区切られる2つの側では長軸に対し異なる側にボアグの欠落が生じているようである[10矢印]．条線を構成する胞紋列は縦溝殻，無縦溝殻ともに10μmあたり約60個．両殻とも胞紋は内側が薄皮によって閉塞される[14]．胞紋の殻外面の開口は丸もしくは長軸方向に長い楕円[9,13]．殻套部には両殻とも胞紋はない[10,12]．

半殻帯 開放型の無紋の帯片よりなる[11]．3枚．
葉緑体 板状，1枚．
有性生殖 不明．
生活形 付着性．
汚濁耐性 中汚濁耐性種．下水処理場の二次処理水中に出現する場合もある．
出現地 阿寒湖（北海道），多摩川（東京都），東京学芸大学構内の防火用水池（東京都）．

ノート 小林・吉田（1984）によると，本種は学校などにあるコンクリート製の池にも頻繁に出現する．

… Plate 154

Achnanthidium gracillimum (Meister) Mayama comb. nov.

Basionym: *Microneis gracillima* Meister 1912. Die Kieselalgen der Schweiz. p. 97. f. 12.
Synonym: *Achnanthes minutissima* var. *gracillima* (Meister) Lange-Bertalot in Krammer & Lange-Bertalot 1989. Bacill. 4. p. 104 ; *Achnanthes alteragracillima* Lange-Bertalot 1993. Biblioth. Diatomol. **27**: 2 ; *Achnanthidium alteragracillimum* (Lange-Bertalot) Round & Bukhtiyarova 1996. Diatom Research **11**: 349.
Dimension: L. 19-31.5 μm, W. 3-4 μm, Str. ca. 22 in 10 μm (center), ca. 36 in 10 μm (ends), Ar. ca. 40 in 10 μm.

被殻・殻 縦溝殻と無縦溝殻より構成される．殻は狭皮針形，殻端は頭状となる(1-8)．帯面は中央部が屈曲した長方形で(9)，縦溝殻が凹状，無縦溝殻が凸状に屈曲するが，無縦溝殻は殻端部で若干反り返る(12)．殻長 19〜31.5 μm，殻幅 3〜4 μm，縦溝殻，無縦溝殻共に殻外面は平滑で，角ばった殻肩に無紋域を生じる(10-13)．

軸域・縦溝 軸域は縦溝殻ではやや紡錘状皮針形〜線形(1,3-5,7)，無縦溝殻では線形(2,6,9)．縦溝は糸状縦溝．殻外面では縦溝枝の中心末端は細長い中心孔を形成して終わる(10)．極末端は両縦溝枝とも同じ側に曲がって終わる(7,10)．殻内面では両方の中心末端は反対方向へ曲がる(14)．

条線・胞紋 条線は縦溝殻，無縦溝殻共に殻全体にわたってほとんど平行に配列し(1-8)，中央部では 10 μm あたり約 22 本，殻端部では約 36 本ある．条線を構成する胞紋列は殻内面では浅い溝の中に配置する(14)．縦溝殻，無縦溝殻ともに 10 μm あたり約 40 個．両殻とも胞紋は内側が薄皮によって閉塞される(14)．薄皮の穿孔は丸ではなく，短い線形で，薄皮の周縁にのみ放射配列する．胞紋の殻外面の開口は多かれ少なかれ短軸方向に細長いスリット状となるが(10-13)，この傾向は殻中央部付近で著しい．また殻外面では明らかにスリット状の胞紋開口が癒合したものも見られる(13)．殻套部には両殻とも細長い1列の胞紋が存在する(12)．

半殻帯 開放型の無紋の帯片よりなる(12)．3枚．
葉緑体 不明．
有性生殖 不明．
生活形 付着性．
汚濁耐性 弱汚濁耐性種．
出現地 阿寒湖（北海道），千曲川（川上，長野県），成出ダム下流（庄川，岐阜県），大戸川（滋賀県），三段峡（柴木川，広島県）など，河川とくに山間部の渓流にしばしば出現する．

ノート 従来，*Achnanthes microcephala*，*A. minutissima* var. *cryptocephala*，また *A. minutissima* var. *jackii* などと同定されていた種類．殻中央部で条線はやや粗くなるが殻端では密になること，また殻端が顕著な頭状になることが光顕的特徴である．電顕的には細い開口をもつ胞紋が他の類似種との識別点となる．Round & Bukhtiyarova(1996)によって作られた新組み合わせである *Achnanthidium alteragracillimum* は本来引用すべき基礎異名 *Microneis gracillima* Meister を引用しておらず，正式出版とみなされない（国際植物命名規約第33条）．

Achnanthidium japonicum (H. Kobayasi) H. Kobayasi

Kobayasi, H. 1997. Nova Hedwigia **65**: 156.
Basionym: *Achnanthes japonica* H. Kobayasi in Kobayasi, H. *et al*. 1986. Diatom **2**: 85, 86. f. 19-21, 27-36, 44-50, 55-58.
Dimension: L. 9-23 μm,　W. 4-5 μm,　Str. 16-20 in 10 μm (RV, center), 30-36 in 10 μm (RV, ends), ca. 18 in 10 μm（ARV),　Ar. ca. 40 in 10 μm.

被殻・殻　縦溝殻と無縦溝殻より構成される．殻は共に狭楕円形から線状皮針形で，殻端は広円(1-10)．帯面は中央部が屈曲した長方形で，縦溝殻が凹側，無縦溝殻が凸側になる．殻長9〜23 μm，殻幅4〜5 μm．縦溝殻，無縦溝殻共に殻外面は平滑．殻肩は縦溝殻では無紋域を生じるが(11,13)，無縦溝殻では丸みを帯び無紋域はない(16)．

軸域・縦溝　軸域は縦溝殻では狭皮針形(1,3,5,7,9)，無縦溝殻では線形(2,4,6,8,10)．縦溝は糸状縦溝．殻外面では縦溝枝の中心末端は直線的で膨れて終わる(11)．両縦溝枝の極末端はボアグ欠落のある側にゆるやかに曲がる(11,13)．殻内面では両方の中心末端は反対方向へ曲がる(12)．極末端は蝸牛舌に終わる(12)．

条線・胞紋　条線は縦溝殻の中央部ではほぼ平行で，10 μmあたり16〜20本，殻端部では強い放射状となり30〜36本ある(1,3,5,7,9)．無縦溝殻の条線は殻全域にわたりほぼ平行で(2,4,6,8,10)，10 μmあたり約18本．ボアグの欠落は両殻とも電顕の使用により観察できるが(11)，すべての個体に見つけることができるわけでなはい．条線を構成する胞紋列は殻内面では浅い溝の中に配置する(12,15)．胞紋は縦溝殻，無縦溝殻ともに10 μmあたり約40個．両殻とも胞紋は内側が薄皮によって閉塞され(12,15,17)，殻外面では内側よりやや直径の小さい丸い開口をもつ(11,13,14,16)．薄皮の穿孔は丸ではなく，短い線形で，中心整列状である(17,18)．縦溝殻の殻套には細長い1列の胞紋が存在する(11,13)．

半殻帯　開放型の無紋の帯片よりなる．3枚．
葉緑体　板状，1枚．
有性生殖　不明．
生活形　付着性．
汚濁耐性　弱汚濁耐性種．
出現地　荒川（埼玉県），多摩川（東京都），由良川（京都府）など，各地の河川の清水域．

（ノート）縦溝殻の殻端部で強く放射状となる密な条線がある，無縦溝殻の殻肩に無紋域がないことが本種のおもな識別形質である．本種と *Achnanthidium convergens* (H. Kobayasi) H. Kobayasi の最初の記載は1965年であったが（Kobayasi 1965），タイプの指定がなかったことから後年にタイプの指定がされて正式出版となった(Kobayasi *et al*. 1986)．本種の微細構造は Kobayasi *et al*.(1986) および Kobayasi(1997) に詳細に記されている．

Plates 156, 157

Achnanthidium minutissimum (Kützing) Czarnecki

Czarnecki, D. B. 1994. In: Kociolek, J. P. (ed.) 11th Int. Diat. Symp. p. 157.
Basionym: *Achnanthes minutissima* Kützing 1833. Alg. Aq. Dulc. Grem. Exs. Dec. 8. No. 75; Linnaea **8**: p. 578. f. 54.
Synonym: *Achnanthidium microcephalum* Kützing 1844. Bacil. p. 75 ; *Achnanthes minutissima* var. *macrocephala* Hustedt 1937. Arch. Hydrobiol. Suppl. **15**: 193 ; *Achnanthes lineariformis* H. Kobayasi 1965. Journ. Jap. Bot. **40**: 374 (non. *Achnanthes lineariformis* Lange-Bertalot 1993. Biblioth. Diatomol. **27**: 7).
Dimension: L. 7-25 μm, W. 2.5-4 μm, Str. ca. 22-26 in 10 μm (center), 28-32 in 10 μm (ends), Ar. 40-50 in 10 μm.

被殻・殻 縦溝殻と無縦溝殻より構成される．殻は狭皮針形，殻端は形態変異の幅が広く，弱頭状，鈍形，ややくさび形のものを生じる[1-34]．帯面は中央部が屈曲した長方形で[6,9]，縦溝殻が凹側[35]，無縦溝殻が凸側に屈曲する．殻長7〜25 μm，殻幅2.5〜4 μm，縦溝殻，無縦溝殻共に殻外面は平滑で，角ばった殻肩には無紋域を生じる[35,36,39,41,42]．

軸域・縦溝 軸域は縦溝殻ではやや紡錘状皮針形〜線形[13,17,28-31]，無縦溝殻では線形[14,18,32-34]．縦溝は糸状縦溝．殻外面では縦溝枝の中心末端は直線的に膨れて終わる[36,39]．極末端は両縦溝枝とも直線的に終わる[36,41]．殻内面では両方の中心末端は反対方向へ曲がる[37,38]．極末端は蝸牛舌に終わる[37,38]．

条線・胞紋 条線は縦溝殻では中央部で若干粗くなるが，横帯が形成されることはない．縦溝殻，無縦溝殻ともに条線は殻全体にわたって放射状に配列し，中央部では10 μmあたり22〜26本，殻端部では28〜32本ある．縦溝殻では殻端部で若干放射の角度が強くなることが電顕観察でわかる[28-31]．明瞭なボアグの欠落は電顕的にもあまり観察できない．条線を構成する胞紋列は殻内面では浅い溝の中に配置する[37,38,40]．縦溝殻，無縦溝殻ともに10 μmあたり40〜50個．両殻とも胞紋は内側が薄皮によって閉塞される[37,38,40]．胞紋の殻外面の開口は丸いが，まれに殻面の一番外側の開口がスリット状になる場合がある[31,34]．縦溝殻の殻端部では，条線を構成する個々の胞紋間を仕切る隔壁が狭くなり，その分，胞紋の開口が若干広くなる．その結果，殻端部全体として胞紋の占める割合が大きくなって見える[28-31,36,41]．殻套部には両殻とも細長い1列の胞紋が存在する[37-39,42]．

半殻帯 開放型の無紋の帯片よりなる．3枚[35]．
葉緑体 板状，1枚．
有性生殖 Geitler(1932)が本種の増大胞子形成を報告している．
生活形 付着性．
汚濁耐性 弱汚濁耐性種．
出現地 荒川（埼玉県），多摩川（東京都），千曲川（長野県）など，各地の河川の清水域や，地蔵院沼（埼玉県）など，湖沼，貯水池．

[ノート] 本種は，殻形の変異に富む．図28〜34は4m四方，深さ50cmの防火用水の壁に付着していた個体群であるが，多様な殻形の個体が含まれる．本種はまた，*Achnanthidium saprophilum* に似るが，本分類群では殻端部がより細いこと，無縦溝殻の軸域が線状であること，電顕的には，胞紋の密度が10 μmあたり40〜50個と少ないこと，縦溝殻の条線の放射が強くなる殻端部で胞紋の占める面積が大きくなることから識別できる．また，清水域に出現する点でも異なっている．なお，Lange-Bertalot & Ruppel(1980)は本種のタイプ試料の光顕および透過電顕観察を行っている．

Achnanthidium pusillum (Grunow) Czarnecki

Czarnecki, D. B. & Edlund, M. B. 1995. Diatom Research **10**: 208.
Basionym: *Achnanthes pusilla* Grunow in Cleve & Grunow 1880. Kongl. Svenska Vet.-Akad. Handl. **17** (2): 23.
Synonym: *Rossithidium pusillum* (Grunow) Round & Bukhtiyarova 1996. Diatom Research **11**: 351.
Dimension: L. 10-22 μm, W. 3-4 μm, Str. 20-24 in 10 μm, Ar. ca. 50 in 10 μm.

被殻・殻 縦溝殻と無縦溝殻より構成される．殻は共に狭楕円形で，殻端は広円(1-9)．帯面は中央部が屈曲した長方形で，縦溝殻が凹側に，無縦溝殻が凸側になる．殻長 10～22 μm，殻幅 3～4 μm，縦溝殻，無縦溝殻共に殻外面は平滑．殻肩は縦溝殻では角ばり無紋域を生じる(10,13)が，無縦溝殻については不明である．

軸域・縦溝 軸域は縦溝殻では紡錘状皮針形～線形(1,3,5,6,8)，無縦溝殻では線形(2,4,7,9)．縦溝は糸状縦溝．殻外面では縦溝枝の中心末端は直線的に膨れて終わる(13)．極末端は両縦溝枝とも曲がらずに終わる(10)．殻内面では両方の中心末端は反対方向へ曲がる(12)．

条線・胞紋 条線は縦溝殻，無縦溝殻共に殻全体を通じて弱い放射状である(1-9)．両殻とも 10 μm あたり 20～24 本．ボアグの欠落は電顕でのみ観察が可能であるが，すべての個体に見つけることができるわけでなはい．条線を構成する胞紋列は殻内面では浅い溝の中に配置する(12)．縦溝殻，無縦溝殻ともに 10 μm あたり約 50 個．両殻とも胞紋は内側が薄皮によって閉塞される(12)．殻外面では内側より若干直径の小さい丸い開口をもつ(10,11,13)．薄皮には丸い穿孔が多数存在する(14)．

半殻帯 開放型の無紋の帯片よりなる．3 枚．
葉緑体 不明．
有性生殖 不明．
生活形 付着性．
汚濁耐性 弱汚濁耐性種．
出現地 竜返しの滝（長野県），芦ノ湖（神奈川県），池田湖（鹿児島県），大峰沼（群馬県），青木湖（長野県），支笏湖（北海道），湯ノ湖（栃木県），木崎湖（長野県）など．

ノート 殻の外形は *Achnanthidium japonicum* に似るが，本種は，縦溝殻，無縦溝殻とも条線の傾きがたいへん弱く，密度も粗いことから識別できる．また，胞紋を閉塞する薄皮の穿孔が，規則散在する丸い孔である点も異なっている．本種の原記載には図版がないので，Grunow in Van Heurck（1880, pl. 27. f. 33, 34）が初出の図となる．

Plates 159, 160

Achnanthidium pyrenaicum (Hustedt) H. Kobayasi

Kobayasi, H. 1997. Nova Hedwigia **65**: 148.
Basionym: *Achnanthes pyrenaica* Hustedt 1939. Ber. Deutsch. Bot. Ges. **56**: 554. f. 5-10.
Synonym: *Achnanthes biasolettiana* Grunow sensu Lange-Bertalot & Krammer 1989. Bibl. Diatomol. **18**: 26., Krammer & Lange-Bertalot 1991. Bacillo. 4. p. 62, 63.
Dimension: L. 11-20 μm, W. 3.5-4.5 μm, Str. ca. 12 in 10 μm (RV, center), ca. 32 in 10 μm (RV, ends), ca. 20 in 10 μm (ARV, center), ca. 23 in 10 μm (ARV, ends), Ar. ca. 45 in 10 μm.

被殻・殻 縦溝殻と無縦溝殻より構成される．殻はやや不相称の皮針形，殻端はくさび形(1-7)．帯面は中央部が屈曲した長方形で，縦溝殻が凹状，無縦溝殻が凸状に屈曲する(12)．殻長11〜20 μm，殻幅3.5〜4.5 μm，縦溝殻，無縦溝殻共に殻外面は平滑で，角ばった殻肩には無紋域を生じる(9-13)．

軸域・縦溝 軸域は縦溝殻，無縦溝殻共にやや紡錘状皮針形〜線形(1-7)．縦溝は糸状縦溝．殻外面では縦溝枝の中心末端は中心孔を形成して終わる(10)．極末端は両縦溝枝とも同じ側にほぼL字に屈曲して終わる(8,9,11)．殻内面では両方の中心末端は反対方向へ曲がる(15,17)．極末端では蝸牛舌を形成する(16,18)．

条線・胞紋 条線は縦溝殻，無縦溝殻共に中央部ではほとんど平行であるが，殻端部ではゆるく放射状(1-7)．また中央部で粗く殻端部に近づくにつれ密になる．縦溝殻の中央部で10 μmあたり約12本，殻端部で約32本，無縦溝殻の中央部で10 μmあたり20本，殻端部で約23本である．ボアグの欠落はそれが観察できる場合は極裂が曲がる側にある(1)．条線を構成する胞紋は縦溝殻，無縦溝殻とも10 μmあたり約45個．両殻とも胞紋は内側が薄皮によって閉塞される(16-18)．胞紋の殻内面の開口は短軸方向にやや長く(11)，それらを閉塞する薄皮は胞紋間の壁によって明瞭には分断されない(16-18)．殻外面の開口は殻面では丸い小孔である(9-11,13)．また，両殻の殻套部にはスリット状の開口をもつ1列の胞紋が存在する(12,13)．

半殻帯 開放型の無紋の帯片よりなる．数枚．
葉緑体 板状，1枚．
有性生殖 不明．
生活形 付着性．
汚濁耐性 弱汚濁耐性種．
出現地 多摩川（東京都），日原川（東京都），員弁川（三重県），帝釈川（広島県）．

[ノート] Lange-Bertalot & Krammer(1989)は，本種のレクトタイプ（選定基準標本）を指定したが，それ以前にLange-Bertalot & Ruppel(1980)は，Grunowの試料からホロタイプ（正基準標本）の観察を行い，それが*Achnanthes gibberula*であることを明記している．したがって，後に指定されたレクトタイプは国際植物命名規約上は非合法のものとなり，*A. pyrenaica* Hustedtが本種の基礎異名となる (Kobayasi 1997)．日本では，1989年以前に採集された河川珪藻試料中には出現していなかった種であったが，今日では比較的多くの河川で見られるようになった．冬から春にかけて，また夏期でも水温の低い河川で優占的に出現することが知られている（小林・石田 1996）．日本産のものは殻の左右がやや不相称のものが多いが，Hustedtのタイププレパラートから撮影された個体群の殻は左右相称である(Simonsen 1987)．1990年代になり初めて日本で観察されるようになった種は，本種の他，*Nitzschia sigmoidea* (Ehrenberg) W. Smithおよび*Nitzschia vermicularis* (Kützing) Hantzschが知られている (Mayama *et al.* 2004)．

Achnanthidium saplophilum (H. Kobayasi & Mayama) Round & Bukhtiyarova

Round, F. E. & Bukhtiyarova, L. 1996. Diatom Research **11**: 349.
Basionym: *Achnanthes minutissima* var. *saprophila* H. Kobayasi & Mayama 1982. Jap. J. Phycol. **30**: 195. f. 2a-h.
Dimension: L. 8-15 μm, W. 3-3.5 μm, Str. 25-28 in 10 μm (center), ca. 32 in 10 μm (ends), Ar. 50-60 in 10 μm.

被殻・殻　縦溝殻と無縦溝殻より構成される．殻は共に狭楕円形で，殻端は弱頭状(1-15)．帯面は中央部が屈曲した長方形で，縦溝殻が凹側，無縦溝殻が凸側に屈曲する．殻長 8〜15 μm，殻幅 3〜3.5 μm，縦溝殻，無縦溝殻共に殻外面は平滑，殻肩は角ばり無紋である(14-16)．

軸域・条線・胞紋　条線は縦溝殻では狭皮針形の軸域（縦溝中肋）から側方に向かって伸び(8-10, 14)，無縦溝殻でも狭皮針形の軸域（中肋）から側方へと伸びているが(11-13, 15)，この中肋は細胞分裂後の殻の形成過程でいったん作られた縦溝中肋が，殻の成熟とともに二次的に消失してできることが確かめられている（Mayama & Kobayasi 1989）．条線は縦溝殻，無縦溝殻でともに弱く放射する．殻中央部では 10 μm あたり 25〜28 本，殻端部では細かく約 32 本．ボアグの欠落は電顕的に両殻で認めることができる．条線を構成する胞紋列は殻内面で浅い溝の中に配置する．10 μm あたり 50〜60 個．殻套部には 1 列の胞紋のみが存在する．胞紋は内側が薄皮によって閉塞されている(16)．薄皮の穿孔は六角整列を示す(17)．

縦　　溝　糸状縦溝．殻外側面では縦溝枝の中心末端は直線的に膨れて終わる(14)．極末端もほとんど直線的に終わる(14, 16)．殻内面では両方の中心末端はかすかに反対方向へそれる．極末端は小さな蝸牛舌におわる．

半殻帯　開放型，3 枚．
葉緑体　板状，1 枚．
有性生殖　不明．
生活形　付着性．
汚濁耐性　強汚濁耐性種．
出現地　南浅川（東京都），多摩川（東京都，神奈川県），空堀川（東京都），柳瀬川（東京都，埼玉県），残堀川（東京都），花見川（千葉）など，各地河川の汚濁水域．

ノート　殻端の形と無縦溝殻の狭皮針形の中肋は共に，光学顕微鏡下において本変種を承名変種から識別する形質である．電子顕微鏡的には本変種の胞紋は長軸方向に長く，承名変種のものは短軸方向に長いものが多い．このため，本変種の 10 μm あたりの胞紋数は約 10 個ほど多い．また sapro + philum（汚濁 + 好き：の意味）の名前のとおり，汚濁域に特徴的に出現する点も承名変種とは異なる（Mayama & Kobayasi 1984）．本種では，無縦溝殻の形成過程の初期には縦溝殻と同様に縦溝スリットが形成されるが，次第にスリットがシリカによって埋められて，最終的に縦溝をもたない殻が完成することが知られている（Mayama & Kobayasi 1989）．

Plates 162, 163

Achnanthidium subhudsonis (Hustedt) H. Kobayasi comb. nov.

Basionym: *Achnanthes subhudsonis* Hustedt 1921. Hedwigia **63**: 144. f. 9-12.
Dimension: L. 8-17 μm,　W. 3.5-4.5 μm,　Str. 18-20 in 10 μm,　Ar. ca. 30-34 in 10 μm.

被殻・殻　縦溝殻と無縦溝殻より構成される．殻は狭楕円形～皮針形，殻端はくさび形(1-12)．帯面は中央部が屈曲した長方形で(17,18)，縦溝殻が凹状(35)，無縦溝殻が凸状に屈曲する．殻長8～17 μm，殻幅3.5～4.5 μm．縦溝殻，無縦溝殻共に殻外面は平滑で，角ばった殻肩には無紋域を生じる(14,19,20)．縦溝殻，無縦溝殻共に殻套の縁に沿っていくつかのプラークが形成されることがある(14,18)．

軸域・縦溝　軸域は縦溝殻では狭皮針形(1,13)，無縦溝殻では紡錘状皮針形(2,21)．縦溝は糸状縦溝．殻外面では縦溝枝の中心末端は中心孔を形成する(13,14)．極末端は両縦溝枝とも釣針状に曲がり殻套部で終わる(13)．殻内面では両方の中心末端は同じ方向へ曲がる(16)．極末端は小さな蝸牛舌を形成する(15)．無縦溝殻の殻外面の軸域には殻形成過程中に，珪酸質による埋め込みのため二次的に消失したと思われる縦溝の痕跡が認められる個体も観察される(19,20,22)．

条線・胞紋　条線は縦溝殻，無縦溝殻共に中央部でゆるく放射状(1-12)．条線は縦溝殻，無縦溝殻ともに10 μmあたり18～20本．電顕ではボアグの欠落が極列の曲がる側にあることが観察される(13)．胞紋は殻内面，殻外面共に開口部はほぼ丸い(13,15,21,22)．胞紋を閉塞する薄皮は殻外面に向かってゆるいドーム状をしており(14,21)，胞紋壁の中央部よりやや外側表面近くに位置することが，殻套部胞紋の断面からわかる(15 矢印)．胞紋は縦溝殻，無縦溝殻ともに10 μmあたり約30～34個．殻套部には両殻とも1または2列の胞紋が存在する(14,19)．

半 殻 帯　開放型の無紋の帯片よりなる．4 枚(17,18,20)．
葉 緑 体　不明．
有性生殖　不明．
生 活 形　付着性．
汚濁耐性　弱汚濁耐性種．
出 現 地　湯ノ湖（栃木県），荒川（埼玉県），川越の湧泉（埼玉県），桜川支流中沢（茨城県つくば市小和田），多摩川（東京都），玉川上水（東京都），千曲川（長野県），児野沢（長野県木曽郡）など各地の河川．

ノート　Simonsen(1987)によって示されたHustedtの用いたタイプスライド中の個体の写真と比較すると，それらは本邦産のものに比べ，やや大形であり，縦溝殻の中心域がより広く，無縦溝殻の条線が中央部のかなりの範囲にわたって平行に配列するなどの違いが見られる．またLange-Bertalot & Krammer(1989)によって，わずか1枚示された*Achnanthes subhudsonis*のSEM写真の個体は殻肩が丸みを帯び，条線の胞紋列が殻套へ連続して配列するように見えており，本邦のものと微妙な相違をもっている．さらなる研究が本種に関する分類について必要であろう．

Lemnicola hungarica (Grunow) Round & Basson

Round, E. F. & Basson, P. W. 1997. Diatom Research **12**: 77.
Basionym: *Achnanthidium hungaricum* Grunow 1863. Verh. Kais.-König. Zool.-Bot. Ges. Wien. **13**: 146. pl. 13. f. 8.
Synonym: *Achnanthes hungarica* (Grunow) Grunow in Cleve & Grunow 1880. Kongl. Svenska Vet.-Akad. Handl. **17** (2): 20.
Dimension: L. 12-42.5 μm, W. 6-8 μm, Str. 19-21 in 10 μm, Ar. ca. 50 in 10 μm.

被殻・殻 縦溝殻と無縦溝殻より構成される．殻は線状皮針形，殻端はくさび形であるが(3-9)，長い個体では広円形となる(1,2,10)．帯面は中央部が屈曲した長方形で(12)，縦溝殻が凹に，無縦溝殻が凸に屈曲する．殻長12～42.5 μm，殻幅6～8 μm，縦溝殻の殻外面は平滑で，角ばった殻肩には無紋域を生じる(16,18)．無縦溝殻の殻外面は平滑で，殻肩はある程度角ばるようである（Round & Basson 1997）．

軸域・縦溝 軸域は縦溝殻，無縦溝殻共に線形(1-9,13-15)．縦溝殻の中心域は殻套部まで延びる横帯を形成するが，一方の側が大きいものとなる(1,3,5,7,8)．無縦溝殻でも中心域は短軸方向に延びるが縦溝殻のものほど大きくはならず横帯も形成しない(2,4,6,9)．詳細に観察すると，無縦溝殻の中心域の長軸方向長は，縦溝殻の中心域同様に一方の側で長いことがわかる(2,4,9,14)．縦溝は糸状縦溝．殻外面では縦溝枝の中心末端は小さな中心孔を作って終わる(17)．両縦溝枝の極末端は互いに反対方向へ曲がる(8,13)．この極裂は光学顕微鏡観察でも顕微鏡の焦点操作により確認できるものである(7)．殻内面では両方の中心末端はわずかに反対方向へ曲がる(20)．極末端は蝸牛舌を形成する(15,19)．

条線・胞紋 条線は縦溝殻，無縦溝殻ともにほとんど平行であるが，殻端部ではわずかに放射状になる(1-9)．両殻とも10 μmあたり19～21本ある．ボアグの欠落はたまに観察される．条線を構成する胞紋は二重胞紋列で，縦溝殻の殻内面では極浅い溝の中に配置する(15,19,20,22)．無縦溝殻の殻内面では条線は肋状に発達した間条線に挟まれる(14,21)．条線を構成する胞紋は縦溝殻同様に二重胞紋列よりなる．胞紋は縦溝殻，無縦溝殻ともに10 μmあたり約50個．両殻の胞紋の内側はスリット状の孔をもつ薄皮によって閉塞される(22)．胞紋の殻外面の開口は丸い(16,18)．縦溝殻の殻套部では1列の胞紋が観察されたが，無縦溝殻では不明．

半殻帯 無紋の開放型帯片よりなる．少なくとも2枚存在するが，3枚目の存在は未確認．
葉緑体 板状，1枚(10-12)．
有性生殖 不明．
生活形 付着性．
汚濁耐性 中汚濁耐性種．
出現地 小千谷の水田（新潟県），塩沢の水田（新潟県），宝蔵寺沼（埼玉県），川越の湧泉（埼玉県），玉川上水（東京都），多摩川（東京都），三宝寺池（東京都），小林弘所有の金魚水槽のガラス面（東京都），木崎湖（長野県）など．

ノート 本属は*Planothidium*属と被殻構造が類似するが，条線が*Planothidium*属では多重胞紋列からなるのに対し，*Lemnicola*属では二重胞紋列からなること，また，後者では縦溝殻で片側がより大きくなる横帯を形成すること，無縦溝殻では幅の細い中心域を作ることから区別されている．

Planothidium frequentissimum (Lange-Bertalot) Lange-Bertalot

Lange-Bertalot, H. 1999. Iconogr. Diatomol. **6**: 276.
Basionym: *Achnanthes lanceolata* spp. *frequentissima* Lange-Bertalot 1993. Biblioth. Diatomol. **27**: 4.
Synonym: *Achnantheiopsis frequentissima* (Lange-Bertalot) Lange-Bertalot 1997. Arch. Protistenk. **148**: 207.
Dimension: L. 9.5-25 µm, W. 4.5-7.5 µm, Str. 12-14 in 10 µm, Ar. 70-80 in 10 µm.

被殻・殻 縦溝殻と無縦溝殻より構成される．殻は皮針形，殻端は広円からややくさび形がかった広円となる[1-8]．帯面は中央部が屈曲した長方形で，縦溝殻は凹状（殻内面観では凸状[9]），無縦溝殻は凸状（殻内面観では凹状[14]）に屈曲する．殻長9.5～25 µm，殻幅4.5～7.5 µm．縦溝殻，無縦溝殻共に殻外面は平滑で殻肩は角ばる[11,12]．

軸域・縦溝 軸域は縦溝殻，無縦溝殻共に線形～やや狭皮針形[1-9,11,12]．縦溝殻の殻中央部では小形の不定形の中心域を形成する[1,3,5,7,9,11]．無縦溝殻の中央部は片側のみに無紋域を生じ，そこに馬蹄域[2,4,6,8]が形成される．馬蹄域は光学顕微鏡観察では，二重の弧を描いて見え特徴的である．走査型電子顕微鏡観察では，殻内面ではこの馬蹄域の上にずきん状のフードのような構造が見られる[10,13]．このずきん状構造の入口が，光学顕微鏡で観察される殻套側の弧の正体である．縦溝は糸状縦溝．殻外面では縦溝枝の中心末端は膨らんで終わる[11]．極末端は両縦溝枝とも同じ方向に曲がる[11]．殻内面では双方の縦溝枝の中心末端は直線的に終わり[9]，極末端は蝸牛節を形成して終わる．

条線・胞紋 条線は縦溝殻，無縦溝殻共に殻全体にわたってゆるく放射状に配列し[1-8]，両殻とも10 µmあたり12～14本ある．ボアグの欠落を識別するのは難しい．両殻の条線は殻内面では長胞構造になっている[9,10,13]．長胞の外側は2～4列の小孔状の胞紋列により覆われている[9-13]．1列あたりの胞紋は，10 µmあたり70～80個．それぞれの小孔を閉塞する薄皮構造の有無は確認されていない．両殻とも殻套部には長胞が存在しない[9]．

半 殻 帯 不明．
葉 緑 体 板状，1枚．
有性生殖 不明．
生 活 形 付着性．
汚濁耐性 弱汚濁耐性種．
出 現 地 阿寒湖（北海道），支笏湖（北海道），城沼（群馬県），秋川（東京都），多摩川（東京都），木崎湖（長野県），野々海湿原（長野），大戸川（滋賀県），池田湖（鹿児島県），タロコ峡（太魯閣峡，台湾）など，各地の湖沼，河川．

ノート　*Achnanthes frequentissima* Lange-Bertalot in Lange-Bertalot & Krammer(1989)は本種に対して最初に与えられた名前であるが，基準標本の指定およびラテン語の記載がされておらず，正式出版した名前とみなされない．このため，*Achnanthes lanceolata* spp. *frequentissima* Lange-Bertalot(1993)が本種の基礎異名となる．

Planothidium lanceolatum (Brébisson ex Kützing) Lange-Bertalot

Lange-Bertalot, H. 1999. Iconogr. Diatomol. **6**: 281.
Basionym: *Achnanthidium lanceolatum* Brébisson ex Kützing 1846. Bot. Z. **4**: 247.
Synonym: *Achnanthes lanceolata* (Brébisson ex Kützing) Grunow in Cleve & Grunow 1880. Kongl. Svenska Vet.-Akad. Handl. **17** (2): 23.
Dimension: L. 10-25 μm, W. 5-7.5 μm, Str. 11-13 in 10 μm, Ar. ca. 60-70 in 10 μm.

被殻・殻 縦溝殻と無縦溝殻より構成される．殻は皮針形，殻端は広円からやや広くちばし形がかった広円となる(1-10)．帯面は中央部が屈曲した長方形で，縦溝殻は凹状，無縦溝殻は凸状（殻内面観では凹状(14)）に屈曲する．殻長10〜25 μm（観察された初生殻では33 μm），殻幅5〜7.5 μm（同9 μm）．縦溝殻，無縦溝殻共に殻外面は平滑で殻肩は角ばる(11, 13)．

軸域・縦溝 軸域は縦溝殻，無縦溝殻共に線形〜やや狭皮針形(1-11)．縦溝殻の殻中央部では小形の楕円形〜不定形の中心域を形成する(1,3,5,6,11)．無縦溝殻の中央部は片側のみに無紋域を生じ，そこに馬蹄域(2,4,7,8)が形成される．馬蹄域は光学顕微鏡観察では，その周辺の部分と白黒が逆転して見えるため大変目立つ．走査型電子顕微鏡観察によると，この紋様は殻内面にあるくぼみであることがわかる(10,12,14)．縦溝は糸状縦溝．殻外面では縦溝枝の中心末端は中心孔を形成して終わる(11)．極末端は両縦溝枝とも同じ方向に曲がる(11)．殻内面では双方の縦溝枝の中心末端は直線的に終わる．極末端は蝸牛舌を形成し，直線的に終わる．

条線・胞紋 条線は縦溝殻，無縦溝殻共に殻全体にわたってゆるく放射状に配列し(1-11)，両殻とも10 μmあたり11〜13本ある．ボアグの欠落を識別するのは難しい．両殻の条線は殻内面では長胞構造になっている(12, 14)．長胞の外側は2〜4列の小孔状の胞紋列により覆われている(11, 13)．状態のよい殻を電顕で観察すると，それぞれの小孔の内側はわずかにドーム状に盛り上がる薄皮によって覆われているのがわかる(12)．薄皮には穿孔が規則散在する．両殻とも殻套部には長胞が存在しない(14)．10 μmあたり60〜70個．

半 殻 帯 詳細な観察例はないが開放型のようである．
葉 緑 体 板状，1枚 (Geitler 1932).
有性生殖 配偶子母細胞に，2つずつ配偶子が形成され，それらが接合し2個の増大胞子を作る (Geitler 1932).
生 活 形 付着性．
汚濁耐性 弱汚濁耐性種．
出 現 地 涸沼（茨城県），荒川（埼玉県），仙女ヶ池（埼玉県），宝蔵寺沼（埼玉県），川越の湧泉（埼玉県），狩川（神奈川県），多摩川（東京都），南浅川（東京都），日原川（東京都），酒匂川（神奈川県），狩野川（静岡県），小千谷の水田（新潟県），竜返しの滝（長野県），千曲川（長野県），諏訪湖（長野県），山中湖（山梨県），児野沢（長野県木曽郡），員弁川（三重県）ほか，各地の湖沼，河川に出現．

ノート Round & Bukhtiyarova(1996)による新組み合わせ *Planothidium lanceolatum* は基礎異名の正しい引用をしていないため正式出版とならず，Lange-Bertalot(1999)による新組み合わせが，正式かつ合法的なものとなる．

Plate 168

Planothidium septentrionale (Østrup) Round & Bukhtiyarova ex U. Rumrich *et al.*

Rumrich, U., Lange-Bertalot, H. & Rumrich, M. 2000. Iconogr. Diatomol. **9**: 216.
Basionym: *Achnanthes septentrionalis* Østrup 1910. Danske Diatomeer. p. 215. f. 13, 27.
Synonym: *Achnantheiopsis septentrionalis* (Østrup) Lange-Bertalot 1997. Arch. Protistenk. **148**: 208.
Dimension: L. 13-22 μm, W. 5.5-8 μm, Str. 11-13 in 10 μm, Ar. ca. 70 in 10 μm.

被殻・殻 縦溝殻と無縦溝殻より構成される．殻は広皮針形，殻端部では先端鋭形となる[3-8]．帯面は中央部が屈曲した長方形で，縦溝殻が凹状（殻内面観では凸状[10]），無縦溝殻が凸状に屈曲する[11]．殻長 13〜22 μm．殻幅 5.5〜8 μm．縦溝殻，無縦溝殻共に殻外面は平滑で，殻肩は若干丸みを帯びる[11]．

軸域・縦溝 軸域は縦溝殻ではやや紡錘状皮針形〜線形[3,5,6]，殻中央部では円形〜楕円形の中心域を形成．無縦溝殻では線形であるが殻中央部で若干膨れる[4,7,8]．縦溝は糸状縦溝．殻外面では縦溝枝の中心末端は直線的に膨れて終わるが，極末端の形状は観察されていない．殻内面では両方の中心末端は互いに反対方向へそれる[10]．極末端は蝸牛舌に終わる[9,10]．無縦溝殻の軸域は殻外面では若干くぼむ[11]．

条線・胞紋 条線は縦溝殻では放射状に配列し[3,5,6]，10 μm あたり 11〜13 本，殻端部に近づくに従い放射の角度は強くなり条線密度も増加する傾向にある．明瞭なボアグの欠落は電顕的にも認められない[9]．無縦溝殻の条線は殻中央部付近ではほぼ並行に配列するが，殻端部ではわずかに放射状となる[4,7,8]．両殻の条線は殻内面では長胞によって構成される[9,10]．それぞれの長胞の外側には 3, 4（まれに 2〜5）列の孔状の胞紋が存在する[9]．この胞紋列は無縦溝殻の場合，殻外面から観察すると，間条線の外側部分を含め殻全体に広がる層状構造をなしていることがわかる[11,12]（縦溝殻の殻外面の観察はされていないので不明）．殻套部には両殻とも長胞は存在しない[9-11]．10 μm あたり約 70 個．

半殻帯 開放型の無紋の帯片よりなる．少なくとも 2 枚存在するが，3 枚目の存在は未確認．

葉緑体 Cox(1996, Fig. 16t)が *Achnanthidium delicatulum* と同定した図は，おそらく本種であり，そこには 1 枚の葉緑体が描かれている．

有性生殖 不明．

生活形 付着性．

汚濁耐性 中汚濁耐性種．

出現地 阿寒湖（北海道），尾駮沼（下北半島，青森県），多摩川（羽田，東京都），多摩川（丹波，山梨県），宇川（京都府）など．

ノート *Achnanthes delicatula* (Kützing) Grunow に似るが，*A. delicatula* のレクトタイプ（選定基準標本）では殻端の形が鈍形な点，および縦溝殻の中心域もより小さく不定形である点[1,2]で本種と異なる．Kobayasi & Mayama(1989)が *A. delicatula* と同定し，識別珪藻群のタイプ B（中汚濁耐性種群）に割り振った個体は本種と同定されるべきものである．

Psammothidium helveticum (Hustedt) Bukhtiyarova & Round

Bukhtiyarova, L. & Round, F. E. 1996. Diatom Research **11**: 8.
Basionym: *Achnanthes austriaca* var. *helvetica* Hustedt 1933. Kies. p. 385. f. 831 g-k.
Synonym: *Achnanthes helvetica* (Hustedt) Lange-Bertalot in Lange-Bertalot & Krammer 1989. Biblioth. Diatomol. **9**: 63.
Dimension: L. 9-23 μm, W. 4.5-9 μm, Str. 22-28 in 10 μm, Ar. 45-50 in 10 μm (RV), 30-40 in 10 μm (ARV).

被殻・殻 縦溝殻と無縦溝殻より構成される．殻は楕円形，殻端は広円である(1-12)．帯面は中央部がわずかに湾曲した長方形で，縦溝殻は長軸および短軸方向に凸状に湾曲し(13,16)，無縦溝殻では長軸および短軸方向に凹状に(15)（殻内面から見ると凸状に(14)）湾曲する．殻長9～23 μm，殻幅4.5～9 μm．縦溝殻，無縦溝殻共に殻外面は平滑で，殻肩はわずかに丸みを帯びる(15,16)．

軸域・縦溝 軸域は縦溝殻，無縦溝殻ともに線形(14-16)．縦溝は糸状縦溝．殻外面で縦溝枝の中心末端は直線的に終わる(16)．縦溝枝の中央部から中心末端にかけて，外裂溝は殻表面の浅い溝の中に存在する(13)．両縦溝枝の極末端は互いに反対側に湾曲して終わる(13,16)．無縦溝殻の軸域は殻内面でわずかに肋状に肥厚する(14)．

条線・胞紋 条線は縦溝殻，無縦溝殻ともに殻全体にわたり放射状に配列する(1-12)．明瞭なボアグの欠落は電顕的にも観察されることはまれである．縦溝殻の中央部で条線は短くなり蝶形の中心域を形成する．10 μmあたり22～28本．条線を構成する胞紋は10 μmあたり45～50個で，光学顕微鏡によりそれらを認めることは不可能である(1,3,7,9,11)．無縦溝殻では条線は10 μmあたり22～28本．条線を構成する胞紋は10 μmあたり殻端部では30～40個であるが，殻中央部では20～30個のためよい状態で封入された試料の場合，光学顕微鏡でそれらを認めることができる(2,4,8)．両殻とも胞紋は内側が丸い薄皮によって閉塞される(17)．無縦溝殻の殻肩の内側では薄皮が癒合して長楕円形を示すこともある(17)．胞紋の殻外面の開口は丸い(15,16)．殻套部には両殻とも胞紋は存在しない(13,14,17)．

半殻帯 開放型の無紋の帯片よりなる(16)．少なくとも2枚存在するが，3枚目の存在は未確認．

葉緑体 本種での観察はないが，*Achnanthes austriaca* では1枚（Cox 1996）．

有性生殖 不明．

生活形 付着性．

汚濁耐性 弱汚濁耐性種．

出現地 塘路湖（北海道），洞爺湖（北海道），尾瀬沼（群馬県），仙女ヶ池（埼玉県），小千谷の水田（新潟県），鎌ヶ池（長野県），千曲川（長野県），青木湖（長野県），木崎湖（長野県），五ノ池（乗鞍岳，岐阜県），権現池（乗鞍岳，岐阜県），亀ヶ池（乗鞍岳，岐阜県），琵琶湖（滋賀県）など，高山などの貧栄養性の湖にしばしば出現する．

ノート 本種は，以前は *Achnanthes* 属に分類されていたが，被殻の屈曲(湾曲)方向が逆であるため，*Psammothidium* 属に組み替えられたものである．微細構造について Kobayasi & Sawatari(1986)の詳細な報告がある．

Psammothidium hustedtii (Krasske) Mayama

Mayama in Mayama, S., Idei, M., Osada, K. & Nagumo, T. 2002. Diatom **18**: 90.
Basionym: *Cocconeis hustedtii* Krasske 1923. Bot. Arch. **3**: 193. f. 10a, b.
Synonym: *Achnanthes krasskei* H. Kobayasi et Sawatari 1986. In: Ricard, M. (ed.) 8th Diat. Symp. p. 261 ; non *Achnanthes hustedtii* J. Bílý & Marvan 1959. p. 63.
Dimension: L. 10-14 µm,　W. 5-7 µm,　Str. 20-22 in 10 µm (RV), 18-21 in 10 µm (ARV),
　Ar. ca. 70-80 in 10 µm (RV), ca. 60-70 in 10 µm (ARV).

被殻・殻　縦溝殻と無縦溝殻より構成される．殻は楕円形～狭楕円形，殻端は広円～ややくさび形を帯びた広円[1-6]．帯面は全体がわずかに湾曲した長方形で，縦溝殻は長軸および短軸方向に凸状に湾曲し[7]，無縦溝殻では長軸および短軸方向に凹状に湾曲する[8]．しかし，短い殻長の個体で殻の湾曲性を確認することは容易でない．殻長 10～14 µm（12～17 µm: Krasske 1923），殻幅 5～7 µm．縦溝殻では殻外面は平滑であるが[7,12]，無縦溝殻ではわずかではあるが軸域部分に網目状の隆起を伴う[8,13]．殻肩は両殻とも角ばる[7,8]．

軸域・縦溝　軸域は縦溝殻では線形～やや狭皮針形[1,3,5,7,12]である．無縦溝殻では広皮針形[2,4,6,8,10,13]．縦溝は糸状縦溝．殻外面では縦溝枝の中心末端は直線的に終わる[12]．縦溝枝の中央部から中心末端にかけて，外裂溝は殻表面の浅い溝の中に存在する[7,12]．縦溝枝の極末端は直線的に終わる[12]．殻内面では両縦溝枝の中心末端は，互いに反対方向へわずかに曲がる[11]．また，極末端は小さな蝸牛舌を生じて終わる[9]．

条線・胞紋　条線は縦溝殻，無縦溝殻ともに殻全体にわたり放射状に配列するが，無縦溝殻では各条線は短い[1-6]．ボアグの欠落は電顕的にも観察が難しい．縦溝殻の中央部で条線は短くなり楕円形の中心域を形成する[1,3,5]．10 µm あたり 20～22 本（縦溝殻），18～21 本（無縦溝殻）．縦溝殻の条線は単列もしくは二重の胞紋列よりなるが，ほとんど単列の胞紋から条線が構成される個体から[7]，単列は中心域周辺および軸域に近い部分のみで多くの部分は二重胞紋列よりなる個体まで[12]，その混在の程度は個体により変異する．条線を構成する胞紋は 10 µm あたり約 70～80 個．無縦溝殻では条線は 10 µm あたり 18～21 本．縦溝殻と同様に条線を構成する胞紋は部分的に二重胞紋列になる[8,13]．胞紋は 10 µm あたり約 60～70 個．胞紋は殻内側では薄皮によって閉塞される[11]．胞紋の殻外面の開口は丸もしくは長円である[12,13]．両殻とも殻套部に胞紋は存在しない[8-10]．
半 殻 帯　無紋の帯片よりなる[7,8,12]．枚数は不明．
葉 緑 体　不明．
有性生殖　不明．
生 活 形　付着性．
汚濁耐性　弱汚濁耐性種．
出 現 地　小沼（北海道），大峰沼（群馬県），古沼（群馬県），近藤沼（群馬県），荒川（埼玉県），大血川（埼玉県），川越の湧泉（埼玉県），秋川（東京都），玉川上水（東京都），児野沢（長野県木曽郡）など．

(ノート) 本種は小形で殻の湾曲もわずかなため，電顕観察の角度によっては殻の湾曲方向が逆に見えてしまい *Achnanthidium* 属と見間違えることもあるが，それは錯覚である．

Psammothidium marginulata (Grunow) Bukhtiyarova & Round

Bukhtiyarova, L. & Round, F. E. 1996. Diatom Research **11**: 5.
Basionym: *Achnanthes marginulata* Grunow in Cleve & Grunow 1880. Kongl. Svenska Vet.-Akad. Handl. **17** (2): 21 ; Grunow in Van Heurck 1880. Syn. Diat. Belg. pl. 27. f. 45.
Dimension: L. 8.5-16 μm, W. 4-6 μm, Str. 26-28 in 10 μm, Ar. ca. 50 in 10 μm (RV), ca. 40 in 10 μm (ARV).

被殻・殻 縦溝殻と無縦溝殻より構成される．殻は楕円形〜狭楕円形，殻端は広円形[1-12]．帯面は中央部がわずかに湾曲した長方形である[14]．縦溝殻は長軸および短軸方向に凸状に湾曲[13,14]（殻内面観では凹状に屈曲[16]）し，無縦溝殻では長軸および短軸方向に凹状に湾曲する[17]．殻長8.5〜16 μm，殻幅4〜6 μm．縦溝殻，無縦溝殻共に殻外面は平滑で，殻肩はわずかに丸みを帯びる[13,17]．

軸域・縦溝 軸域は縦溝殻では狭皮針形，中央部で楕円形の中心域となる[9,11,13]．無縦溝殻では菱形[10,12,17]．縦溝は糸状縦溝．殻外面で縦溝枝の中心側および殻端側末端はともに直線的に終わる[13]．殻内面では両方の中心末端は反対方向へわずかに曲がる[15,16]．極末端は蝸牛舌に終わる[16]．

条線・胞紋 両殻とも条線は殻全体にわたって放射状に配列し[1-12]，10 μmあたり26〜28本ある．2個のボアグの欠落は電顕の使用により長軸に対し同じ側に生じることがわかる．条線を構成する胞紋列は殻内面では浅い溝の中に配置し[15,16]．縦溝殻では10 μmあたり約50個，無縦溝では10 μmあたり約40個．胞紋の殻外面の開口は両殻とも丸い[13,17]．また胞紋は内側が薄皮によって閉塞される[15]．縦溝殻では各胞紋を閉ざす薄皮が明瞭な区切れをもたずに癒合する場合もある[15]．胞紋は軸域から殻肩まで存在し殻套には存在しない[13,17]．

半 殻 帯 開放型の無紋の帯片よりなる[13,14]．少なくとも2枚存在するが，3枚目の存在は未確認．
葉 緑 体 板状，1枚．
有性生殖 不明．
生 活 形 付着性．
汚濁耐性 弱汚濁耐性種．
出 現 地 荒川（埼玉県），八丁池（伊豆，静岡県），鎌ヶ池（長野県），亀ヶ池（乗鞍岳，岐阜県），五ノ池（乗鞍岳，岐阜県），琵琶湖（滋賀県），屋久島（鹿児島県）など，清水域に出現する．

ノート 本種は，以前は*Achnanthes*属に分類されていたが，被殻の屈曲（湾曲）方向が逆であるため*Psammothidium*属に組み替えられたものである．

Psammothidium montanum (Krasske) Mayama

Mayama in Mayama, S., Idei, M., Osada, K. & Nagumo, T. 2002. Diatom **18**: 90.
Basionym: *Achnanthes montana* Krasske 1929. Bot. Arch. **27**: 350. f. 8a, b.
Dimension: L. 7-18 μm, W. 5.5-7.5 μm, Str. 22-24 in 10 μm, Ar. ca. 60-65 in 10 μm.

被殻・殻 縦溝殻と無縦溝殻より構成される．殻は広皮針形，殻端はくさび形〜やや広円(1-4)．帯面はゆるく湾曲した長方形で(5)，縦溝殻は長軸および短軸に沿って凸に(5)，無縦溝殻は長軸および短軸に沿って凹に湾曲する(6)．殻長 7〜18 μm，殻幅 5.5〜7.5 μm，縦溝殻の殻外面は平滑で，殻肩に沿って無紋域を生じる(7)．無縦溝殻の殻外面は殻中央部では，埋もれた胞紋の痕跡により凹凸を生じている(6)．

軸域・縦溝 軸域は縦溝殻では紡錘状皮針形〜狭皮針形(1,3,7)，無縦溝殻では皮針形〜広皮針形(2,4,6,8)．縦溝は糸状縦溝．殻外面では縦溝枝の中心末端は直線的に終わる(7)．両縦溝枝の中央部から中心末端にかけて，外裂溝は縦溝に沿った浅い溝の中で，わずかに隆起した縁を伴って存在する(5,7)．極末端は両縦溝枝とも互いに反対方向に曲がって終わる(13)．殻内面では両方の中心末端はほとんど直線的に終わる(9)．極末端は蝸牛舌に終わる(9)．

条線・胞紋 条線は縦溝殻，無縦溝殻共に放射状(1-4)で 10 μm あたり 22〜24 本．縦溝殻では条線は中央部で短くなり，片側が若干大きい楕円〜長方形の中心域を作る(1,3,5,7)．明瞭なボアグの欠落は電顕的にも観察が難しい．条線を構成する胞紋列は殻内面では浅い溝の中に配置する(9)．縦溝殻，無縦溝殻ともに 10 μm あたり約 60〜65 個．両殻とも胞紋は内側が薄皮によって閉塞される(8,9,11,12)．胞紋の殻外面の開口は両殻とも長軸方向に長い楕円である(5-7,10)．薄皮には規則分散する穿孔が多数開いている(12)．殻套部には両殻とも胞紋はない(8,9)．

半 殻 帯 開放型の無紋の帯片よりなる．3 枚(6,7)．
葉 緑 体 不明．
有性生殖 不明．
生 活 形 付着性．
汚濁耐性 弱汚濁耐性種．
出 現 地 大血川コケ表面（埼玉県）．

ノート 本種の縦溝殻，無縦溝殻の湾曲方向は *Psammotidium* 属を特徴づけるものである．

Psammothidium subatomoides (Hustedt) Bukhtiyarova & Round

Bukhtiyarova, L. & Round, F. E. 1996. Diatom Research **11**: 13.
Basionym: *Navicula subatomoides* Hustedt in A. Schmidt *et al.* 1936. Atlas Diat. pl. 404. f. 33-35.
Synonym: *Achnanthes subatomoides* (Hustedt) Lange-Bertalot & R. E. M. Archibald 1985. Biblioth. Diatomol. **9**: 9, 97; *Achnanthes detha* Hohn & Hellerman 1963. Trans. Amer. Mic. Soc. **82**: 274.
Dimension: L. 6.5-9.5 μm, W. 4-5.5 μm, Str. 32 in 10 μm, Ar. ca. 40 in 10 μm.

被殻・殻 縦溝殻と無縦溝殻より構成される．殻は楕円，殻端は広円(1-4)．帯面は全体がわずかに湾曲した長方形で，縦溝殻は長軸および短軸方向に凸状に湾曲し(7)，無縦溝殻では長軸および短軸方向に凹状に湾曲する(8)．しかし，殻の湾曲性は観察する角度により明瞭にわからないことも多い(5,6,9)．殻長6.5〜9.5 μm(6〜15 μm: Krammer & Lange-Bertalot 1991)，殻幅4.5〜5.5 μm(3.5〜6 μm: l.c.)．縦溝殻，無縦溝殻ともに殻外面は平滑である(5,7,8)．殻肩は両殻とも無紋域を生じ，若干丸みを帯びる(5,7,8)．

軸域・縦溝 軸域は縦溝殻では線形〜紡錘状皮針形(5,6)，無縦溝殻では線形(8,9)．縦溝は糸状縦溝．殻外面では縦溝枝の中心末端はやや膨れて直線的に終わる(5,7)．縦溝枝の中央部から中心末端にかけて，外裂溝は殻表面の浅い溝の中に存在する(5)．縦溝枝の極末端は直線的に終わる(5,7)．殻内面では両縦溝枝の中心末端は，互いに反対方向へわずかに曲がり，極末端はかすかな蝸牛舌を生じて終わる(6)．縦溝殻の軸域は殻内面でわずかに肋状に肥厚する(6)．無縦溝殻の軸域は中央部では若干肥厚するが，極側付近では軸域の中央部がわずかにくぼんで見える(9矢印)．これは殻形成時に一時的に生じた縦溝が殻の完成と共に珪酸質により埋められた痕跡と考えられる．

条線・胞紋 条線は縦溝殻，無縦溝殻ともに殻全体にわたってゆるく放射状に配列する(1-4)．ボアグの欠落は必ずしも明瞭ではないが電顕観察により認めることができる．縦溝殻，無縦溝殻ともに条線は中央部で短くなり，左右非対称の長方形（片側が狭い）の中心域を形成する(5,6,8,9)．中心域の広い側が殻形成における2次側で，ボアグの欠落はこちら側にある．縦溝殻の条線は10 μmあたり32本．条線を構成する胞紋は10 μmあたり約40個．無縦溝殻では条線は10 μmあたり32本．胞紋は10 μmあたり約40個．胞紋は殻内側では薄皮によって閉塞される(11)．薄皮の穿孔は中央部が規則分散で，周辺部がスリット状の穿孔が放射整列する(12)．胞紋の殻外面の開口は丸もしくは楕円(10)．両殻とも殻套部に1列の胞紋をもつ(5,8)．

半 殻 帯 帯片は無紋の開放型．確認されている帯片は2枚まで(8)，3枚目の存在は未確認．
葉 緑 体 不明．
有性生殖 不明．
生 活 形 付着性．
汚濁耐性 弱汚濁耐性種．
出 現 地 尾瀬沼（群馬県），塩原大沼（栃木県）．

ノート 本種の縦溝殻および無縦溝殻の湾曲方向は*Achnanthidium*属に見られるものと逆である．Kobayasi & Sawatari(1986)が微細構造の詳細を示した*Achnanthes detha* Hohn & Hellerman は本種の異名である．

Nupela lapidosum (Krasske) Lange-Bertalot

Lange-Bertalot, H. 1999, Iconogr. Diatomol. **6**: 280.
Basionym: *Achnanthes lapidosa* Krasske 1929. Bot. Arch. **27**: 350. f. 9.
Synonym: *Achnanthidium lapidosum* (Krasske) H. Kobayasi in Mayama *et al.* 2002. Diatom **18**: 89.
Dimension: L. 7-25 μm, W. 3-5 μm, Str. 22-26 in 10 μm, Ar. ca. 45-55 in 10 μm.

被殻・殻 縦溝殻と無縦溝殻より構成される．殻は狭皮針形～皮針形，殻端はやや丸みを帯びたくさび形(1-13)．帯面は中央部が屈曲した長方形で，縦溝殻が凹状(13)（殻内面観では凹状(16)），無縦溝殻が凸状（殻内面観では凹状(14)）に屈曲する．殻長7～25 μm，殻幅3～5 μm．殻外面の様子は不明．

軸域・縦溝 軸域は縦溝殻では線形(1,3,5,7,9,13)である．殻中央部で丸い中心域を形成する．しばしば，中心域の片側は殻套部まで続くことがある(3,5)．軸域は無縦溝殻では狭皮針形(2,4,6,8,10-12)，殻中央部では縦溝殻同様に片側で中心が殻套部まで続く場合(2)もあり変異に富む．縦溝は糸状縦溝．殻内面では縦溝枝の中心末端はT字形に終わる(16,18)．極末端は蝸牛舌を生じて終わる(16,17)．無縦溝殻の殻内面には，殻の形成過程で二次的に埋められ消失した縦溝の痕跡が，線形の浅いくぼみとなって残っている(14矢印)．この痕跡は光学顕微鏡観察では，薄黒い線状の模様として認めることができる(8,10-12)．

条線・胞紋 両殻とも条線は殻全体にわたり放射状に配列し(1-13)，10 μmあたり22～26本．明瞭なボアグの欠落の確認は電顕的にも容易でない．条線を構成する胞紋列は殻内面では浅い溝の中に配置する(14,16)．胞紋の内側開口は小さく円形だが(15,17,18)，外側開口はいくぶん大きめである（Lange-Bertalot & Krammer 1989: *Achnanthes lapidosa* として）．縦溝殻，無縦溝殻とも10 μmあたり約45～55個．本種の胞紋は本属の他種同様に，外側開口部が薄皮によって閉塞されるようだが，Lange-Bertalot & Krammer(1989)，Krammer & Lange-Bertalot(1991)の写真でも，薄皮がとれた個体の写真が示されているのみである．

半殻帯 接殻帯片の幅が広いことは本種の特徴である(16)．帯片内接部には1列の胞紋がある．帯片全体の枚数は不詳．

葉緑体 不明．

有性生殖 不明．

生活形 付着性．

汚濁耐性 弱汚濁耐性種．

出現地 庄内川（山形県），荒川（埼玉県），桜川支流の中沢（茨城県つくば市），滄浪泉園の小流（東京都小金井市），多摩川おいらん淵（山梨県），小菅川（山梨県），竜返しの滝（長野県），木崎湖（長野県），銅山川（愛媛県）．

ノート *Nupela* Vyverman & Compère(1991)は，*N. giluwensis* 1種で設立された属である．*Nupela*属は*Brachisira*属との近縁性が重視されたが，その特徴には不明な点も多かった．その後，本属に帰属する種が増えるにつれ，多様な特徴をもつことが次第に明らかになった．本種すべてに共通な形質は，小形であること，殻外面で胞紋が薄皮により閉塞されること，胞紋の外側開口が内側開口より目立って大きいこと，帯片に胞紋列をもつことである．上下殻はともに完全な縦溝をもつ種が多いが，片側の殻の縦溝が短くなる種（*N. tenuistriata* (Hustedt) Lange-Bertalot，*N. jahniae-reginae* Lange-Bertalotなど），片側の殻の縦溝がシリカによって埋められ縦溝の痕跡だけを残し単縦溝珪藻になった種（本種や*N. rumrichorm*など）もある．また，内裂溝の中心側末端も，T字形以外に，片側に曲がるもの，直線的に終わるものなどがある．

Diatomella balfouriana Greville

Greville, R. K. 1855. Ann. Mag. Nat. Hist. ser. 2. **15**: 259. pl. 9. f. 10-13.
Basionym: *Grammatophora balfouriana* Simth, W. 1856. Syn. Brit. Diat. vol. 2. p. 43. pl. 61. f. 383.
Dimension: L. 9-18 μm,   W. 4.5-5 μm,   Str. 16-18 in 10 μm,   Ar. ca. 50 in 10 μm.

被殻・殻　殻面は狭楕円形～やや線形で，殻端は広円(1-3,5)．殻は長方形(4)．殻長9～18 μm（12～52 μm: Krammer & Lange-Bertalot 1986），殻幅4.5～5 μm（6～8 μm: l.c.）．殻面は平滑で，殻肩は角ばり無紋域を生じる(7)．

軸域・縦溝　軸域は殻幅のおよそ1/2を占める，狭皮針形(1,5)．縦溝は光学顕微鏡観察では糸状に見える．殻外面では縦溝枝の中心末端は中心孔を形成して終わる(7)．極末端は曲がって終わる(7)．殻内面では縦溝枝の中心末端は，互いに同じ方向へわずかに曲がって終わる(9)．極末端は蝸牛舌に終わる．

条線・胞紋　殻面における条線は中心部，殻端部を問わずたいへん短く周辺部のみにある(1,5,7)．これに対し，殻套部にはより長い条線が存在する(6,7)．条線の密度は10 μmあたり16～18個（18～21 in Krammer & Lange-Bertalot 1986）．条線を構成する胞紋は二重胞紋列であるが，他属の種でしばしば見られるように，それぞれの列の胞紋同士が互い違いに配列することはなく，本種では，2胞紋が隣り合って配列するという特徴をもつ(6-8)．殻内面では殻套域の間条線は肋状に発達しているようで，それらに挟まれて胞紋列が存在する(8)．胞紋の殻外面の開口は小孔状である(6-8)．胞紋は10 μmあたり約50個．

半殻帯　接殻帯片には3穴が開いた隔壁がある(8,9)．隔壁の中央部の橋渡しの部分にはジグザグした継ぎ目状構造が観察できる(8)．接殻帯片以外の帯片は開放型で胞紋をもたない．4枚(6)．

葉緑体　不明．
有性生殖　不明．
生活形　付着性．
汚濁耐性　弱汚濁耐性種．
出現地　竜返しの滝コケ表面（長野県）．

ノート　本属名は国際植物命名規約で保存名とされているものである．

Diploneis elliptica (Kützing) Cleve

Cleve, P. T. 1891. Acta Soc. Fauna Fl. Fennica **8** (2): 42.
Basionym: *Navicula elliptica* Kützing 1844. Bacill. p. 98. pl. 30. f. 55.
Dimension: L. 28-55 µm, W. 16-33 µm, Str. 7-9 in 10 µm, Ar. 8-12 in 10 µm.

- **被殻・殻** 楕円形から広楕円形(1-4)．殻は殻面から殻套に向かってかなり湾曲する(5)．殻長28〜55 µm，殻幅16〜33 µm．
- **縦走管・縦溝** 縦走管は狭く，中央で強く弓状となる(1-4)．その外壁には条線の胞紋より小さな1列（中央で2列）の胞紋がある．縦溝は縦走管に囲まれた溝の中にあるため，光顕では縦溝は太く，特に中央側では膨らんで終わるように見える(1-4)．実際には両方の外中心裂溝は，同じ側にわずかにくの字に折れ，極裂も同じ側に折れ曲がる(5)．内裂溝は両側とも真っ直ぐに終わる．
- **条線・胞紋** 条線は全体にやや放射状で粗く，10 µm あたり7〜9本．条線は長胞で，内面では薄皮によって完全に閉ざされている(7)．条線は明瞭な1列の胞紋から成り，10 µm あたり8〜12個．それぞれの胞紋は，外面で横走肋と縦走肋によって完全に分離され，ほぼ四角か円形で凹み，多孔師板によって閉塞されている(5,6)．多孔師板は単純ではなく，内面から樹状の細い肋によって裏打ちされている．
- **半殻帯** 開放型．内接部に深裂状の切れ込みをもつ接殻帯片と，無紋で小舌をもつ1枚の帯片．
- **葉緑体** 縁が切れ込む葉状，2枚．
- **有性生殖** 不明．
- **汚濁耐性** 弱汚濁耐性種．
- **生活形** 付着性．
- **出現地** 黒山（埼玉県），塩原（栃木県），羽村（東京都），雲取山（東京都）など，清水のコケに付着して出現する．また，神流湖（群馬県），河口湖（山梨県），青木湖（長野県）など，各地の湖沼にも出現する．

[ノート] 淡水に出現する本属の種類の中では特に胞紋が粗い．

Diploneis ovalis (Hilse) Cleve

Cleve, P. T. 1891. Acta Soc. Fauna Fl. Fennica **8** (2): 44.
Basionym: *Pinnularia ovalis* Hilse, in Rabenhorst, G. L. 1861. Alg. Eur. no. 1025.
Dimension: L. 15-52 μm,　W. 9-23 μm,　Str. 10-14 in 10 μm,　Ar. 16-20 in 10 μm.

被殻・殻　楕円形(1-4). 殻は殻面から殻套に向かってなだらかに湾曲する(6). 殻長 15～52 μm, 殻幅 9～23 μm.

縦走管・縦溝　縦走管は狭く, 中央で弓状となる(1-3). 縦走管の外壁には小さな 2 列の胞紋がある. 縦溝は縦走管に囲まれた溝の中にあり, 光顕では縦溝はやや太く, 特に中央側では膨らんで終わるように見える(1-3). 実際には, 外中心裂溝は小さな丸いくぼみの中に真っ直ぐに終わり(7), 両極裂は同じ側に折れ曲がる(4). 内裂溝は両側とも真っ直ぐに終わる.

条線・胞紋　条線は全体にやや放射状で, 10 μm あたり 10～14 本. 条線は長胞で, 内面では薄皮によって完全に閉ざされている(6,8). 条線は明瞭な 1 列の胞紋から成り, 10 μm あたり 16～20 個. 胞紋の外面は多孔師板によって閉塞されているように見えるが, それほど単純な構造ではなく, 縦横に肥厚した小肋が裏打ちしている. 電顕では 1 つ 1 つの胞紋の仕切りが, 外面からは区別しにくいが(5), 内面に肥厚した縦肋によって区画されている(8). また, 内面に肥厚した横肋により, 胞紋が 2 列に見えることもある.

半殻帯　開放型. 内接部に鋸状の切れ込みをもつ接殻帯片と無紋で小舌をもつ 1 枚の帯片.
葉緑体　縁が切れ込む葉状, 2 枚.
有性生殖　不明.
生活形　付着性.
汚濁耐性　弱汚濁耐性種.
出現地　池田湖 (鹿児島県), 鵜ノ池 (愛知県), 小田内沼 (青森県), 仙女ヶ池 (埼玉県), 牛久沼 (茨城県), 北浦 (茨城県), 八郎潟 (秋田県) など, 各地の湖沼.

ノート　本属では淡水に最も広く出現する種類である. 同じ胞紋が 1 列の種類の *Diploneis elliptica* に比べ, 条線も点紋も細かい.

Diploneis smithii (Brébisson ex W. Smith) Cleve

Cleve, P. T. 1894. Kongl. Svebska Vet.-Akad. Handl. **26**: 96.
Basionym: *Navicula smithii* Bréb. ex W. Smith 1856. Syn. Brit. Diat. Vol. 2. p. 92.
Dimension: L. 22-53 µm,　　W. 11.5-23.5 µm,　　Str. 6-8 in 10 µm,　　Ar. 16-20 in 10 µm.

被殻・殻　殻は楕円形で，殻面から殻套に向かってなだらかに湾曲する[1-3,6]．殻長22～53 µm，殻幅11.5～23.5 µm．

縦走管・縦溝　縦走管は狭く，中央でやや広くなり，全体として皮針形となる[1-3]．縦走管の外壁には小さな2列の胞紋がある[4]．縦溝は縦走管に囲まれた溝の中にある．光顕でもわかるように，中心裂溝はやや太く，同じ側に小さく折れ曲がる[4]．極裂は同じ側にゆるく折れ曲がる[7]．内裂溝は両端とも真っ直ぐに終わる．

条線・胞紋　条線は全体に放射状で，10 µmあたり6～8本である．条線は長胞で，内面は薄皮によって完全に閉ざされる．条線は明瞭な交互に並ぶ2列の胞紋から成り，10 µmあたり16～20個[5,6]．胞紋の外面は多孔師板によって閉塞されている[5]．それぞれの胞紋の仕切りは，殻の中央寄りでは明瞭であるが，殻縁に向かうに従って不明瞭となる[5]．長胞を仕切る横走肋は，内側で頭状に膨らむ．

半 殻 帯　開放型．内接部に鋸状の切れ込みをもつ接殻帯片と無紋で小舌をもつ1枚の帯片．

葉 緑 体　縁が切れ込む葉状，2枚．

有性生殖　同型配偶により2個の増大胞子を作る．

生 活 形　付着性．

汚濁耐性　弱汚濁耐性種．

出 現 地　湖山池（鳥取県），涸沼（茨城県），霞ヶ浦（茨城県），小田内沼（青森県），十三湖（青森県），八郎潟（秋田県），野尻湖（長野県）など，各地の淡水から汽水の湖沼．

[ノート]　汽水または海産の種類と思われがちだが，淡水の湖沼にも出現する．従来の記載では大きさの範囲が，殻長12～200 µm，殻幅6.5～75 µm（Krammer & Lange-Bertalot 1986）となっているが，パラタイプ（副基準標本）の大きさなどから考えて，100 µmを越えることはないと思われる．

Entomoneis japonica (Cleve) K. Osada

Osada, K. in Mayama, S., Idei, M., Osada, K. & Nagumo, T. 2002. Diatom **18**: 89.
Basionym: *Amphiprora alata* (Ehrenberg) Kützing var. *japonica* Cleve 1894. Kongl. Svenska. Vet.-
 Akad. Handl. **26** (2): 16. pl. 1. f. 2.
Synonym: *Entomoneis alata* (Ehrenberg) Ehrenberg var. *japonica* (Cleve) Osada et Kobayasi 1985.
 Jpn. J. Phycol. **33**: 215-224. f. 1-30.
Dimension: L. 55-155 µm,　W. 18-40 µm,　Str. 11-12 in 10 µm.

被殻・殻　被殻は全体が長軸に対してねじれているが[7]，帯面観では中央が強くくびれたバイオリン形を呈する[1,4]．殻は殻面観において弱狭さくバイオリン形で，殻端は鋭先形[2]．殻の両側には盾（scutum）をもつ[5 矢印, 7 矢印]．殻長 55～155 µm，殻幅 18～40 µm．せり上がった竜骨は S 字状に強くカーブし，中心節と極の間では翼状に発達する[5-7]．竜骨の頂上部は縦溝と縦溝中肋から成り，その内側には縦溝管[8 rc] をもつ．縦溝中肋から両側に伸びた横走肋は殻縁まで続く[5,6]．

翼構造・接合線　向かい合った横走肋は帯面観で互いに重なるように走行し，3 種類の小骨（基部小骨：basal fibula，中間小骨：intermediate fibula，縦溝小骨：raphe fibula）で連結される[8]．殻本体と翼の境界は，光顕では縦に走る接合線（junction line）として観察される．各接合線は基部小骨の縦の列によって構成され，大きく 2 回湾曲する[1,3,4-7]．

縦　溝　管状縦溝．裂溝はひだ状で，極めて狭い．両殻端の極裂はそれぞれ反対方向に屈曲するが（Osada & Kobayasi 1985；as *E. alata* var. *japonica*），殻内面の極側末端は発達した蝸牛舌で終わる[6 矢印]．

条線・胞紋　条線は縦溝中肋から接合線を横切って殻縁に達するが[1,3,4]，殻の本体（竜骨を除く部分）では，10 µm あたり 11～12 本の密度で，横軸に対してほぼ平行に配列する[2]．各条線は 2 列の孔状胞紋で構成される．胞紋は，細長い小孔を伴うドーム型の薄皮によって外側表面が閉ざされて，殻の内側に開口する[8]．この小孔の配列様式は散在型で（Osada & Kobayasi 1985），中心整列型を示す *E. alata* (Ehrenberg) Ehrenberg（Mann 1981）のものとは異なる．

半殻帯　通常，5 枚の開放型帯片で構成される．すべての帯片には円形状あるいは細長い胞紋が配列する．殻から最も遠い帯片の胞紋は 1 列であるが，その他の胞紋は 2 列で帯片中肋の両側に配列する．殻側から 2, 3, 4 番目の各帯片では，通常，殻に近い方の胞紋列を構成する胞紋は殻から遠い方の胞紋列のものに比べて明らかに短い．帯片の胞紋は外側表面にある薄皮によって閉ざされる．また，これらの薄皮は，中央部で散在する円形の小孔と胞紋の縁辺に沿って 1 列に配列する細長い小孔で穿孔されている（Osada & Kobayasi 1985）．

葉緑体　板状，2 枚．各葉緑体は殻帯側を向き，長軸方向に縦列する．
有性生殖　不明．
生活形　付着性（底生）．
汚濁耐性　不明．
出現地　野付湾（北海道），尾駮沼（下北半島，青森県），多摩川河口（神奈川県），江奈湾（三浦半島，神奈川県），七尾北湾（能登半島，石川県），中海（島根県），名蔵川河口（石垣島，沖縄県）など，本邦では各地の汽水域によく見られる．

ノート　本分類群はこれまで *E. alata* (Ehrenberg) Ehrenberg の変種として扱われてきたが，殻形，条線密度および接合線の形状などが *E. alata* のものとは明らかに異なる．

Entomoneis paludosa (W. Smith) Reimer

Reimer, C. W. in Patrick, R. M. & Reimer, C. W. 1975. p. 4, 5. pl. 1. f. 1.
Basionym: *Amphiprora paludosa* W. Smith 1853. p. 44. pl. 31. f. 269.
Dimension: L. 37.5-72.5 μm, W. 7-12 μm, Str. 20-24 in 10 μm.

被殻・殻 被殻はバイオリン形で，長軸に対してわずかにねじれる[1,2]．殻は，殻面観では両側が平行な線状皮針形で先端が微突頭となる[3]．殻長37.5～72.5 μm，殻幅7～12 μm．竜骨は二葉形で翼状に隆起し，S字状にゆるやかにカーブする[1-3]．

翼構造・接合線 翼は，頂上部にある二重の管；縦溝管[6 rc]と付随管（additional canal）[6 ac]，翼棒（wing bar）[6 wb]および翼棒間の薄皮で構成される[5]．縦溝管と付随管は，向かい合った横走肋を連結する1列の縦溝小骨[6 rc]によって隔てられている[6]．これらの二重管と殻の本体は中実の翼棒によって連結され[5,6]，翼棒の間の部分は，帯面観で重なって見える2枚の細長い薄皮によって閉塞される．接合線は殻端付近に角を伴って滑らかにカーブする[1,2,4]．

縦　　溝 管状縦溝をもつ．裂溝は極めて狭く，ひだ状．中心側および極側末端は殻の内外面で共に単純に終わる．

条線・胞紋 条線は殻の本体と翼の二重管の部分にあり，10 μmあたり20～24本の密度でほぼ平行に配列する．各条線は1列の胞紋で構成される．胞紋は基本的に楕円形であるが，しばしば胞紋内部の微小な棒状体によって二分されていることもある[4,5]．胞紋を閉塞する薄皮は殻外面にあり，規則的散在型（Mann 1981）の小孔を伴う（Osada & Kobayasi 1990）．また，翼棒の間を向かい合って閉塞する薄皮も同様な小孔をもつ[5]．

半殻帯 少なくとも5枚または6枚の開放型帯片で構成される．各帯片は帯片中肋の両側に1列の胞紋をもつが，帯片内接部に近い胞紋列は円形の胞紋，他方の胞紋列は多少細長い胞紋で構成される．いずれの胞紋も外側表面が規則的散在型の小孔を伴った薄皮によって閉塞される（Osada & Kobayasi 1990）．

葉 緑 体 板状，1枚．中央にくびれをもち，殻帯側を向く．

有性生殖 不明．

生 活 形 付着性（底生）．

汚濁耐性 不明．

出 現 地 藻興部川河口（北海道），尾駮沼（下北半島，青森県），涸沼（茨城），荒川（東京都），多摩川河口（東京都，神奈川県），など本邦各地の汽水域や河口域によく見られる．

欧文解説と図版

Plate 1   *Melosira moniliformis* (O. Müller) C. Agardh

Figs 1-3.   LM.

Fig. 1.   Girdle view of the filamentous living cells with many discoid chloroplasts. ×1000.
Fig. 2.   Valve view. ×2000.
Fig. 3.   Girdle view showing a valve mantle and a cingulum. ×2000.

Locality
Figs 1-3.   Mokoto-ko (Lake Mokoto), Hokkaido.

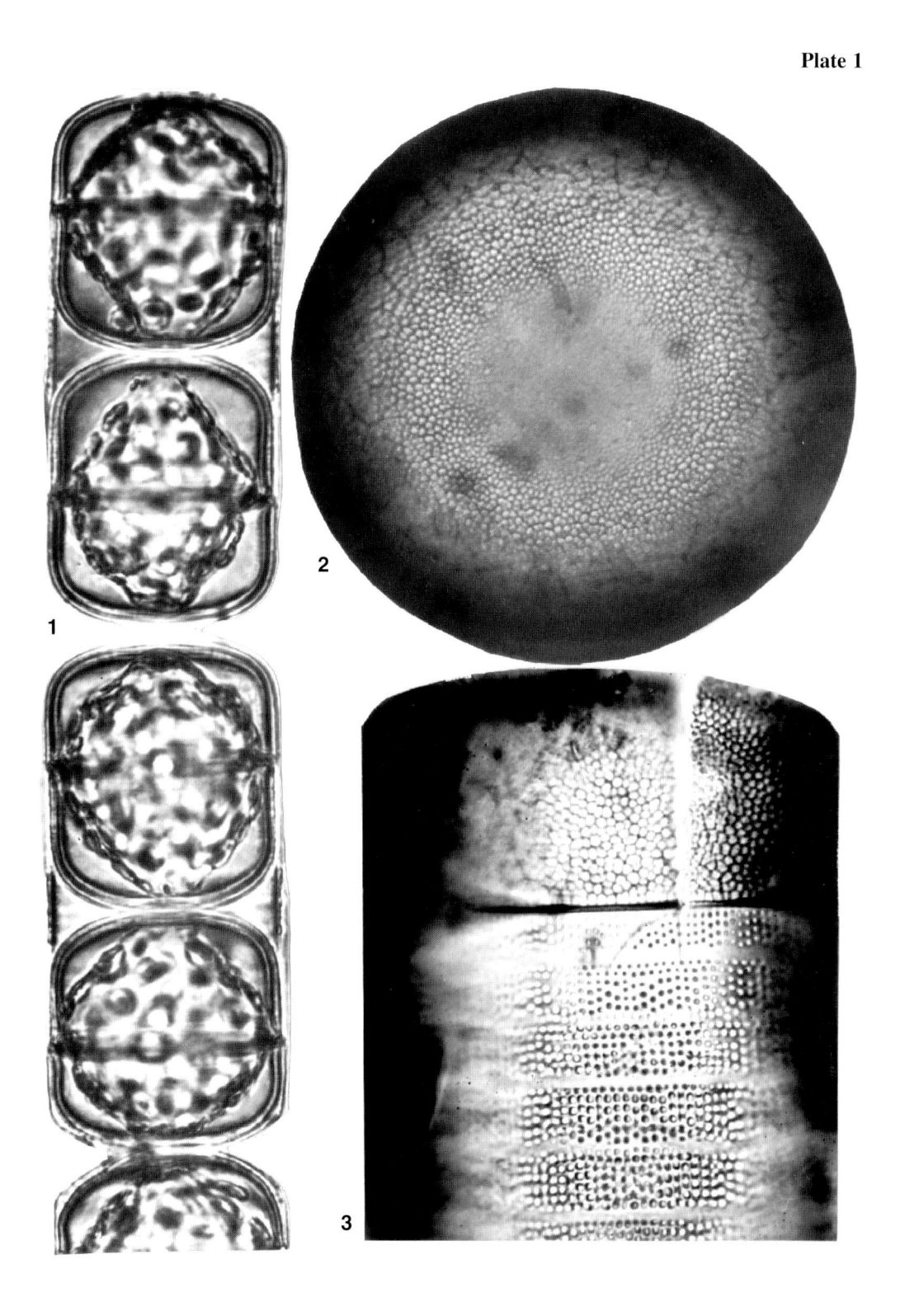

Plate 1

Plate 2 *Melosira moniliformis* (O. Müller) C. Agardh

Figs 4, 5.   SEM.

Fig. 4.   External oblique view of the valve showing. ×2000.

Fig. 5.   External view of the valve face showing numerous granules, starlike processes, and the openings of the labiate process. ×8000.

Locality

Figs 4, 5.   Hinuma-gawa (Hinuma River), Ibaraki.

Plate 2

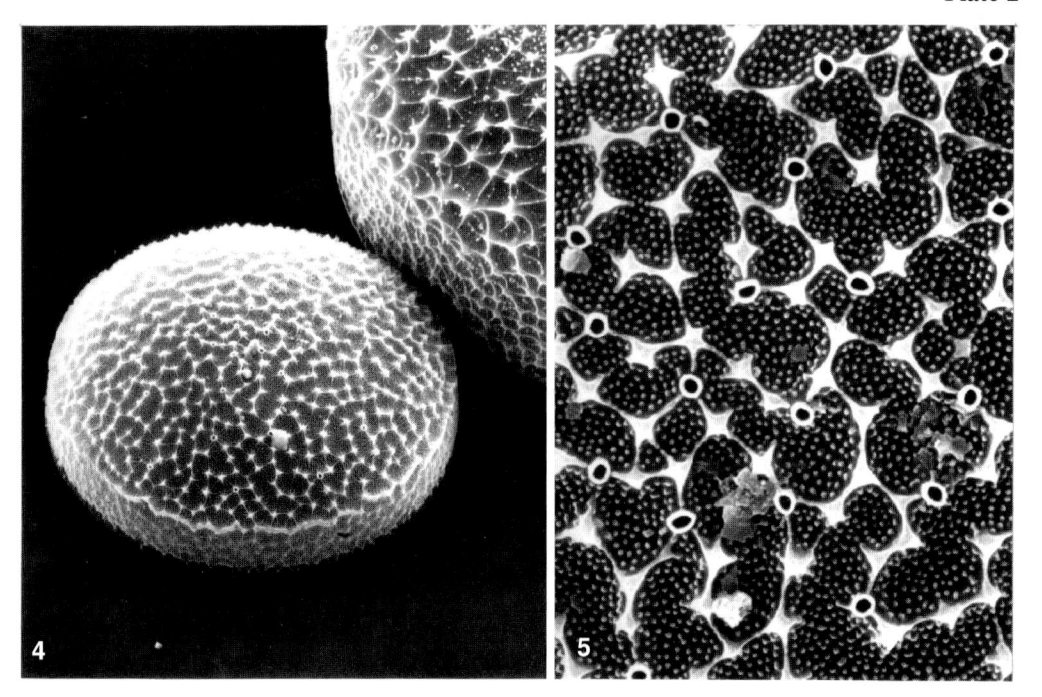

Plate 3 *Melosira nummuloides* C. Agardh

Figs 1-3, 8.   LM. ×2000.
Figs 4-7.   Drawings after Hustedt's Kieselalgen.
Fig. 9.   SEM.

Figs 1, 2, 4, 6.   Girdle view.
Figs 3, 7, 8.   Valve view.
Fig. 5.   Auxospore with a mother valve.
Fig. 9.   External view of the sibling cells showing the collar of each epivalve. ×2000.

Localities
Fig. 1.   Sennyoga-ike (Sennyo Pond), Saitama.
Figs 2, 3, 8.   Leipzip, Germany.
Fig. 9.   Aono-gawa (Aono River), Shizuoka.

Plate 3

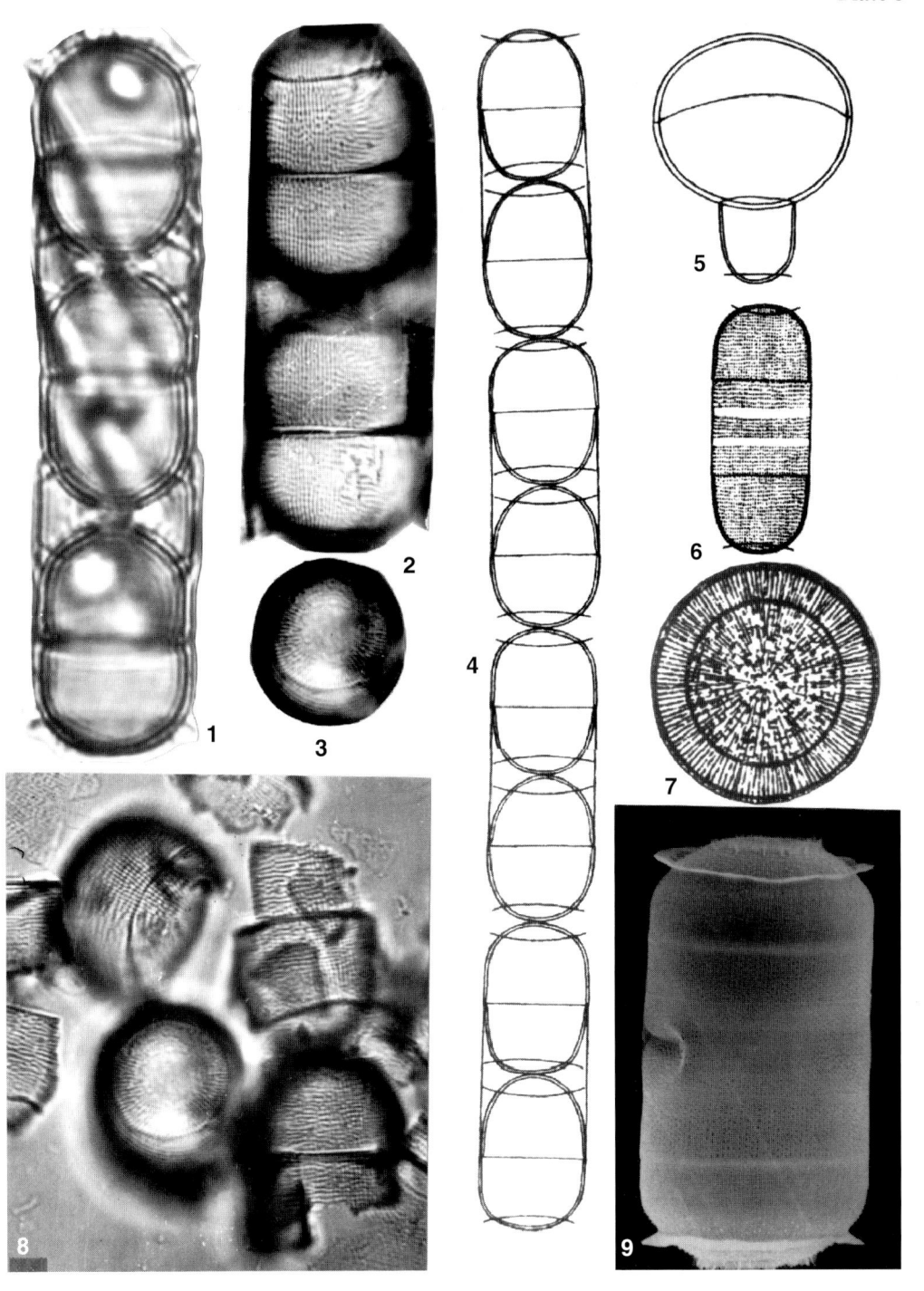

153

Plate 4   *Melosira nummuloides* C. Agardh

Figs 10-12.   SEM.

Fig. 10.   External view of a whole frustule showing the collars and the openings of submarginal labiate processes. ×5000.

Fig. 11.   External oblique view of the valve face showing the openings of central labiate processes, a collar and a band. ×5000.

Fig. 12.   External oblique view of the valve showing an opening of the submarginal labiate process, squared areolae with cribrum, and the granules arranged in the four corners of the areola. ×30000.

Locality

Figs 10-12.   Aono-gawa (Aono River), Shizuoka.

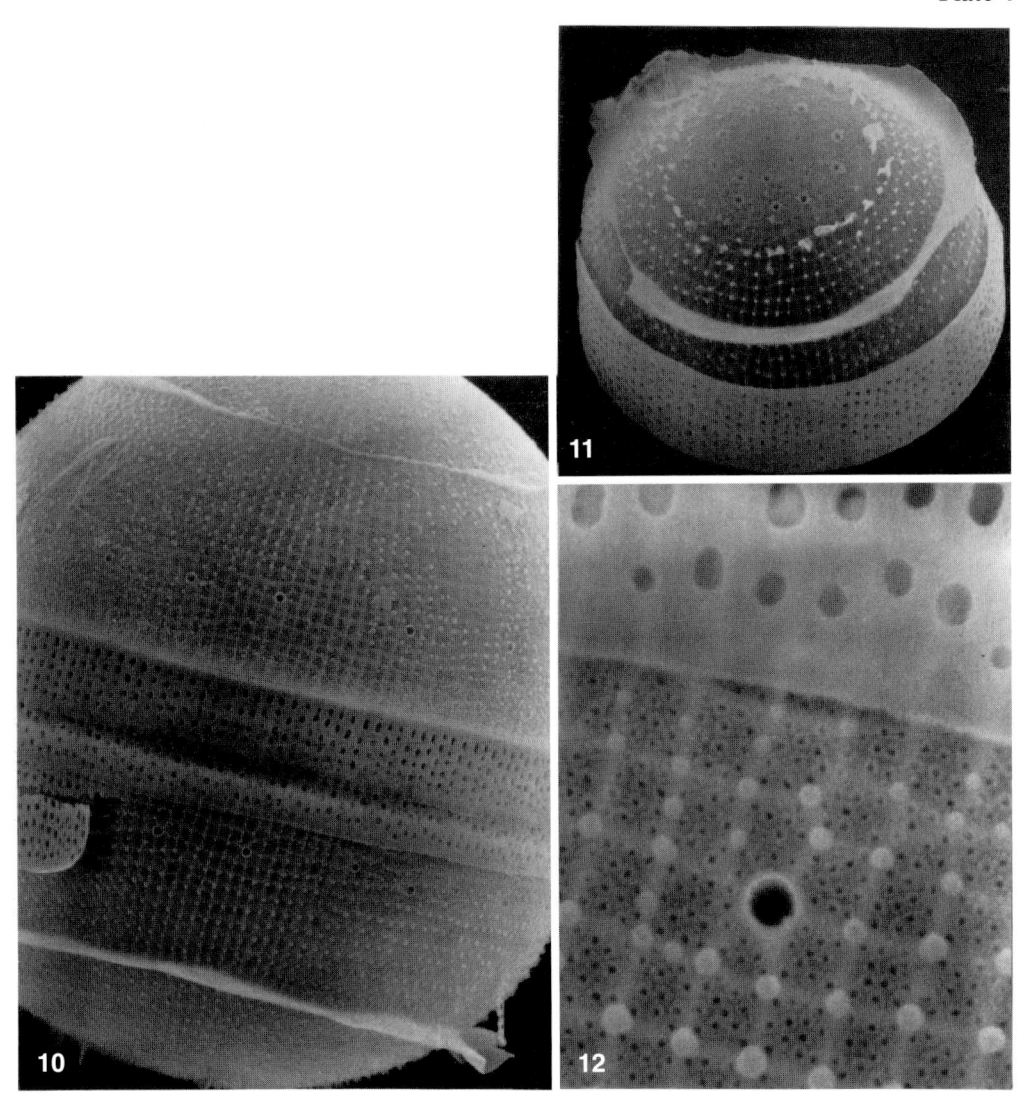

Plate 4

Plate 5 *Melosira varians* C. Agardh

Figs 1-3.　Drawings after Hustedt's Kieselalgen.　×1500.
Figs 4-9.　LM.　×2000.

Fig. 1.　Auxospore with a mother valve.
Figs 2, 7, 8.　Valve view.
Figs 3-6.　Girdle view of the filamentous living cells with many discoid chloroplasts.
Fig. 9.　Girdle view.

Localities
Fig. 4.　A paddy field in Sarugakyo, Gunma.
Figs 5, 6, 9.　Matsumae, Hokkaido.
Figs 7, 8.　A paddy field in Shiozawa, Niigata.

Plate 5

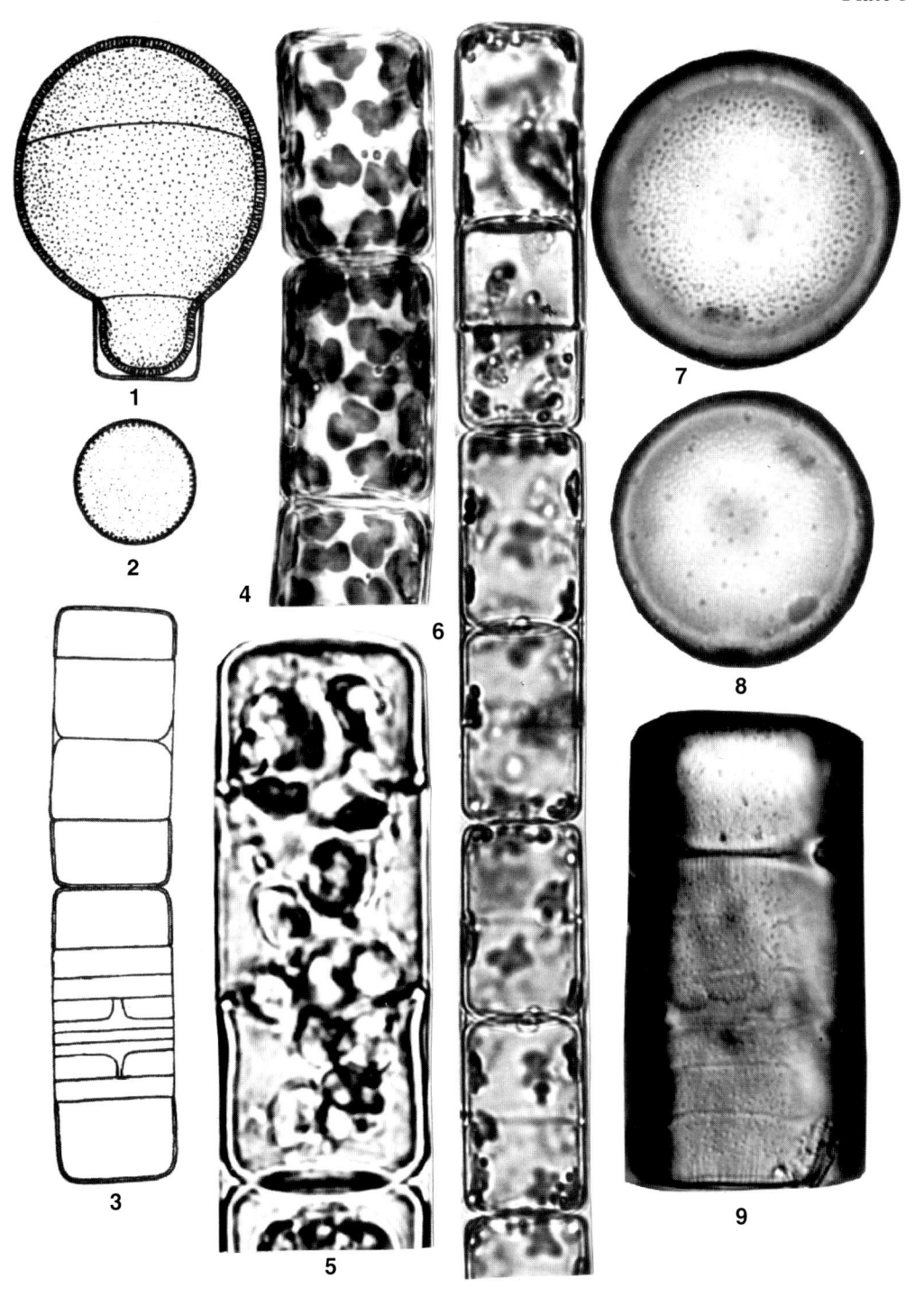

157

Plate 6 *Melosira varians* C. Agardh

Figs 10-14. SEM.

Fig. 10. A whole frustule. ×2000.

Fig. 11. External view of the valve face showing numerous pores, irregular spines, and external openings of the labiate processes. ×15000.

Fig. 12. External oblique view of the valve. ×3000.

Fig. 13. External view of the valve mantle showing the openings of the labiate processes and granules. ×3000.

Fig. 14. Internal oblique view of the valve with lots of labiate processes. ×3000.

Locality

Figs 10-14. Matsumoto-jo Hori (The moat of Matsumoto Castle), Nagano.

Plate 6

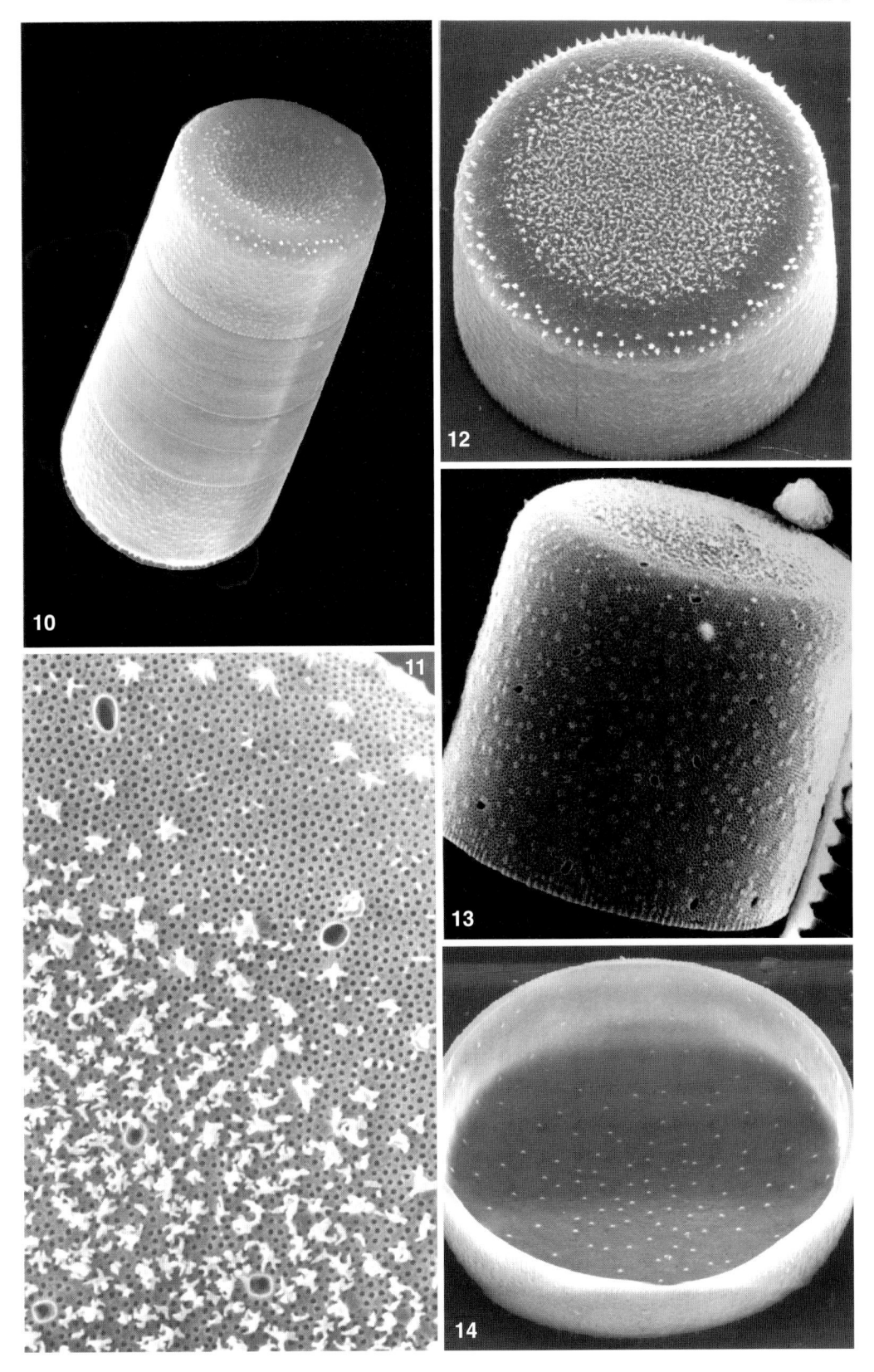

Plate 7 *Ellerbeckia arenaria* f. *teres* (Brun) R. M. Crawford

Figs 1, 2.   LM. ×800.
Figs 3-6.   SEM.

Fig. 1.   Image focused on the girdle surface. Note the tube processes appearing as black dots (arrows).

Fig. 2.   Image focused on the valve center.

Fig. 3.   External oblique view of a whole valve with convex valve face. ×600.

Fig. 4.   External oblique view of the mantle showing many rows of areolae arrayed along the pervalvar axis and the thick mantle edge with many grooves. Note the outer openings of the tube processes (arrows). ×5000.

Fig. 5.   Internal oblique view of the girdle showing occluded areolae and the tube processes. ×5000.

Fig. 6.   Internal oblique view of a partly broken valve. The valve face is plain and the mantle edge is striated by shallow grooves. ×1500.

Localities

Figs 1, 2, 6.   Akan-ko (Lake Akan), Hokkaido.
Figs 3-5.   Onogami (fossil diatom), Gunma.

Plate 7

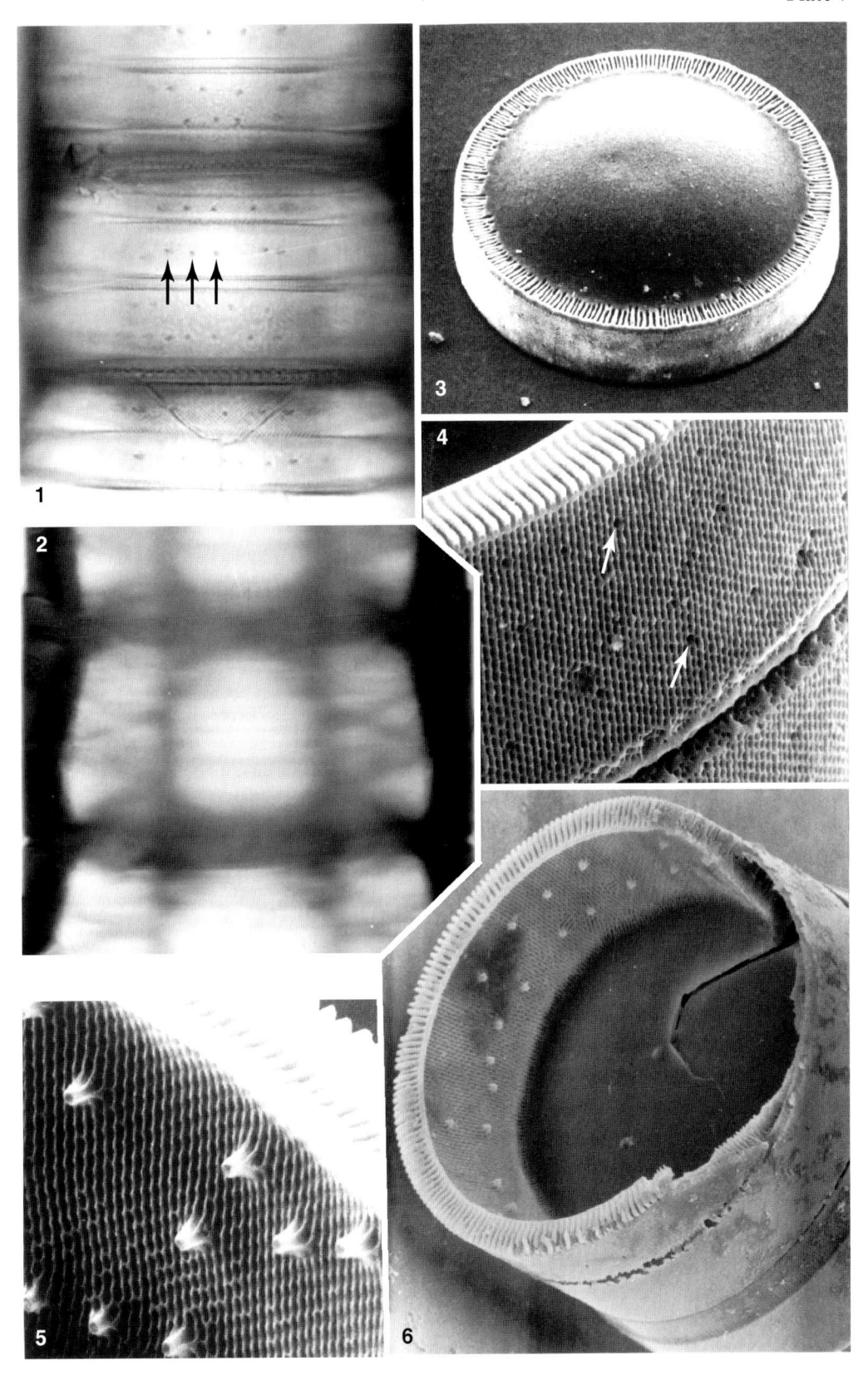

Plate 8 *Aulacoseira ambigua* (Grunow) Simonsen

Figs 1, 2.   LM. ×1000.
Figs 4-13.   LM. ×2000.
Fig. 3.   Drawing. ×2000.

Fig. 1.   A fixed filamentous colony with shrunk chloroplasts.
Figs 2, 4.   Small cells before auxosporulation and enlarged post-initial cells.
Figs 5, 6.   Valve view.
Figs 7-13.   Girdle view of the filamentous colony after acid cleaning.
Fig. 7.   Optical section of the same colony as in Fig. 8.
Figs 7, 8, 11-13.   Linear colony.
Figs 9, 10.   Curved colony.

Localities

Figs 1, 12.   Biwa-ko (Lake Biwa), Shiga.
Figs 2, 11.   A pond in Musashi-kyuryo Shinrin Park, Saitama.
Figs 5-8.   Oze-numa (Oze Pond), Gunma.
Figs 9, 10.   Suwa-ko (Lake Suwa), Nagano.
Fig. 13.   Kondo-numa (Kondo Pond), Gunma.

Plate 8

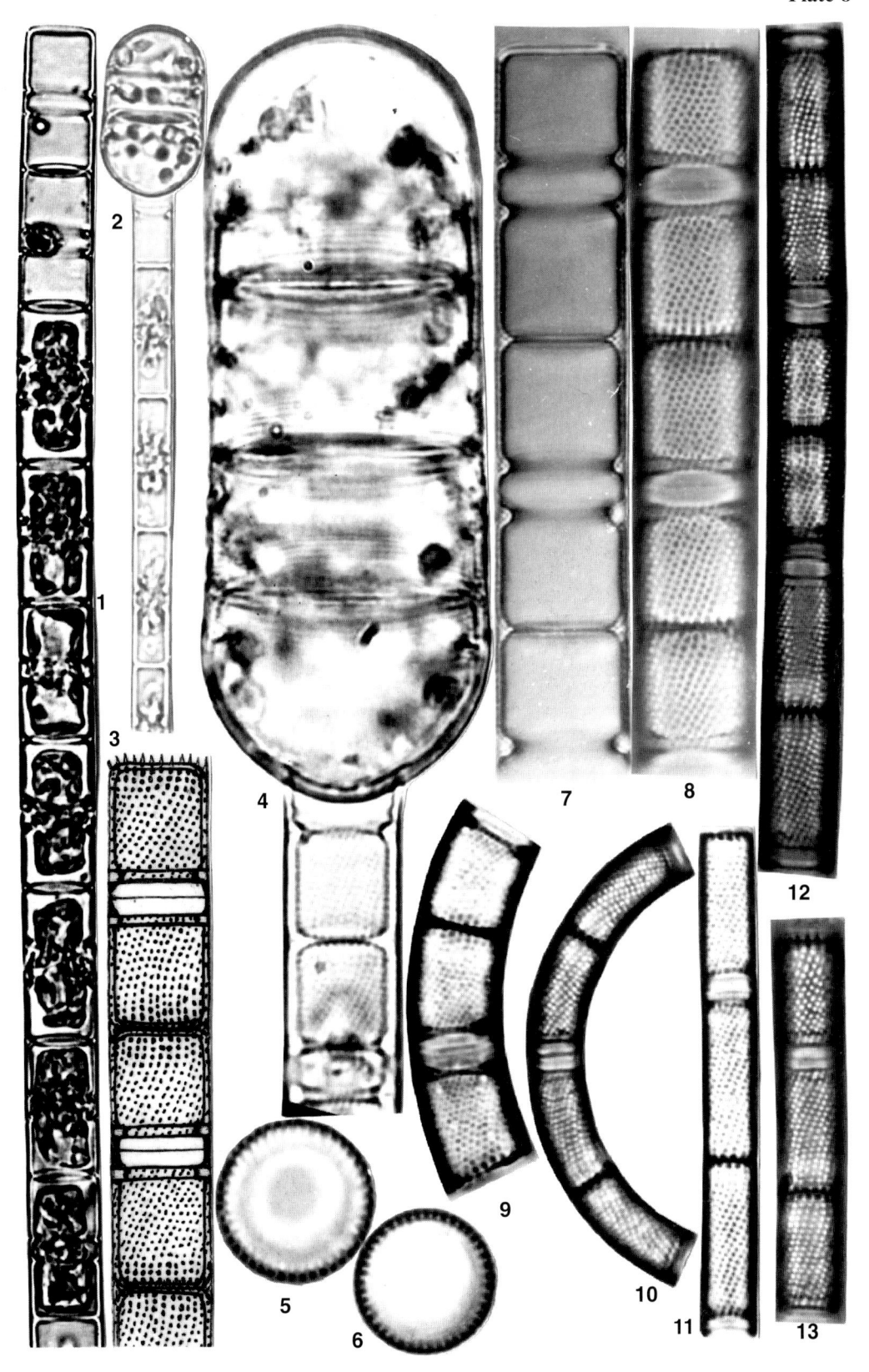

163

Plate 9 *Aulacoseira ambigua* (Grunow) Simonsen

Figs 14-19. SEM.

Fig. 14. A separation valve with acute spines. ×8000.

Fig. 15. Broken valve showing a tubular ring-costa near the mantle edge. ×8000.

Fig. 16. Internal view of a fragmented mantle edge showing the tubular ring-costa with a labiate process (arrow). ×10000.

Fig. 17. External view of the mantle edge with a outer opening of labiate process (arrow) and a valve copula. ×8000.

Fig. 18. Enlargement of the linkage between separation valves. ×10000.

Fig. 19. Enlargement of the linkage between connecting valves. ×10000.

Localities

Figs 14, 15, 17-19.   Biwa-ko (Lake Biwa), Shiga.

Fig. 16. Oze-numa (Oze Pond), Gunma.

Plate 9

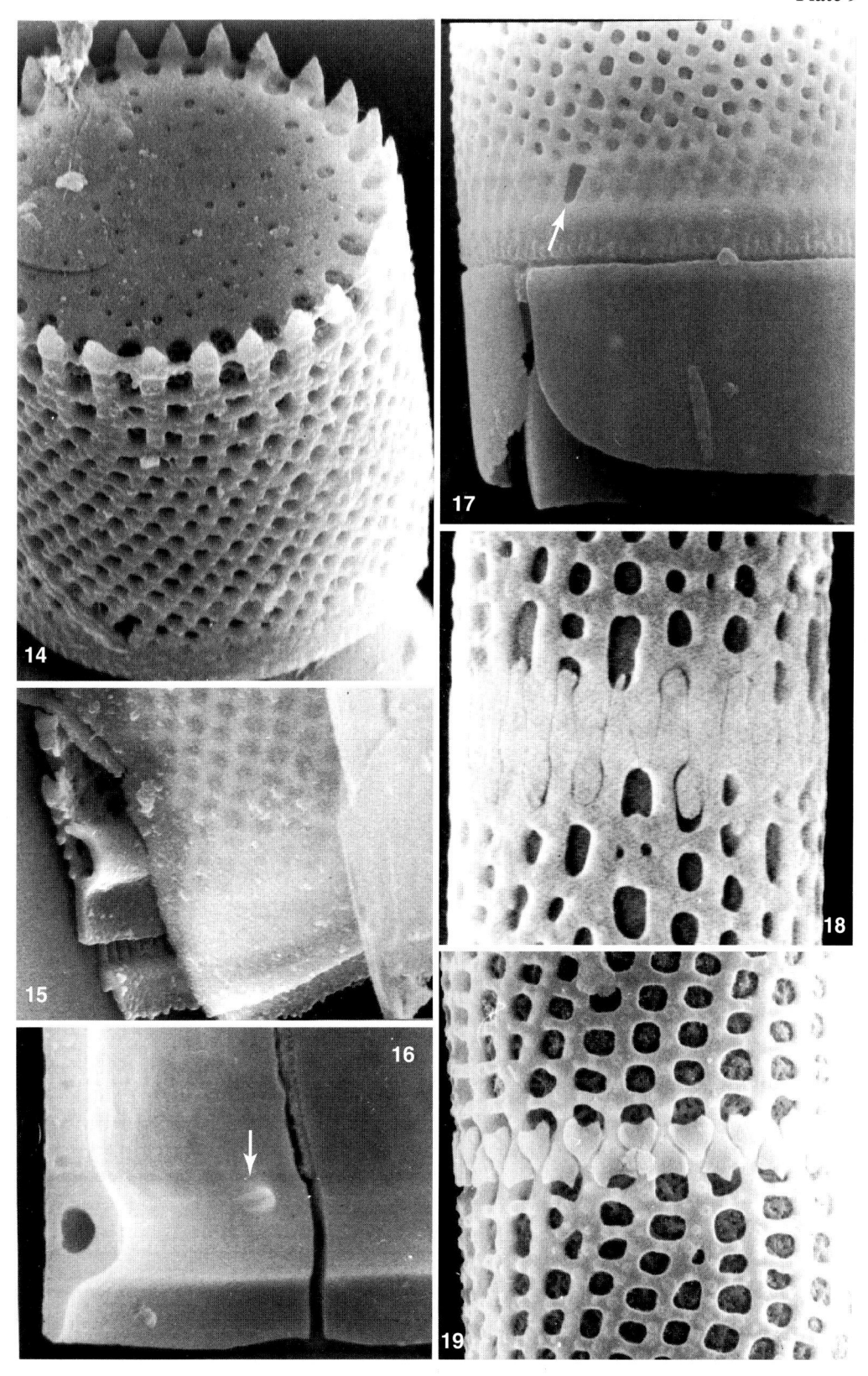

165

Plate 10 *Aulacoseira distans* (Ehrenberg) Simonsen

Figs 1-3.   Drawings. ×2000.
Figs 4, 5.   LM. ×2000.
Figs 6-9.   SEM.

Fig. 1.   Girdle view. Drawing after Van Heurck 1882. pl. 86. f. 21.
Fig. 2.   Valve face.
Fig. 3.   Ring-costa with labiate processes in valve view.
Figs 4, 5.   Topotype. Coll. Grunow 167 Polischifer von Bilin. (W).
Fig. 6.   Oblique view from valve bottom showing developed ring-costa. ×2000.
Fig. 7.   Valve interior. ×2000.
Fig. 8.   Valve face with areolae. ×4000.
Fig. 9.   Oblique view showing small linking spines and rows of pervalvar areolae in the girdle. ×4000.

Localities

Figs 4, 5.   Polischifer von Bilin.
Figs 6-9.   Noto Peninsula, Ishikawa (fossil).

Plate 10

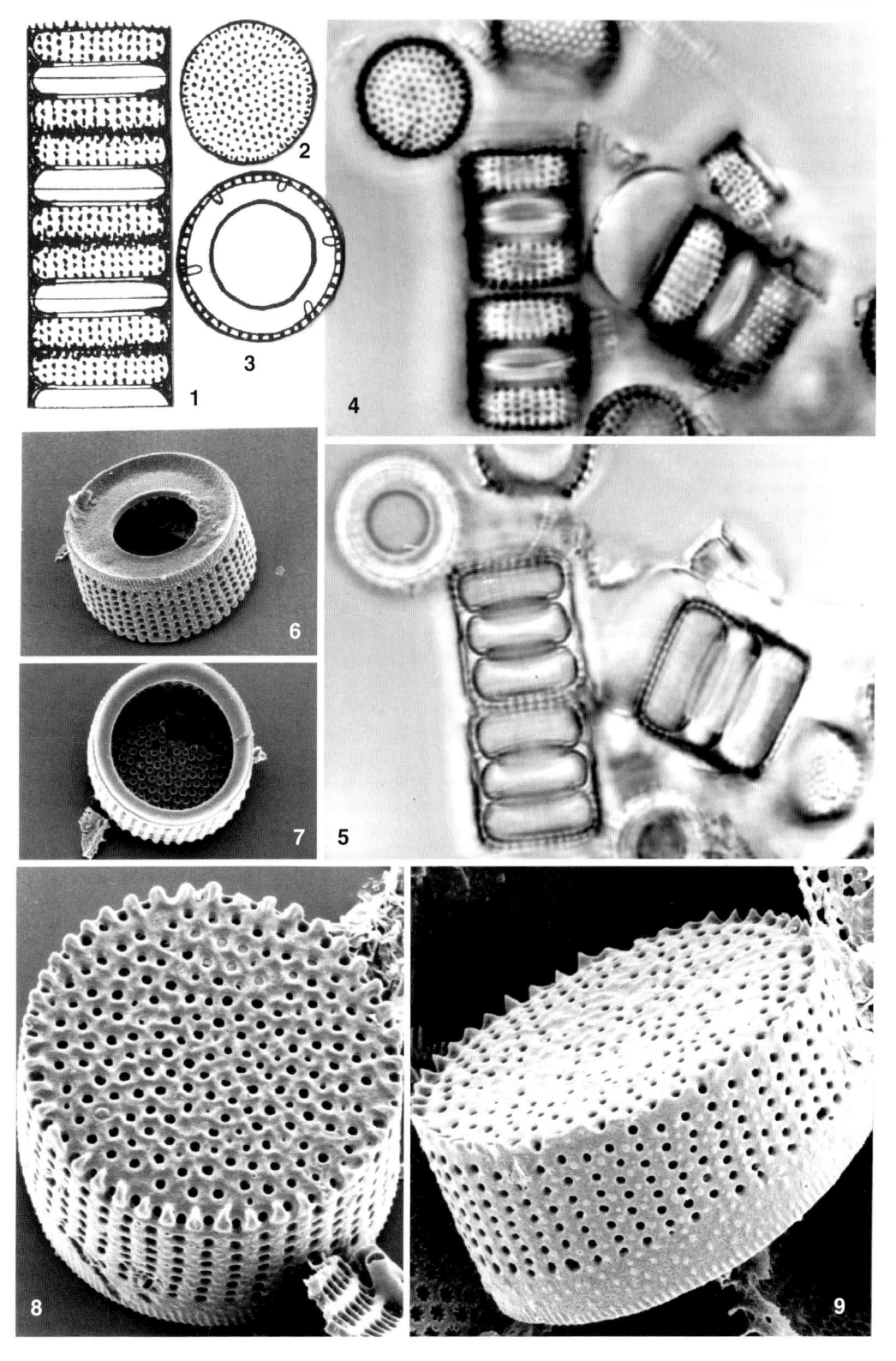

Plate 11 *Aulacoseira granulata* (Ehrenberg) Simonsen

Figs 1, 2, 4, 7.   LM. ×2000.
Figs 3, 5, 8.   Drawings. ×2000.
Fig. 6.   SEM. ×2000.

Fig. 1.   A filamentous colony embedded in a low reflectance medium, Eukit, showing an optical section.
Fig. 2.   Fixed filamentous colony with shrunk chloroplasts.
Fig. 3.   Valve view.
Figs 4, 7.   Girdle view of filamentous colony embedded in a high reflectance medium, Pleurax.
Fig. 5.   Round auxospore. After Van der Werff & Huls (1963).
Fig. 6.   The filamentous colony with a separation valve bearing long spines.
Fig. 8.   The filamentous colony composed of valves with different areolation.

Locality
Fig. 4.   Akan-ko (Lake Akan), Hokkaido.

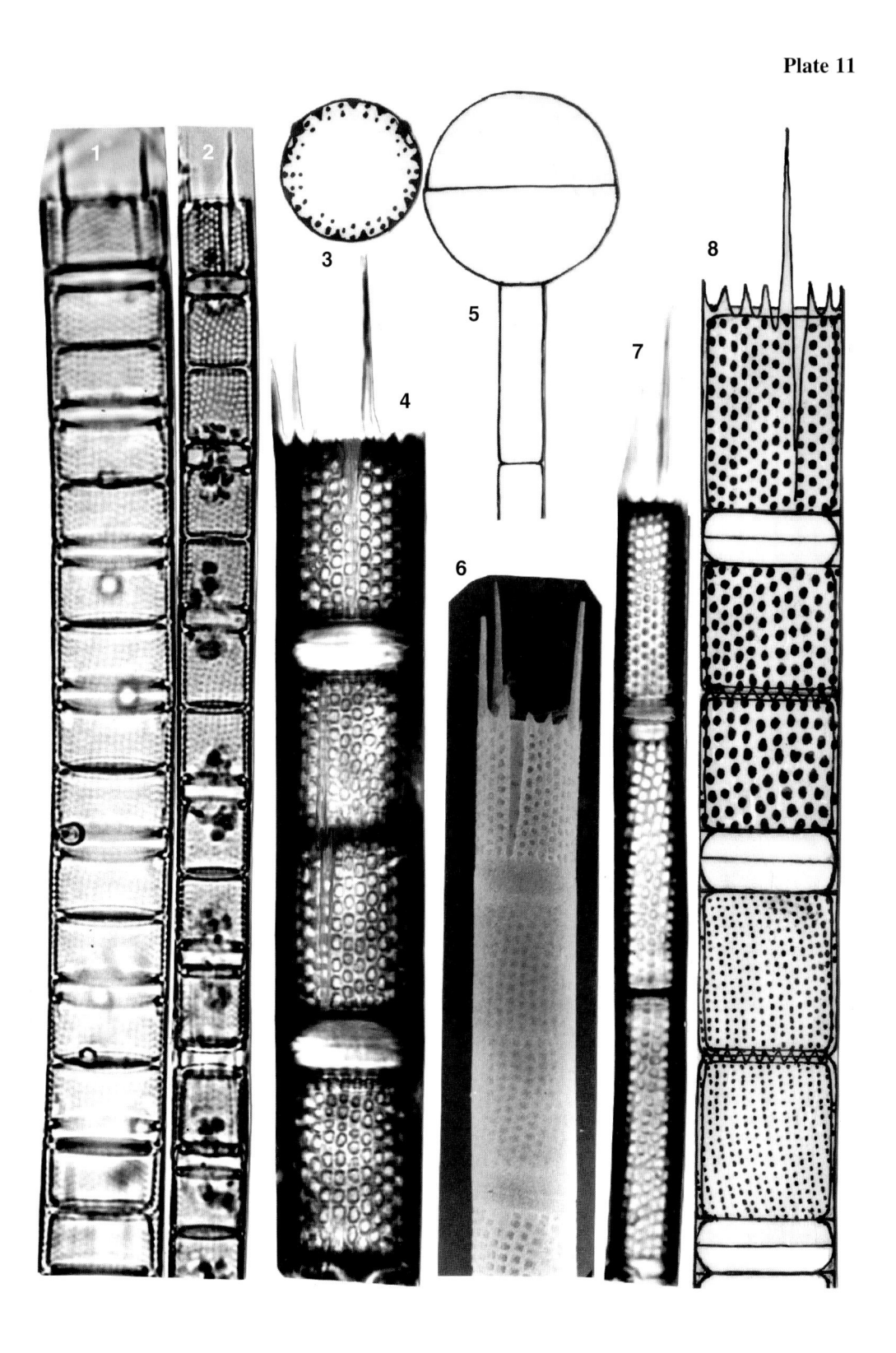

Plate 11

Plate 12 *Aulacoseira granulata* (Ehrenberg) Simonsen

Figs 9-11.   SEM.

Figs 9, 10.   The separation valve with long spines. ×5000.
Fig. 11.   Enlargement of the mantle surface. ×8000.

Locality
Figs 9-11.   Biwa-ko (Lake Biwa), Shiga.

Plate 12

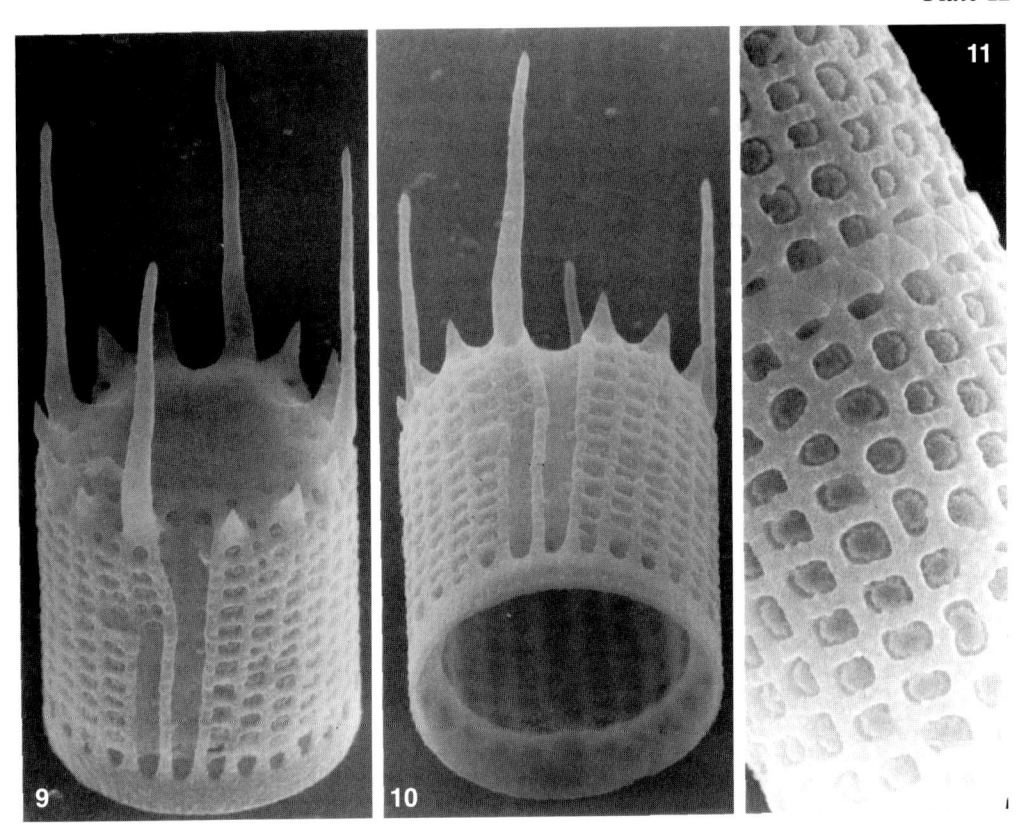

Plate 13 *Aulacoseira italica* (Ehrenberg) Simonsen

Figs 1, 3-6. LM. ×2000.
Fig. 2. Drawing. ×2000.
Figs 7, 8. SEM.

Figs 1-5. Girdle view.
Fig. 6. Valve view.
Fig. 7. Girdle view showing daughter valves connected by linking spines. The mantle edge with impression (arrow). ×2200.
Fig. 8. Internal view of the fragmented daughter valves connected by the linking spines (arrow). ×2000.

Localities
Fig. 1. BM 26775 'V.H. Type 464, Belgique.'
Fig. 2. Shikotsu-ko (Lake Shikotsu), Hokkaido.
Figs 3-8. Ryugakubo (Ryugakubo Pond), Niigata.

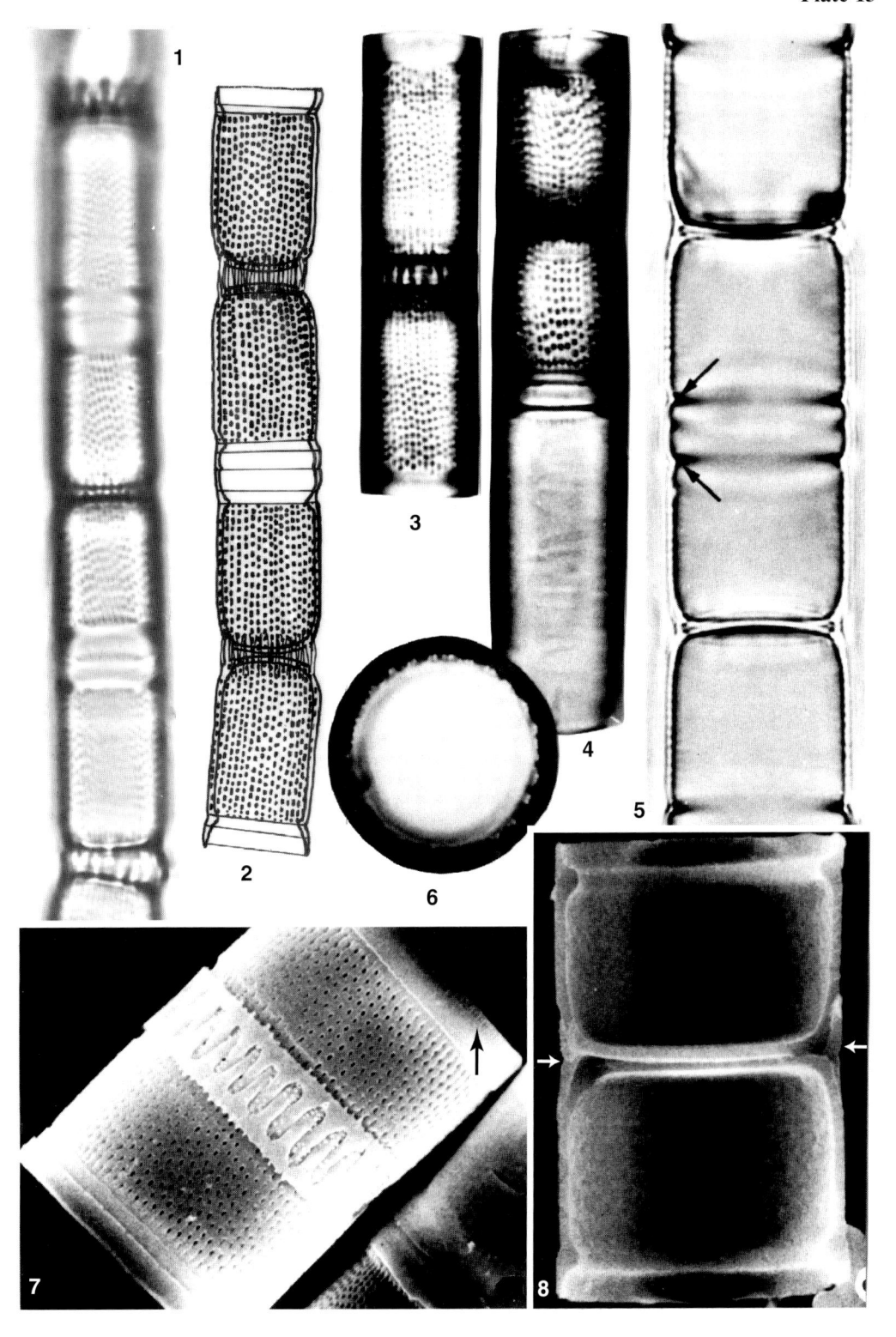

Plate 13

Plate 14 *Aulacoseira italica* (Ehrenberg) Simonsen

Figs 9, 11-14. SEM.
Fig. 10. Drawing.

Fig. 9.   Girdle view showing daughter valves connected by linking spines. ×2000.
Fig. 10. Copy traced from an SEM photograph showing frustule elements. EV: epivalve, EC: epicingulum, HV: hypovalve, VC: valvocopula, C: copulae, P: pleura.
Fig. 11. Oblique view of the valve showing the hyaline valve face, the partly broken linking spines and the areolated mantle. ×5000.
Fig. 12. Enlargement of the interlocked linking spines. Note their expanded apices and granules on the basal parts. ×8000.
Fig. 13. Detailed outer openings of areolae. Note volae visible inside. ×30000.
Fig. 14. Internal view of the fragmented mantel showing ring-costa with a labiate process (arrow).

Locality
Figs 9, 11-14.   Ryugakubo (Ryugakubo Pond), Niigata.

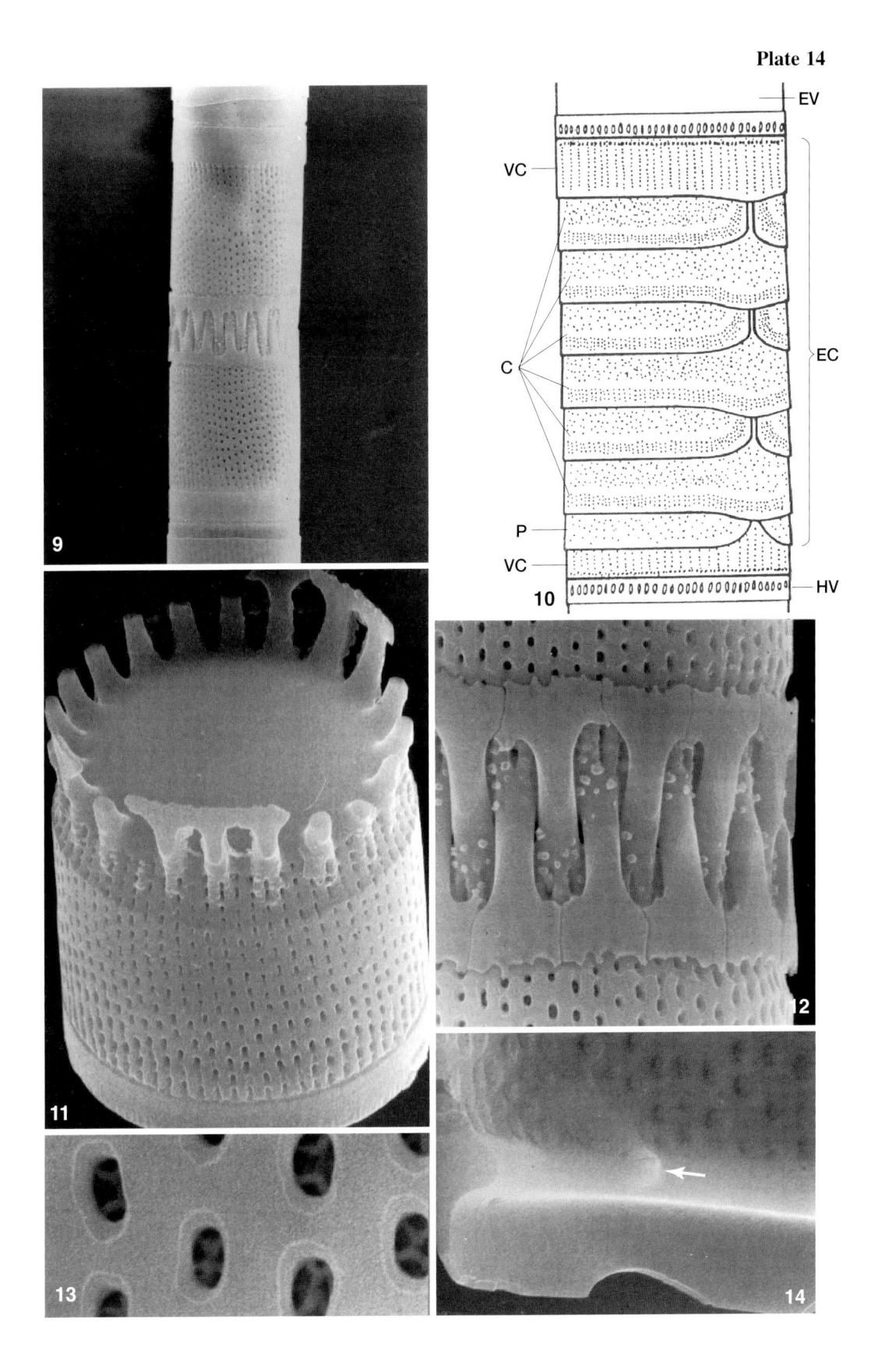

Plate 14

Plate 15 *Aulacoseira longispina* (Hustedt) Simonsen

Figs 1-5, 8-10. LM. ×2000.
Fig. 6. Drawing. ×2000.
Fig. 7. SEM. ×2000.

Figs 1-3, 5, 7, 8. Girdle view of cleaned colony.
Fig. 4. Living colony showing the discoid plastids.
Figs 9, 10. Valve views of the same specimen but at different focal planes. The valve face is shown in Fig. 9 and the ring-costa with elongated labiate processes in Fig. 10.

Localities
Fig. 1. Coll. Hustedt A2/21 (BRM), Chuzenji-See (Chuzenji-ko, Lake Chuzenji), Tochigi.
Figs 2-5, 7-10. Chuzenji-ko (Lake Chuzenji), Tochigi.

Plate 15

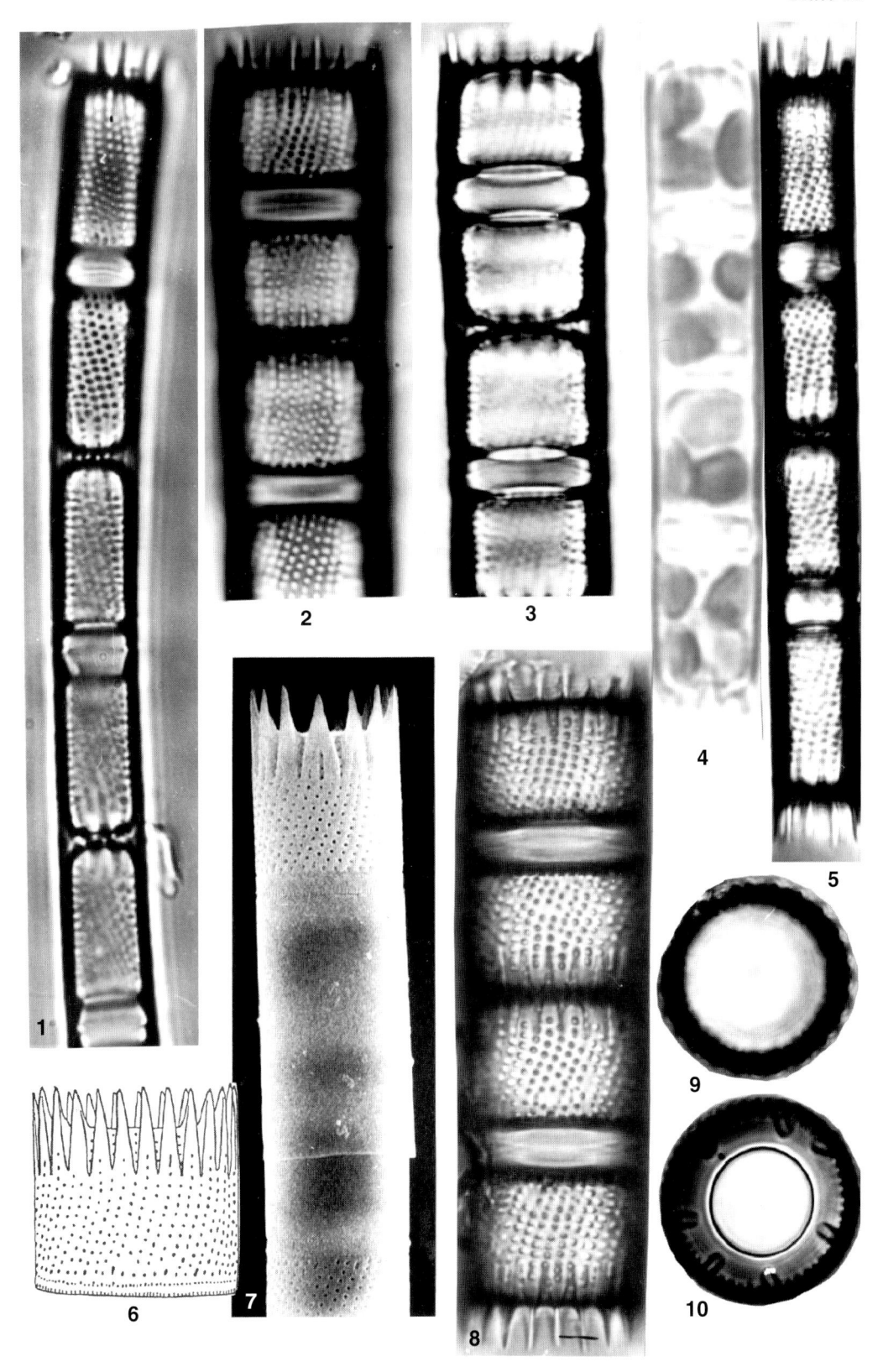

Plate 16 *Aulacoseira longispina* (Hustedt) Simonsen

Figs 11-16. SEM.

Fig. 11. External oblique view of the valve showing plain valve face, acute linking spines and spiral areolation in the girdle. ×3000.

Fig. 12. Internal oblique view of the valve showing well developed ring-costa slightly inside of the mantle edge. ×3000.

Fig. 13. Internal view of the fragmented valve showing the areolar occlusion and labiate processes (arrows) placed on the thick ring-costa. ×5000.

Fig. 14. Fragmented mantle showing the section of the labiate process (arrow) associated with a tube penetrating the mantle wall (arrowhead). ×20000.

Fig. 15. Enlarged external view of the areolae showing internal volae. ×50000.

Fig. 16. Details of the acute linking spines, which interlock together. ×8000.

Locality

Figs 11-16. Chuzenji-ko (Lake Chuzenji), Tochigi.

Plate 16

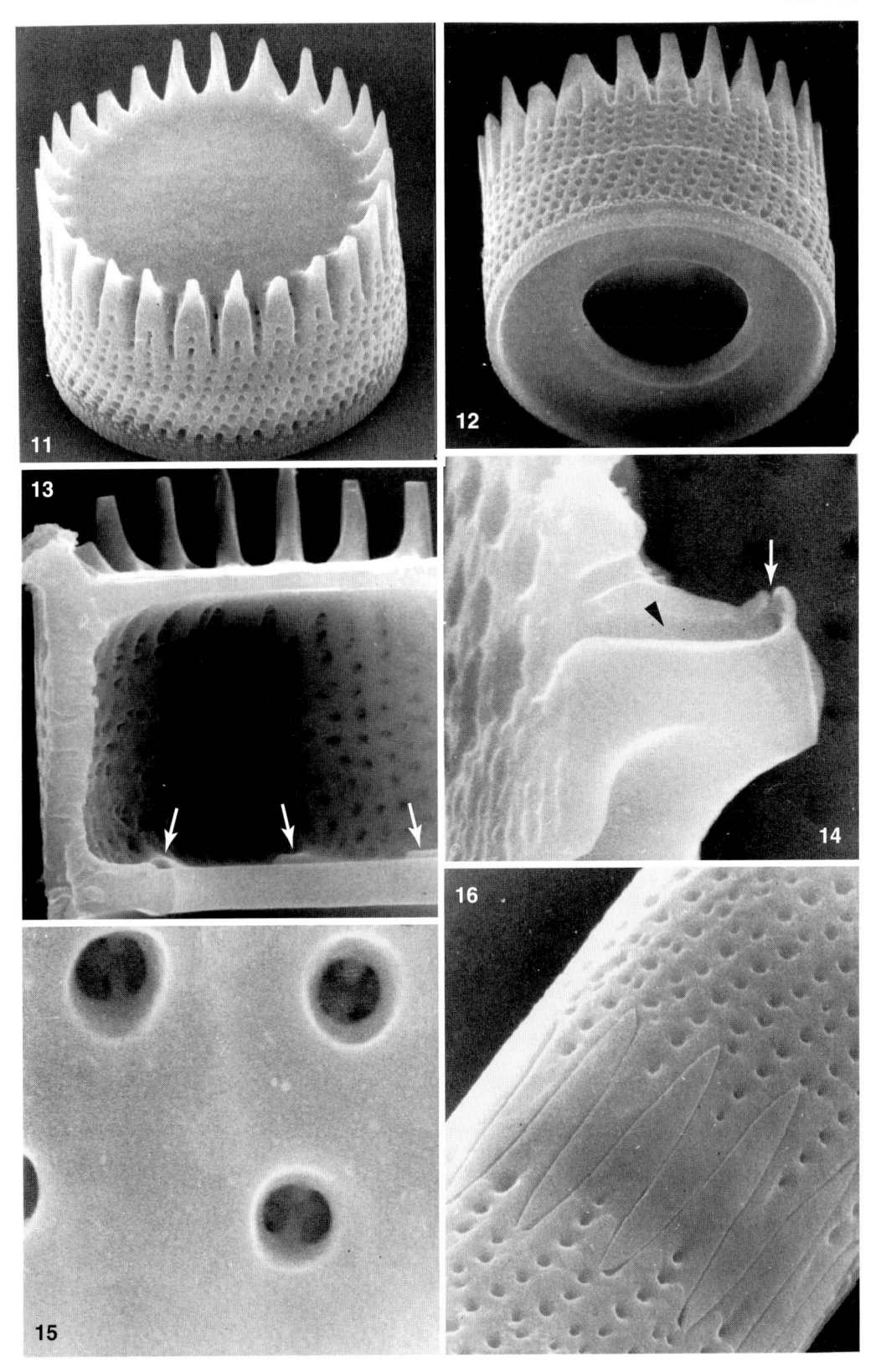

179

Plate 17 *Aulacoseira nipponica* (Skvortsov) Tuji

Figs 1, 6, 7. Drawings. ×2000.
Figs 2-5, 8-10. LM. ×2000.
Figs 11, 12. SEM.

Figs 1-5, 10. Girdle view.
Figs 6-9. Valve view.
Fig. 8. Image focused on a ring-costa.
Fig. 9. Image focused on the valve face.
Fig. 11. Whole view of a frustule connected to a single valve. ×3000.
Fig. 12. Internal view of fragmented mantle showing a ring-costa, on which four labiate processes, partly broken, are visible (arrows). ×8000.

Locality
Unindentified.

Plate 17

Plate 18 *Aulacoseira nipponica* (Skvortsov) Tuji

Figs 13, 14.　SEM.

Fig. 13.　External oblique view of a whole valve showing plane valve face, partly broken linking spines and mantle areolation. ×5000.

Fig. 14.　Internal view of the fragmented sibling valves showing the volate occlusion of the areolae, and ring-costa on which labiate processes extend (arrows). Note the ring-costa thicker than the valve face. ×5000.

Locality
Unindentified.

Plate 18

Plate 19 *Aulacoseira subarctica* (O. Müller) Haworth

Figs 1, 2, 4-14.   LM. ×2000.
Fig. 3.   Drawing from SEM photograph.
Figs 15, 16.   SEM.

Figs 1-3, 5-7, 10-14.   Girdle view of the filamentous colony. Figs 2, 7, 11 are the same specimen as Figs 1, 6, 10, respectively, but taken at different focal planes.
Figs 4, 8.   Valve face.
Fig. 9.   Ring-costa from at the same specimen as in Fig. 8. Note the inconspicuous labiate processes (arrows).
Figs 13, 14.   Girdle view of the post-initial frustules.
Fig. 15.   External oblique view of the theca showing the valve face with scattered areolae, linking spines, girdle areolation, and some bands. ×5000.
Fig. 16.   Details of the acute linking spines, which interlock the sibling valves. ×10000.

Localities
Figs 1, 2, 5-12, 15.   Akan-ko (Lake Akan), Hokkaido.
Figs 4, 16.   Panke-to (Panke Pond), Hokkaido.

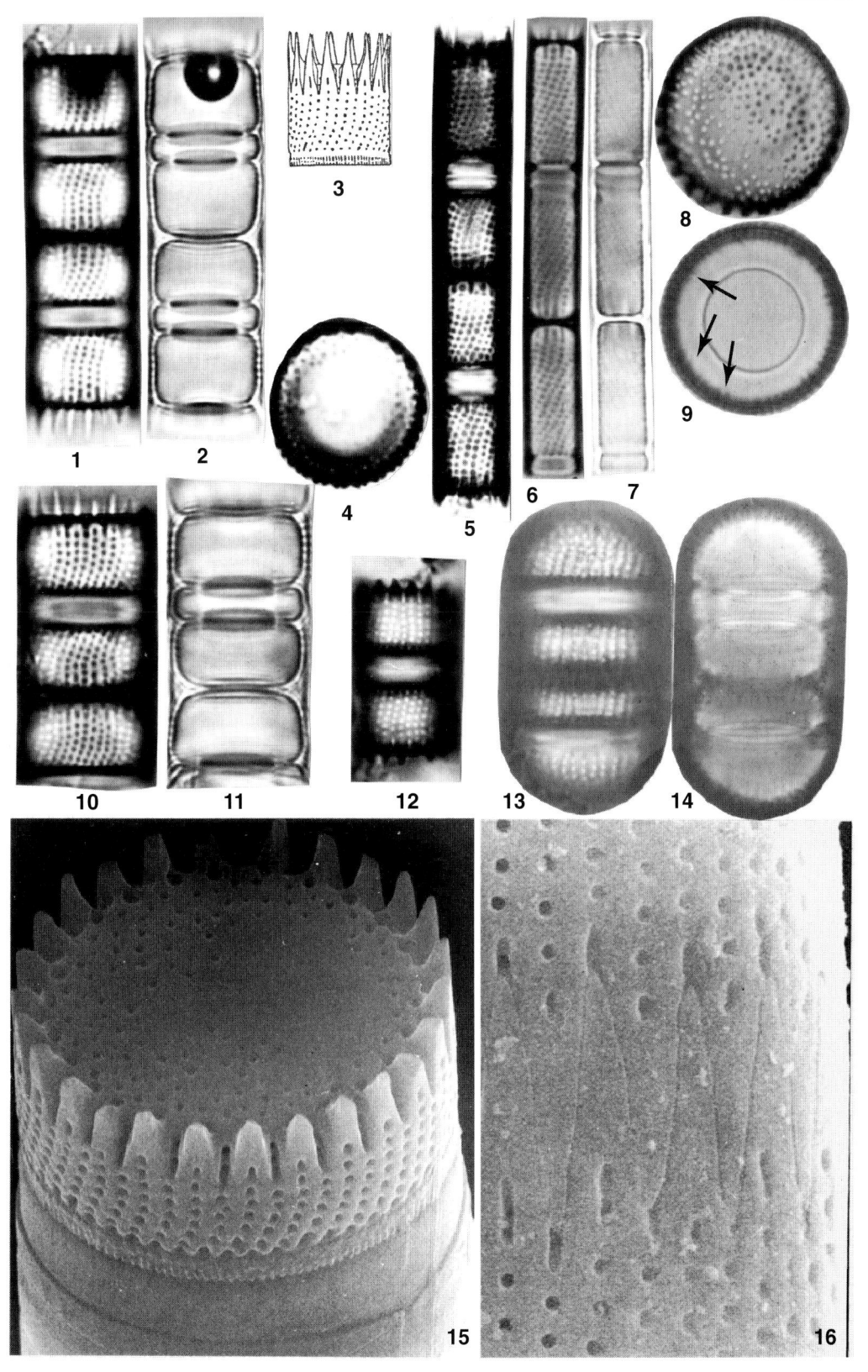

Plate 19

Plate 20 *Aulacoseira subarctica* (O. Müller) Haworth

Figs 17-21. SEM.

Fig. 17. Internal oblique view of the fragmented valve. ×8000.

Fig. 18. Internal fragmented valve showing areolae occluded by volae. Compare the thickness of the valve face and the ring-costa on which two labiate processes (arrows) are located. ×10000.

Fig. 19. Enlargement of the labiate process (arrow) located at the bottom of Fig. 17. ×20000.

Fig. 20. External oblique view of a whole initial frustule showing the areolated semispherical valve and epibands. ×3000.

Fig. 21. Oblique view of the initial theca associated with the post-initial valve, which shows bottom view of the ring-costa. ×3000.

Locality

Figs 17-21. Akan-ko (Lake Akan), Hokkaido.

Plate 20

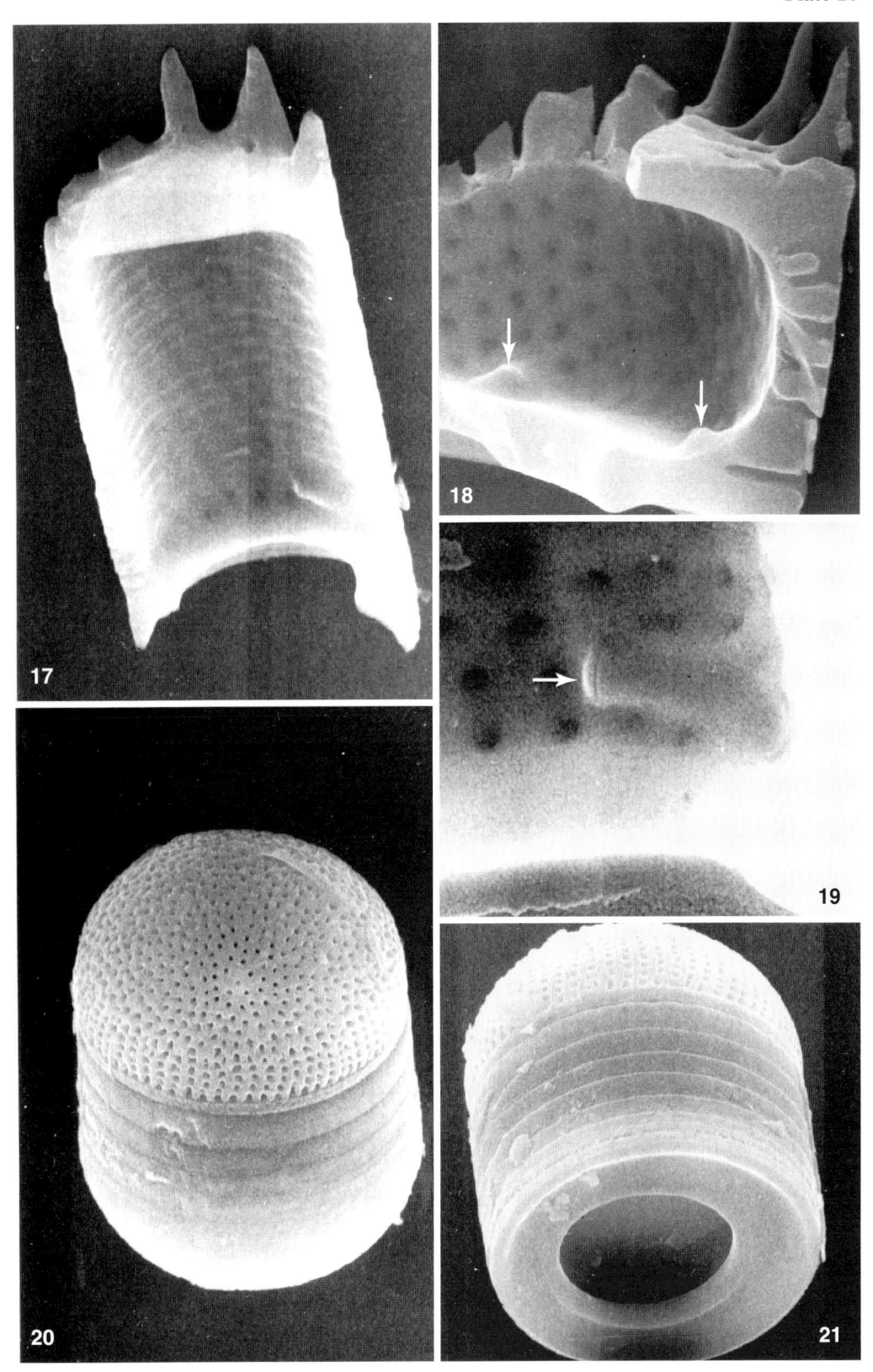

Plate 21 *Aulacoseira tenuis* (Hustedt) H. Kobayasi

Figs 1-8. LM. ×2000.
Figs 9, 10. Drawings. ×2000.

Figs 1-4. Girdle view of cleaned colony.
Fig. 5. Fixed colony showing somewhat shrunken plastids.
Figs 6-8. Valve view.
Fig. 6. Image focused on the valve face.
Fig. 7. Image focused on the labiate process.
Fig. 8. Image focused on the ring-costa.
Figs 9, 10. Valves with acute linking spines.

Localities
Figs 1-3. Coll. Hustedt A2/22. (BRM), Juno-See (=Yuno-ko (Yu Pond)), Tochigi.
Figs 4-10. Yuno-ko (Yu Pond), Tochigi.

Plate 21

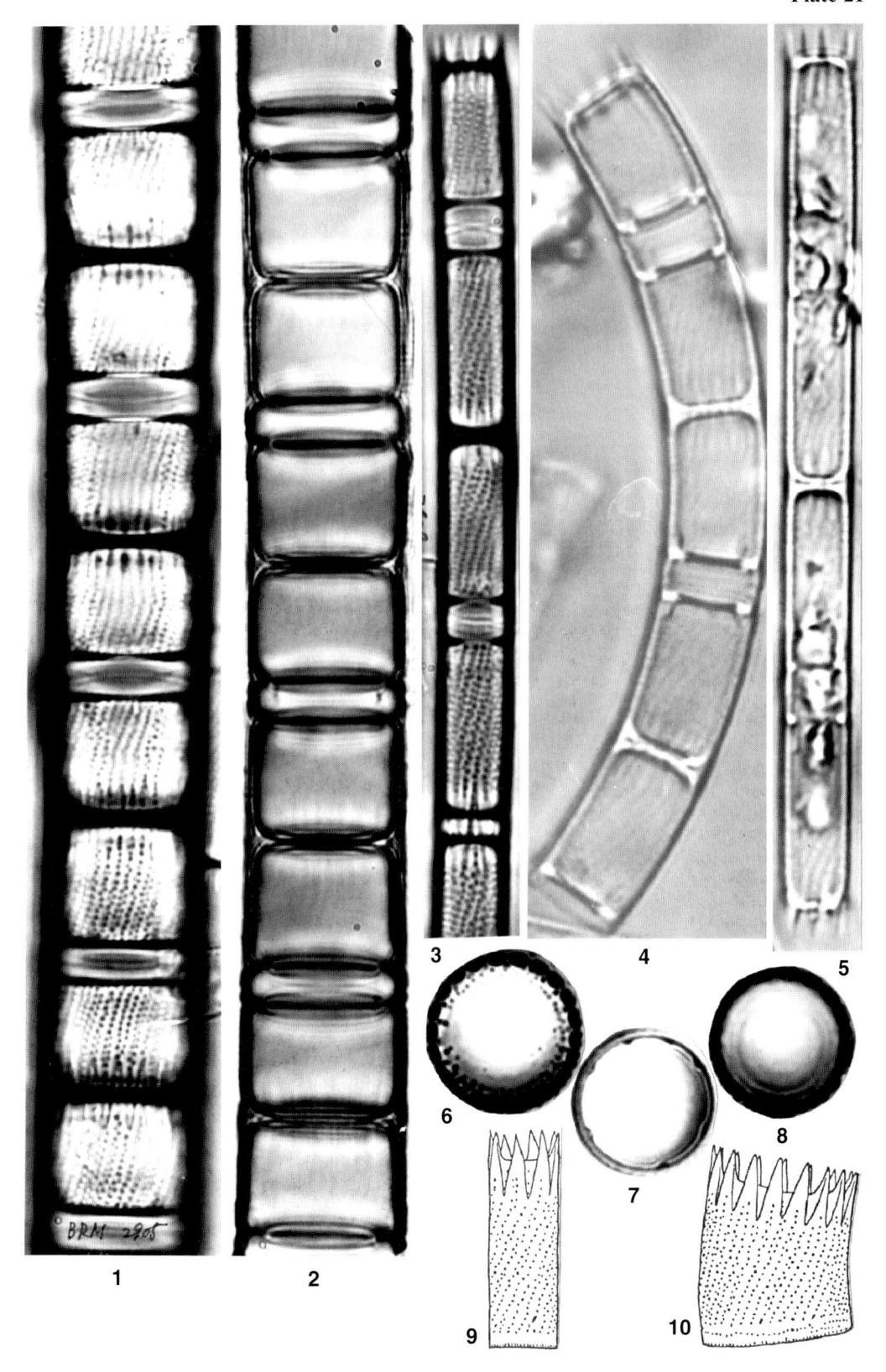

189

Plate 22 *Aulacoseira tenuis* (Hustedt) H. Kobayasi

Figs 11-16.　SEM.

Figs 11, 12.　External view of the valves, which form a curved filamentous colony. Each series of spiral areolation consists of a single and/or double row of areolae. ×5000.

Fig. 13.　External view of the valves, which form a straight filamentous colony. Note the mixture of single and double rows of areolae. Outer opening of the labiate process is visible in the left specimen (arrow). ×5000.

Fig. 14.　Internal view of a fragmented valve showing the labiate process with a helicoid tube on the mantle slightly above the ring-costa. ×5000.

Fig. 15.　Enlarged fragment of the mantle showing areolae occluded internally by volae. ×30000.

Fig. 16.　Details of the labiate process with the helicoid tube and the volae around it. ×30000.

Locality

Figs 11-16.　Yuno-ko (Yu Pond), Tochigi.

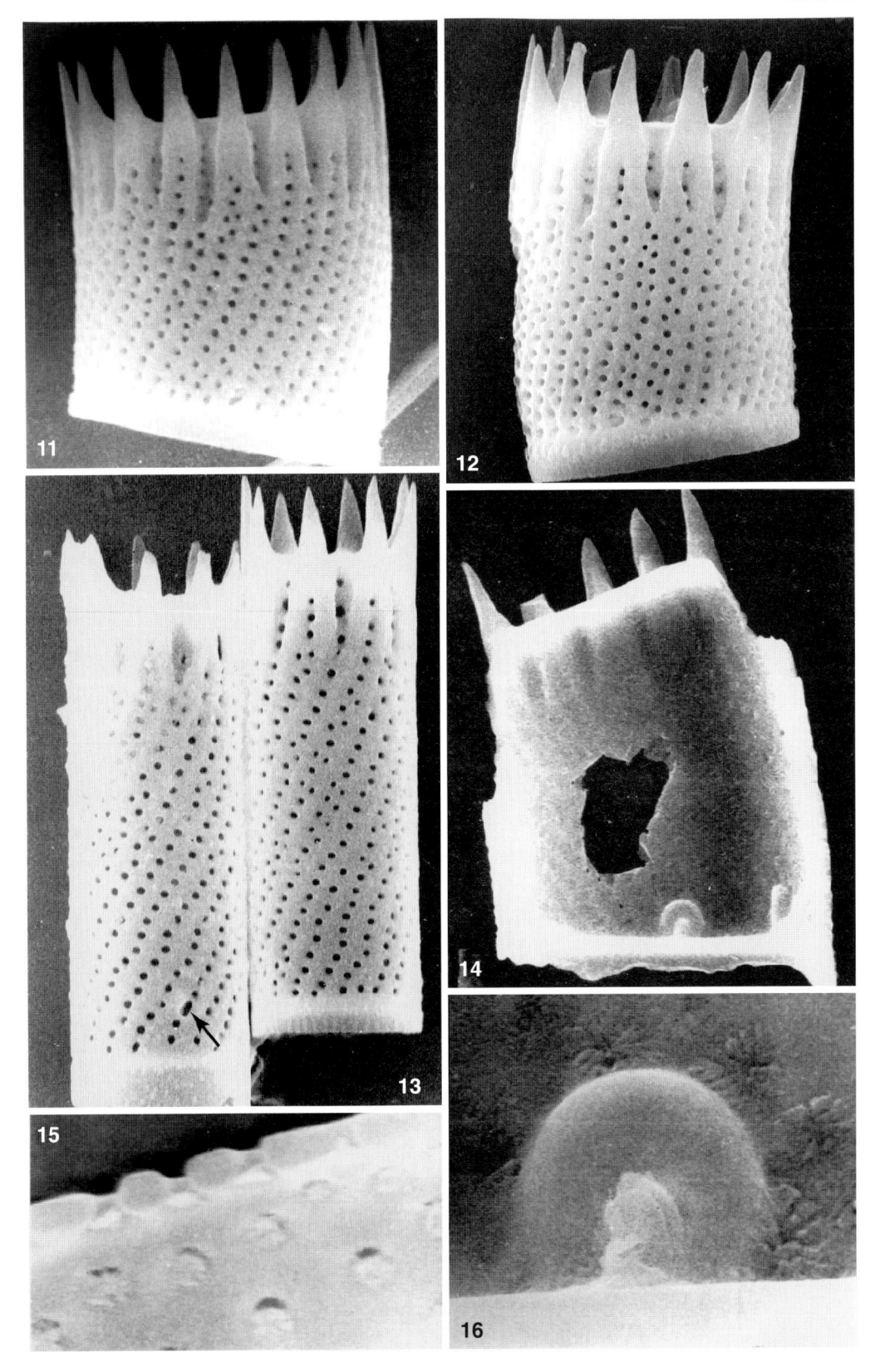

Plate 22

Plate 23　*Aulacoseira valida* (Grunow) Krammer

Figs 1-8, 11, 12.　LM.　×2000.
Figs 9, 10.　Drawings.　×2000.

Figs 1-9.　Girdle view.
Figs 1-3.　Coll. Grunow 2827 Lac de Gerardmer, Vosges (W).
Figs 4, 5.　Coll. Hustedt A1/69 '*M. italica* var. *valida* Grunow. R.S. A69. Norwegen Feforvatn. 1930' (BRM).
Figs 10-12.　Valve view.

Localities

Figs 1-3.　Lac de Gerardmer, Vosges.
Figs 4, 5.　Norwegen Feforvatn.
Fig. 6.　A pond in Ohzutsumi, Fukushima.
Fig. 7.　Oze-numa (Oze Pond), Gunma.
Fig. 8.　A pond in Musashi-kyuryo Shinrin Park, Saitama.
Figs 11, 12.　Shimosueyoshi Formation, Kanagawa (fossil).

Plate 23

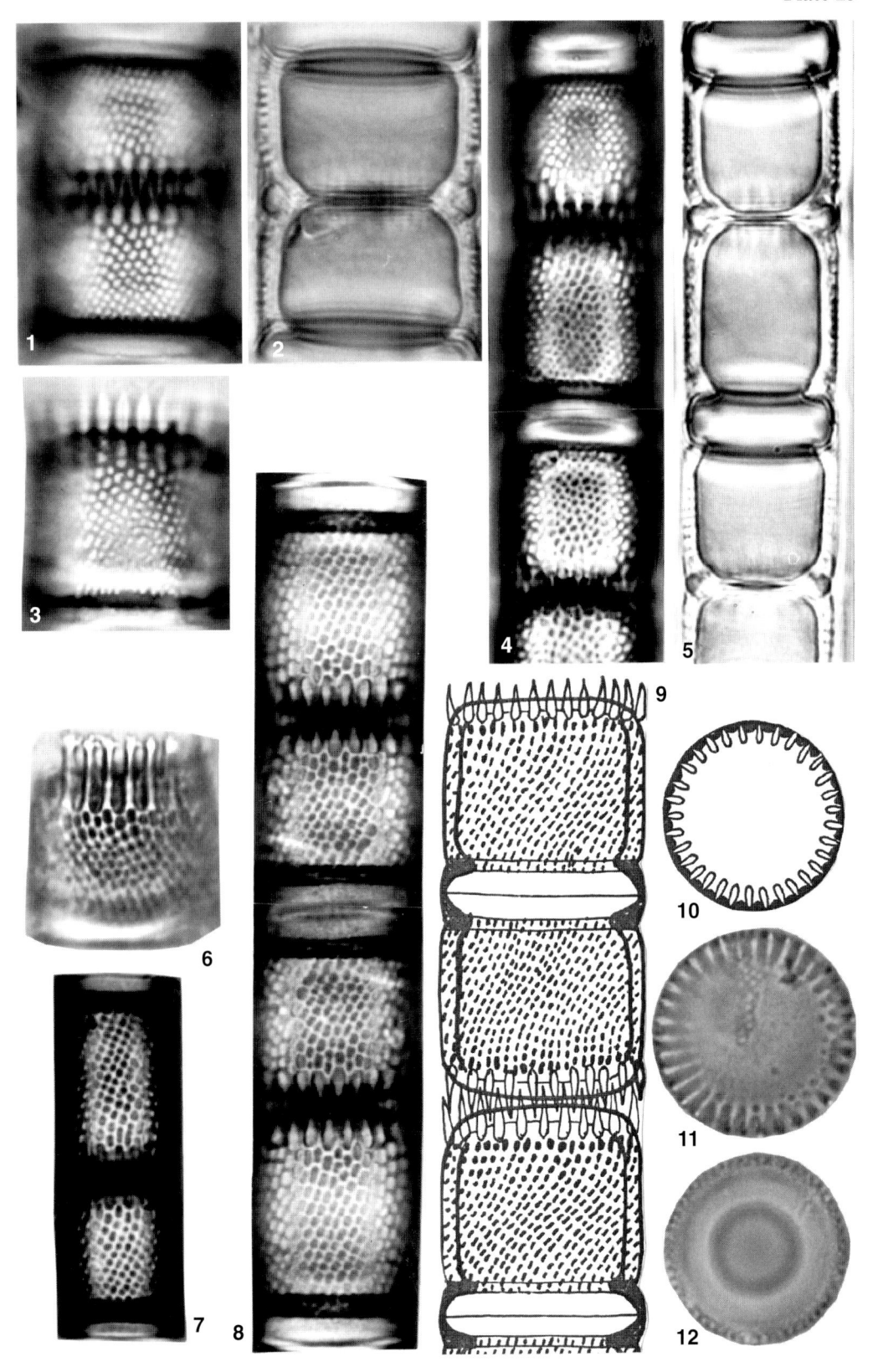

193

Plate 24 *Aulacoseira valida* (Grunow) Krammer

Figs 13-18. SEM.

Fig. 13. Oblique view of the valve showing a ring of centrifugally expanded areolae, partly broken linking spines and spiral areolation in the mantle. ×5000.

Fig. 14. Oblique view of connected daughter valves showing a ring-costa. ×2500.

Fig. 15. Detailed linking spines interlocked together showing their expanded apices. ×10000.

Fig. 16. Internal view of a fragmented daughter cell. ×2000.

Fig. 17. Enlargement of the fragmented internal mantle edge. Note the wall consisting of two structurally different spongy layers. ×20000.

Fig. 18. Details of the bottom mantle edge of the same specimen in Fig. 16 showing a labiate process (arrow) on the ring-costa. ×30000.

Localities

Figs 13, 16-18. A pond in Ohzutsumi, Fukushima.

Figs 14, 15. Oze-numa (Oze Pond), Gunma.

Plate 24

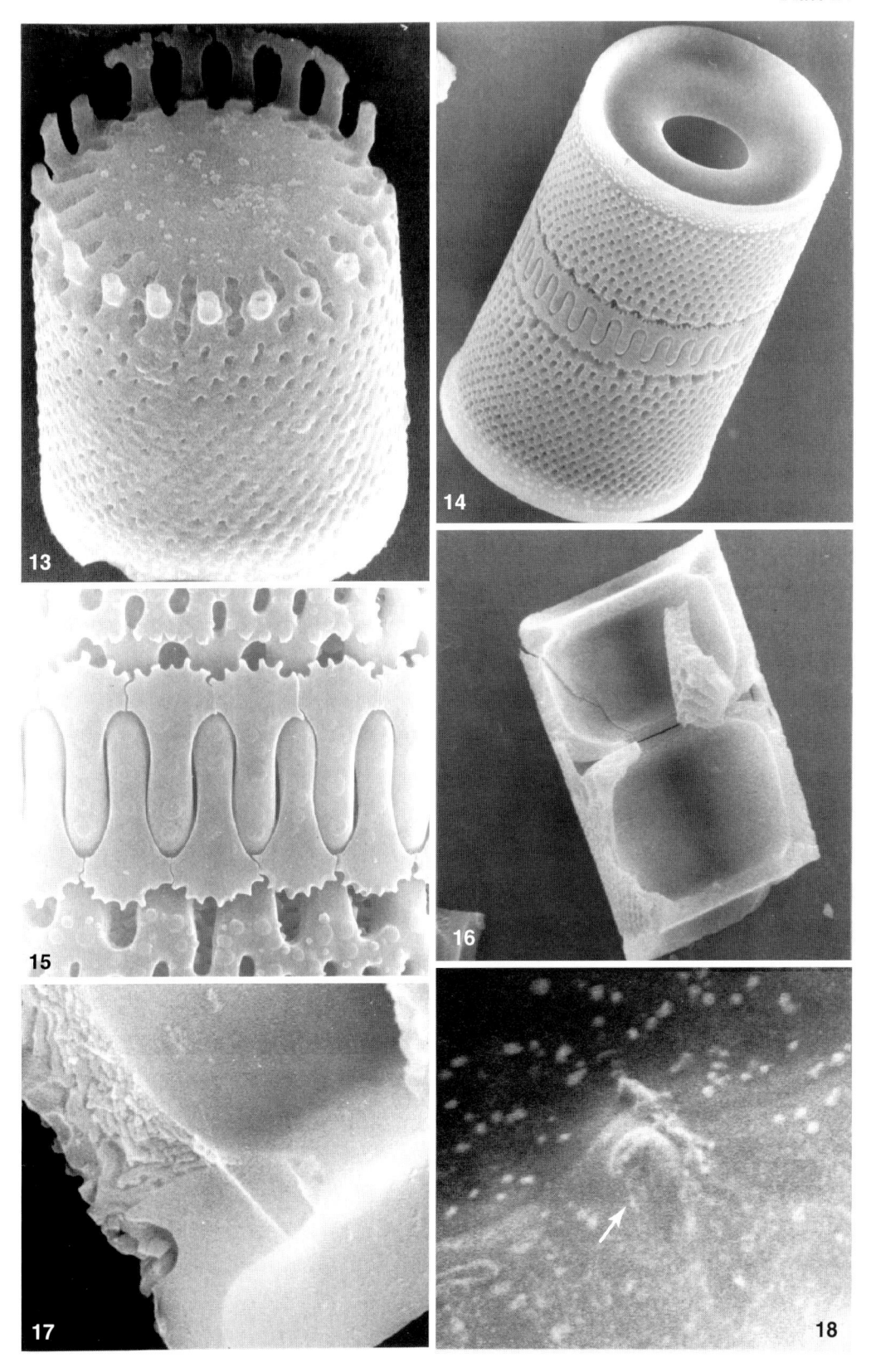

13

14

15

16

17

18

195

Plate 25 *Orthoseira asiatica* (Skvortsov) H. Kobayasi

Figs 1, 3.   Drawings. ×2000.
Figs 2, 4, 6.   LM. ×2000.
Figs 5, 7-9.   SEM.

Fig. 5.   External oblique view of a whole valve showing three "carinoportulae" in the central area of the valve face and well developed triangular plate-like spines on the valve shoulder, i.e. the valve face and girdle juncture. ×1900.

Fig. 7.   External oblique view of a whole frustule, which has an epivalve with a small central area as seen in Fig. 6. ×1800.

Fig. 8.   Frustule showing a hypovalve with four "carinoportulae" in the center. ×1800.

Fig. 9.   External oblique view of the valve with a large central area as seen in Fig. 4. ×2000.

Locality

Figs 1, 3-9.   On moss at Hashidate Limestone Cave, Saitama.

Plate 25

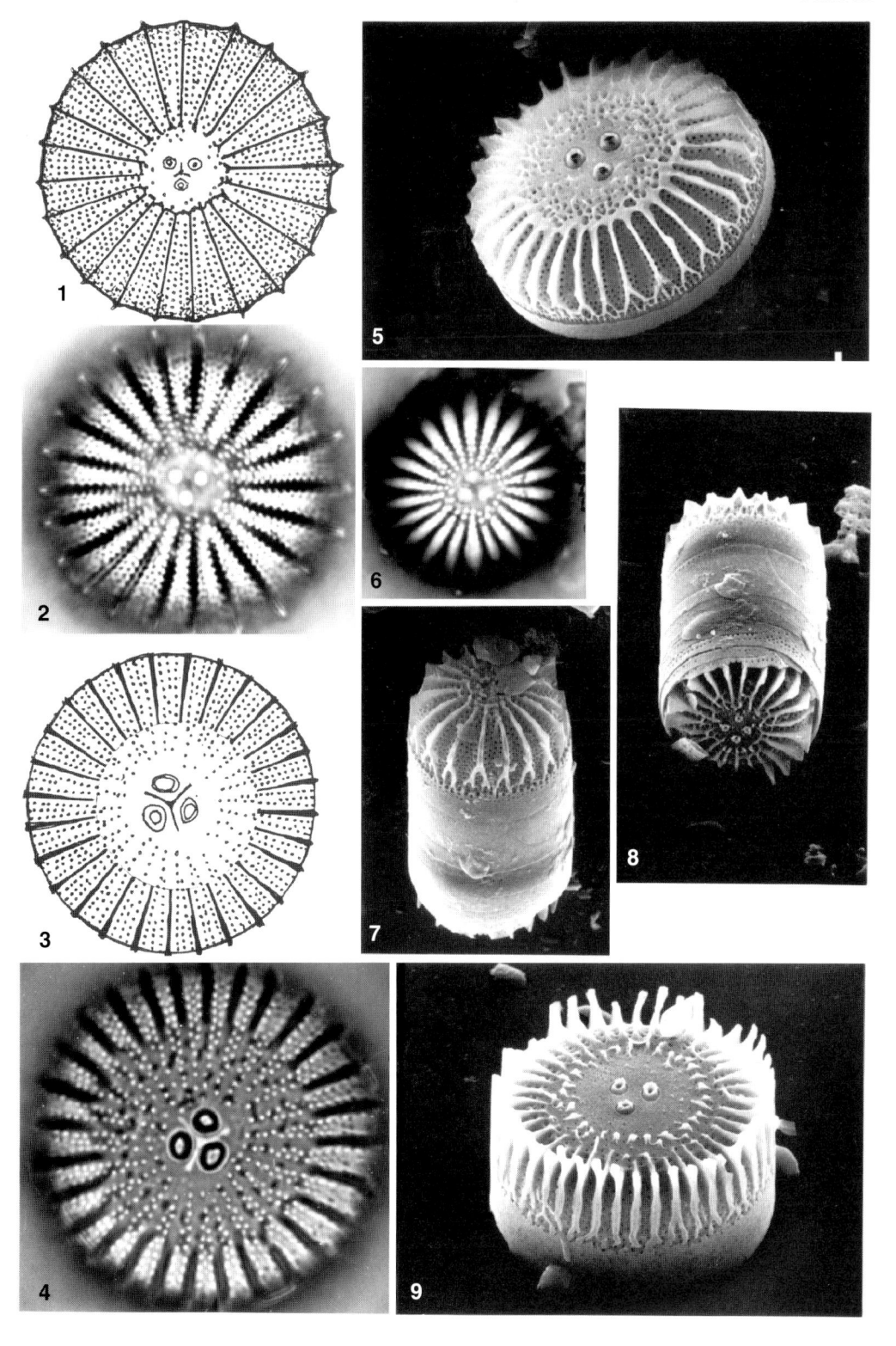

Plate 26 *Orthoseira epidendron* (Ehrenberg) H. Kobayasi

Fig. 1.   Drawing. ×2000.
Figs 2-4.   LM. ×2000.
Figs 5-8.   SEM.

Figs 1, 3, 4.   Valve view. Specimens with three "carinoportulae."
Fig. 2.   Valve view. Specimen with two "carinoportulae."
Fig. 5.   External view of the theca showing the corrugated valve face and mantle, and the triangular plate-like spines on the valve shoulder, i.e. the valve face and mantle juncture. There are two "carinoportulae" at the valve center. ×3000.
Fig. 6.   Internal view of the valve showing inner openings of the "carinoportulae" and the areolae without volae due to erosion. ×5000.
Fig. 7.   External view of a whole frustule. ×2000.
Fig. 8.   Enlarged oblique view of the valve face and mantle. Outer opening of each areola is a simple pore. The base of the triangular plate-like spine is often bifurcated. ×10000.

Locality
Figs 2-8.   On moss at Hashidate Limestone Cave, Saitama.

Plate 26

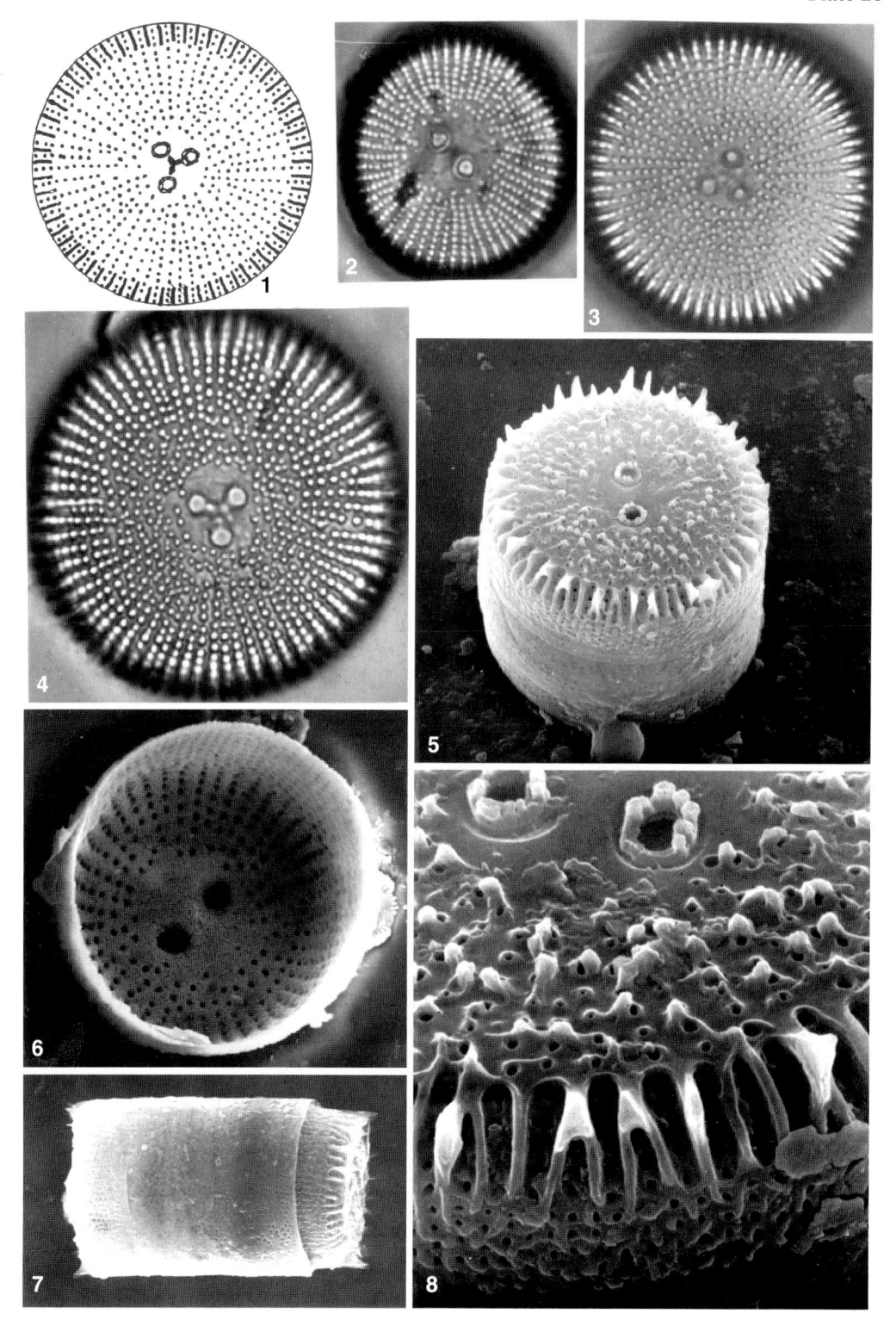

Plate 27 *Brebisira arentii* (Kolbe) Krammer

Fig. 1.   Drawing. ×2000.

Figs 2-5.   LM. ×2000.

Figs 6, 7.   SEM.

Figs 1-3.   Valve view.

Figs 4, 5.   Girdle view of two frustules with difference focuses.

Fig. 6.   External view of valve showing numerous granules on valve face and spines at the valve mantle. ×5000.

Fig. 7.   Internal view of valve showing striation and marginal spines. ×5000.

Localities

Figs 2, 4-7.   Omine-numa (Omine Pond), Gunma.

Fig. 3.   Yashimaga-ike (Yashima Pond), Nagano.

Plate 27

Plate 28 *Thalassiosira allenii* Takano

Figs 1-4.   LM. ×2000.
Figs 5, 6.   Drawings. ×2000.
Figs 7, 8.   SEM.

Figs 1-5.   Valve view.
Fig. 6.   Girdle view.
Figs 7, 8.   External view of frustule showing outer opening of central strutted process, marginal strutted processes and outer tube of marginal labiate process (arrow). ×5000.

Locarities
Figs 1-4.   Estuary of Tama-gawa (Tama River), Tokyo.
Figs 7, 8.   Osaka-wan (Osaka Bay), Osaka.

Plate 28

Plate 29 *Thalassiosira eccentrica* (Ehrenberg) Cleve

Fig. 1. Drawing. ×2000.

Fig. 2. LM. ×2000.

Figs 3-6. SEM.

Figs 1, 2. Valve view.

Fig. 3. External view of frustule showing marginal spines and outer tube of labiate process (arrow). ×2000.

Fig. 4. Detail of valve face showing outer openings of alveoli and central strutted processes (arrows). ×10000.

Fig. 5. External view of a valve showing marginal spines and outer tube of labiate process (arrow). ×2000.

Fig. 6. Detail of valve internally showing openings of strutted processes with four struts (arrows). ×10000.

Locality

Figs 2-6. Estuary of Tama-gawa (Tama River), Tokyo.

Plate 29

Plate 30 *Thalassiosira faurii* (Gasse) Hasle

Fig. 1. Drawing. ×2000.
Figs 2-4. LM. ×2000.
Fig. 5. TEM. ×2000.
Figs 6-8. SEM.

Figs 1-5. Valve view. Arrows showing central strutted processes.
Fig. 6. External view of whole frustule showing flat valve face and outer tubes of marginal strutted processes. ×3000.
Fig. 7. Internal view of valve showing grouped central strutted processes. ×5000.
Fig. 8. Details of valve margin internally showing internal openings of marginal strutted processes with four struts and labiate processes. ×10000.

Localities
Fig. 1. Tama-gawa (Tama River), Tokyo.
Figs 2-8. Gake-gawa (Gake River), Adachi, Tokyo.

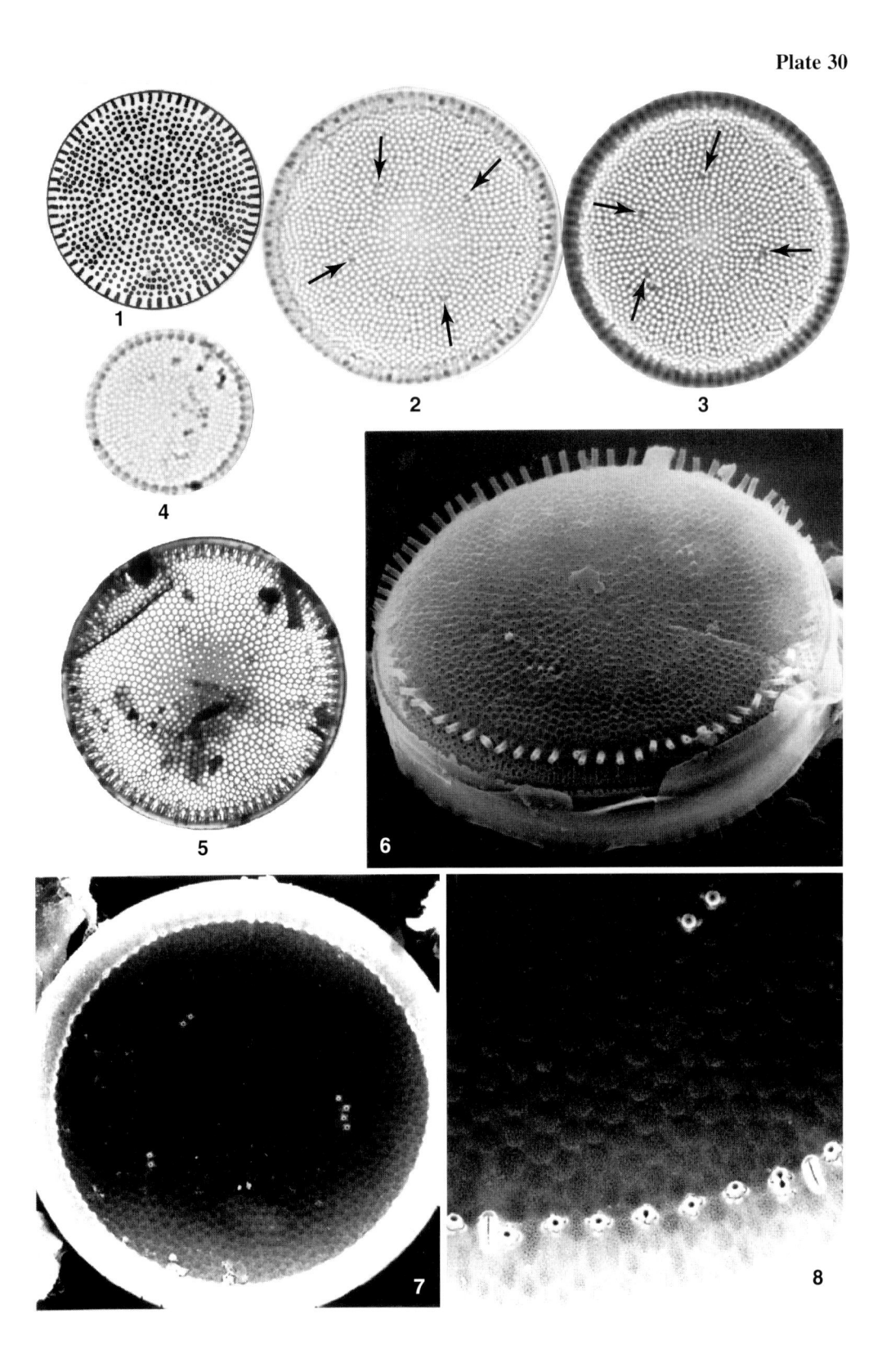

Plate 30

Plate 31 *Thalassiosira guillardii* Hasle

Fig. 1.   Drawing. ×2000.
Figs 2-4.   LM. ×2000.
Figs 5-8, 11.   SEM.
Figs 9, 10.   TEM.

Figs 1-6.   Valve view.
Fig. 7.   External view of valve showing outer tube of labiate process (arrow) and outer tube of marginal strutted processes. ×5000.
Fig. 8.   Internal view of valve showing marginal strutted processes and labiate process. ×8000.
Fig. 9.   Whole valve showing marginal strutted processes and labiate prosses (arrow). ×8000.
Fig. 10.   Detail of valve margin showing marginal strutted processes, labiate process and central strutted process (arrow). ×12000.
Fig. 11.   Detail of valve margin internally showing marginal strutted processes and labiate process. ×20000.

Localities
Figs 2-4.   Hi-numa (Lake Hi), Ibaraki.
Figs 5-11.   Matsumoto-jo Hori (The moat of Matsumoto Castle), Nagano.

Plate 31

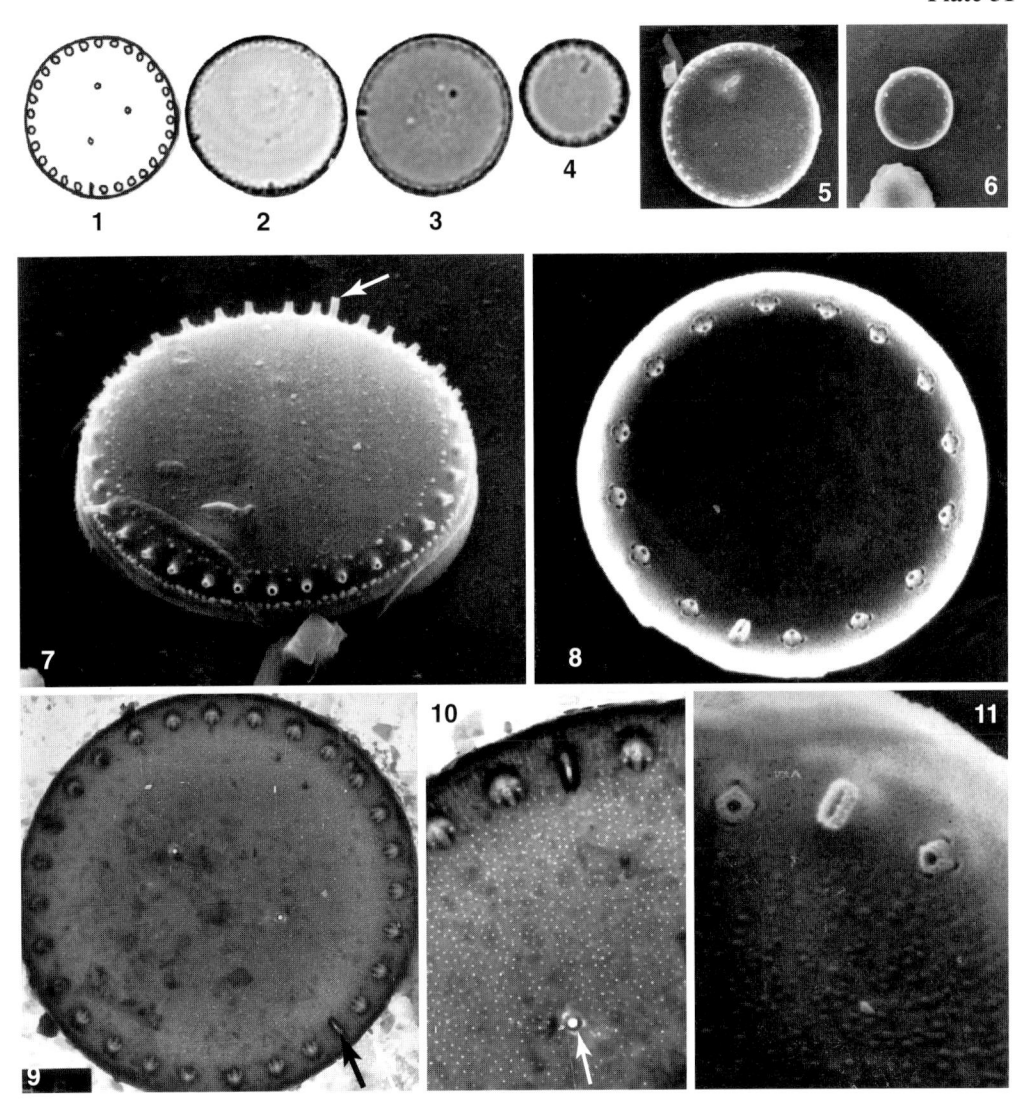

Plate 32 *Thalassiosira lacustris* (Grunow) Hasle

Fig. 1.   Drawing. ×2000.
Figs 2, 3.   LM. ×2000.
Figs 4-7.   SEM.

Figs 1-3.   Valve view.
Fig. 4.   Detail of central strutted processes with four struts. ×30000.
Fig. 5.   External view of a valve showing strongly undulated central portion and marginal spines.
Fig. 6.   Internal view of whole valve showing marginal strutted processes and a labiate process (arrow). ×3000.
Fig. 7.   Detail of mantle externally showing spines and external opening of strutted process. ×10000.

Localities
Figs 2, 3.   Tama-gawa (Tama River), Tokyo.
Figs 4, 6.   Panke-to (Panke Pond), Hokkaido.
Figs 5, 7.   Yamanaka-ko (Lake Yamanaka), Yamanashi.

Plate 32

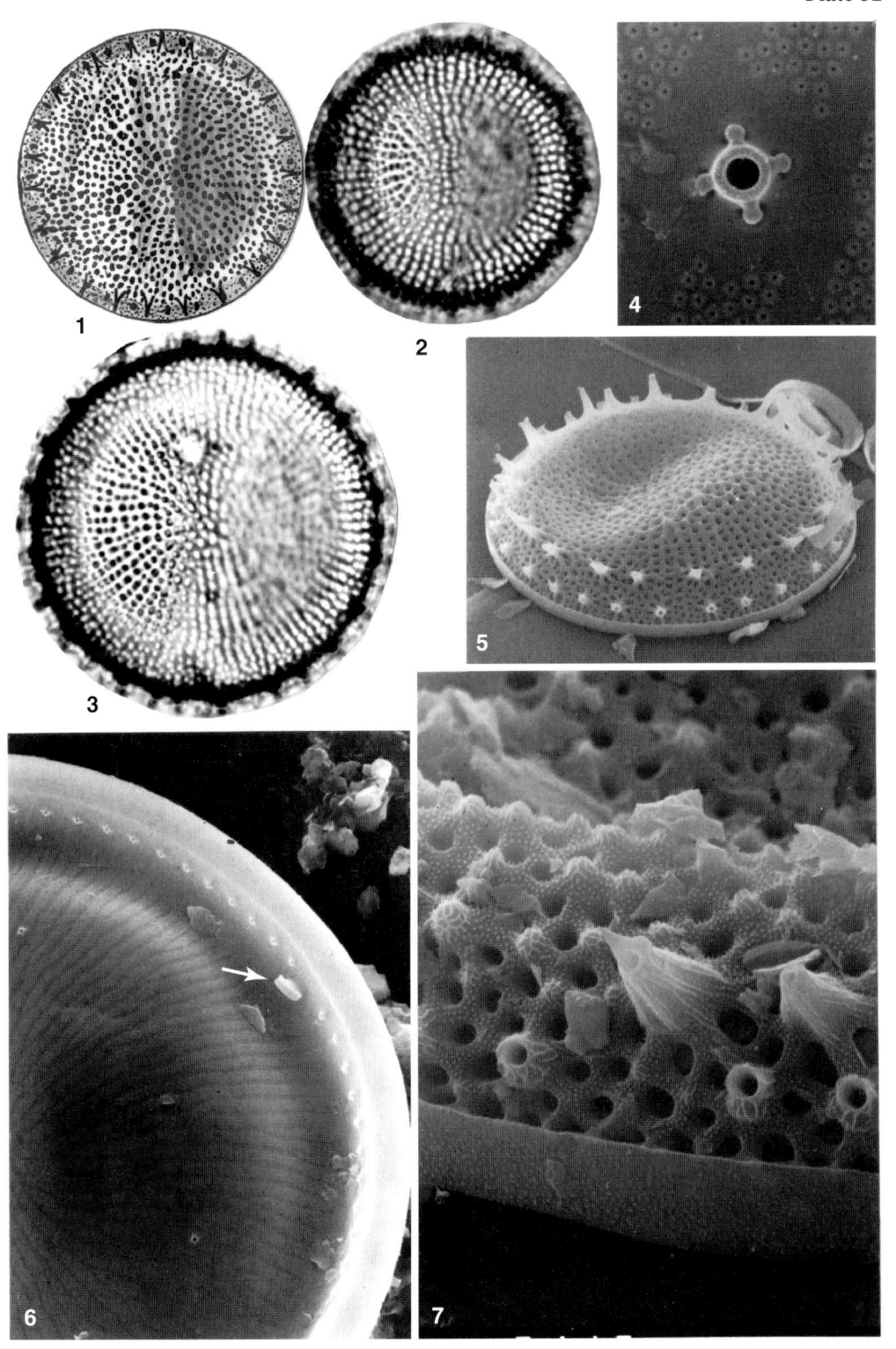

Plate 33 *Thalassiosira nordenskioeldii* Cleve

Figs 1, 2.   Drawings. ×2000
Fig. 3.   LM. ×2000.
Fig. 4.   LM. ×800.
Figs 5, 6.   SEM

Fig. 1.   Girdle view.
Figs 2, 3.   Valve view.
Fig. 4.   Colony joined with mucilage.
Fig. 5.   External view frustule showing outer tube of central strutted process with muscilage (arrow). ×1500.
Fig. 6.   Oblique view of frustule externally showing well developed outer tube of marginal strutted processes and outer tube of central strutted process with mucilage (arrow). ×3000.

Localities
Fig. 3.   Estuary of Tama-gawa (Tama River), Tokyo.
Figs 4-6.   Tokyo-wan (Tokyo Bay), Tokyo.

Plate 33

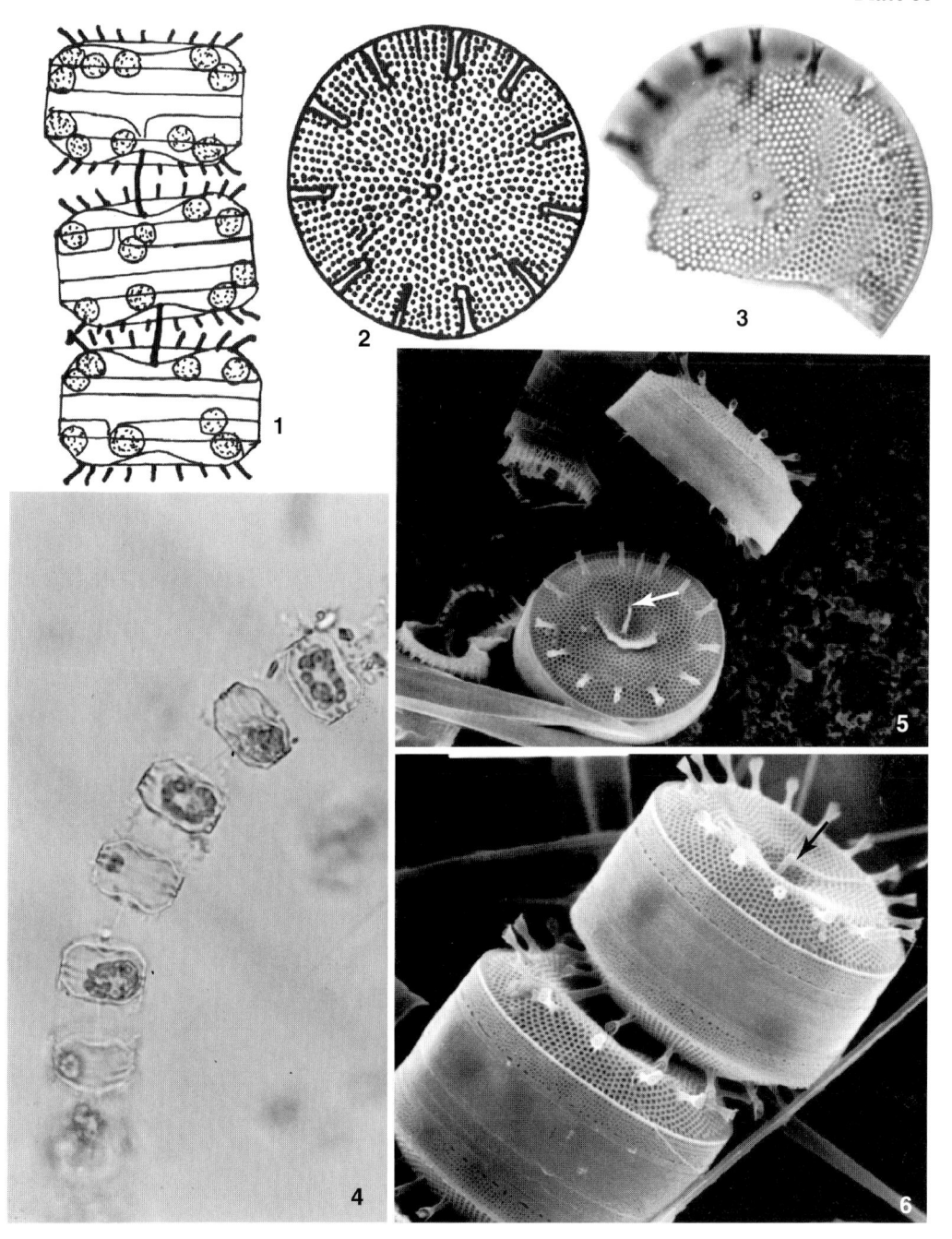

Plate 34 *Thalassiosira pseudonana* Hasle & Heimdal

Figs 1-5.　Drawings.　×2000.
Figs 6-8.　LM.　×2000.
Figs 9, 11.　TEM.
Figs 10, 12, 13.　SEM.

Figs 1-9.　Valve view.
Fig. 10.　External and internal view of frustule showing epivalve, bands and hypovalve. ×10000.
Fig. 11.　Whole valve showing marginal strutted processes and areolation. ×15000.
Fig. 12.　External view of valve showing external opening of strutted processes. ×10000.
Fig. 13.　Detail of valve margin internally showing labiate process (arrow) and marginal strutted processes. ×30000.

Lacalities
Fig. 8.　Hachiro-gata (Lake Hachiro), Akita.
Figs 10, 11.　Gake-gawa (Gake River), Tokyo.
Figs 6, 7, 9, 12, 13.　Hinuma-gawa (Hinuma River), Ibaraki.

Plate 34

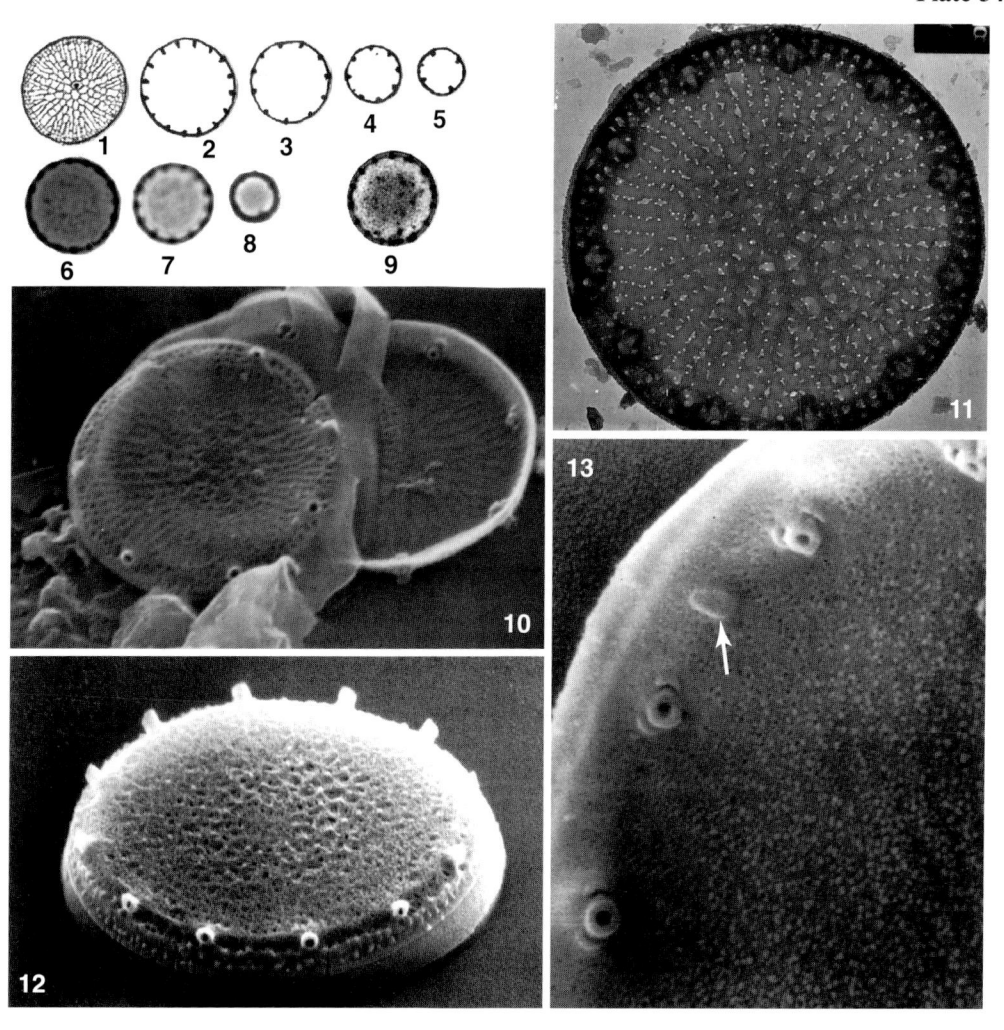

Plate 35 *Thalassiosira tenera* Proshkina-Lavrenko

Fig. 1.   Drawing. ×2000.
Figs 2-5.   LM. ×2000.
Figs 6-10.   SEM.

Figs 1-5.   Valve view.
Fig. 6.   External view of a valve showing outer opening of central strutted process (arrow). ×4000.
Fig. 7.   Oblique view of frustule showing outer opening of central strutted process (arrow) and strongly undulated marginal spines. ×4000.
Fig. 8.   External view of a valve. ×3000.
Fig. 9.   Girdle view of whole frustule showing strongly undulated marginal spines. ×4000.
Fig. 10.   Oblique view of frustule with granules showing outer opening of central strutted process (arrow) and strongly undulated marginal spines. ×5000.

Localities
Figs 2, 3.   Estuary of Tama-gawa (Tama River), Tokyo.
Figs 4, 5.   Obuchi-numa (Obuchi Pond), Aomori.
Figs 6, 10.   Tokyo-wan (Tokyo Bay), Tokyo.
Figs 7, 9.   Matsukawa-ura (Matsukawa Lagoon), Fukushima.
Fig. 8.   Mikawa-wan (Mikawa Bay), Aichi.

Plate 35

Plate 36 *Thalassiosira weissflogii* (Grunow) G. A. Fryxell & Hasle

Fig. 1.   Drawing. ×2000.
Figs 2-4.   LM. ×2000.
Figs 5-7, 9.   SEM.
Fig. 8.   TEM.

Figs 1-4.   Valve view.
Fig. 5.   External view of a valve showing numerous external openings of marginal strutted processes. ×3000.
Fig. 6.   Detail of central strutted processes with four struts. ×20000.
Fig. 7.   Internal view of a valve showing central portion and marginal spines. ×8000.
Fig. 8.   Detail of central strutted processes with four struts and marginal strutted processes. ×10000.
Fig. 9.   Detail of mantle internally showing a labiate process and internal openings of marginal strutted processes. ×20000.

Locarities

Figs 2-4.   Estuary of Tama-gawa (Tama River), Tokyo.
Figs 5-7, 9.   Hachiro-gata (Lake Hachiro), Akita.
Fig. 8.   Funada-ike (Funada Pond), Chiba.

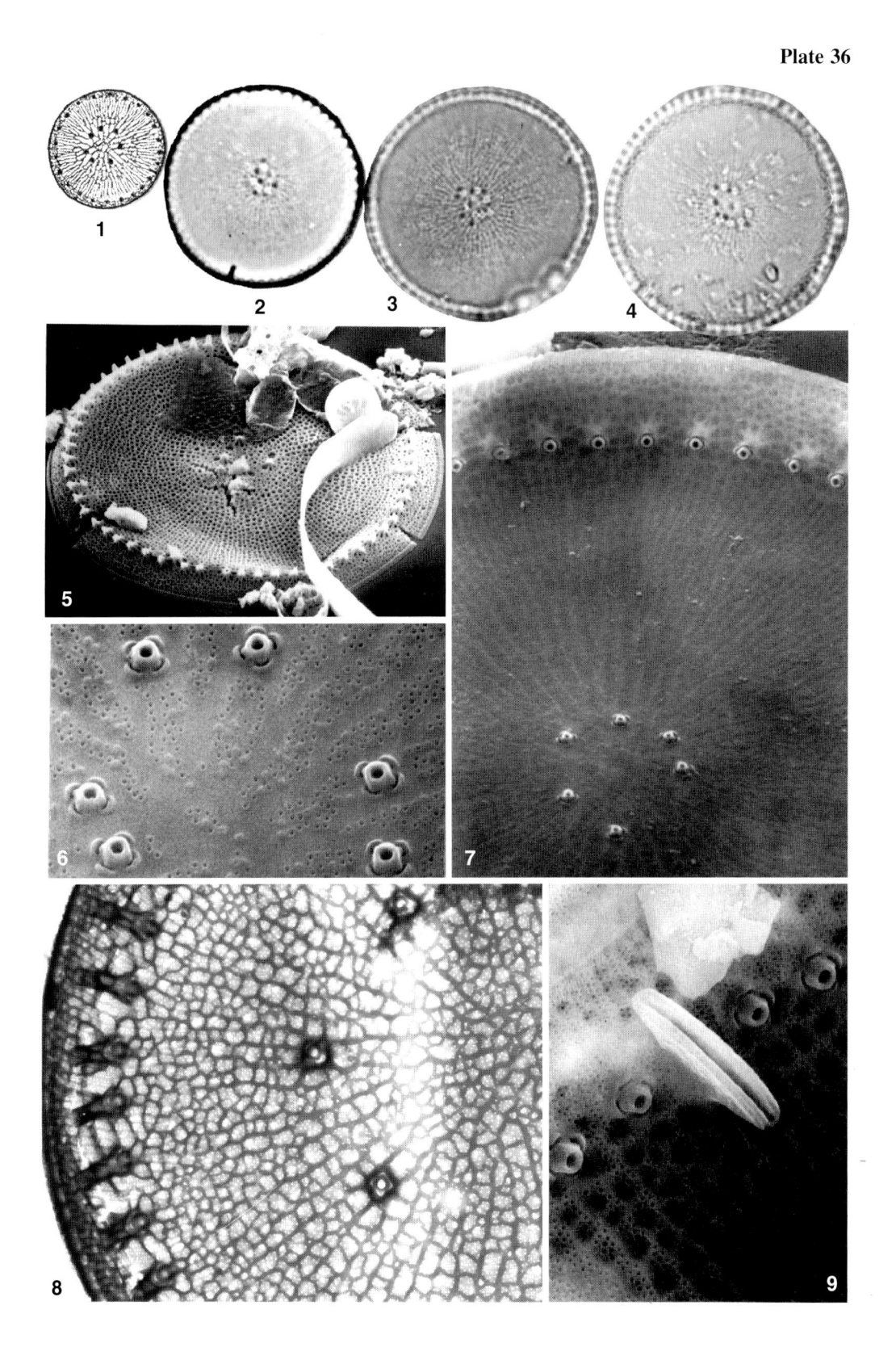

Plate 36

Plate 37 *Cyclostephanos dubius* (Fricke) Round

Figs 1-6, 8-10.   LM.   ×2000.
Fig. 7.   Drawing.   ×2000.
Figs 11-15.   SEM.

Figs 1-8.   Valve view.
Figs 9, 10.   Girdle view.
Fig. 11.   External view of a frustule showing acute marginal spines and marginal strutted processes secreting mucilage.   ×5000.
Fig. 12.   Internal view of the valve.   ×2000.
Fig. 13.   External view of the valve showing the fascicles and the interfascicles.   ×8000.
Fig. 14.   Internal view of the valve showing the areolae occluded by the domed cribra and central strutted processes with bisatellite pores.   ×20000.
Fig. 15.   Internal view of the valve showing the central strutted process (arrow) and marginal fascicles.   ×2000.

Localities
Figs 1-6, 9, 10, 11, 15.   Suwa-ko (Lake Suwa), Nagano.
Fig. 8.   Tama-gawa (Tama River), Tokyo.
Figs 12-14.   Kita-ura (Lake Kita), Ibaraki.

Plate 37

Plate 38 *Cyclostephanos dubius* (Fricke) Round

Figs 16-21.   SEM.

Fig. 16.   External view of a valve.
Fig. 17.   Enlargement of the valve margin showing the openings of the areolae, the marginal spines and the opening of the labiate process. ×20000.
Fig. 18.   The cingulum with a wide valvocopula and four open bands. ×20000.
Fig. 19.   Internal view of the valve showing the central strutted processes and the marginal interfascicles. ×5000.
Figs 20, 21.   Internal view of the valve margin showing the marginal strutted processes, a labiate process and the domed cribra of the areolae. ×20000.

Locality
Figs 16-21.   Kita-ura (Lake Kita), Ibaraki.

Plate 38

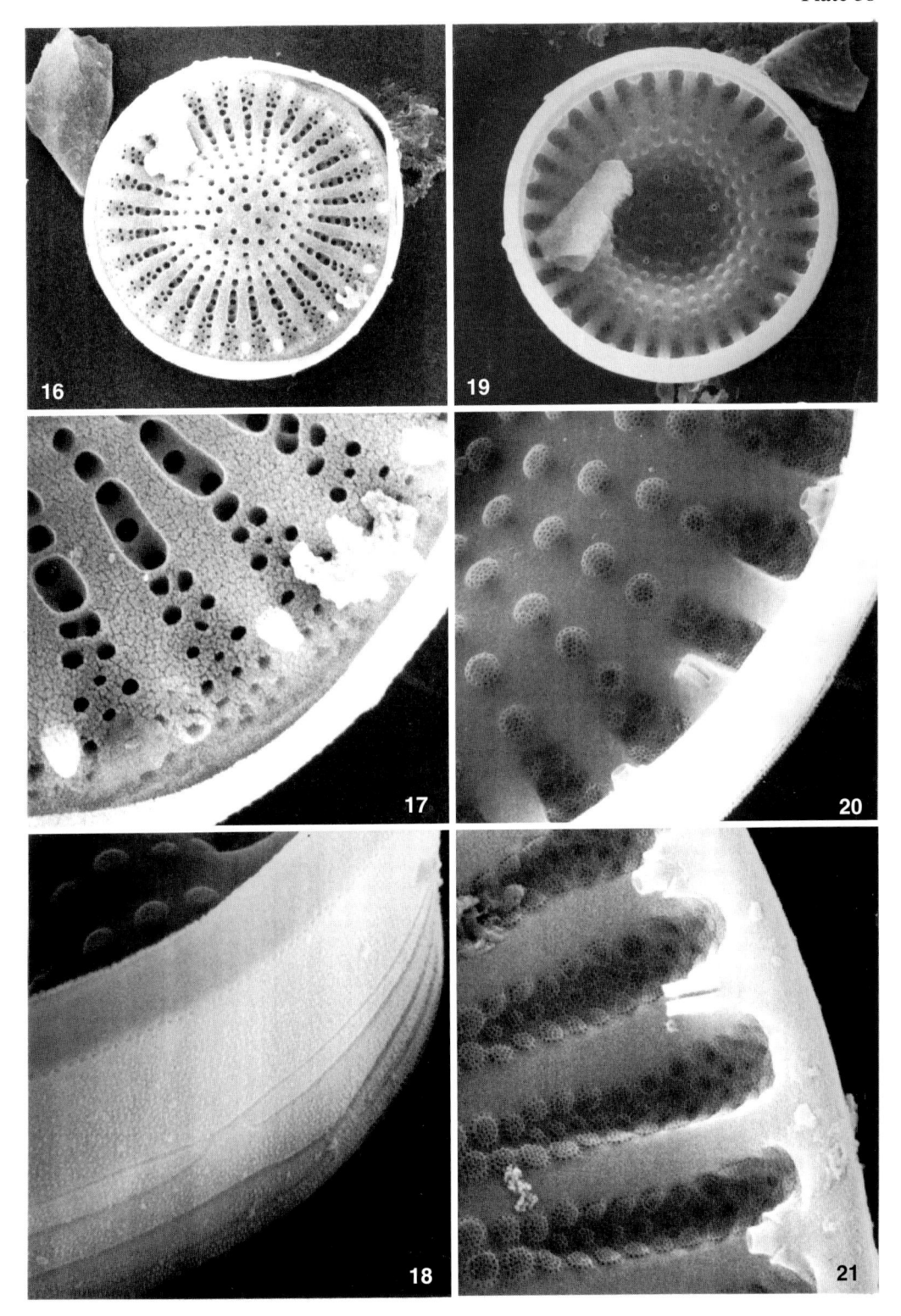

Plate 39 *Cyclostephanos invisitatus* (Hohn & Hellerman) Theriot *et al.*

Figs 1-4. LM. ×2000.
Fig. 5. Drawing.
Figs 6-8. SEM.
Fig. 9. TEM.

Figs 1-4. Valve view.
Fig. 5. Diagramatic representation of the features: a) marginal spines, b) marginal strutted process, c) flange, d) outer opening of labiate process, e) bifurcation of the interfascicle, f) central strutted process, g) pattern center, h) interfascicle, i) fascicle.
Fig. 6. Girdle view of frustule showing acute marginal spines. ×8000.
Fig. 7. Internal view of valve showing central strutted process, marginal strutted processes and labiate process (arrow). ×8000.
Fig. 8. External view of strongly silicified valve showing openings of areolae and opening of central strutted process. ×8000.
Fig. 9. Valve view of weakly silicified valve. ×6000.

Localities
Figs 1, 2. Naka-gawa (Naka River), Tokyo.
Figs 3, 4, 6-9. Waku-ike (Waku Pond), Nagano

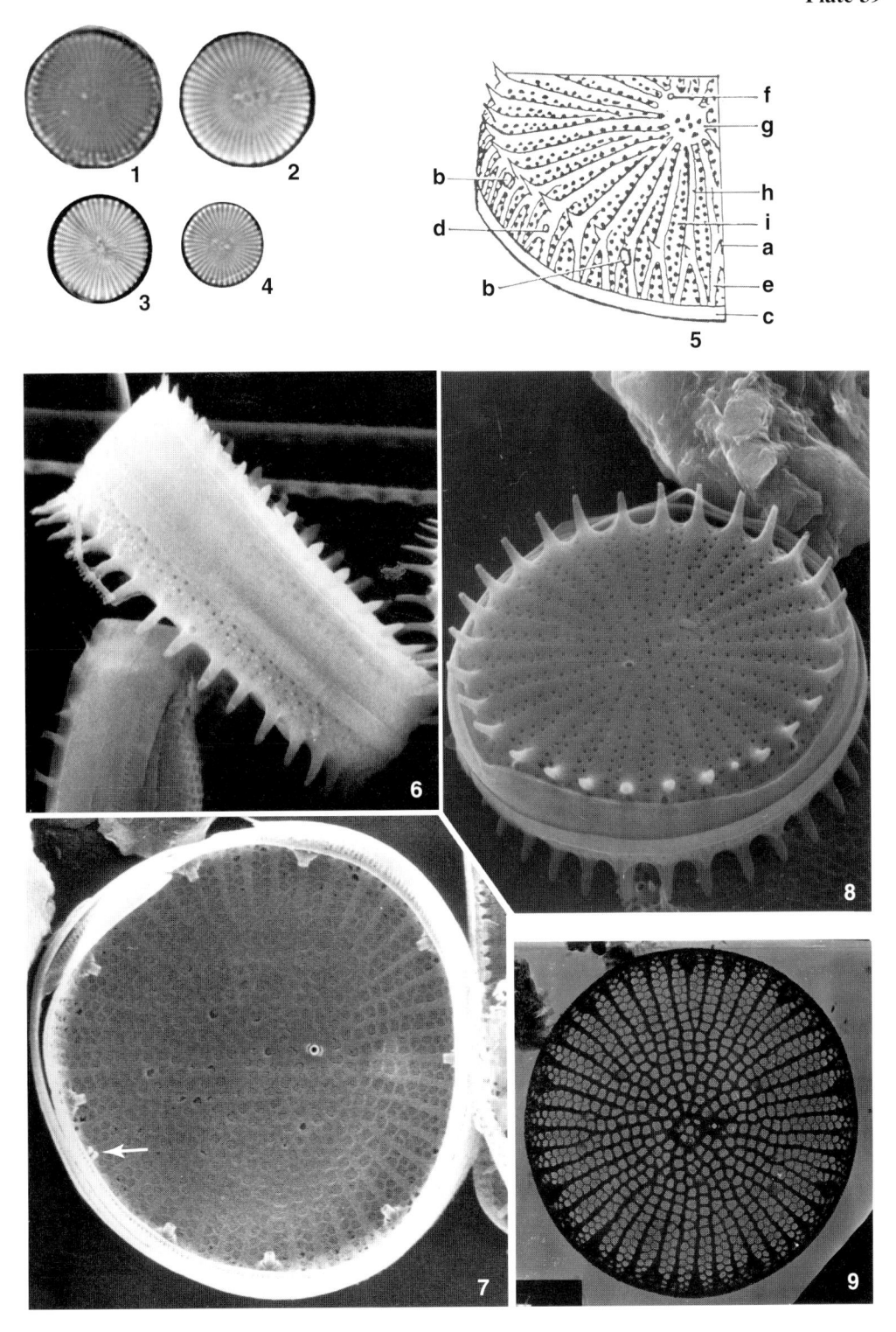

Plate 39

225

Plate 40 *Cyclostephanos invisitatus* **(Hohn & Hellerman) Theriot *et al.***

Figs 10, 11. SEM.

Fig. 10. Internal view of valve showing areolae occluded by cribra and marginal strutted processes. ×20000.

Fig. 11. External view of valve showing marginal spines, openings of marginal strutted processes, bifurcations of interfascicles and a central strutted process (arrow). ×8000.

Locality

Figs 10, 11. Waku-ike (Waku Pond), Nagano.

Plate 40

Plate 41 *Cyclotella atomus* Hustedt

Figs 1, 2.   Drawings. ×2000.
Figs 3-6.   LM. ×2000.
Figs 7-9.   TEM. ×2000.
Fig. 10.   TEM. ×8000
Figs 11-13. SEM.

Figs 1-10.   Valve view.
Fig. 11.   Internal view of whole valve showing marginal strutted processes with two struts and labiate process (arrow). ×10000.
Fig. 12.   External view of a frustule showing outer openings of marginal strutted processes.
Fig. 13.   Detail of marginal striae showing perforations of the pore occlusion and a external opening of labiate process (arrow), and external openings of strutted processes. ×30000.

Localities
Figs 3, 4, 7-13.   Naka-gawa (Naka River), Tokyo.
Figs 5, 6.   Estuary of Tama-gawa (Tama River), Tokyo.

Plate 41

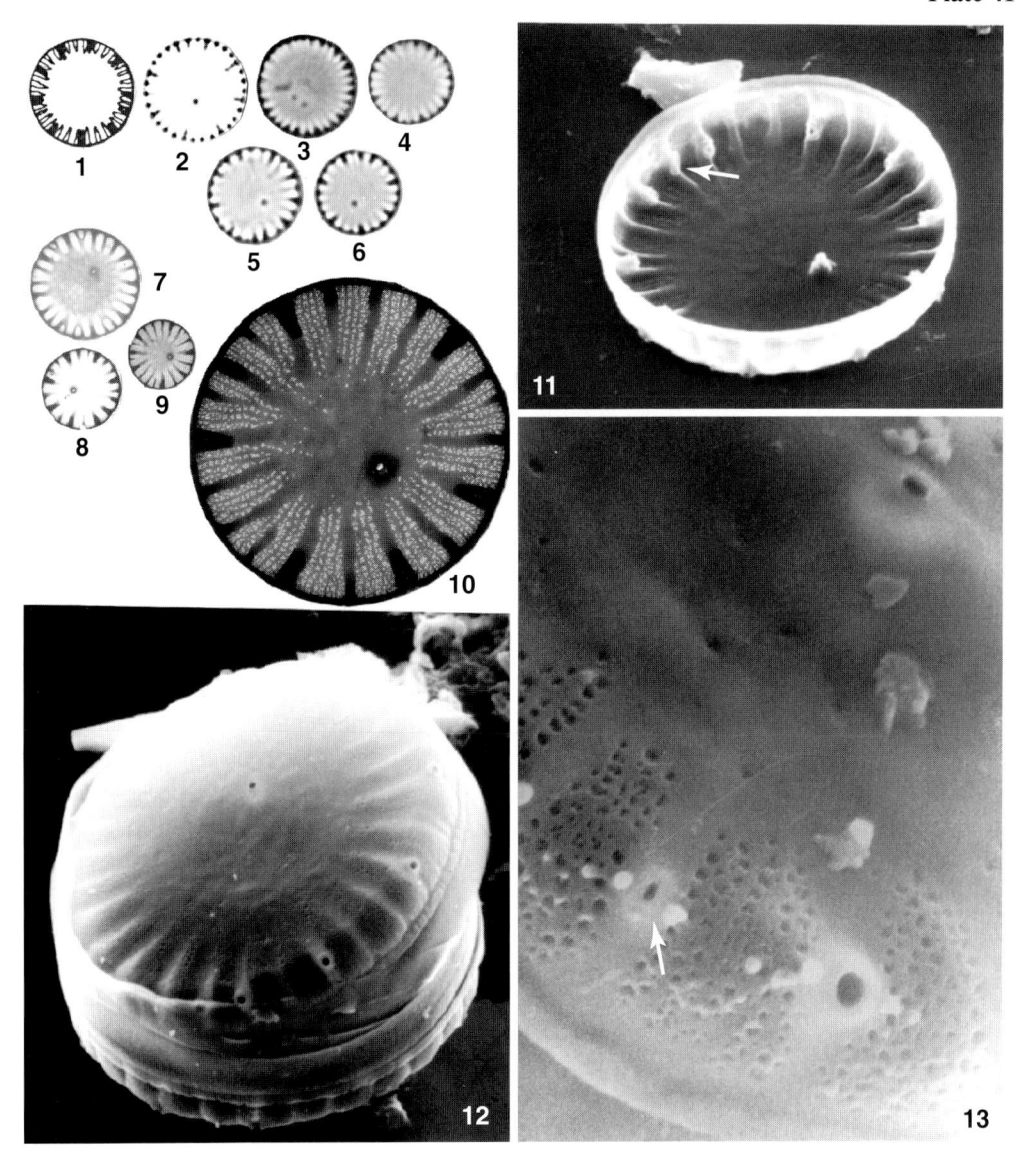

229

Plate 42 *Cyclotella atomus* Hustedt var. *gracilis* Genkal & Kiss

Fig. 1. Drawing. ×2000.
Figs 2-4, 6. LM. ×2000.
Figs 5, 7. TEM.
Figs 8-10. SEM.

Figs 1-5. Valve view. ×2000.
Fig. 6. Girdle view. ×2000.
Fig. 7. Whole valve. ×8000.
Fig. 8. External view of a frustule showing outer openings of marginal strutted processes and labiate process (arrow). ×10000.
Fig. 9. Internal view of whole valve showing marginal strutted processes with two struts and labiate process (arrow). ×10000.
Fig. 10. Detail of broken valve along the marginal stria. ×13000.

Locality
Figs 2-10. Inohana-ko (Lake Inohana), Aichi.

Plate 42

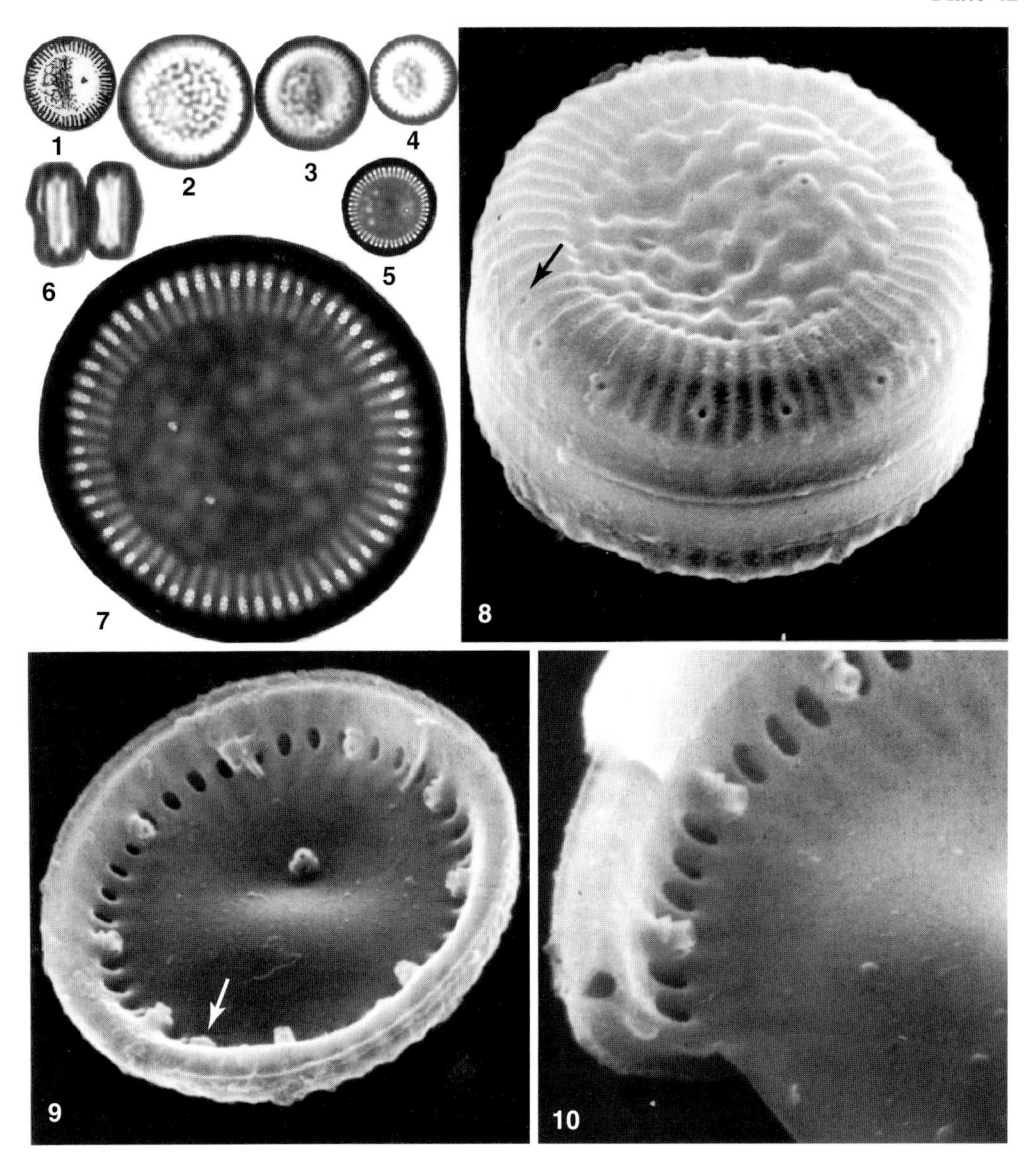

Plate 43 *Cyclotella criptica* Reimann *et al.*

Fig. 1. Drawing. ×2000.
Figs 2, 3. LM. ×2000.
Figs 4, 5, 7. TEM.
Figs 6, 8. SEM.

Figs 1-5. Valve view.
Fig. 5. Whole valve. ×6000.
Fig. 6. External view of whole valve showing outer opening of central strutted process (arrow) and outer openings of marginal strutted processes and a labiate process (arrowhead). ×6000.
Fig. 7. Detail of marginal striae showing perforations of the pore occlusion and marginal strutted processes. ×30000.
Fig. 8. Internal view of whole valve showing marginal strutted processes with two struts and central strutted process (arrow) and labiate process (arrowhead). ×10000.

Localities
Figs 4, 5, 8. Waku-ike (Waku Pond), Nagano.
Figs 2, 3, 6, 7. Naka-gawa (Naka River), Tokyo.

Plate 43

Plate 44 *Cyclotella litoralis* Lange & Syvertsen

Fig. 1.   Drawing. ×2000.
Figs 2-4.   LM. ×2000.
Fig. 5.   SEM.

Figs 1-4.   Valve view.
Fig. 5.   Detail of central strutted processes with three struts. ×20000.

Locality
Figs 2-5.   Arakawa-lowland, Saitama.

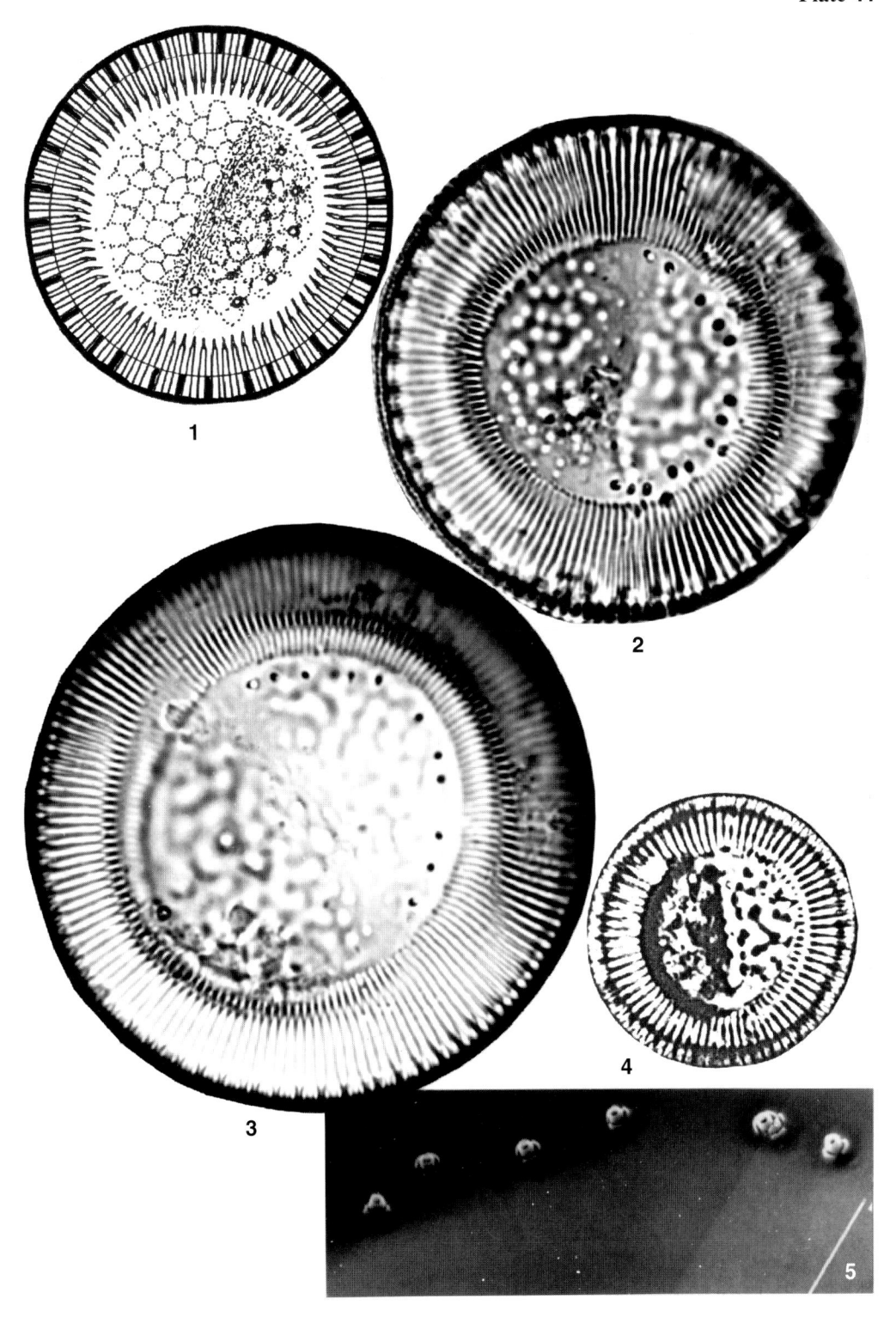

Plate 44

Plate 45 *Cyclotella litoralis* Lange & Syvertsen

Figs 6-9.　SEM.

Fig. 6.　External view of a valve showing strongly undulated central portion. ×2200.

Fig. 7.　Detail of valve margin showing strongly undulated valve face, areolae rows and outer openings of marginal strutted processes. ×15000.

Fig. 8.　Internal view of whole valve showing marginal strutted processes and labiate process (arrow), and a lined strutted processes with three struts. ×3000.

Fig. 9.　Details of marginal strutted processes with two struts and a labiate process. ×15000.

Locality

Figs 6-9.　Arakawa-lowland, Saitama.

Plate 45

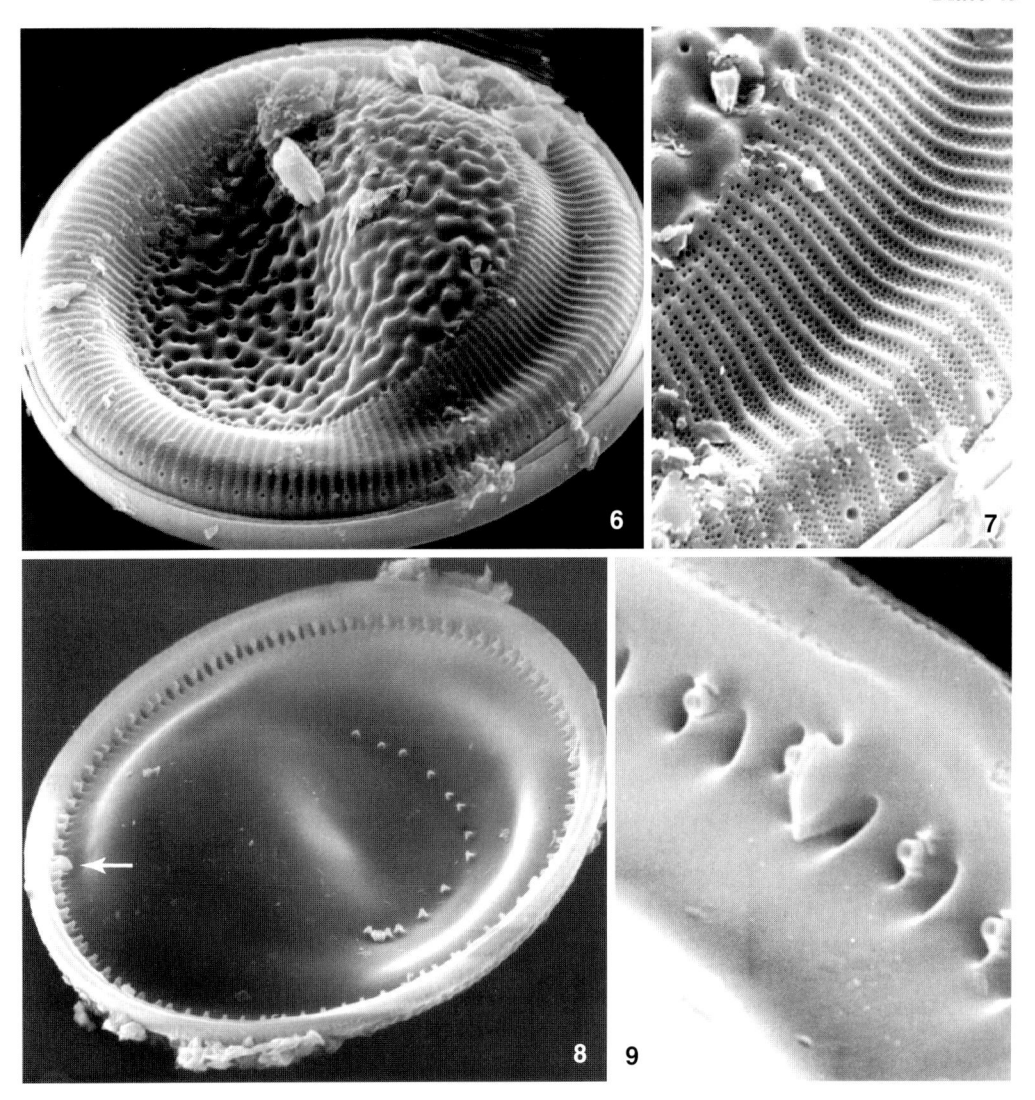

Plate 46 *Cyclotella meduanae* Germain

Fig. 1.   Drawing. ×2000.
Figs 2-4, 6.   LM. ×2000.
Figs 5, 7, 10.   TEM.
Figs 8, 9.   SEM.

Figs 1-6.   Valve view.
Fig. 7.   Whole valve. ×8000.
Fig. 8.   Internal view of whole valve showing marginal strutted processes with three struts and labiate process (arrow). ×12000.
Fig. 9.   External view of a valve showing outer openings of marginal strutted processes and outer opening labiate process (arrow), and many small granules on the mantle surface. ×12000.
Fig. 10.   Detail of marginal striae showing perforations of the pore occlusions. ×30000.

Localities
Figs 2, 3, 4.   Hachiro-gata (Lake Hachiro), Akita.
Figs 5-10.   Naka-gawa (Naka River), Tokyo.

Plate 46

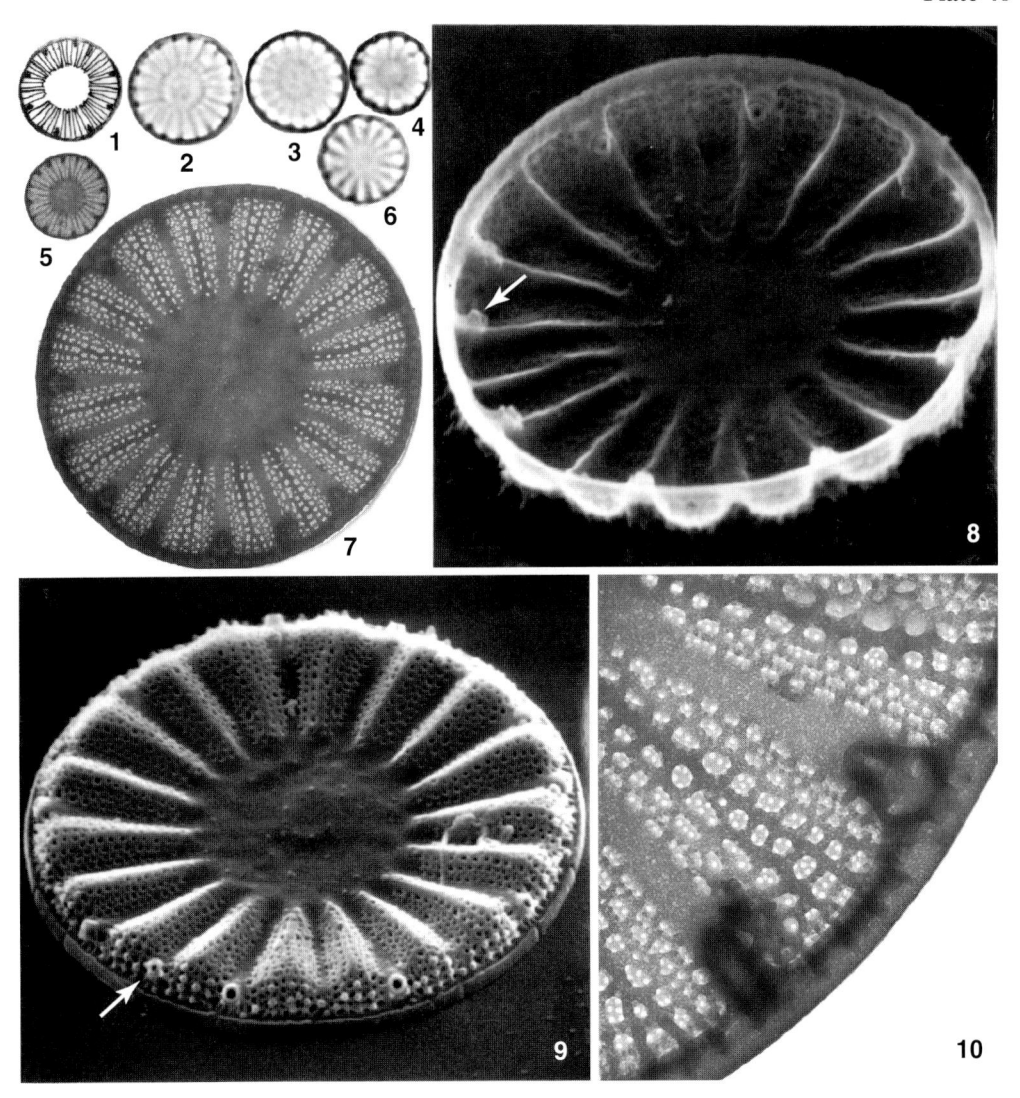

Plate 47　*Cyclotella meneghiniana* Kützing

Fig. 1.　Drawing. ×2000.
Figs 2-5.　LM. ×2000.
Figs 6-9, 11.　SEM.
Fig. 10.　TEM.

Figs 1-5.　Valve view.
Fig. 6.　External view of whole valve showing strongly undulated central portion and marginal spines. ×3000.
Fig. 7.　Internal view of whole valve showing marginal strutted processes with three struts (arrow) and labiate process. ×3000.
Fig. 8.　Detail of marginal strutted processes with three struts. ×30000.
Fig. 9.　Detail of central strutted processes with three struts . ×50000.
Fig. 10.　Whole valve. ×3000.
Fig. 11.　Detail of valve mantle showing perforations of the pore occlusion and external opening of strutted process. ×10000.

Localities
Fig. 2.　A paddy field in Ojiya, Niigata.
Fig. 3.　Tokyo-wan (Tokyo Bay), Tokyo.
Fig. 4.　Tama-gawa (Tama River), Tokyo.
Fig. 5.　Hozoji-numa (Hozoji Pond), Saitama.
Figs 6-9, 11.　Matsumoto-jo Hori (The moat of Matsumoto Castle), Nagano.
Fig. 10.　Ichigaya-bori (Ichigaya Moat), Tokyo.

Plate 47

Plate 48 *Cyclotella ocellata* Pantocsek

Figs 1-3, 5.   LM. ×2000.
Fig. 4.   Drawing. ×2000.
Figs 6-10.   SEM.

Figs 1-5.   Valve view.
Fig. 6.   Girdle view of frustule showing cinglum. ×4000.
Fig. 7.   Oblique view of frustule showing valve face with depression and outer openings of marginal strutted processes. ×4000.
Fig. 8.   Detail of external valve margin showing outer openings of marginal strutted processes. ×20000.
Fig. 9.   Oblique view of valve internaly showing a central strutted process with two struts and labiate process (arrow). ×5000.
Fig. 10.   Oblique view of valve with cinglum internaly showing central strutted processes with two struts and labiate process (arrow). ×5000.

Locarity
Figs 1-3, 5-10.   Miyajima fomation (Shiobara group), Tochigi.

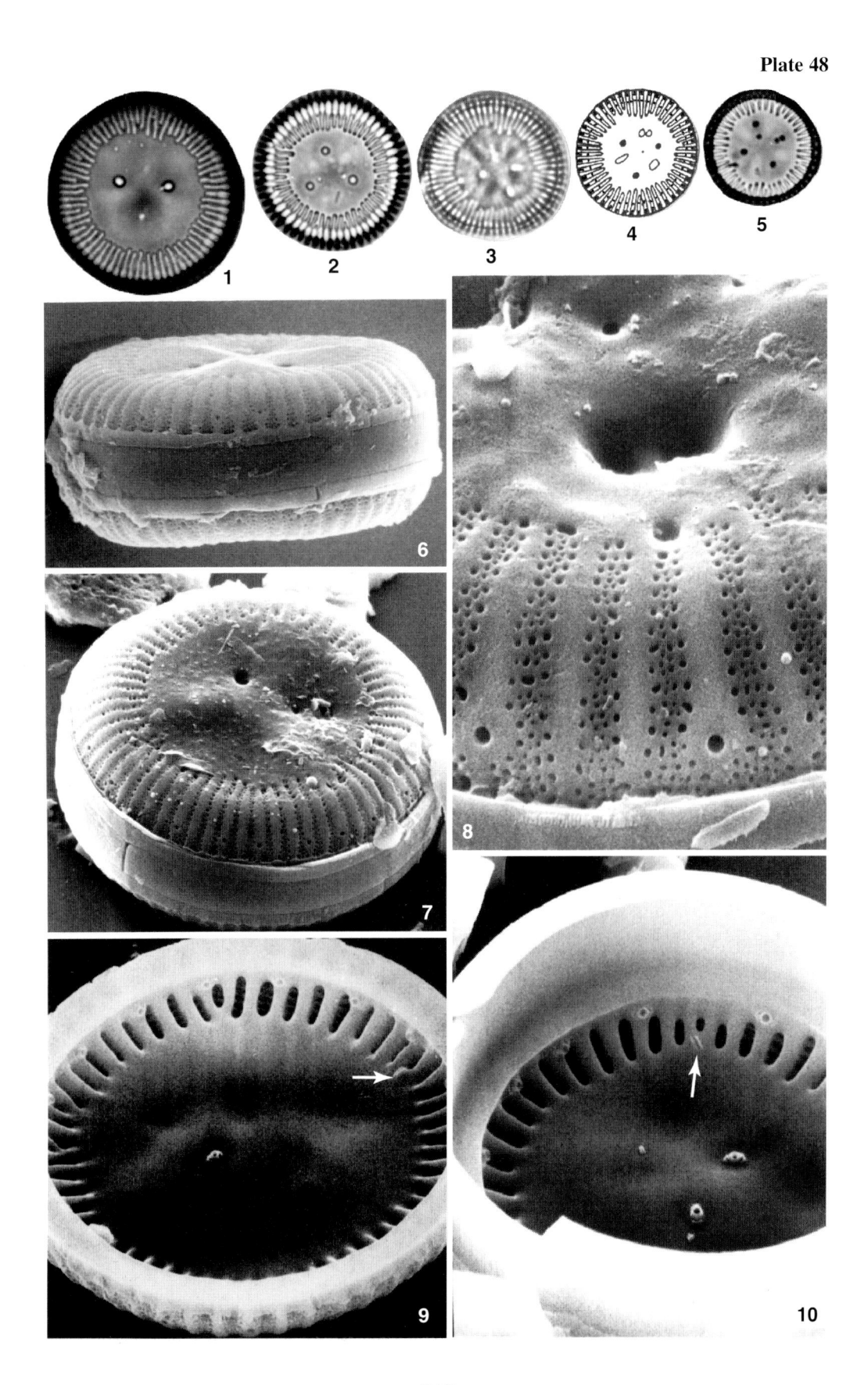

Plate 48

243

Plate 49 *Cyclotella pantaneliana* Castracane

Fig. 1. Drawing. ×2000.
Figs 2, 3. LM. ×2000.
Figs 4, 5. SEM.

Figs 1-3. Valve view.
Fig. 4. Oblique view of central area externally showing slightly concave valve face and external opening of marginal labiate process (arrow). ×6000.
Fig. 5. Detailed view of external valve margin showing outer opening of labiate process (black arrow) and marginal strutted process (white arrow). ×10000.

Locality
Figs 2-5. Tama-gawa (Tama River), Tokyo.

Plate 49

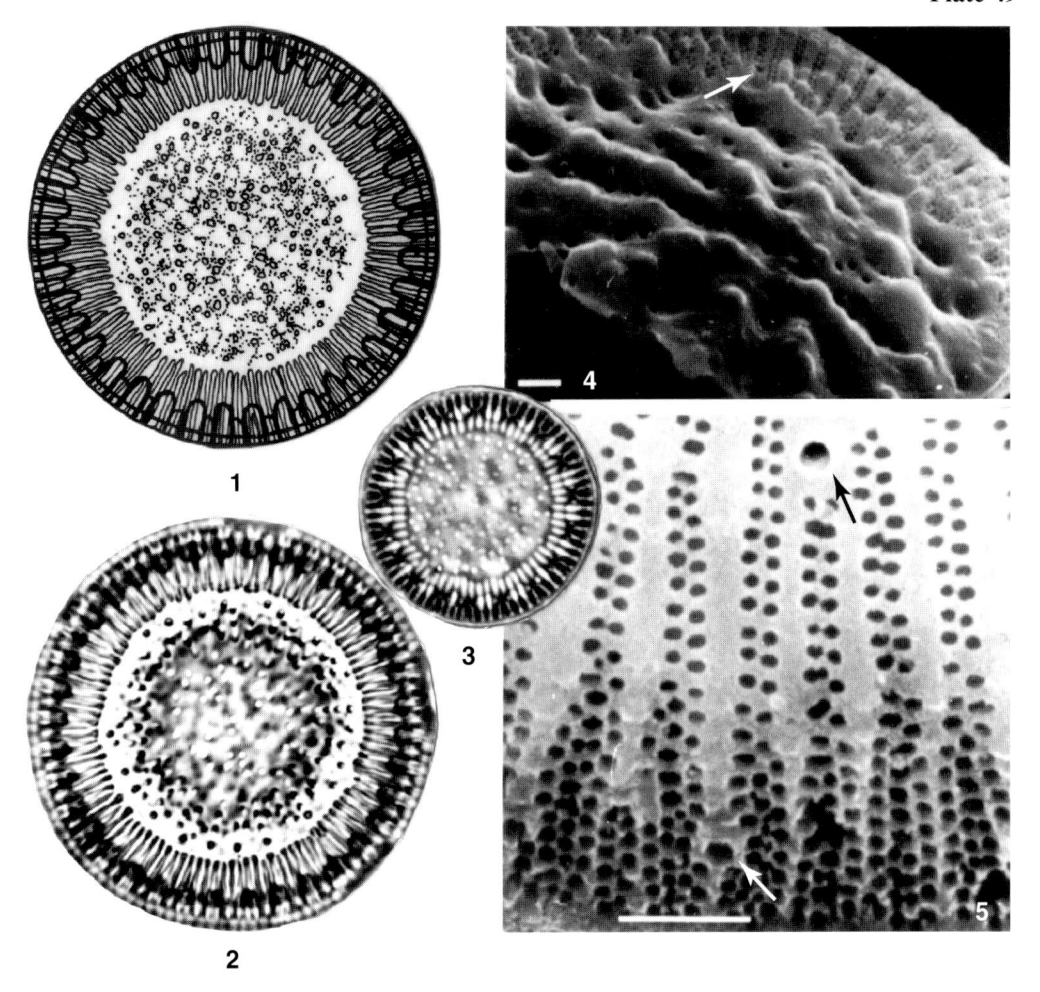

Plate 50   *Cyclotella striata* (Kützing) Grunow

Fig. 1.   Drawing. ×2000.
Figs 2, 3.   LM. ×2000.
Figs 4-6.   SEM.

Figs 1-3.   Valve view.
Fig. 4.   Detail of marginal striae showing perforations of the pore occlusion and external openings of marginal strutted processes. ×20000.
Fig. 5.   External view of a valve showing strongly tangentially folded colliculate central zone and external openings of central strutted processes. ×2000.
Fig. 6.   Internal view of whole valve showing marginal strutted processes and labiate process (arrow). ×2000.

Localities
Figs 2, 3, 6.   Hinuma-gawa (Hinuma River), Ibaraki.
Figs 4, 5.   Hi-numa (Lake Hi), Ibaraki.

Plate 50

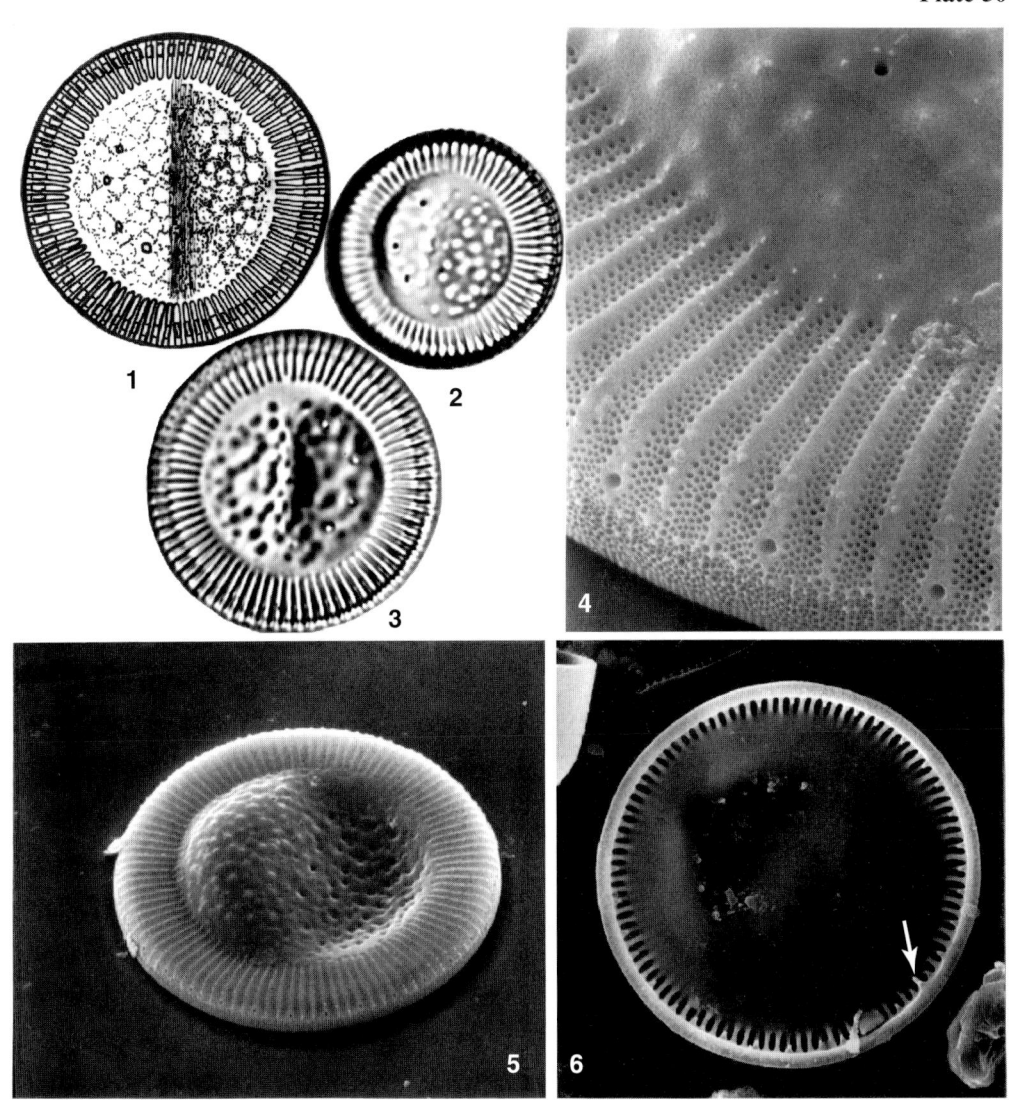

247

Plate 51 *Discostella asterocostata* (Lin *et al.*) Houk & Klee

Figs 1-4. LM. ×2000.
Fig. 5. Drawing. ×2000.
Figs 6-8. SEM.

Figs 1-4. Valve view.
Fig. 6. Internal view of a valve showing marginal strutted processes with two struts. ×2000.
Fig. 7. Detail of marginal striae internally, showing marginal strutted processes with two struts and labiate process (arrow). ×6000.
Fig. 8. Internal view of a valve showing arrangement of marginal strutted processes and internal openings of areolae. ×5000.

Localities
Figs 1, 2. Miyako-gawa (Miyako River), Chiba.
Figs 3, 4, 6-8. Kita-ura (Lake Kita), Ibaraki.

Plate 51

Plate 52 *Discostella pseudostelligera* (Hustedt) Houk & Klee

Figs 1, 2.   Drawings.   ×2000.
Figs 3-5, 13-15, 18. 20.   TEM.
Figs 6-12, 16, 17.   LM.   ×2000.
Figs 19, 21.   SEM.

Figs 3-12, 16, 17.   Valve view.

Figs 13-15, 18.   Detail of valves showing variation of the central area and striation, and labiate process (arrow in Fig. 15).   ×8000.

Figs 16, 17.   Lectotypes. BRM. 51/1, Ems, bei Papenburg. 197.

Fig. 19.   External view of a frustule showing outer openings of marginal strutted processes.   ×3000.

Fig. 20.   Detail of a frustule showing marginal striae and perforations of the pore occlusions and external openings of strutted processes.   ×8000.

Fig. 21.   Detail of a frustule showing flat surface, external openings of strutted processes and labiate process (arrow).   ×7000.

Localities

Figs 3, 4, 9-14, 19, 20.   Waku-ike (Waku Pond), Nagano.
Figs 5, 18.   Inokashira-ike (Inokashira Pond), Tokyo.
Fig. 6.   Tama-gawa (Tama River), Tokyo.
Figs 7, 21.   Naka-gawa (Naka River), Tokyo.
Fig. 8.   Ueda-jo Hori (The moat of Ueda Castle), Nagano.
Fig. 15.   Kano-gawa (Kano River), Shizuoka.

Plate 52

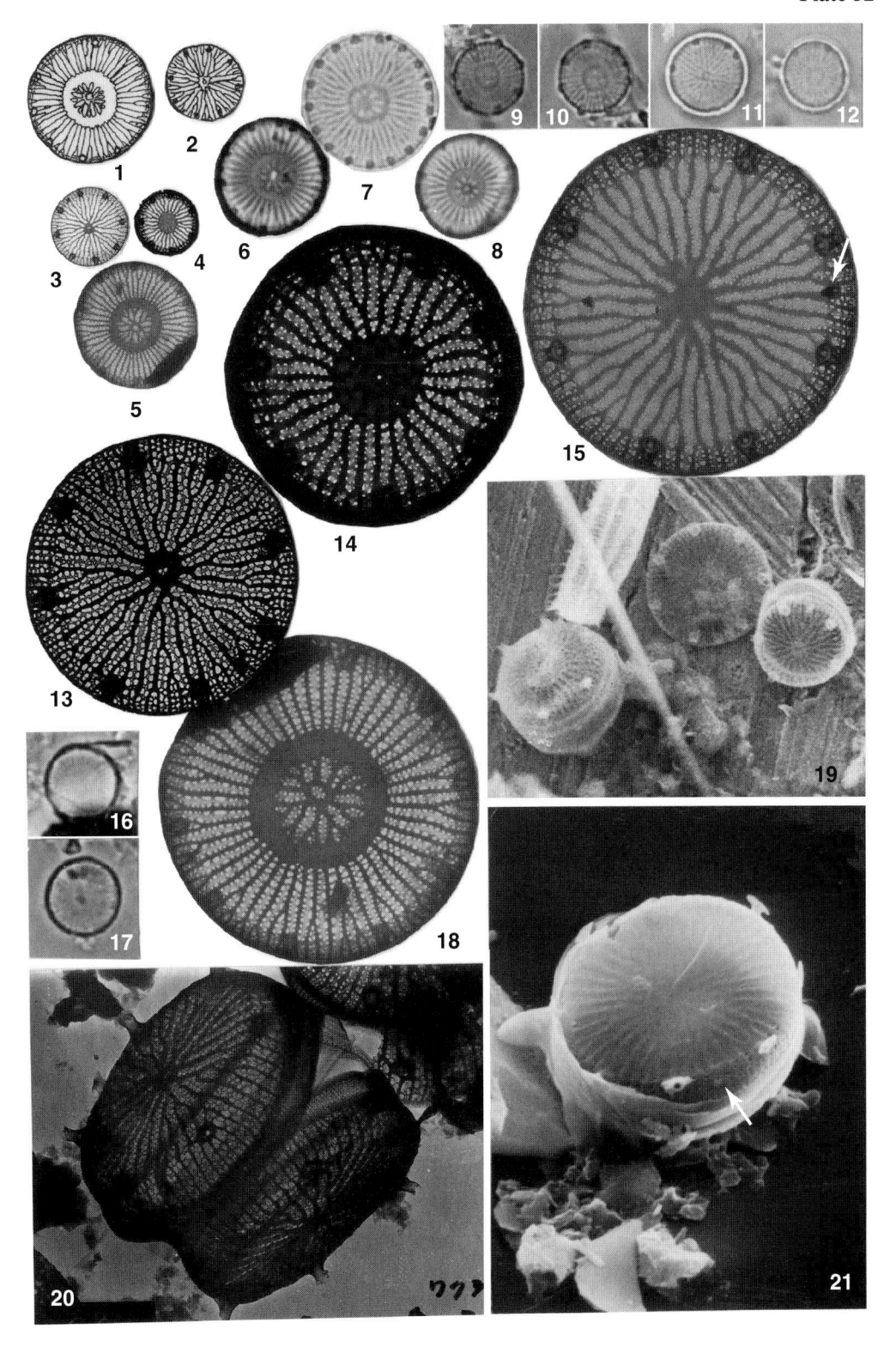

251

Plate 53 *Discostella stelligera* (Ehrenberg) Houk & Klee

Figs 1, 4.   Drawings. ×2000.
Figs 2, 3, 5-10.   LM. ×2000.
Figs 11, 13, 14.   SEM.
Fig. 12.   TEM.

Figs 1-10.   Valve view.
Fig. 11.   Oblique view of valve showing convex central area and outer openings of marginal strutted processes. ×5000.
Fig. 12.   Valve view showing detail arrangement of central and marginal stria. ×4000.
Fig. 13.   External view of valve. ×5000.
Fig. 14.   Internal view of valve showing marginal strutted processes and a labiate process (arrow). ×5000.

Localities

Figs 2, 3, 11.   A pond in Shinrin-koen (Shinrin Park), Saitama.
Figs 5, 8.   Yamanaka-ko (Lake Yamanaka), Yamanashi.
Figs 6, 7, 9, 10, 12-14.   A pond in Tokyo Gakugei University, Tokyo.

Plate 53

Plate 54 *Discostella stelligera* (Ehrenberg) Houk & Klee

Fig. 15.　TEM.
Fig. 16.　SEM.

Fig. 15.　Detail of valve margin showing marginal strutted processes and marginal striae. ×15000.

Fig. 16.　Detail of valve margin internally showing marginal strutted process and a labiate process. ×30000.

Locality

Figs 15, 16.　A pond in Tokyo Gakugei University, Tokyo.

Plate 54

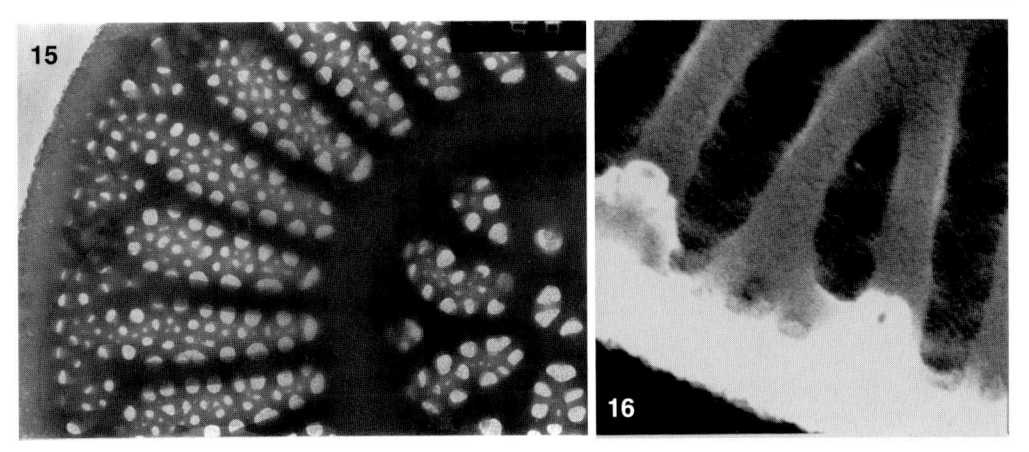

Plate 55 *Puncticulata praetermissa* (Lund) Håkansson

Figs 1, 5.   Drawings. ×2000.

Figs 2-4.   LM. ×2000.

Figs 6-9.   SEM.

* All figures are *P. praetermissa* except for Fig. 4, which is *Puncticulata radiosa* (Grunow) Håkansson.

Figs 1-5.   Valve view.

Fig. 6.   External view of valve showing marginal striation and openings of strutted processes (arrows). ×4000.

Fig. 7.   Internal view of valve showing a labiate process and strutted processes. ×4000.

Fig. 8.   Internal view of valve showing strutted processes with two struts at marginal, with three struts in central and a labiate process. ×10000.

Fig. 9.   Detail of marginal striae showing perforations of the pore occlusion and a external opening of labiate process (arrowhead), and external openings of strutted processes on central area of valve face (arrow). ×8000.

Localities

Figs 2, 3.   Estuary of Tama-gawa (Tama River), Tokyo.

Fig. 4.   Tama-gawa (Tama River), Tokyo.

Figs 6-9.   Estuary of Tama-gawa (Tama River), Tokyo.

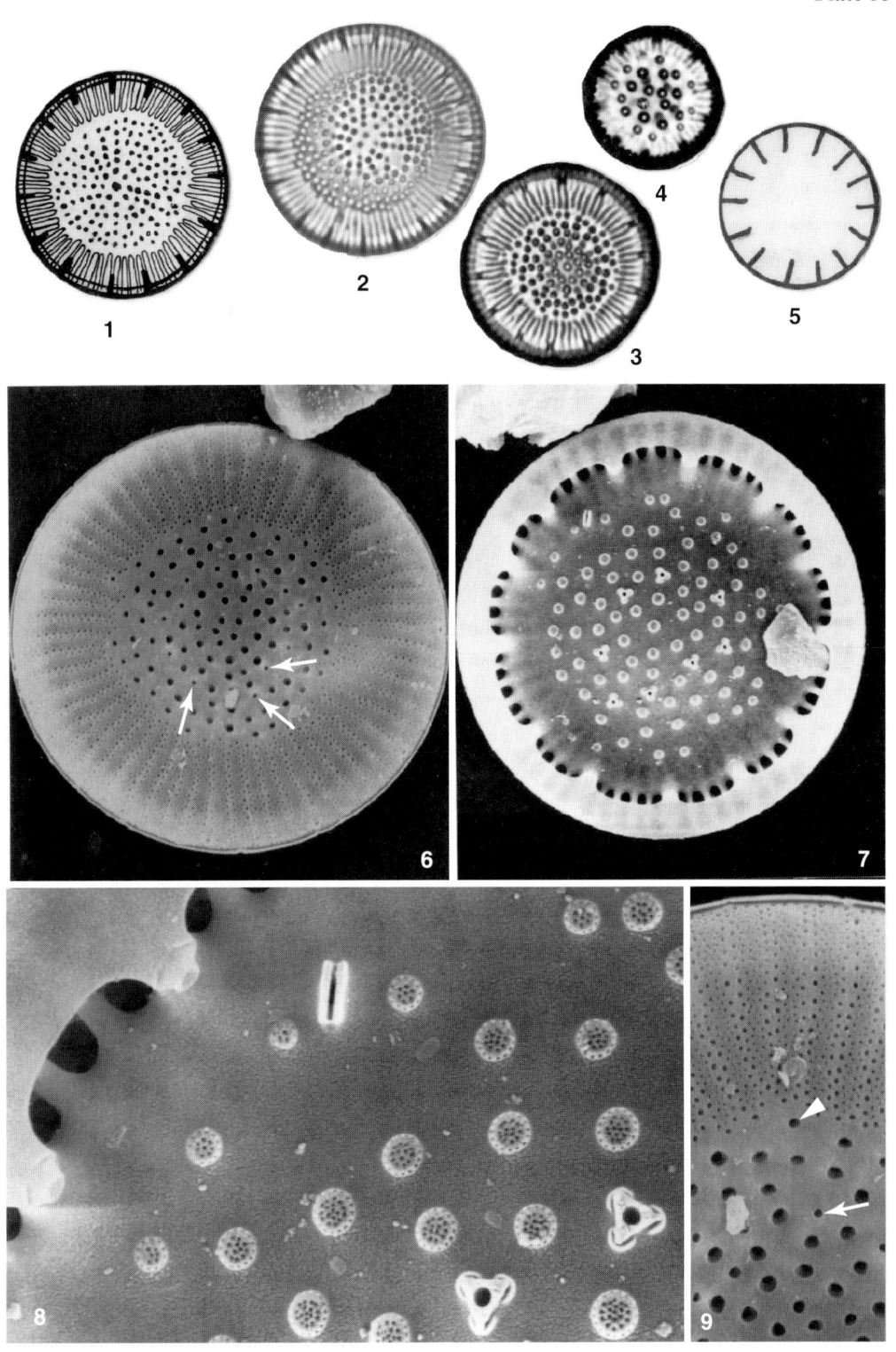

Plate 55

Plate 56 *Puncticulata shanxiensis* (S. Q. Xie & Y. Z. Qi) Nagumo comb. nov.

Fig. 1.   Drawing. ×2000.
Figs 2-5.   LM. ×2000.
Figs 6-10.   SEM.

Figs 1-5.   Valve view.
Fig. 6.   Internal view of valve showing marginal strutted processes with two struts. ×5000.
Fig. 7.   External view of a valve face showing outer openings of alveolus. ×5000.
Fig. 8.   Detail of broken valve showing marginal strutted processes with two struts. ×15000.
Fig. 9.   Detail of valve margin showing external openings of marginal strutted processes (arrows). ×30000.
Fig. 10.   Detail of marginal striae internally, showing marginal strutted processes with two struts and a labiate process (arrow). ×15000.

Locality
Figs 2-10.   Inabe-gawa (Inabe River), Mie.

Plate 56

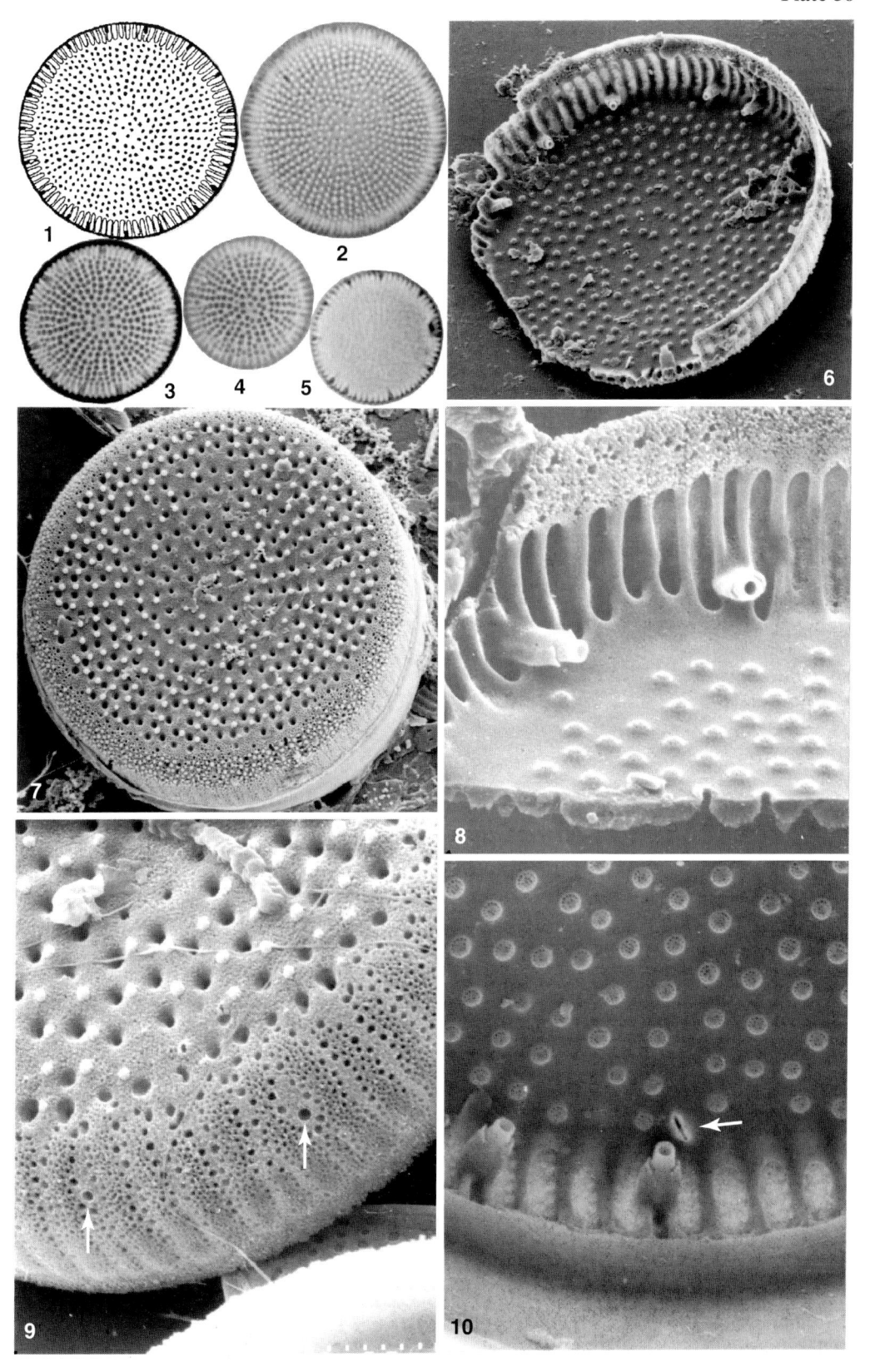

Plate 57 *Stephanodiscus hantzschii* f. *tenuis* (Hustedt) Håkansson & Stoermer

Figs 1, 2, 4-7. LM. ×2000.
Fig. 3. TEM. ×2000.
Fig. 8. Drawing.
Figs 9-12. SEM.

Figs 1-7. Valve view.
Fig. 8. Diagramatic representation of the features. a) marginal spines, b) marginal strutted process, c) vertical slit-like marking of the flange, d) outer opening of labiate process, e) areolar rows on the valve mantle, f) flange, g) pattern center, h) interfascicle, i) fascicle.
Fig. 9. External view of valve showing acute marginal spines and marginal strutted processes. ×5000.
Fig. 10. External view of thick valve. ×5000.
Fig. 11. Internal view of areolae occluded by domed cribra. ×50000.
Fig. 12. Internal view of valve showing marginal strutted processes and a labiate process. ×8000.

Localities
Figs 1-4, 6, 7, 9-12. Waku-ike (Waku Pond), Nagano.
Fig. 5. Naka-gawa (Naka River), Tokyo.

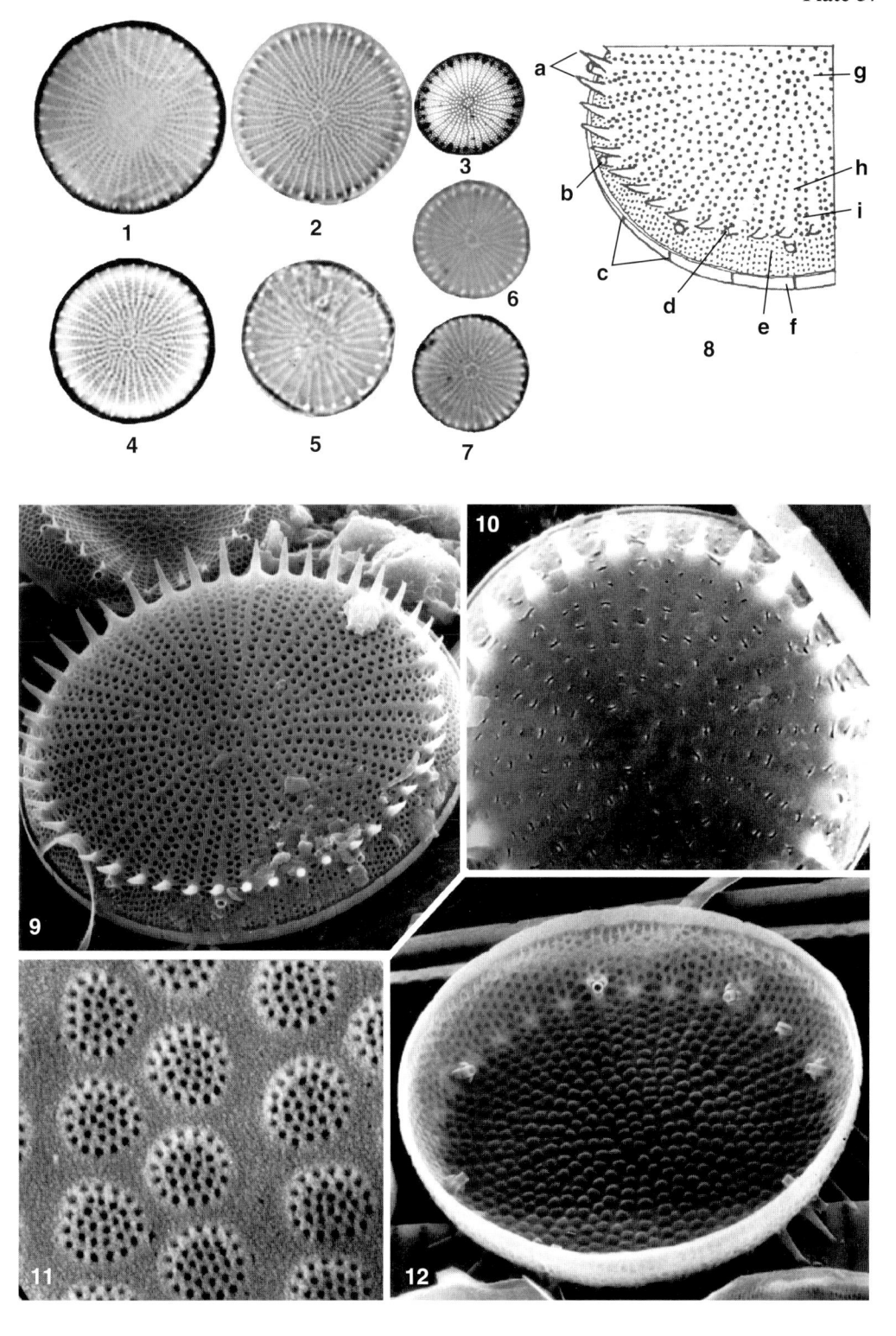

Plate 57

Plate 58 *Stephanodiscus hantzschii* f. *tenuis* (Hustedt) Håkansson & Stoermer

Figs 13,14. LM. ×2000.
Fig. 15. SEM.
Figs 16, 17. TEM.

Figs 13, 14. Type of Hustedt.
Fig. 15. Side view of frustule showing acute and truncate spines. ×8000.
Fig. 16. Valve view. ×7000.
Fig. 17. Areolae occluded by cribra. ×20000.

Localities
Figs 13,14. Hustedt Collection, slide 51/1.
Figs 15-17. Waku-ike (Waku Pond), Nagano.

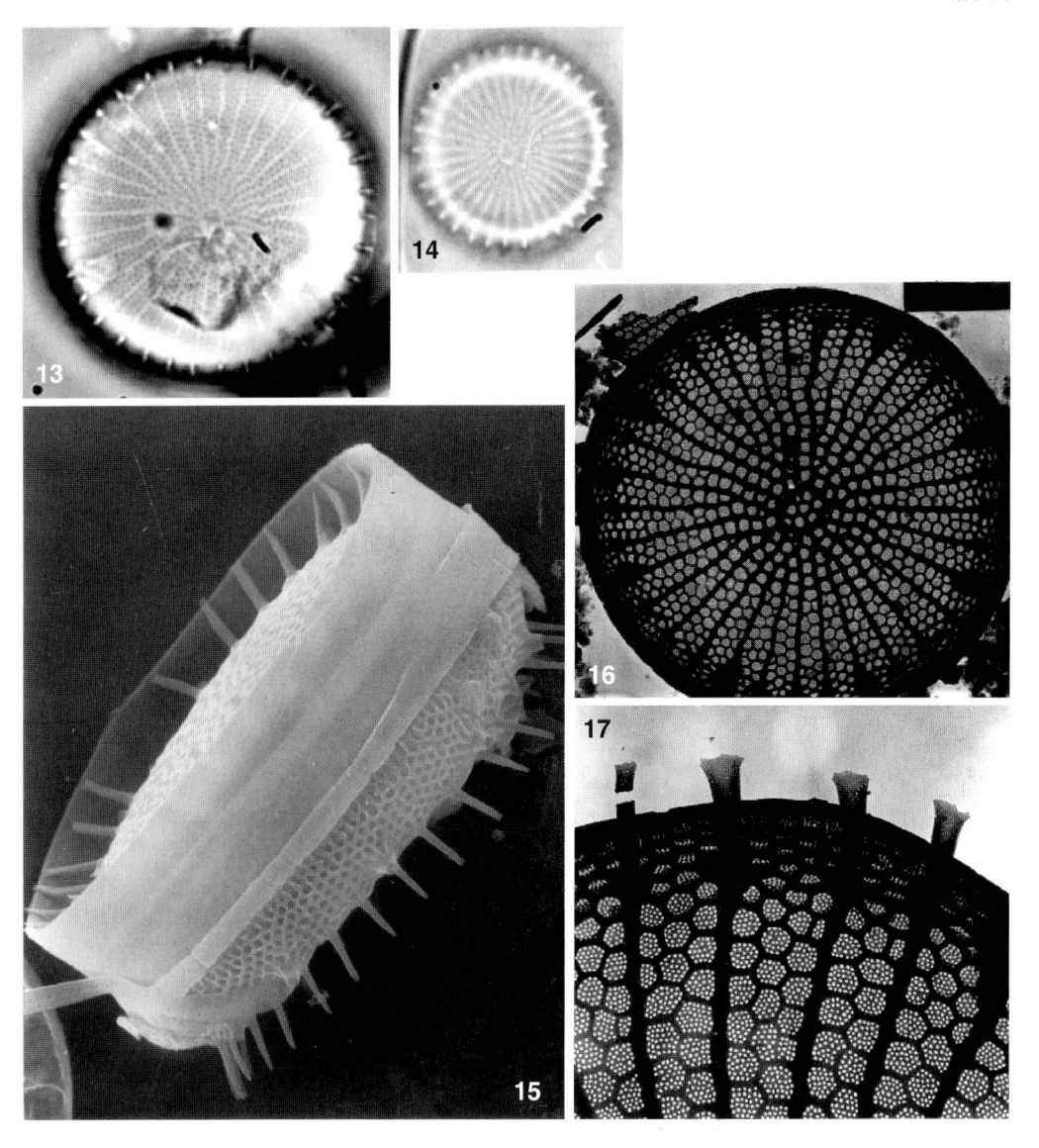

Plate 58

Plate 59 *Stephanodiscus minutulus* (Kützing) Round

Figs 1-5.   LM. ×2000.
Figs 6, 11.   TEM. ×2000.
Fig. 7.   Drawing.
Figs 8-10.   SEM.

Figs 1-6.   Valve view.
Fig. 7.   Diagrammatic representation of the features: a) marginal spines, b) marginal strutted process, c) vertical slit-like marking of the flange, d) outer opening of labiate process, e) areolar rows on the valve mantle, f) flange, g) central strutted process, h) pattern center, i) interfascicle, j) fascicle, k) asteroid arrangement of pores.
Fig. 8.   External view of heavily silicified valve showing acute marginal spines, central strutted process (arrow) and marginal strutted processes. ×10000.
Fig. 9.   External view of valve with flat surface. ×10000.
Fig. 10.   Internal view of valve showing marginal strutted processes, central strutted process and labiate process (arrow). ×10000.
Fig. 11.   Valve view. ×8000.

Localities
Figs 1-3, 8-10.   Hime-numa (Hime Pond), Rishiri Island, Hokkaido.
Fig. 4.   Wakkanai Oh-numa (Wakkanai Oh Pond), Hokkaido.
Figs 5, 6, 11.   Hachiro-gata (Lake Hachiro), Akita.

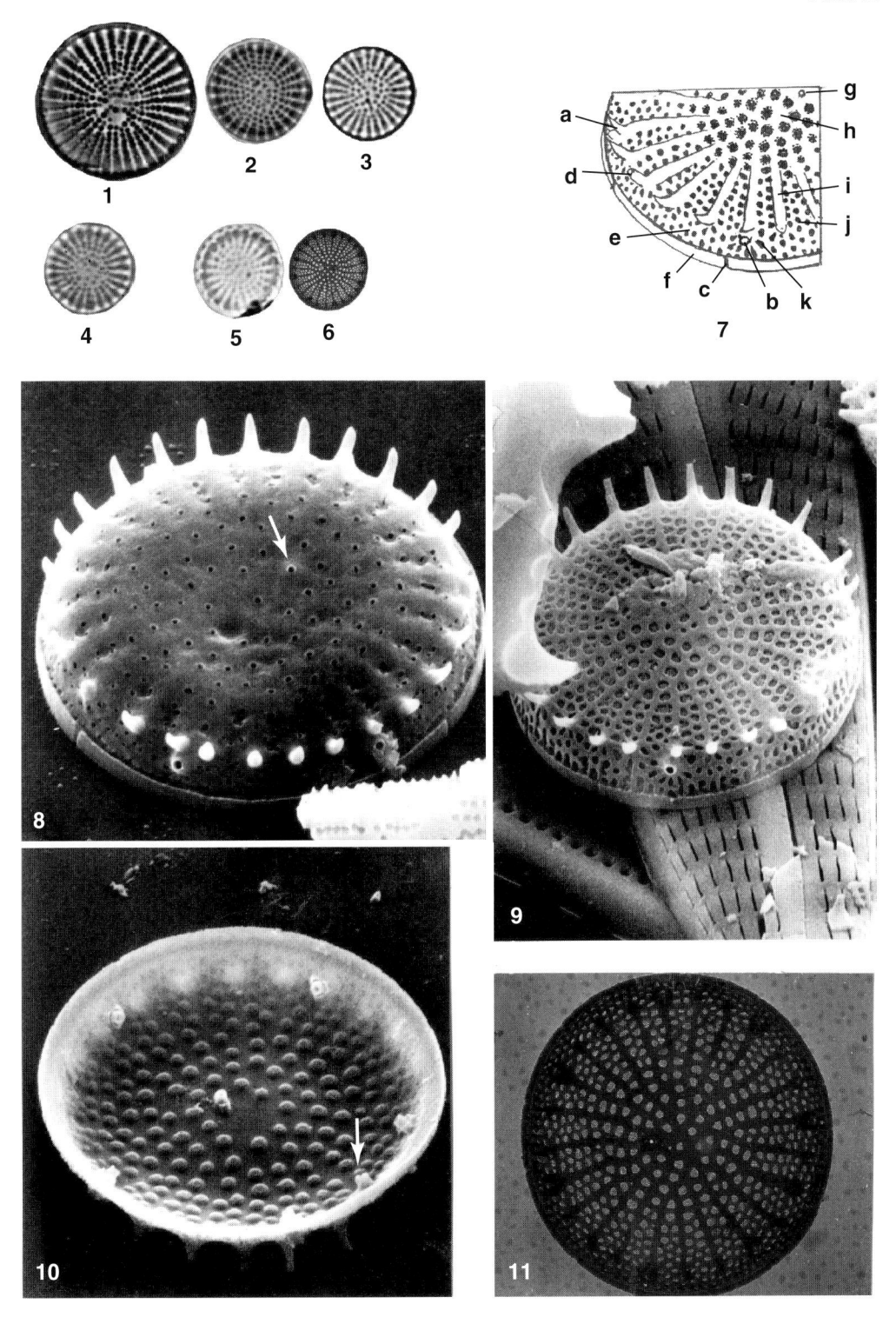

Plate 59

Plate 60 *Stephanodiscus minutulus* (Kützing) Round

Figs 12-18. LM. ×2000.
Figs 19, 24. TEM. ×2000.
Fig. 20. Drawing.
Figs 21-23. SEM.

Figs 12-19. Valve view.
Fig. 20. Diagrammatic representation of the exterior valve structure.
Figs 21, 22. External view of heavily silicified valve with undulate surface. ×10000.
Fig. 23. Internal view of valve showing marginal strutted processes, central strutted process and labiate process. ×10000.
Fig. 24. Valve view. ×10000.

Localities

Figs 12, 13, 15, 22, 23. Hime-numa (Hime Pond), Rishiri Island, Hokkaido.
Figs 14, 21. Wakkanai Oh-numa (Wakkanai Oh Pond), Hokkaido.
Figs 16-18, 24. Hachiro-gata (Lake Hachiro), Akita.
Fig. 19. Waku-ike (Waku Pond), Nagano.

Plate 60

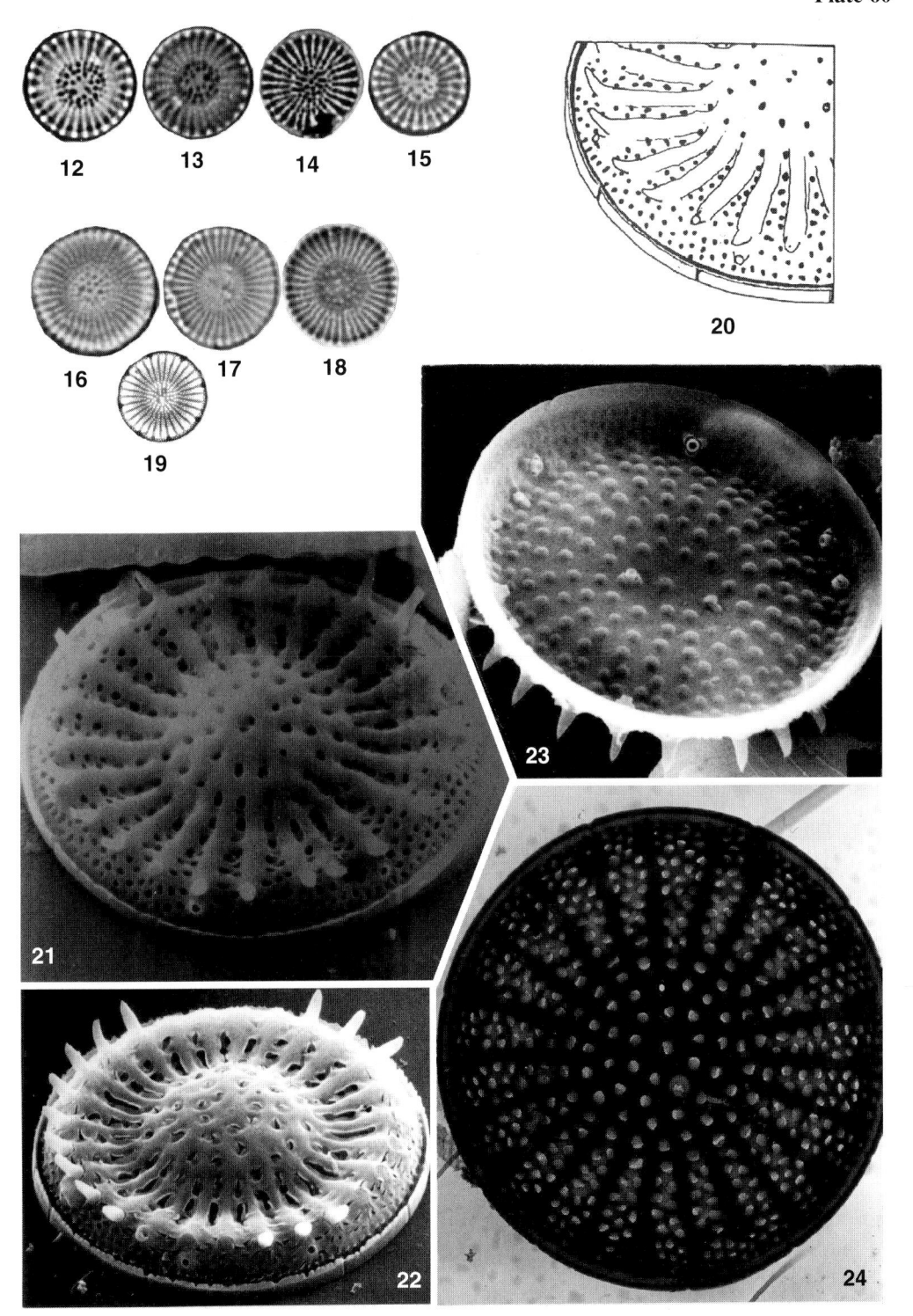

Plate 61 *Stephanodiscus rotula* (Kützing) Hendey

Figs 1, 3.   LM. ×2000.
Figs 2, 4-6.   SEM.

Figs 1, 3.   Valve view.
Fig. 2.   Internal view of valve showing striations and marginal strutted processes. ×1000.
Fig. 4.   External view of a frustule showing undulate valve surface and marginal spines. ×2000.
Fig. 5.   Internal view of the valve showing areolation on the valve face and mantle, marginal strutted processes with three satellite pores and a labiate process. ×8000.
Fig. 6.   Enlargement of the labiate process and the strutted processes. ×10000.

Localities

Fig. 1.   A micrograph from type slide, BM 17997, Coll. Kützing.
Figs 2-6.   Nakanojo, Gunma.

Plate 61

Plate 62 *Eucampia zodiacus* Ehrenberg

Fig. 1.　Drawing. ×2000.
Figs 2-4.　LM. ×2000.
Figs 5-8.　SEM.

Fig. 1.　Colony of girdle view.
Fig. 4.　Living cells.
Fig. 5.　Oblique view of frustule showing perforations of areorae occlusion and external opening of labiate process. ×1000.
Fig. 6.　Oblique view of frustule showing bands and labiate processes. ×1500.
Fig. 7.　Internal view of valve showing perforations of areorae in radially and labiate process. ×2000.
Fig. 8.　Internal view of valve center showing labiate process. ×5000.

Locality
Figs 2-7.　Coast of Shimoda, Shizuoka.

Plate 62

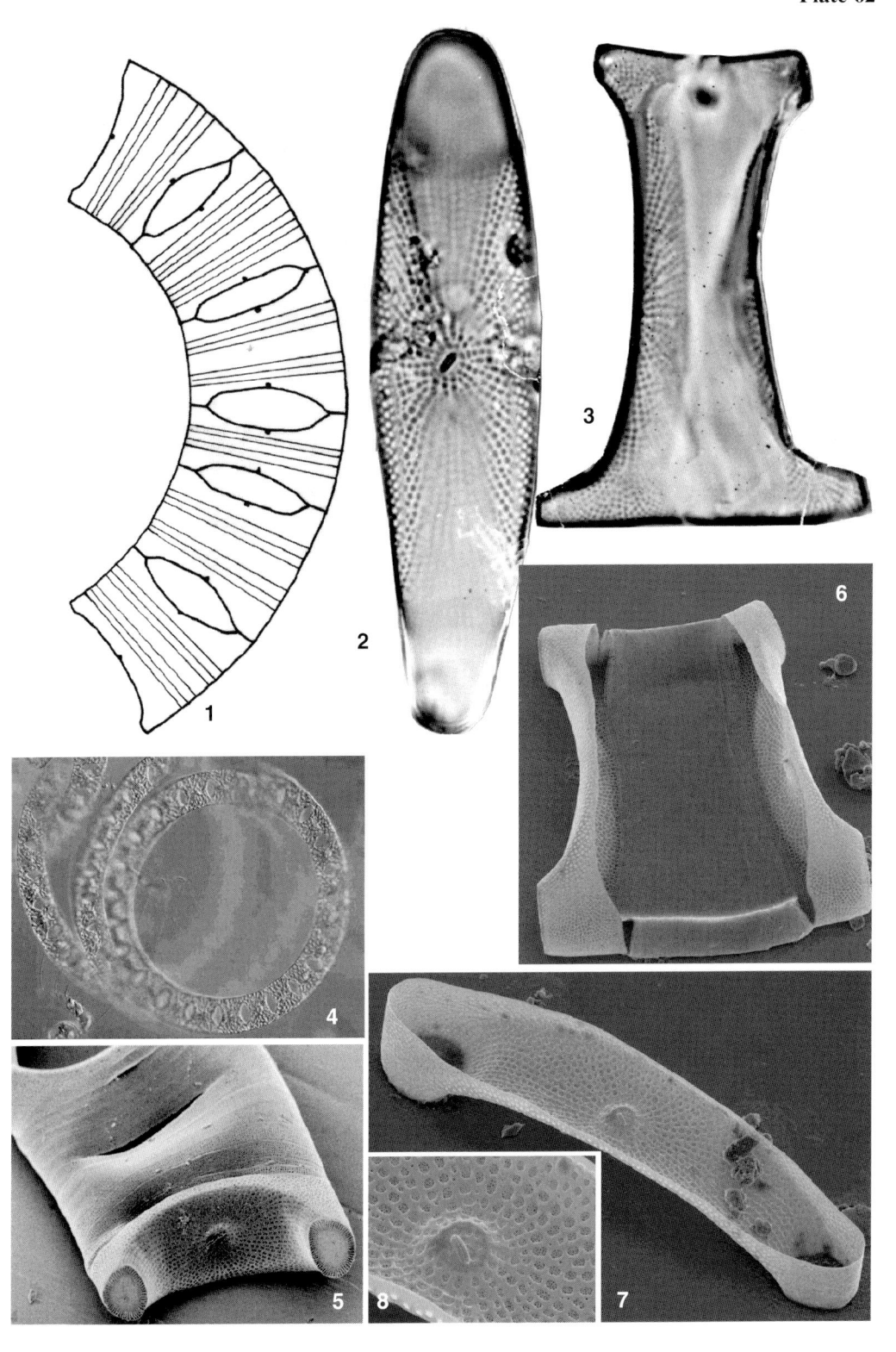

271

Plate 63 (Figs 1a-e) *Asterionella formosa* Hassall

Figs 1a-c. LM. ×2000.
Fig. 1d. Drawing. ×2000.
Fig. 1e. LM. ×300.

Figs 1a-d. Valve view.
Fig. 1e. Stellate colonies formed by the frustules joined at their footpoles.

Localities
Figs 1a, 1b, 1e. Yuno-ko (Yu Pond), Tochigi.
Fig. 1c. Biwa-ko (Lake Biwa), Shiga.

Plate 63 (Figs 2a-g) *Asterionella gracillima* (Hantzsch) Heiberg

Figs 2a-d. LM. ×2000.
Figs 2e, 2g. Drawings. Fig. 2e. ×2000.
Fig. 2f. LM. ×500.

Fig. 2a. Girdle view of sibling frustules.
Figs 2b-e. Valve view of the valve.
Fig. 2f. Zigzag colony formed by cells with small, plate-like chloroplasts.
Fig. 2g. Drawing of a zigzag colony.

Locality
Figs 2a-d, 2f. Biwa-ko (Lake Biwa), Shiga.

Plate 63

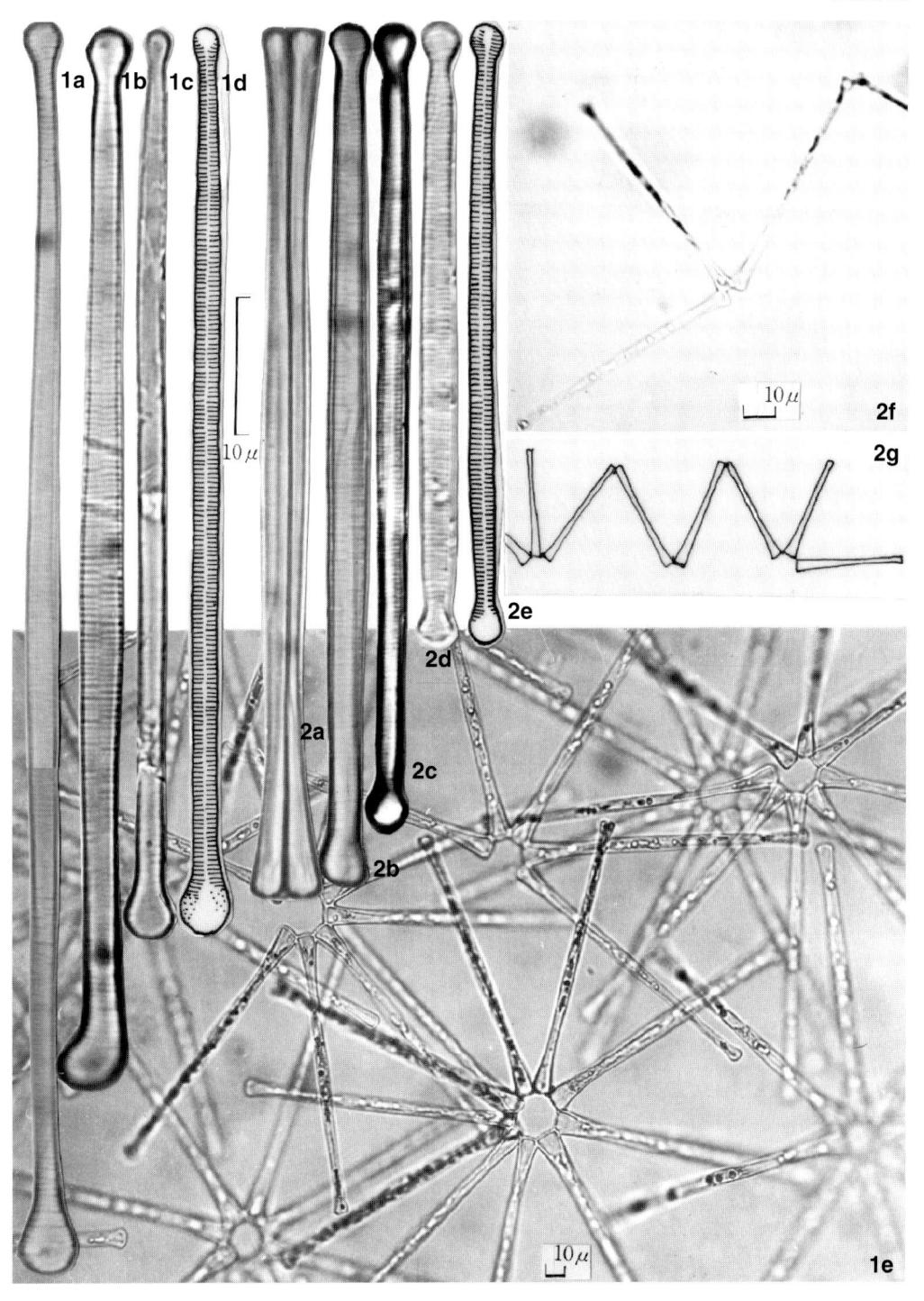

273

Plate 64　*Asterionella ralfsii* W. Smith

Figs 1-3.　LM.　×2000.
Fig. 6.　LM.　×1000.
Figs 4, 5.　Drawings.　×2000.

Figs 1, 2, 4, 5.　Valve view of the valve.
Fig. 3.　Girdle view of sibling frustules.
Fig. 6.　Stellate colony.

Locality
Figs 1-3, 6.　Bogs of Naeba-san (Mt. Naeba), Niigata.

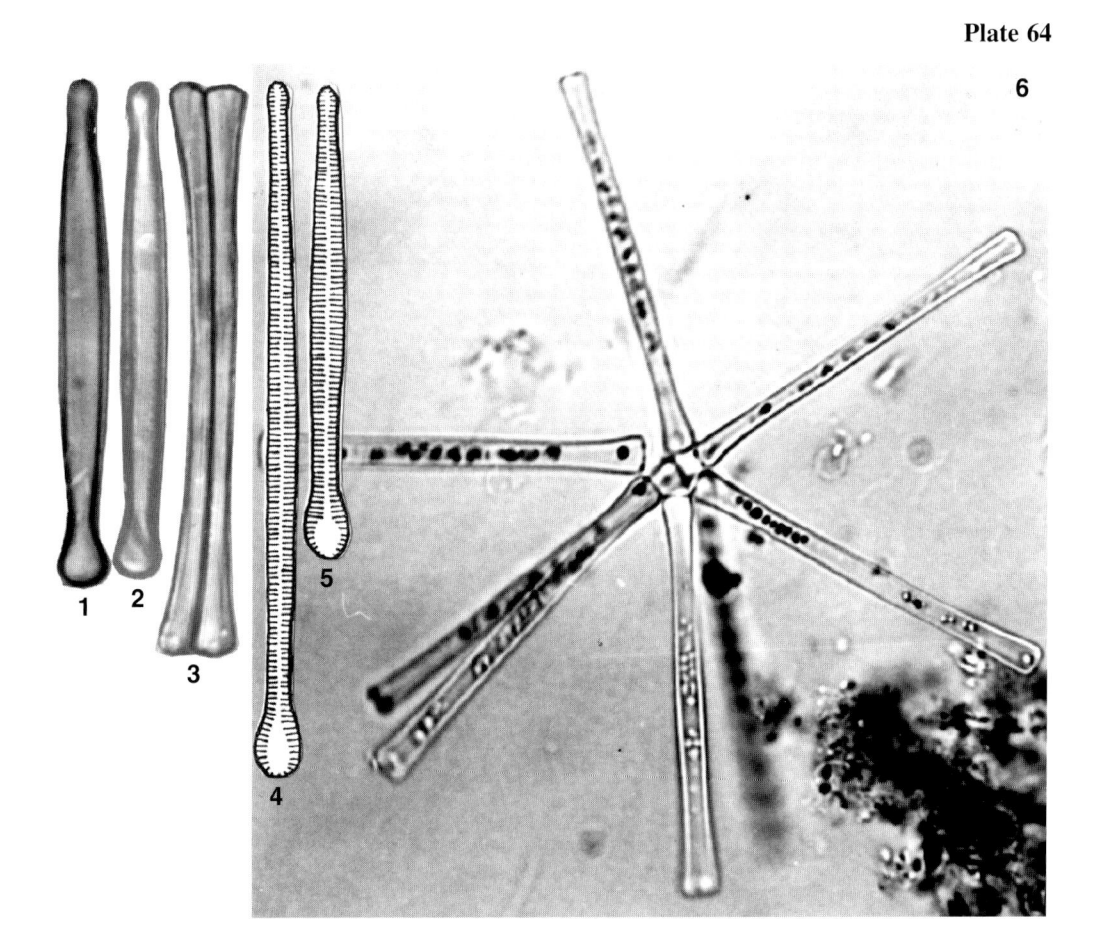

Plate 64

Plate 65 *Asterionellopsis glacialis* (Castracane) Round

Fig. 1. LM. ×200.
Fig. 2. LM. ×750.
Figs 3-5. SEM.
Fig. 6. Drawing. ×2000.

Fig. 1. Twisting filamentous colonies.
Fig. 2. Enlargement of a colony, showing girdle view of cells with two chloroplasts.
Fig. 3. Colonies formed by the frustules jointed at their footpoles. ×1250.
Fig. 4. Girdle view of headpole showing pointed spines arranged in the margin of the valve. ×10000.
Fig. 5. External view of footpole showing the areolae and the narrow sternum (arrow) in the valve (v), the joint (arrowhead) between ocelli, and the valvocopula (vc) with rows of small areolae. ×9000.

Locality
Figs 1-5. Tokyo-wan (Tokyo Bay), Tokyo.

Plate 65

Plate 66 *Catacombas obtusa* (Pantocsek) Snoeijs

Fig. 1.   Drawing. ×2000.
Figs 2, 3.   LM. ×2000.
Figs 4-6.   SEM.

Fig. 4.   External polar view of frustule showing ocellulimbus, complete epicingulum, an opening of labiate process and unoccluded pores nearest ocellulimbus. ×6000.
Fig. 5.   External view of valve showing areolation interrupted by the valve shoulder. ×10000.
Fig. 6.   Internal view of valve showing apertures of the striae. ×15000.

Localities
Fig. 2.   Sumida-gawa (Sumida River), Ryogoku, Tokyo.
Figs 3-6.   Ara-kawa (Ara River), Kasai, Tokyo.

Plate 66

Plate 67 *Ctenophora pulchella* (Rakfs ex Kützing) D. M. Williams & Round

Figs 1-4. LM. ×2000.
Fig. 5. Drawing. ×2000.
Figs 6-9. SEM.

Fig. 6. Internal view of valve center showing hyaline central area and striation. ×5000.
Fig. 7. External view of valve pole showing areolae, opening of labiate process and ocellulimbus. ×10000.
Fig. 8. Internal view of valve showing areolation and labiate process. ×10000.
Fig. 9. Internal view of valve center. ×8000.

Localities
Figs 3, 4. BM 18310, Ralfs 193.
Figs 6-9. Tenryu-gawa (Tenryu River), Shizuoka.

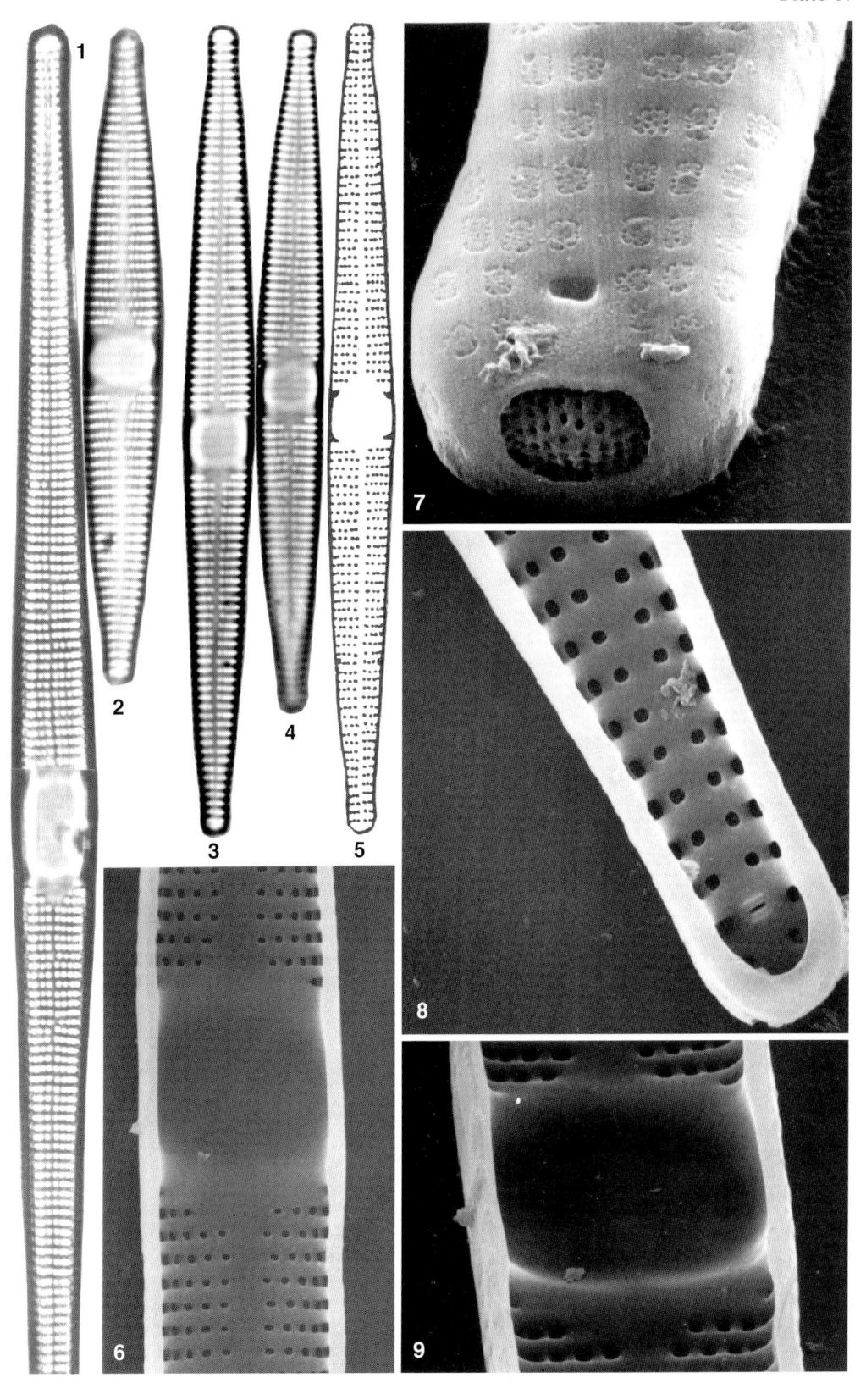

Plate 67

Plate 68 *Diatoma mesodon* (Ehrenberg) Kützing

Figs 1, 3-8.   LM. ×2000.
Figs 2, 12.   Drawings.
Figs 9-11, 13.   SEM.

Figs 1-5.   Valve view. ×2000.
Fig. 3.   Living filamentous colony.
Fig. 7.   Girdle view of normal vegetative frustules.
Fig. 8.   Girdle view of the frustule with an inner valve (arrow).
Fig. 9.   Internal oblique view showing striae composed of poroid areolae, transverse costae, labiate process and apical pore field. ×5000.
Fig. 10.   External oblique view showing the spinules on the valve shoulder, i.e. the valve face and mantle juncture. Note the outer opening of the labiate process (arrow). ×5000.
Fig. 11.   External view of the valve without spinule on the valve shoulder showing apical pore field at both ends and the outer opening of the labiate process. ×5000.
Fig. 12.   Whole frustule. A copy traced from the specimen of Fig. 13. ×3250.
Fig. 13.   Whole frustule showing bands curved extremely at both ends of the frustule. ×5000.

Localities
Figs 1, 5, 10, 11.   Ara-kawa (Ara River), Saitama.
Figs 3, 4, 7, 8.   Yuno-ko (Yu Pond), Tochigi.
Fig. 6.   A pond on Mt. Naeba, Niigata.
Fig. 9.   Kanna-gawa (Kanna River), Shimokubo Dam, Gunma.
Figs 12, 13.   Chikuma-gawa (Chikuma River), Kawakami, Nagano.

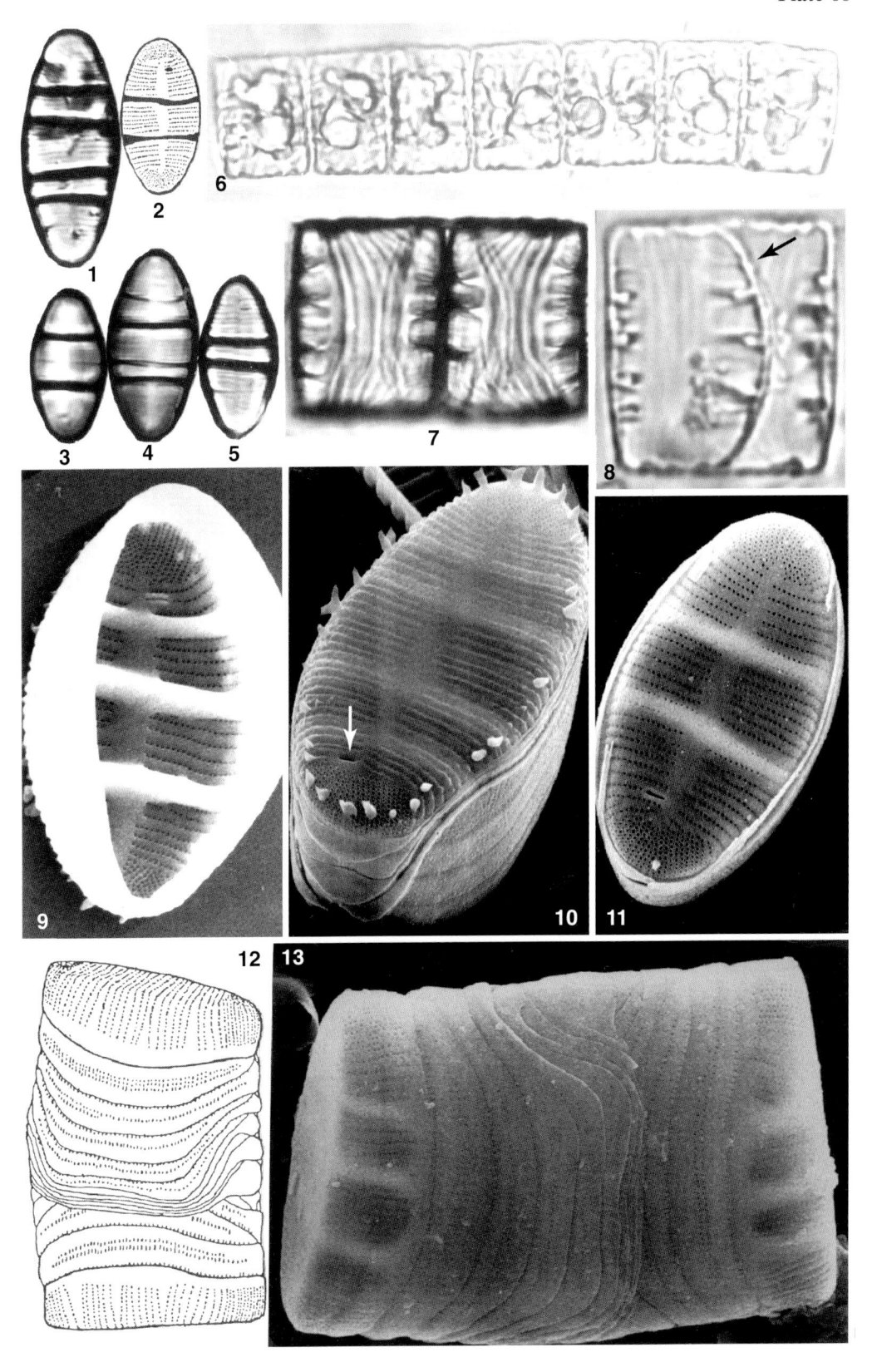

Plate 68

283

Plate 69   *Diatoma tenuis* C. Agardh

Figs 1, 2, 4.   LM. ×2000.
Fig. 3.   Drawing. ×2000.
Figs 5-9.   SEM.

Figs 1-4.   Valve view.
Fig. 5.   Internal oblique view of the valve showing narrow axial area, inner openings of poroid areolae, transverse costae, labiate process (arrow) and apical pore field. ×10000.
Fig. 6.   External oblique view of the valve showing flat valve surface and round valve shoulder. ×10000.
Fig. 7.   Enlarged valve end showing the apical pore field and the outer opening of the labiate process (arrow). ×20000.
Fig. 8.   Internal view of the valve end showing apical pore field. ×20000.
Fig. 9.   Opposite end of the valve in Fig. 8 showing the labiate process (arrow) on the transverse costa. ×20000.

Localities

Fig. 1.   Kasumiga-ura (Lake Kasumigaura), Ibaraki.
Fig. 2.   Yuno-ko (Yu Pond), Nikko, Tochigi.
Fig. 4.   Shikotsu-ko (Lake Shikotsu), Hokkaido.
Figs 5, 8, 9.   Chikuma-gawa (Chikuma River), Kamiyamada, Nagano.
Figs 6, 7.   Chikuma-gawa (Chikuma River), Koumi, Nagano.

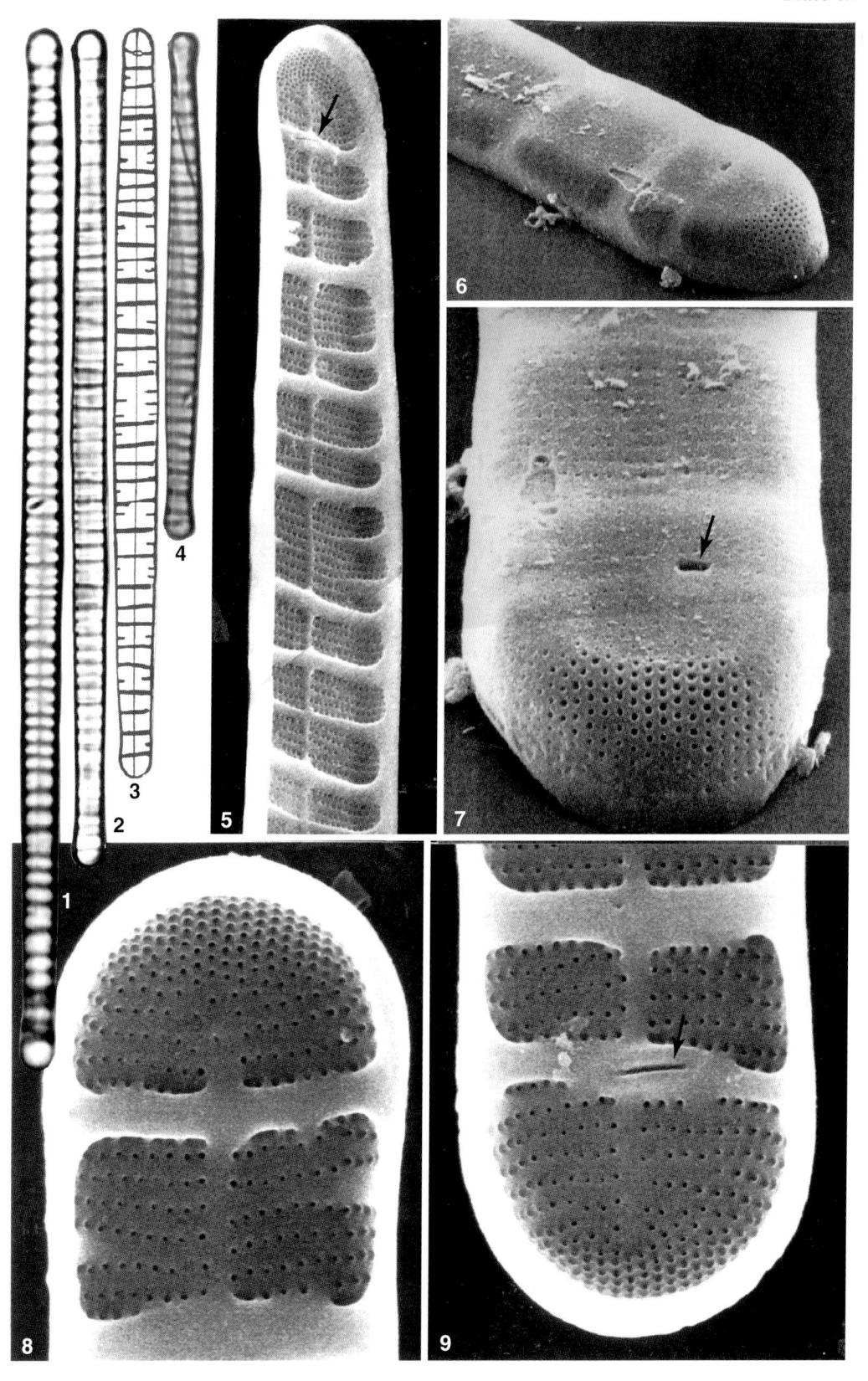

Plate 69

Plate 70 *Diatoma vulgaris* Bory

Figs 1-5. LM.
Figs 6, 7. SEM.

Figs 1-4. Valve view. ×2000.
Fig. 5. Living cells forming zigzag colony. ×500.
Fig. 6. Oblique view of the frustules and valves. ×1000.
Fig. 7. Details of the internal valve end showing apical pore field, transverse costae, poroid areolae and labiate process. ×8000.

Localities
Figs 1, 5, 6. Kawaguchi-ko (Lake Kawaguchi), Yamanashi.
Fig. 2. Suwa-ko (Lake Suwa), Nagano.
Figs 3, 4. Kinu-gawa (Kinu River), Tochigi.
Fig. 7. Tama-gawa (Tama River), Ohme, Tokyo.

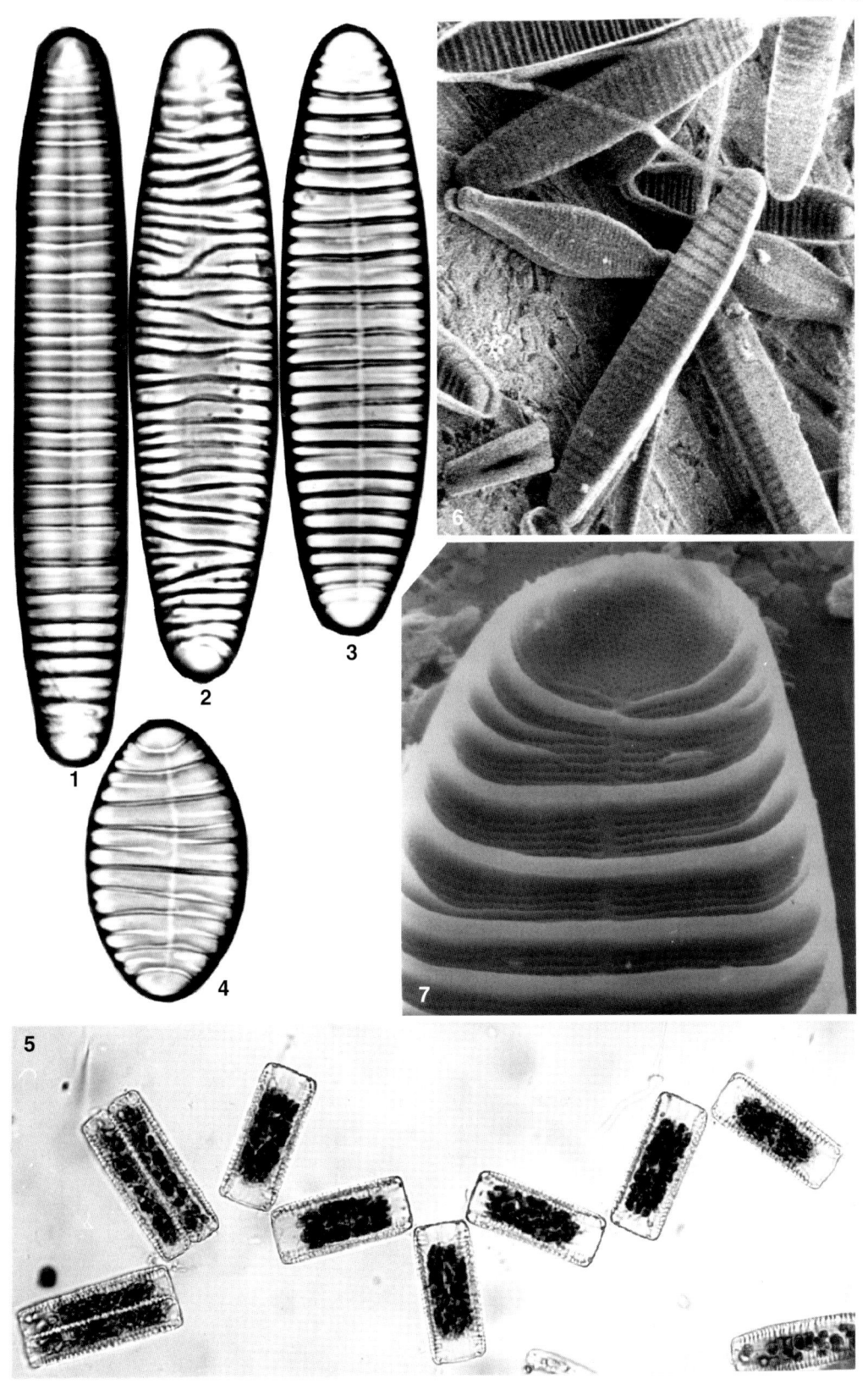

Plate 70

Plate 71 *Fragilaria capitellata* (Grunow) J. B. Petersen

Figs 1-5, 7-9. LM. ×2000.
Fig. 6. Drawing. ×2000.
Figs 10-12. SEM.

Figs 1-5, 7, 8. Valve view.
Fig. 9. Girdle view of two frustules.
Fig. 10. External polar view of valve showing areolation and ocellulimbus. ×20000.
Fig. 11. Internal view of valve with a labiate process. ×3000.
Fig. 12. External view of frustule showing an opening of labiate process, ocellulimbi and cingulum. ×20000.

Localities
Figs 1, 2. Akan-gawa (Akan River), Hokkaido.
Figs 3, 5, 7, 8. Kinu-gawa (Kinu River), Tochigi.
Figs 4, 9-12. Ara-kawa (Ara River), Saitama.

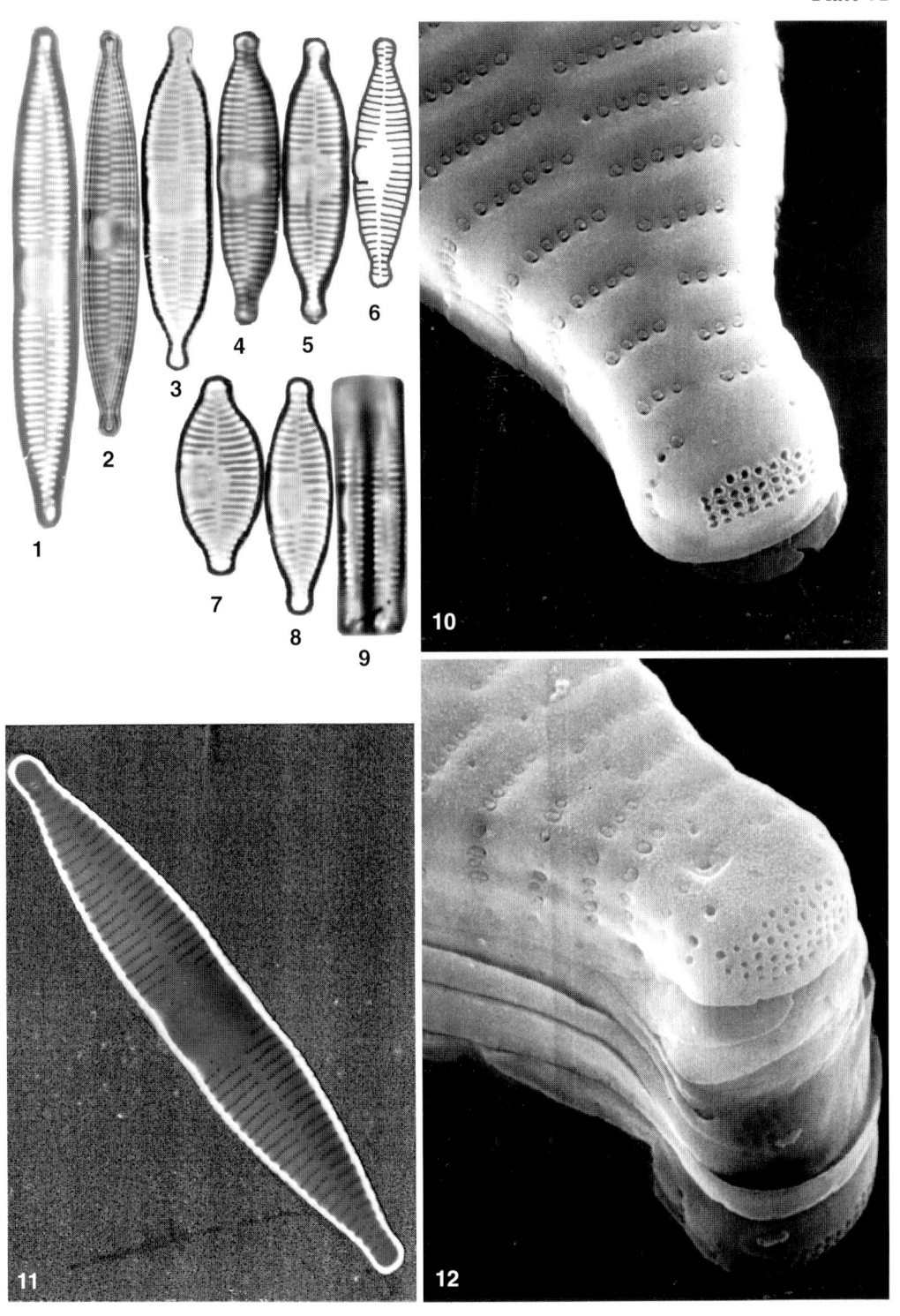

Plate 71

Plate 72 *Fragilaria capitellata* (Grunow) J. B. Petersen

Figs 13, 14.   Drawings. ×2000.
Figs 15-21.   SEM.

Figs 13, 14.   Valve view. ×2000.
Figs 15, 16.   External view of sigmoid valve. ×2000.
Fig. 17.   External polar view of valve showing areolae and an opening of labiate process. ×20000.
Fig. 18.   External polar view of valve showing areolation, ocellulimbus and an opening of labiate process. ×20000.
Fig. 19.   External central view of valve showing areolation and central area. ×10000.
Fig. 20.   External view of areolae with rota occlusion. ×20000.
Fig. 21.   Enlargement of areolae with rota occlusion. ×50000.

Locality
Figs 15-21.   Towada-ko (Lake Towada), Aomori.

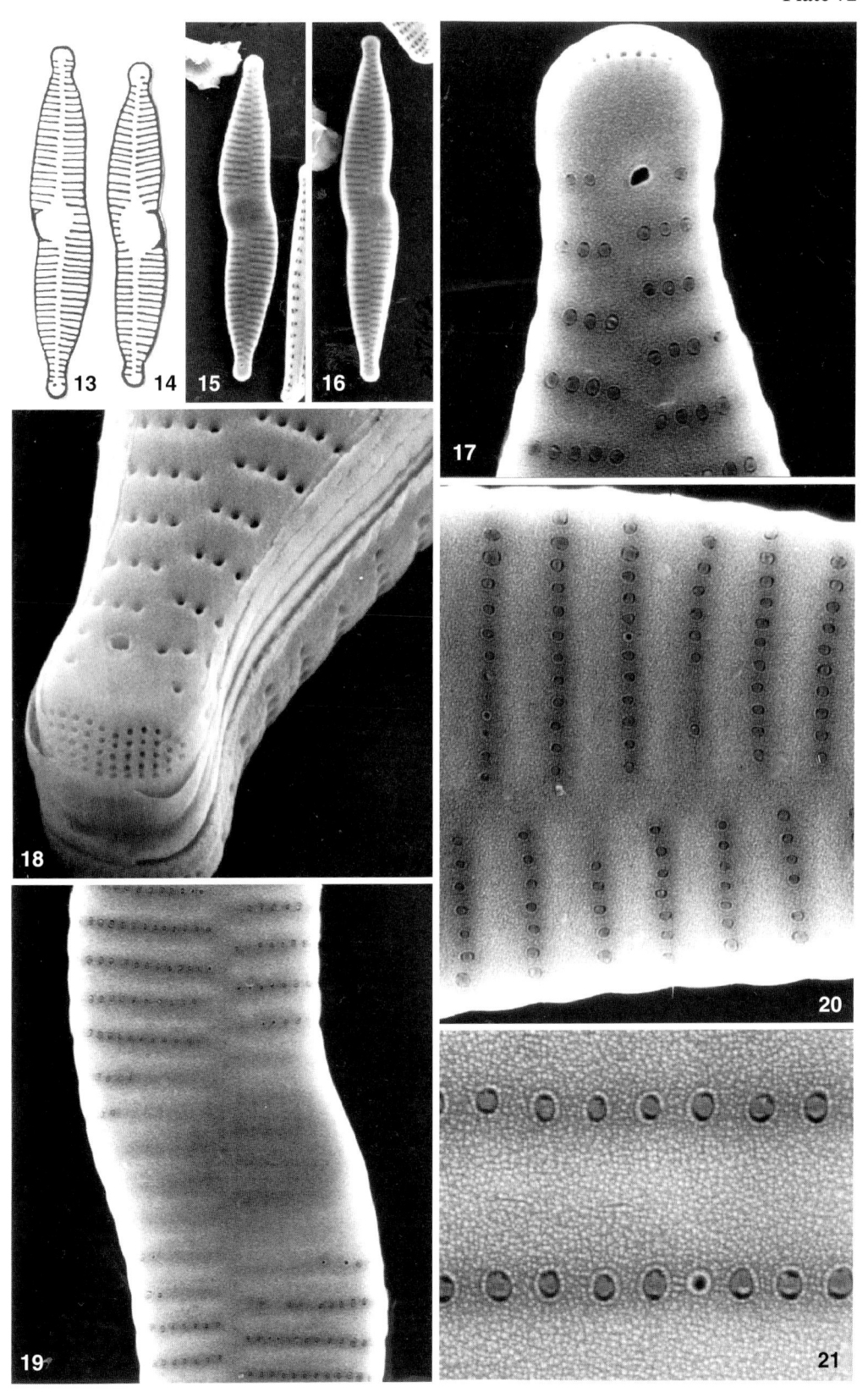

Plate 72

Plate 73 *Fragilaria crotonensis* Kitton

Figs 1, 3-10. LM. ×2000.
Fig. 2. Drawing. ×2000.
Fig. 11. SEM.

Fig. 1. Girdle view of colony.
Figs 2-9. Valve view.
Fig. 10. Girdle view of colony.
Fig. 11. External and internal polar view showing an opening of labiate process, ocellulimbi and two terminal spines. ×20000.

Localities

Figs 1, 4, 6, 11. Yuno-ko (Yu Pond), Tochigi.
Fig. 3. Kizaki-ko (Lake Kizaki), Nagano.
Fig. 5. Yamanaka-ko (Lake Yamanaka), Yamanashi.
Fig. 7. Nojo (Fields), Tokyo Gakugei Univ., Tokyo.
Fig. 8. Mookoppe-gawa (Mookoppe River), Hokkaido.
Figs 9, 10. Tsukui-ko (Lake Tsukui), Kanagawa.

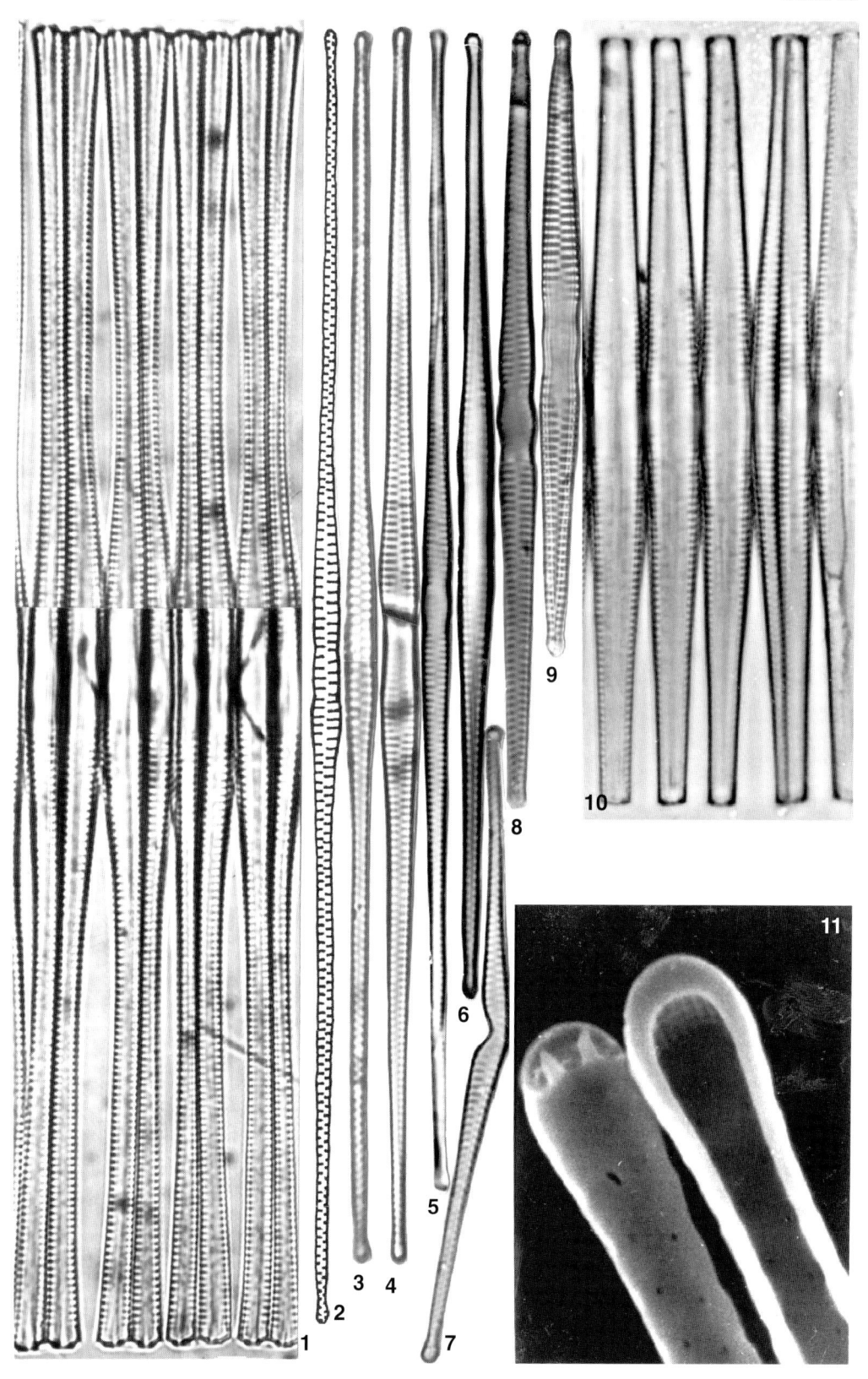

Plate 73

Plate 74 *Fragilaria crotonensis* Kitton

Figs 12-16.   LM. ×2000.
Figs 17-20.   SEM.

Figs 12-16.   Valve view.
Fig. 17.   Colony. ×500.
Fig. 18.   External view of frustules showing linking spines. ×2600.
Fig. 19.   External polar view showing areolation and cingula. ×10000.
Fig. 20.   External view of frustules showing central linking spines. ×10000.

Localities
Figs 12-18, 20.   Tsukui-ko (Lake Tsukui), Kanagawa.
Fig. 19.   Yuno-ko (Yu Pond), Tochigi.

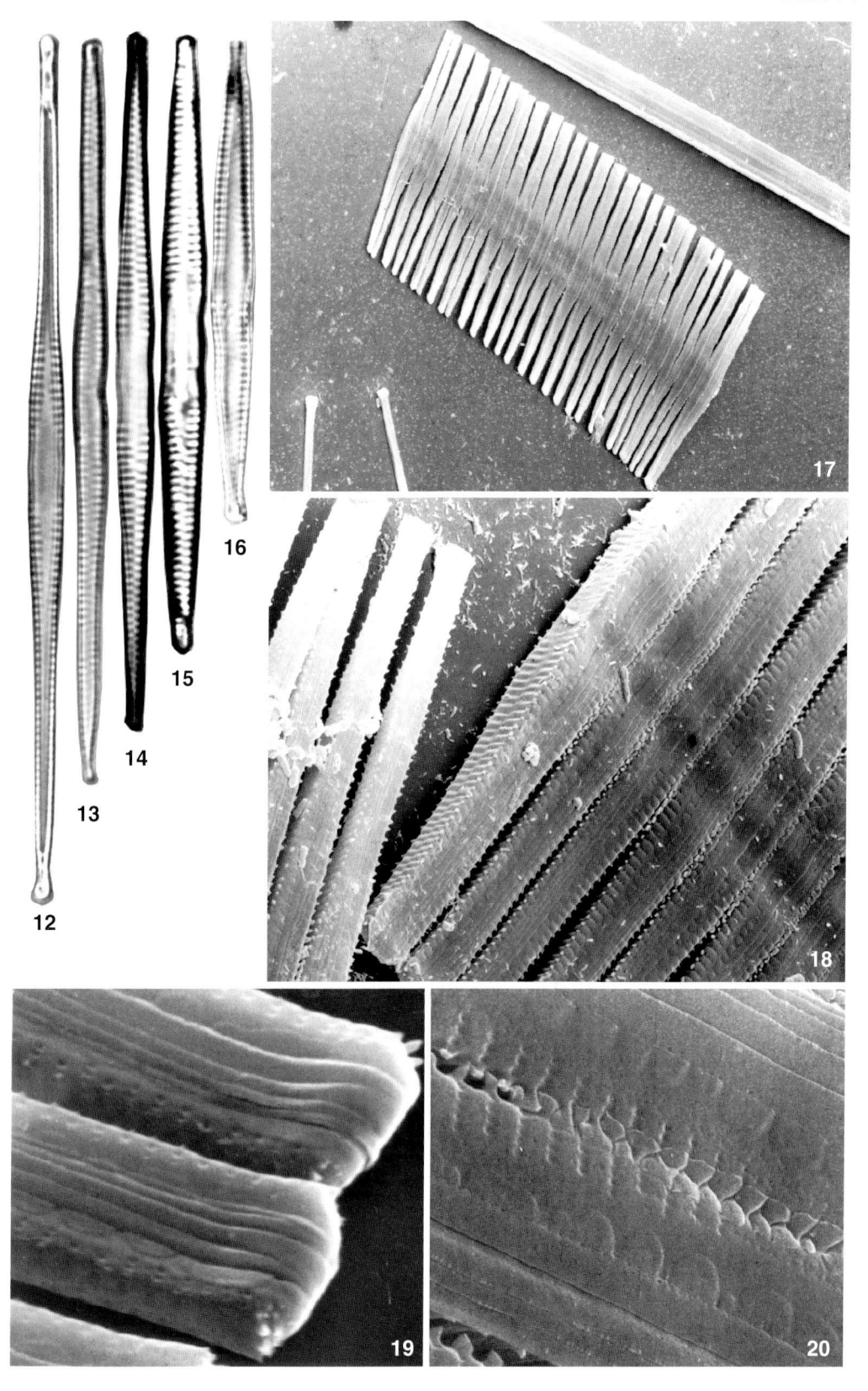

Plate 74

Plate 75 *Fragilaria mesolepta* Rabenhorst

Figs 1-3, 5, 6.   LM.   ×2000.
Fig. 4.   Drawing.   ×2000.
Figs 7-10.   SEM.

Figs 1-5.   Valve view.
Fig. 6.   Girdle view.
Fig. 7.   Internal view of valve center showing striation and central area with ghost striae. ×10000.
Fig. 8.   Internal polar view of linking valves showing a labiate process (arrow) located on the valve mantle. ×12000.
Fig. 9.   Internal polar view showing a labiate process (arrow) and areolation. ×20000.
Fig. 10.   External polar view showing cingulum, linking spines and ocellulimbus. ×36000.

Localities

Figs 1-3, 6, 7, 9, 10.   Lake Päijänne, Finland.
Fig. 5.   Biwa-ko (Lake Biwa), Shiga.
Fig. 8.   Akan-ko (Lake Akan), Hokkaido.

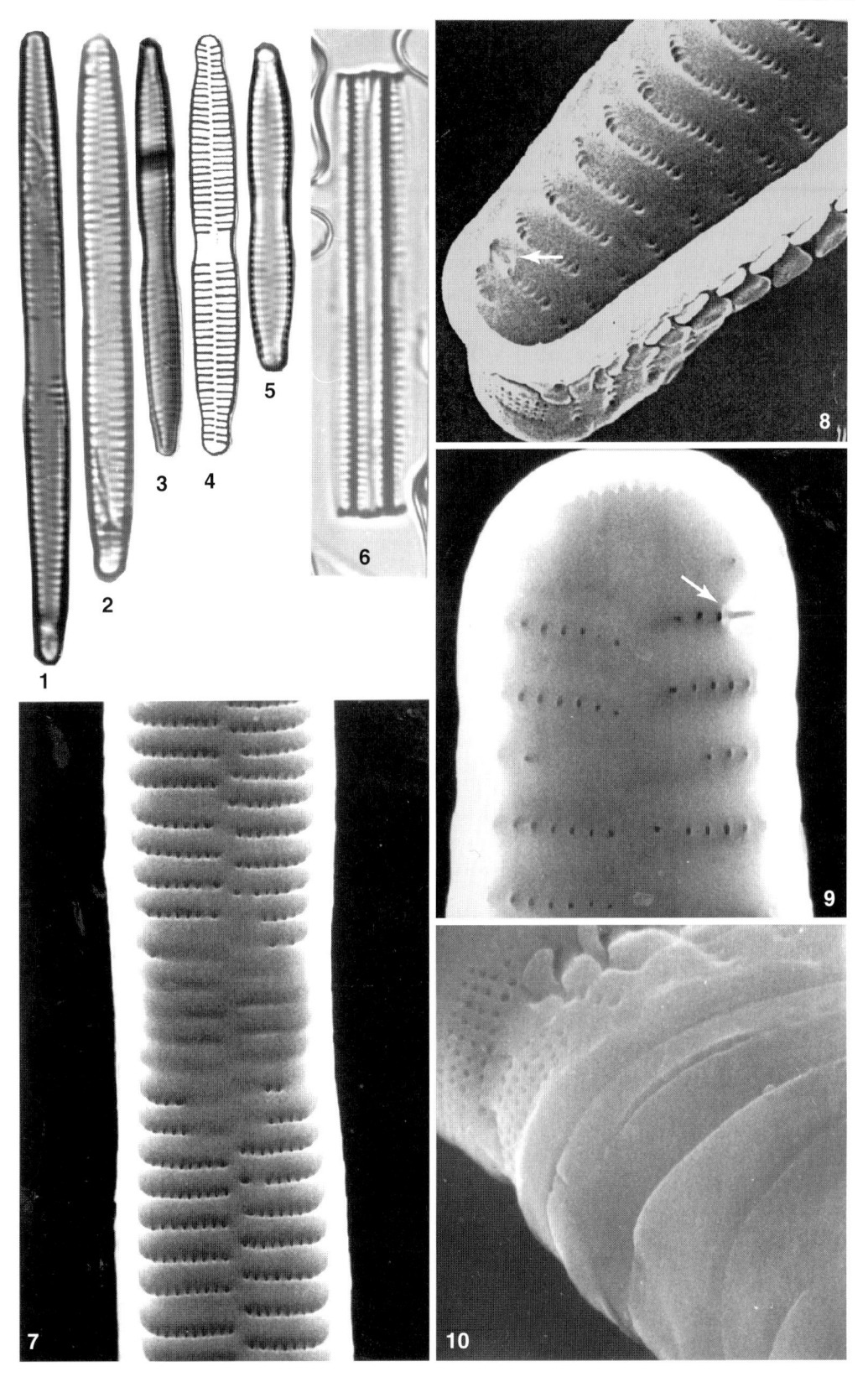

Plate 75

Plate 76 *Fragilaria neoproducta* Lange-Bertalot

Figs 1-3, 5, 6. LM. ×2000.
Fig. 4. Drawing. ×2000.
Figs 7-11. SEM.

Figs 1-5. Valve view.
Fig. 6. Girdle view of colony.
Fig. 7. External view of valve showing striation and marginal linking spines. ×3100.
Fig. 8. Internal view of valve with sibling valve. ×3100.
Fig. 9. External view of linking valves showing linking spines and cingulum. ×5000.
Fig. 10. External view of areolation and linking spines. ×10700.
Fig. 11. External polar view showing ocellulimbus and striation. ×13000.

Locality

Figs 1-11. Akan-ko (Lake Akan), Hokkaido.

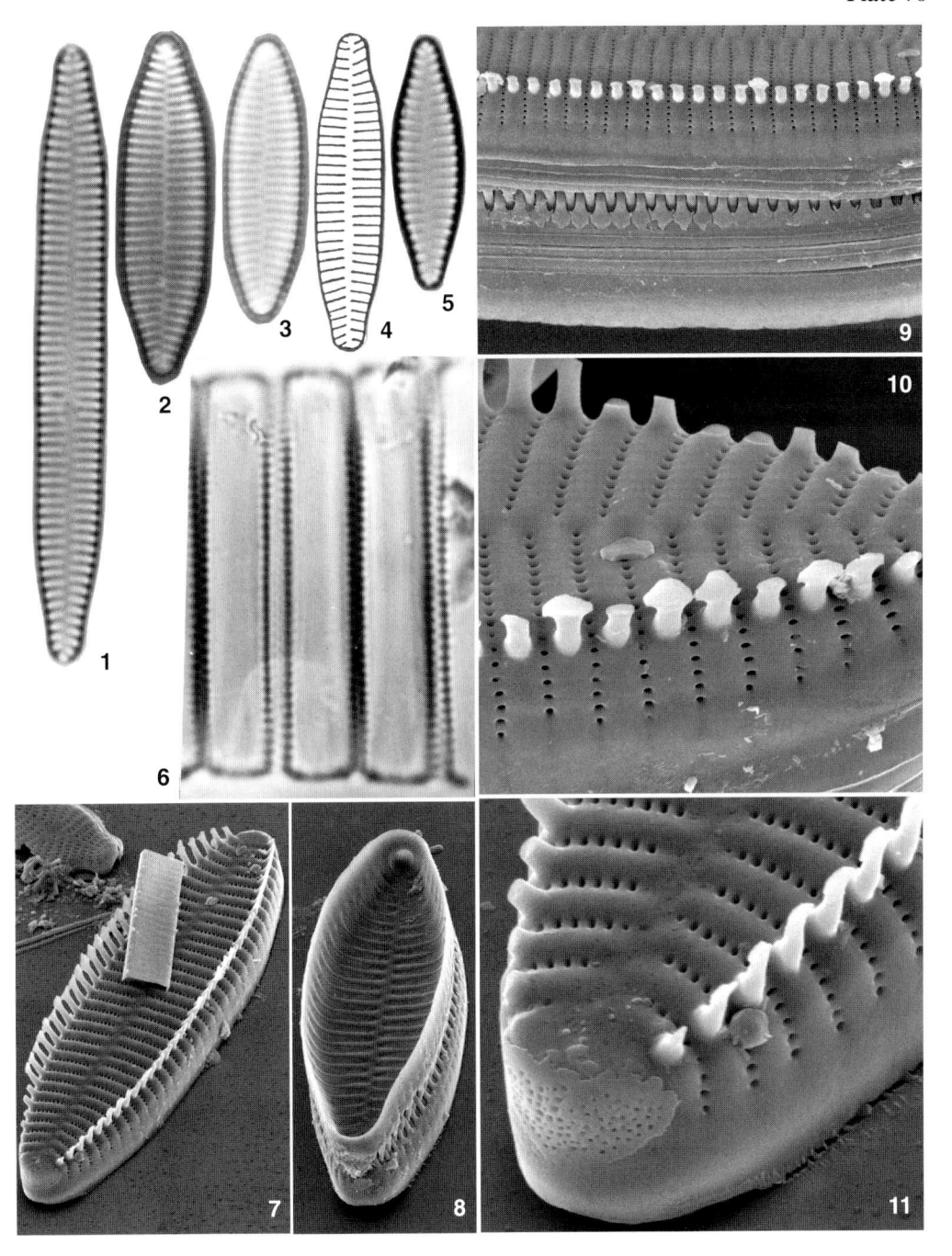

Plate 76

Plate 77 *Fragilaria perminuta* (Grunow) Lange-Bertalot

Figs 1-8.   LM.   ×2000.
Figs 9-11.   SEM.

Fig. 9.   External view of valve showing areolae with rota occlusion.   ×20000.
Fig. 10.   External view of valve showing undulated valve surface and ocellulimbus.   ×10000.
Fig. 11.   External central view showing hyaline area and ghost striae.   ×10000.

Localities
Figs 2-5, 9-11.   Shikotsu-ko (Lake Shikotsu), Hokkaido.
Figs 6-8.   Biwa-ko (Lake Biwa), Shiga.

Plate 77

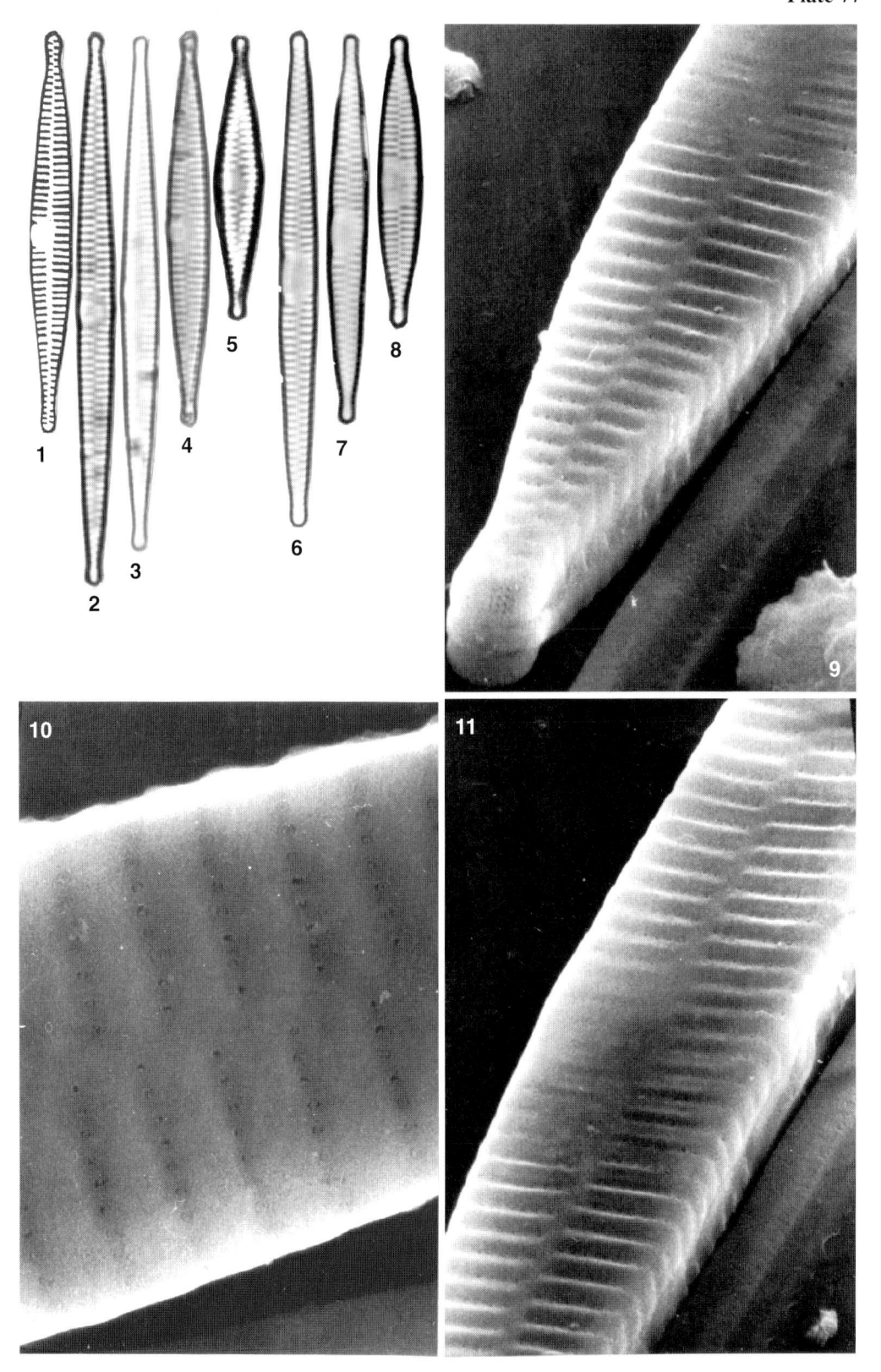

Plate 78 *Fragilaria vaucheriae* (Kützing) J. B. Petersen

Fig. 1. Drawing. ×2000.
Figs 2-9. LM. ×2000.
Figs 10-12. SEM.

Figs 1-6, 8, 9. Valve view.
Fig. 7. Girdle view of colony cells.
Fig. 10. Internal view of whole valve showing a depression of central area and a broken valvocopula. ×5000.
Fig. 11. External view of valve center showing an epicingulum composed of four bands. ×10000.
Fig. 12. External view of valve showing central area, ocellulimbus, striation and remains of spines. ×10000.

Localities

Figs 2, 4, 10-12. Ara-kawa (Ara River), Saitama.
Figs 3, 6, 8. A paddy field in Kawagoe, Saitama.
Figs 5, 7. A paddy field in Shiozawa, Niigata.
Fig. 9. Kita-ura (Lake Kita), Ibaraki.

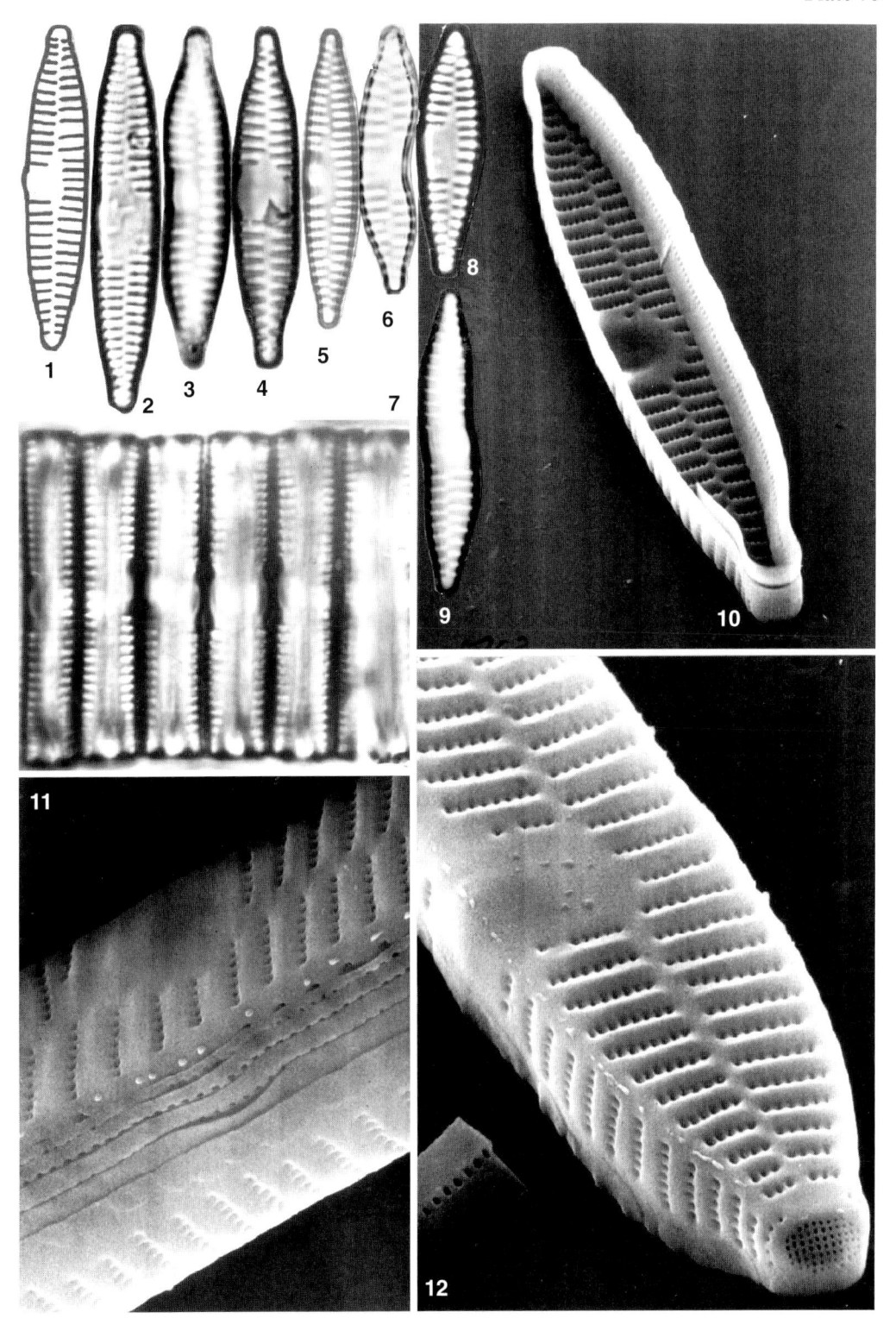

Plate 78

Plate 79 *Fragilariforma bicapitata* (A. Mayer) D. M. Williams & Round

Figs 1, 2.   Drawings. ×2000.
Figs 3-5.   LM. ×2000.
Fig. 6.   LM. ×1000.
Figs 7, 8.   SEM.

Figs 1-5.   Valve view.
Fig. 6.   Zigzag colony in girdle view.
Fig. 7.   External view of colonial frustules. ×2000.
Fig. 8.   Girdle view of frustules in colony showing marginal spines, striation and bands. ×5000.

Localities
Figs 3-5.   Asahi-ike (Asahi Pond), Niigata.
Figs 6-8.   Yufutsu-gawa (Yufutsu River), Hokkaido.

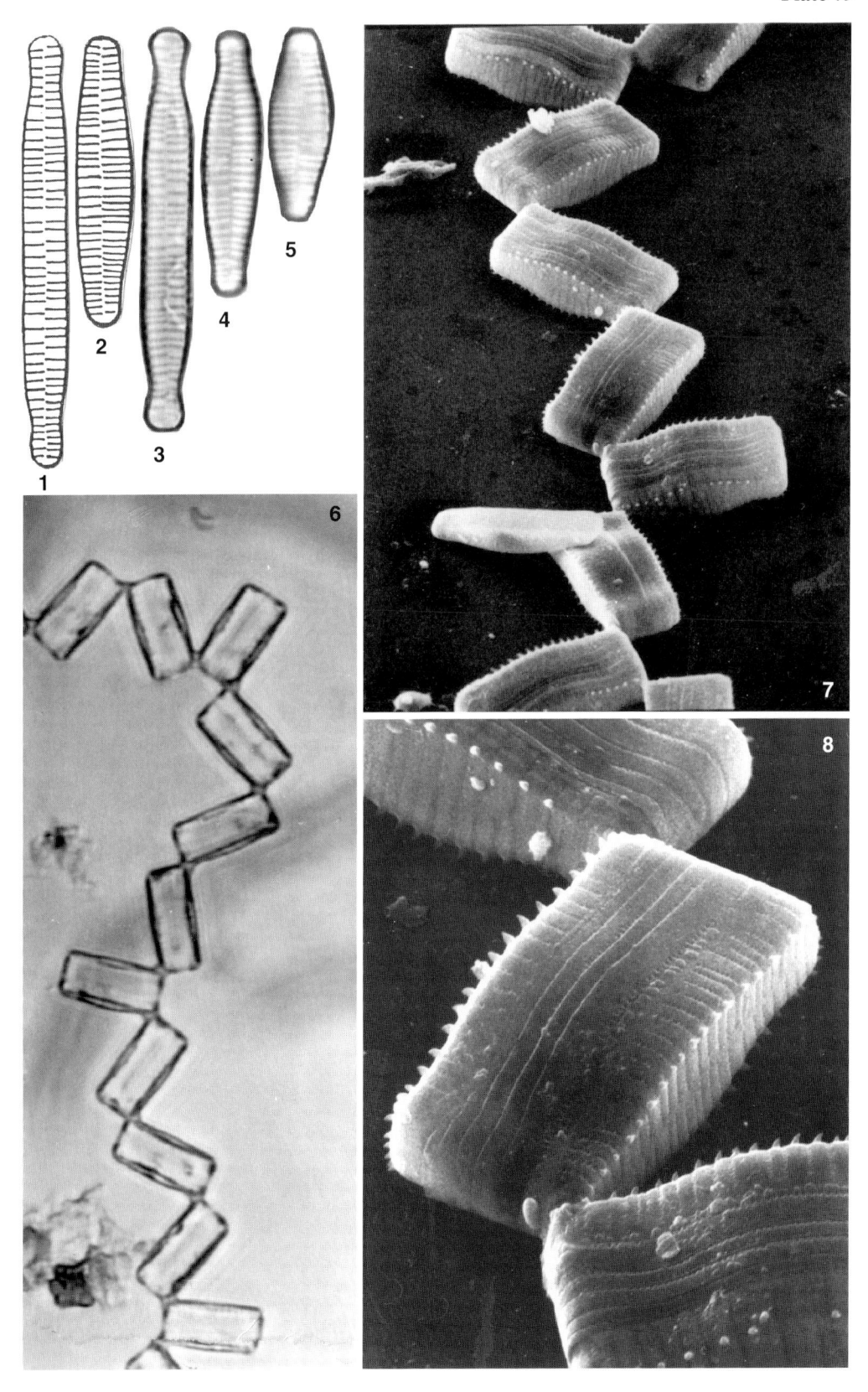

Plate 79

Plate 80 *Hannaea arcus* (Ehrenberg) R. M. Patrick

Figs 1-4.   LM. ×2000.
Figs 5-7.   SEM.

Fig. 5.   External view of central area. ×10000.
Fig. 6.   Internal polar view showing a labiate process and ocellulimbus. ×20000.
Fig. 7.   External polar view showing an opening of labiate process and ocellulimbus. ×10000.

Locality
Figs 1-7.   Churui-gawa (Churui River), Hokkaido.

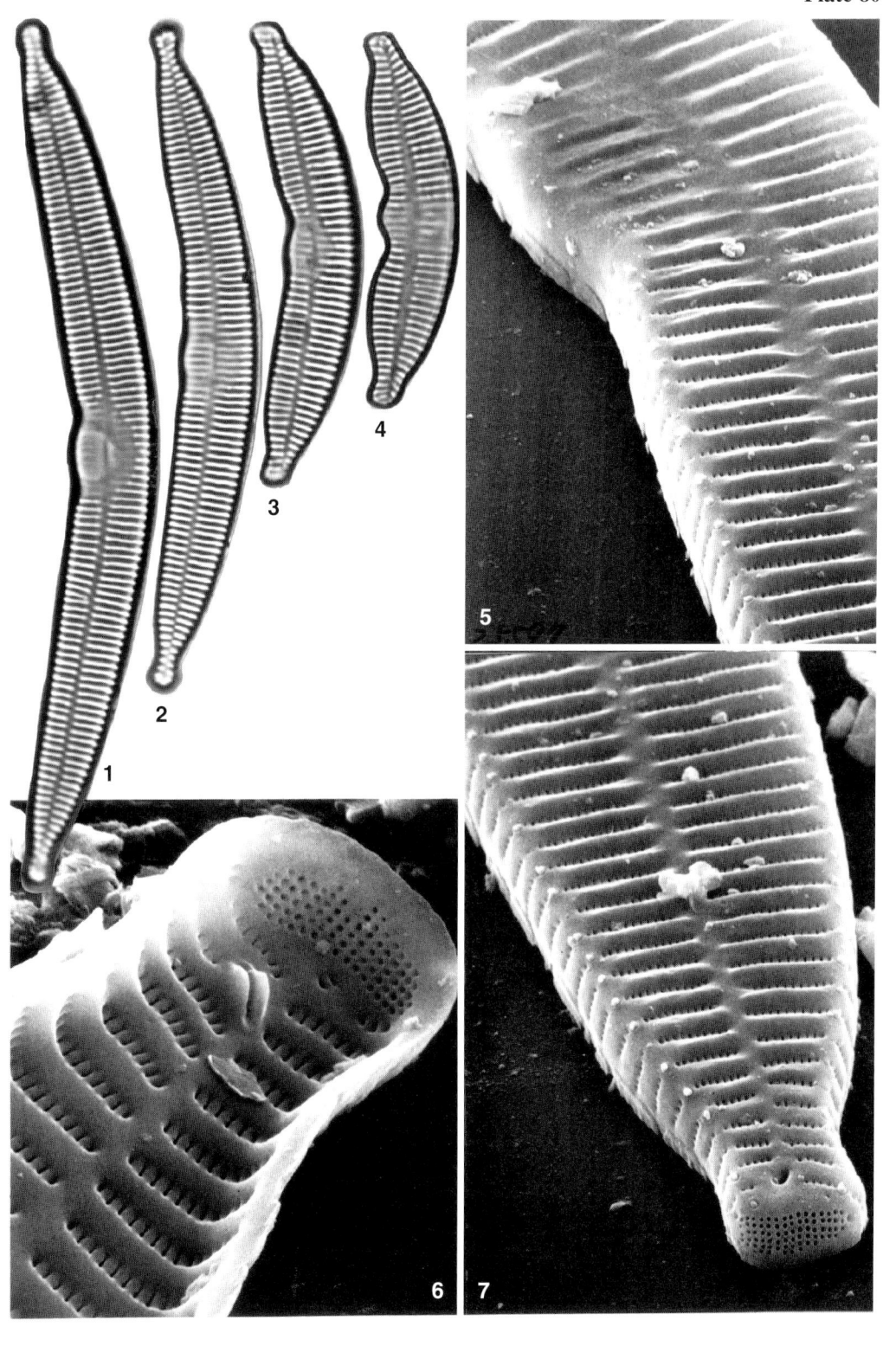

Plate 80

Plate 81 *Hannaea arcus* var. *recta* (Cleve) M. Idei comb. nov.

Figs 1-7.　LM.　×2000.
Figs 8, 9.　SEM.

Fig. 8.　External polar view of colonial frustules showing an opening of labiate process, ocellulimbi and cingulum. ×9000.
Fig. 9.　Girdle view showing linking spines. ×9000.

Locality
Figs 1-9.　Ara-kawa (Ara River), Saitama.

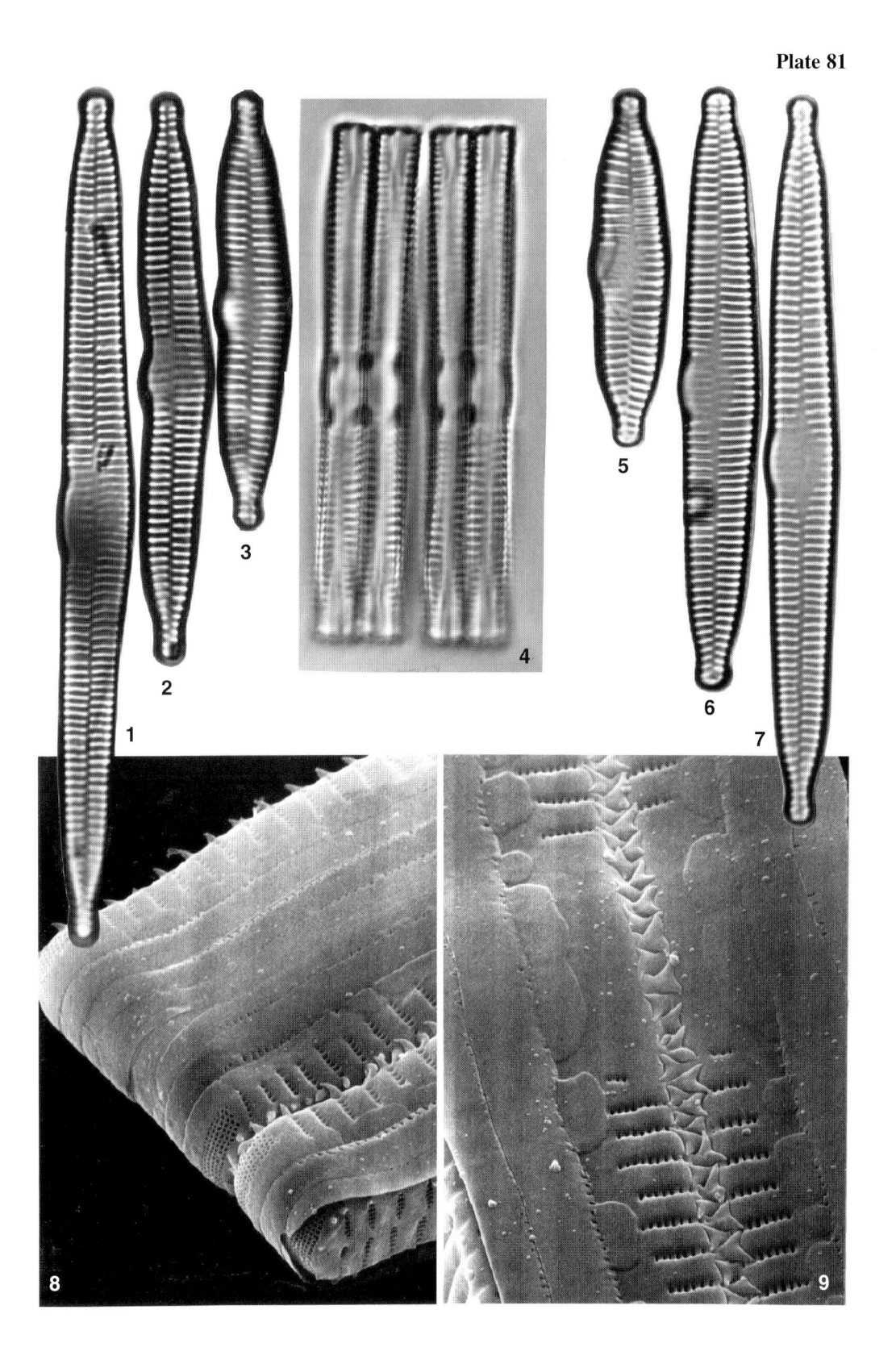

Plate 81

Plate 82 *Martyana martyi* (Héribaud) Round

Figs 1-8.   LM. ×2000.
Figs 9-13. SEM.

Figs 1-3, 5-8.   Valve view.
Fig. 4.   Girdle view.
Fig. 9.   Internal view of valve showing areolae with delicate occlusion and apical pore field. ×15000.
Fig. 10. External view of headpole showing step and striae crossed by bars. ×10000.
Fig. 11. External view of valve. ×2000.
Fig. 12. External view of footpole with apical pore field. ×10000.
Fig. 13. Oblique view of frustule. ×6000.

Locality
Figs 1-13. Akan-ko (Lake Akan), Hokkaido.

Plate 82

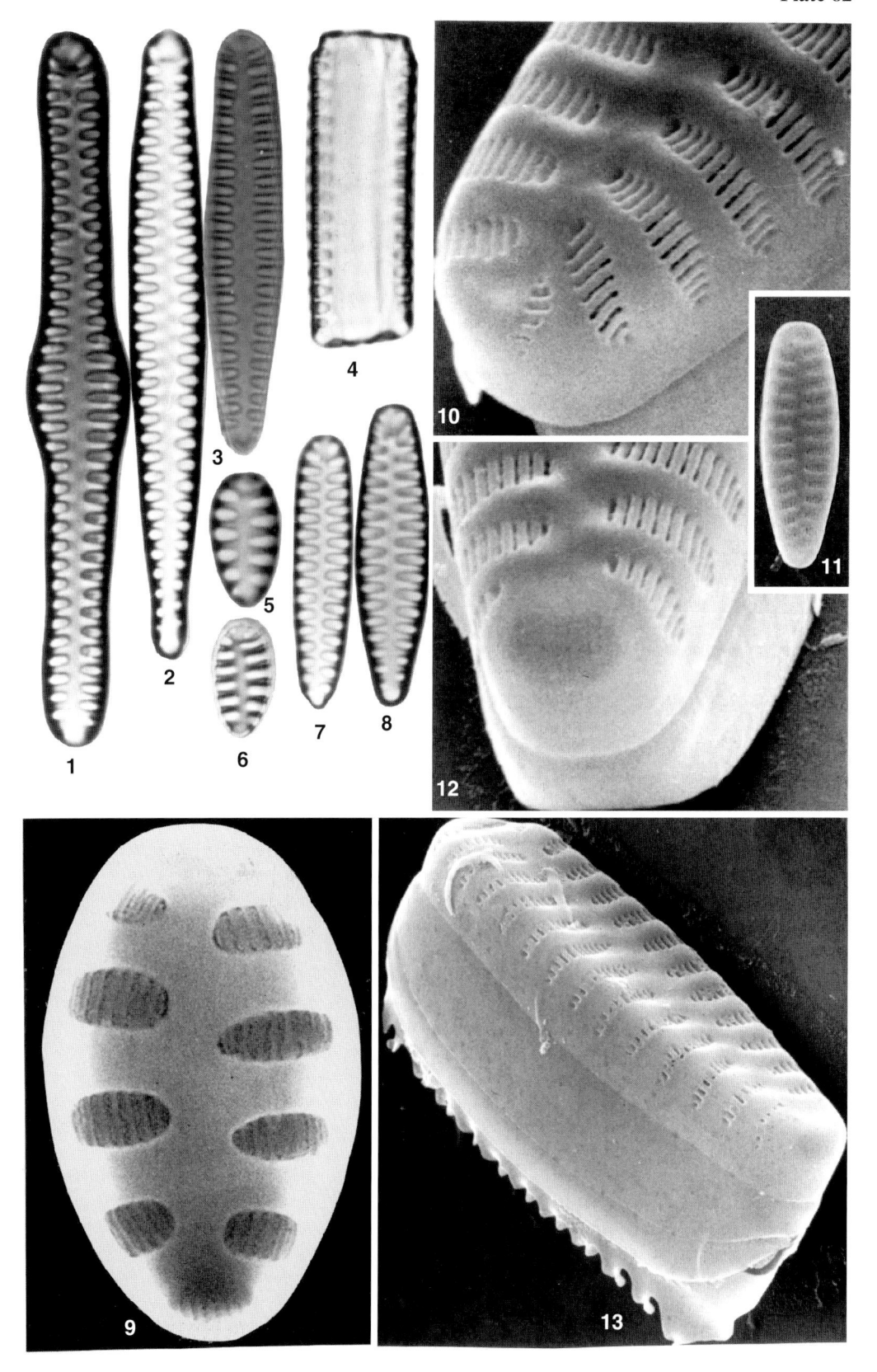

311

Plate 83 *Martyana martyi* (Héribaud) Round

Figs 14-17.   SEM.

Fig. 14.   External view of initial whole valve with central inflation. ×3000.
Fig. 15.   External central view of initial valve showing areolae crossed by bars. ×10000.
Fig. 16.   External polar view of initial valve showing ocellulimbus. ×14000.
Fig. 17.   External view of initial valve. ×5000.

Locality
Figs 14-17.   Akan-ko (Lake Akan), Hokkaido.

Plate 83

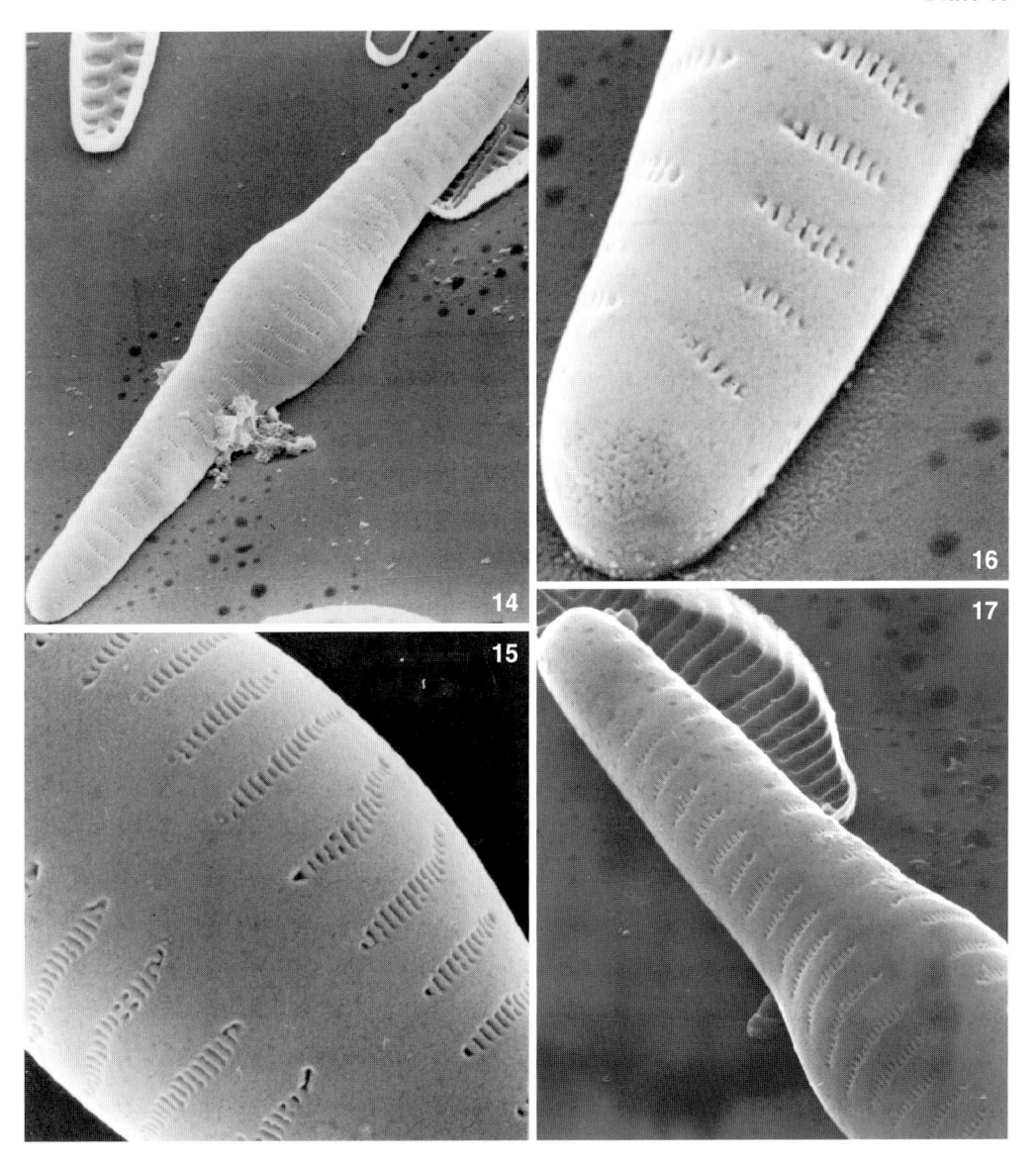

Plate 84 *Meridion circulare* (Greville) C. Agardh

Figs 1-5.   LM. ×2000.
Fig. 6.   Drawing. ×2000.
Figs 7-10.   SEM.

Figs 1-4, 6.   Valve view.
Fig. 5.   Girdle view showing fan-shaped colony with inner valves.
Fig. 7.   External view of the wider valve end (headpole) showing the areolation and outer opening of the labiate process (arrow). ×10000.
Fig. 8.   External oblique view of the headpole showing the apex lacking ocellulimbus and the plaques (arrows) on the mantle edge. ×8000.
Fig. 9.   Opposite end of the valve in Figs 7, 8 (footpole) showing the ocellulimbus at the apex. ×10000.
Fig. 10.   External oblique view showing the plaques (arrows) on the mantle edge. ×8000.

Localities
Fig. 1.   Schlesien Graben.
Figs 3, 4.   Daido-gawa (Daido River), Shiga.
Figs 7-10.   Chikuma-gawa (Chikuma River), Koumi, Nagano.

Plate 84

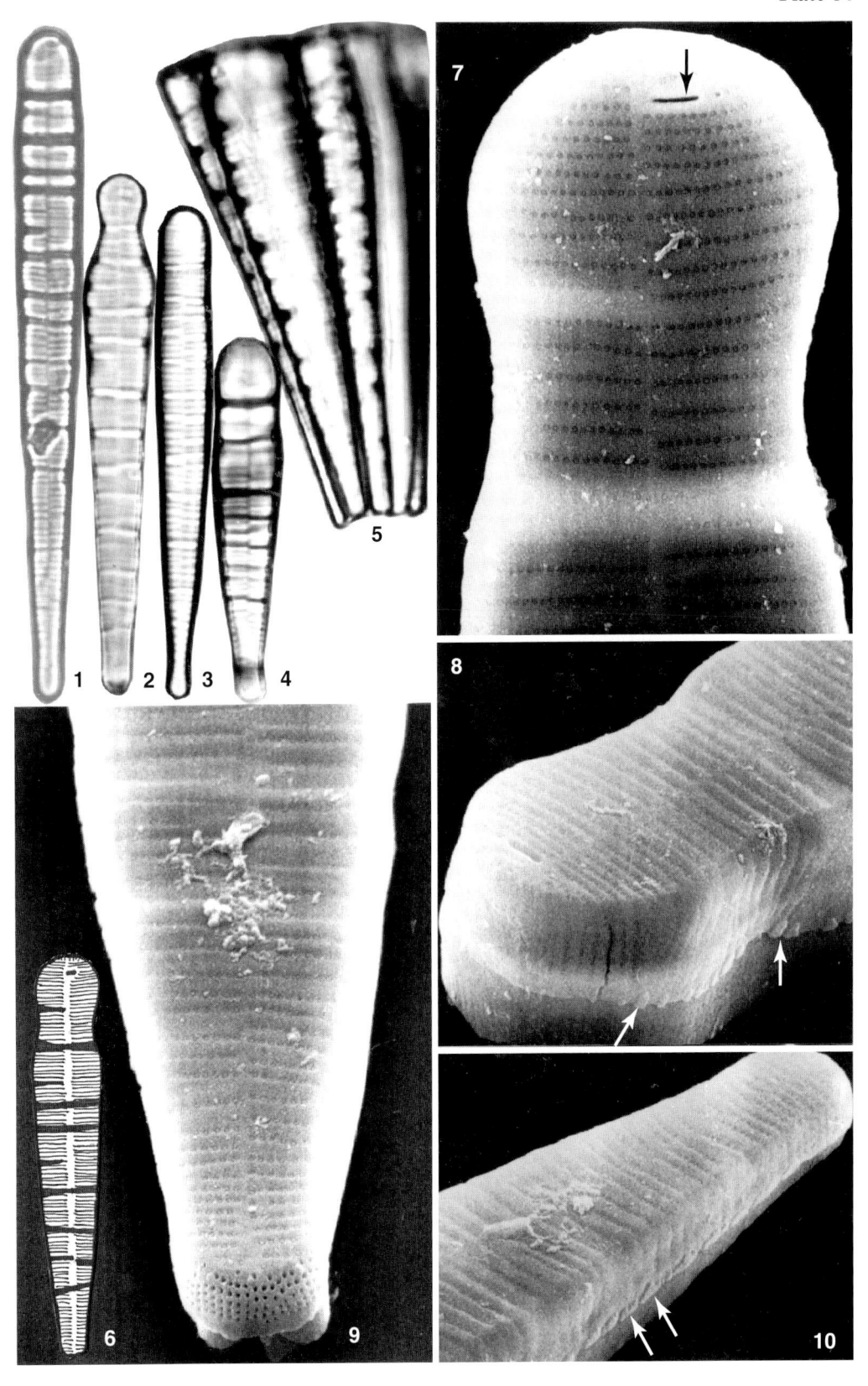

Plate 85 *Pseudostaurosira brevistriata* **(Grunow) D. M. Williams & Round**

Fig. 1.   Drawing. ×2000.
Figs 2-6.   LM. ×2000.
Figs 7-12.   SEM.

Figs 1-6.   Valve view.
Fig. 7.   External view of valve. ×4000.
Fig. 8.   Internal view of valve. ×5500.
Fig. 9.   External view of valve pole showing linking spines, apical pore field, areolae with velum and plaques (arrow) along the mantle edge. ×10000.
Fig. 10.   External view of valve showing marginal striae and spines. ×5000.
Fig. 11.   Internal view of valve showing inner striation. ×8000.
Fig. 12.   External view of mantle showing open valvocopula and five copulae. ×30000.

Localities

Figs 2, 5, 7-11.   Akan-ko (Lake Akan), Hokkaido.
Fig. 3.   Nojiri-ko (Lake Nojiri), Nagano.
Fig. 4.   Yamanaka-ko (Lake Yamanaka), Yamanashi.
Fig. 6.   Fossil deposit, Kanomura (Wamura), Ueda, Nagano.
Fig. 12.   Chikuma-gawa (Chikuma River), Nagano.

Plate 85

317

Plate 86 *Pseudostaurosira brevistriata* var. *nipponica* (Skvortsov) H. Kobayasi

Figs 1, 2.   Drawings. ×2000.
Figs 3-6.   LM. ×2000.
Figs 7-13.   SEM.

Figs 1-6.   Valve view.
Fig. 7.   External and internal view of sibling valves. ×5000.
Fig. 8.   Internal view of valve. ×5000.
Fig. 9.   Internal oblique view of sibling valves showing linking spines and inner occlusions of striae. ×4000.
Fig. 10.   External view of valve showing areolae of valve face and of mantle and marks of spines. ×10000.
Fig. 11.   Internal view of valve showing inner occlusions of striae. ×10000.
Fig. 12.   Valve poles with apical pore fields. ×12000.
Fig. 13.   Enlargement of inner occlusions of areolae. ×16000.

Locality
Figs 3-13.   Akan-ko (Lake Akan), Hokkaido.

Plate 86

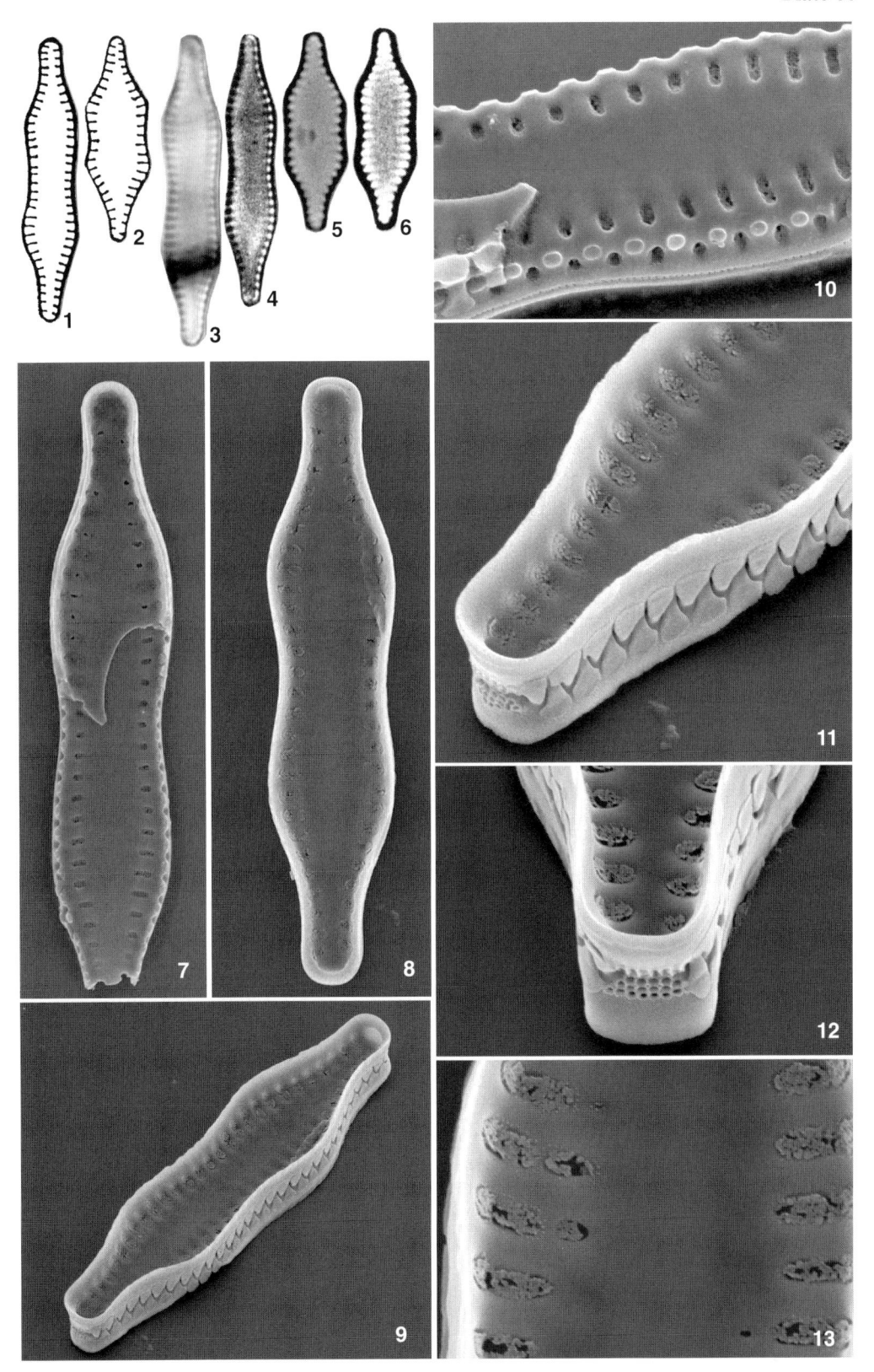

Plate 87 *Pseudostaurosira robusta* (Fusey) D. M. Williams & Round

Fig. 1. Drawing. ×2000.
Figs 2-4. LM. ×2000.
Figs 5-11. SEM.

Figs 1-4. Valve view.
Fig. 5. Internal view of sibling valves. ×4000.
Fig. 6. Internal view of valve showing areolae and sternum. ×4000.
Fig. 7. Oblique view of sibling valves showing areolae in valve face and mantle. ×6000.
Fig. 8. Internal and external view of sibling valves showing inner occlusions and external round openings of the areolae. ×15000.
Fig. 9. Girdle view of frustules showing linking spines and cingula. ×4000.
Fig. 10. Polar view of sibling valves showing inner occlusion of areolae, linking spines and apical pore field. ×13000.
Fig. 11. Girdle view of frustules showing linking spines, cingula and plaques. ×13000.

Locality
Figs 2-11. Akan-ko (Lake Akan), Hokkaido.

Plate 87

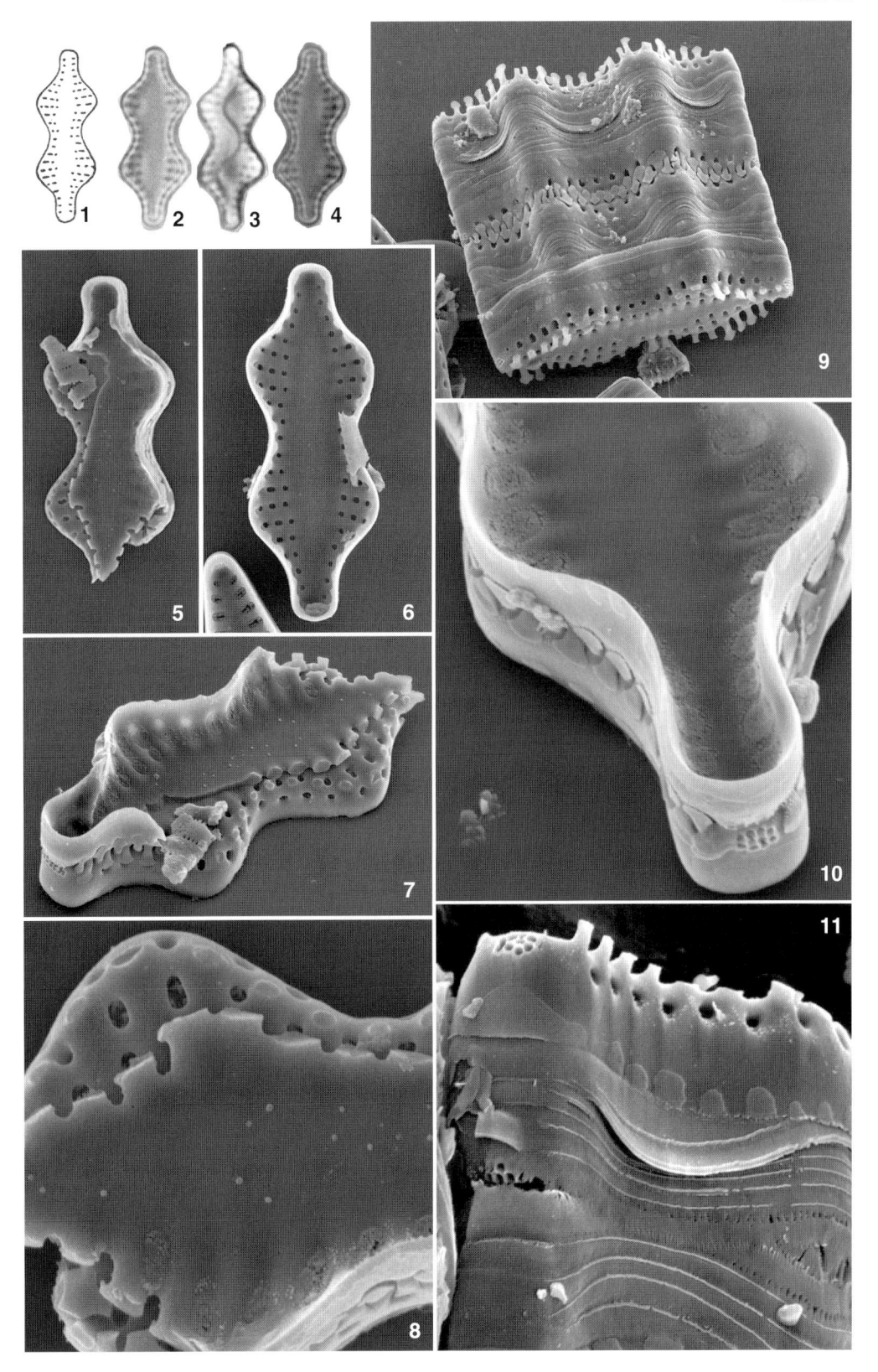

321

Plate 88 *Punctastriata linearis* D. M. Williams & Round

Fig. 1.   Drawing. ×2000.
Figs 2-9.   LM. ×2000.
Figs 10-14.   SEM.

Figs 1-8.   Valve view.
Fig. 9.   Girdle view of linking cells.
Fig. 10.   External oblique view of frustule showing undulation of valve surface. ×7000.
Fig. 11.   External view of valve showing multiseriate striae. ×8000.
Fig. 12.   External view of linking frustules showing multiseriate striae and linking spines. ×8000.
Fig. 13.   External view of valve mantle showing areolation of striae and girdle bands. ×20000.
Fig. 14.   Enlargement of multiseriate striae. ×20000.

Localities

Figs 2, 5.   Kinu-gawa (Kinu River), Tochigi.
Figs 3, 4, 6-9, 11-14.   Manyo-ike (Manyo Pond), Tokyo Gakugei University, Tokyo.
Fig. 10.   Akan-ko (Lake Akan), Hokkaido.

Plate 88

Plate 89 *Staurosira construens* Ehrenberg var. *construens*

Figs 1, 2.   Drawings. ×2000.
Figs 3-5.   LM. ×2000.
Figs 6-8.   SEM.

Figs 1-4.   Valve view.
Fig. 5.   Girdle view of colony.
Fig. 6.   External view of valve showing striation and marginal spines. ×7800.
Fig. 7.   External oblique view of valve showing ocellulimbus and striation. ×6000.
Fig. 8.   External view of valve center showing linking spines and striation. ×10000.
Fig. 9.   External view of colonial frustules showing linking spines and cingula. ×5600.

Localities

Fig. 3.   Sohrohsenen no yusui (Spring at Sohrohsenen), Koganei, Tokyo.
Fig. 4.   Fossil deposit Kanomura (Wamura), Ueda, Nagano.
Fig. 5.   Sanpoji-ike (Sanpoji Pond), Tokyo.
Figs 6, 7.   Hachimantai Oh-numa (Oh Pond), Akita.
Figs 8, 9.   Benten-numa (Benten Pond), Tomakomai, Hokkaido.

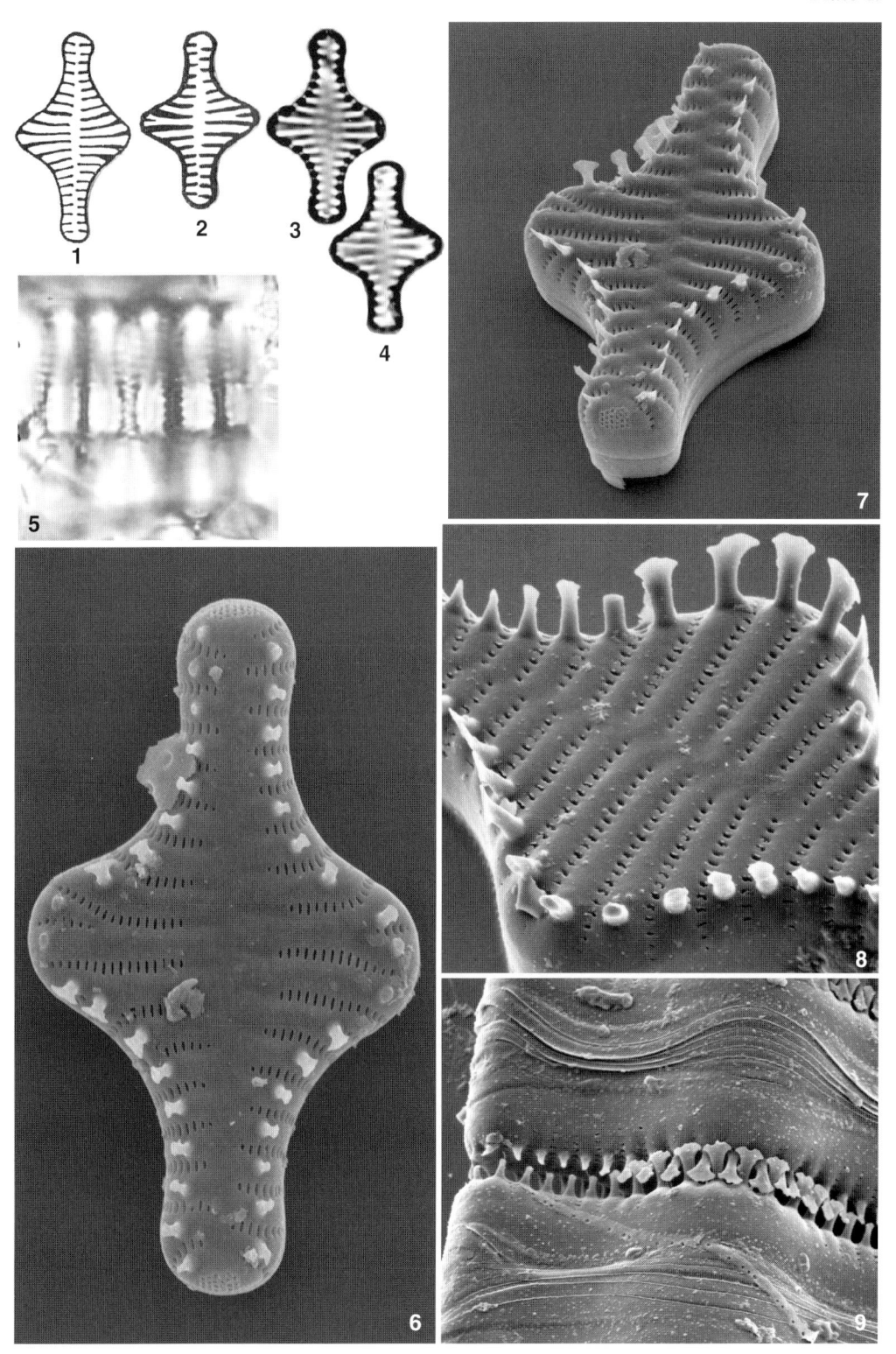

Plate 89

Plate 90 *Staurosira construens* var. *binodis* (Ehrenberg) P. B. Hamilton

Figs 1, 6. Drawings. ×2000.
Figs 2-5. LM. ×2000.
Figs 7-11. SEM.

Figs 1-4. Valve view.
Fig. 5. Girdle view.
Fig. 6. Diagram of frustule.
Fig. 7. Oblique view of sibling valves. ×4000.
Fig. 8. External view of valve showing striae and linking spines. ×13500.
Fig. 9. Oblique view of sibling valves showing linking spines and apical pore field. ×10000.
Fig. 10. Internal view of valve showing inner occlusions of areolae. ×20000.
Fig. 11. External view of frustules showing girdle bands and linking spines. ×8000.

Localities

Figs 2, 4. Oze-numa (Oze Pond), Gunma.
Fig. 3. Akan-ko (Lake Akan), Hokkaido.
Fig. 5. Tatara-numa (Tatara Pond), Gunma.
Figs 7-10. Kizaki-ko (Lake Kizaki), Nagano.
Fig. 11. Tega-numa (Tega Pond), Chiba.

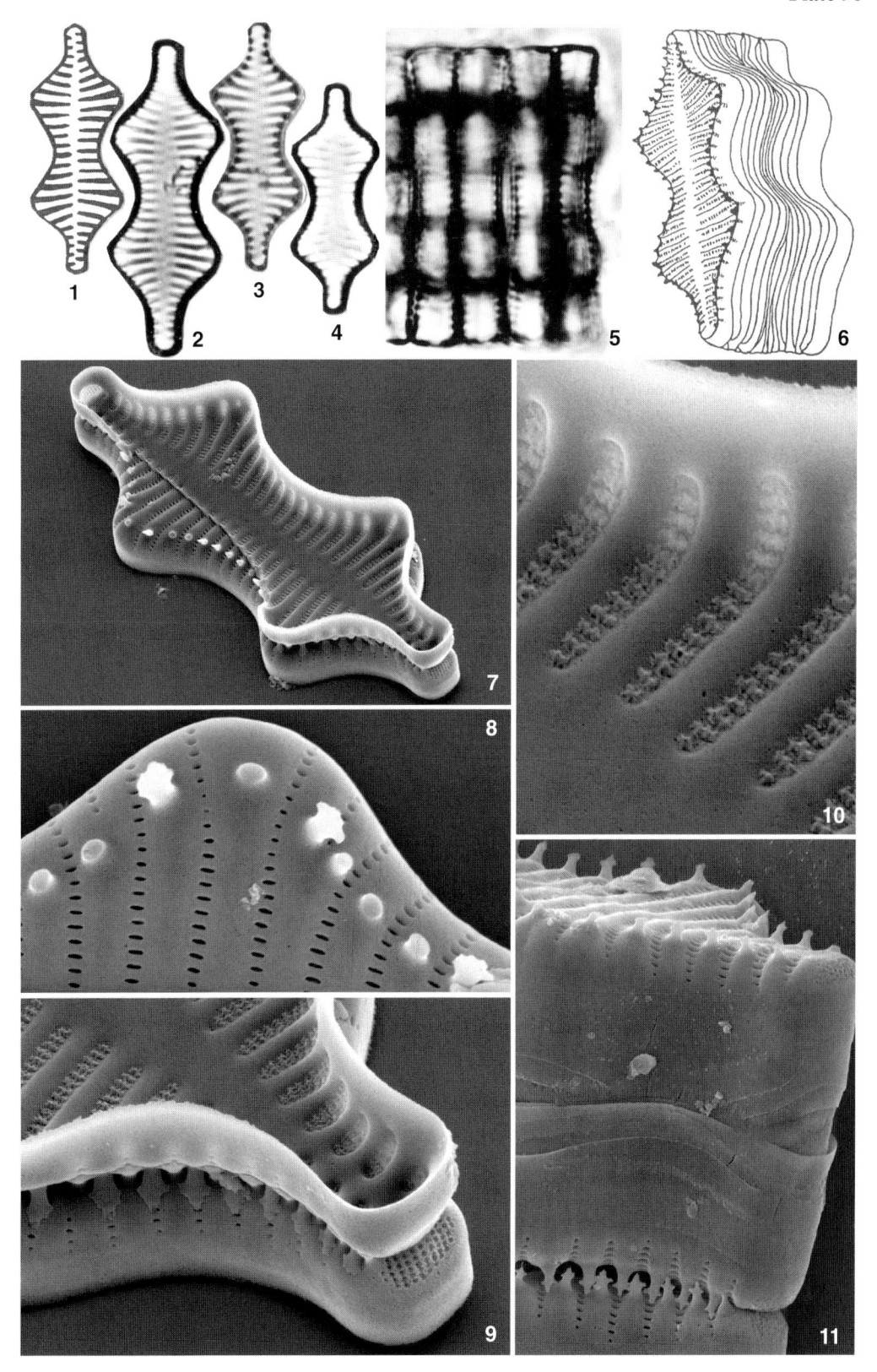

Plate 90

Plate 91 *Staurosira construens* var. *exigua* (W. Smith) H. Kobayasi

Figs 1, 2. Drawings. ×2000.

Figs 3-6. LM. ×2000.

Figs 7-11. SEM.

Figs 1-6. Valve view.

Fig. 7. Internal view of valve. ×6000.

Fig. 8. Oblique view of sibling valves showing striation and linking spines. ×5900.

Fig. 9. Oblique view of valve poles with apical pore field. ×15000.

Fig. 10. Internal view of valve showing inner occlusions of areolae and apical pore field. ×18000.

Fig. 11. Internal and external view of sibling valves showing linking spines and inner occlusions of and external openings of areolae. ×18000.

Locality

Figs 3-11. Hachiro-gata (Lake Hachiro), Akita.

Plate 91

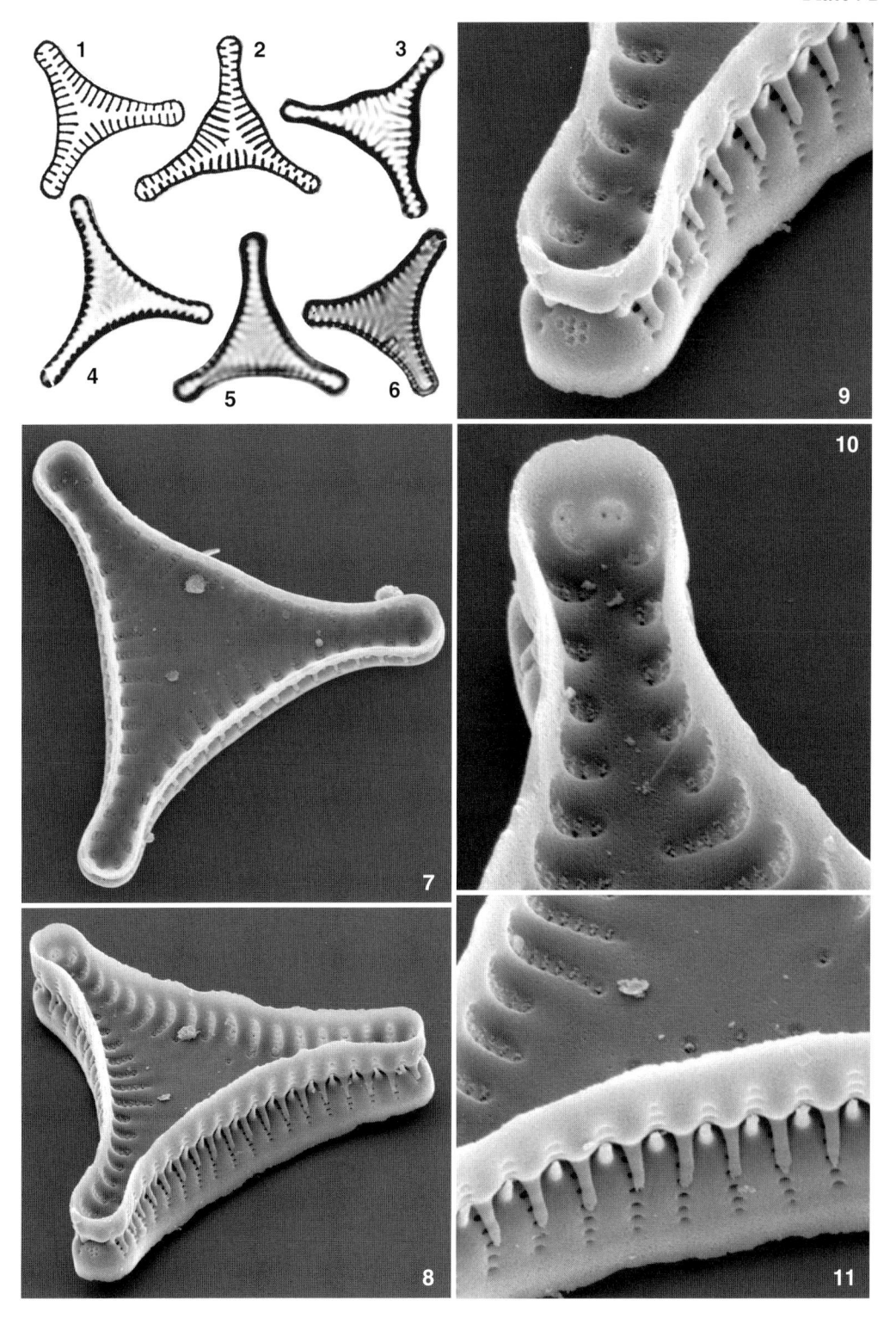

Plate 92 *Staurosira construens* var. *triundulata* (H. Reichelt) H. Kobayasi

Figs 1, 2.   Drawings. ×2000.
Figs 3, 4.   LM. ×2000.
Figs 5-13.   SEM.

Figs 1-4.   Valve view.
Fig. 5.   External and internal views of sibling valves. ×2000.
Figs 6, 7.   Internal view of valve. ×2000.
Fig. 8.   External view of valve showing openings of areolae and linking spines. ×10000.
Fig. 9.   Internal oblique view of valve showing striation and axial area. ×3200.
Fig. 10.   External oblique view of sibling valves showing apical pore field and linking spines. ×10000.
Fig. 11.   Enlargement of linking spines. ×20000.
Fig. 12.   Internal view of valve showing inner occlusions of areolae. ×25000.
Fig. 13.   Girdle view of sibling thecae showing bands and linking spines. ×10000.

Localities

Figs 3, 5-13.   Tega-numa (Tega Pond), Chiba.
Fig. 4.   Biwa-ko (Lake Biwa), Shiga.

Plate 92

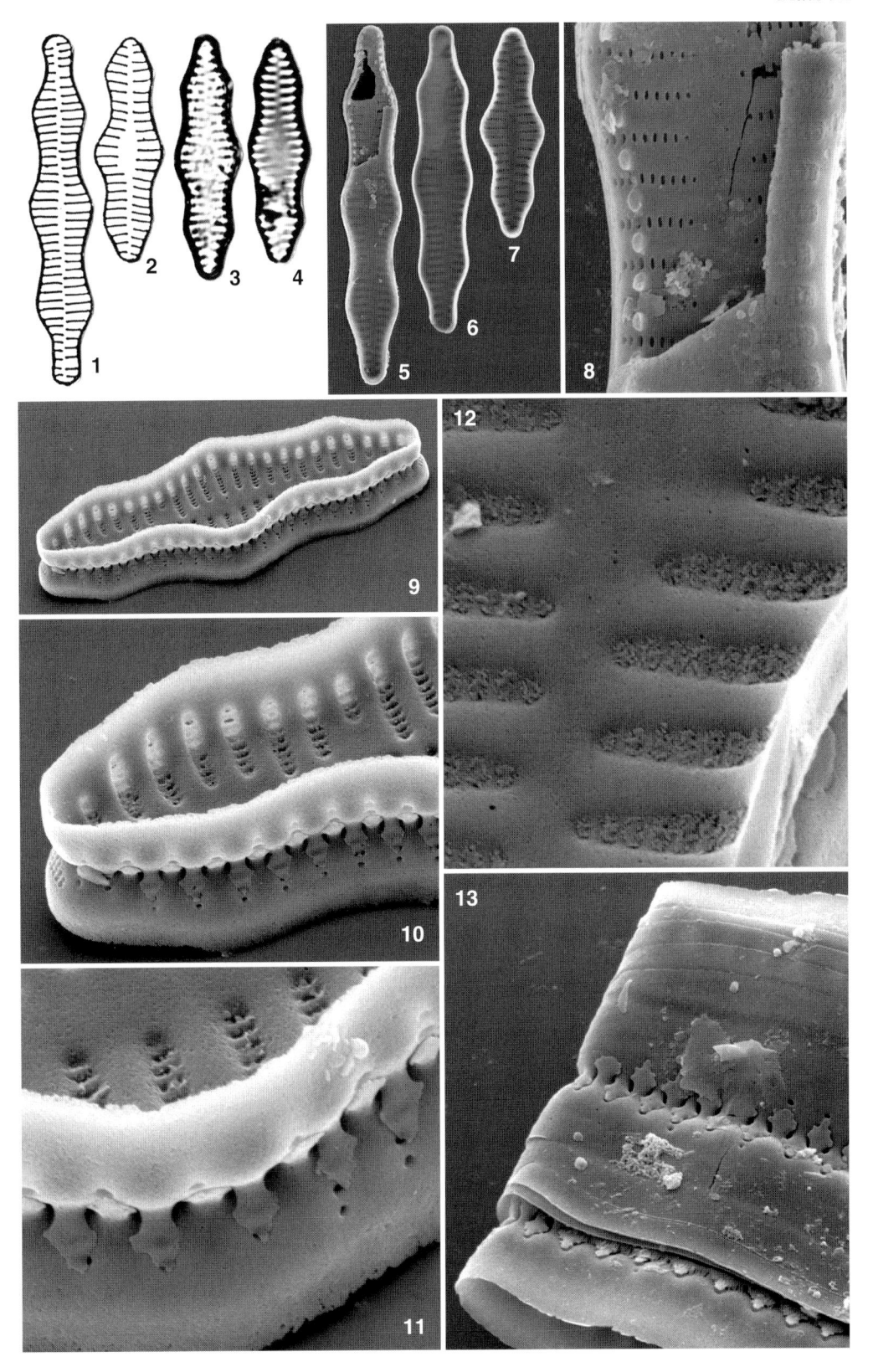

Plate 93 *Staurosira elliptica* (Schumann) D. M. Williams & Round

Figs 1, 2. Drawings. ×2000.
Figs 3-6. LM. ×2000.
Figs 7-10. SEM.

Figs 1-4. Valve view.
Figs 5, 6. Girdle view.
Fig. 7. Filamentous colony. ×2000.
Figs 8, 9. External view of frustule showing areolation and marginal linking spines. ×10000.
Fig. 10. Enlargement of linking spines. ×14000.

Localities

Fig. 3. Akan-ko (Lake Akan), Hokkaido.
Figs 4-7. Sanpoji-ike (Sanpoji Pond), Tokyo.
Figs 8-10. Asahi-ike (Asahi Pond), Niigata.

Plate 93

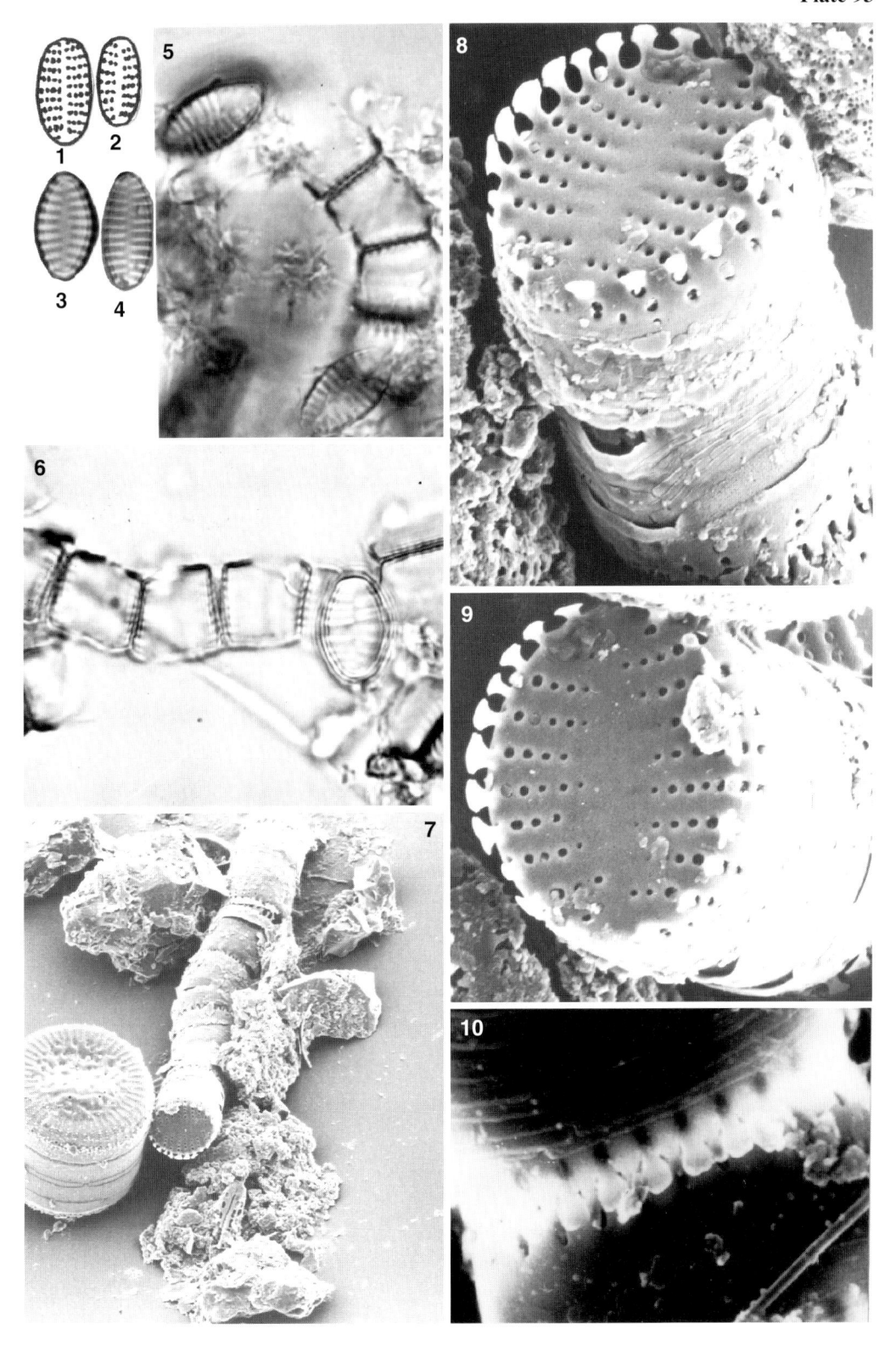

Plate 94 *Staurosira venter* (Ehrenberg) H. Kobayasi

Figs 1-3.   Drawings. ×2000.
Figs 4-9.   LM. ×2000.
Figs 10-16. SEM.

Figs 1-8.   Valve view.
Fig. 9.   Girdle and valve view of colony.
Fig. 10.   External view of valve showing striae and linking spines. ×6000.
Fig. 11.   Oblique view of sibling valves showing inner striae and linking spines. ×6700.
Fig. 12.   Internal and external view of sibling valves showing inner occlusions and outer openings of areolae. ×16000.
Fig. 13.   Internal and external view of valve pole showing apical pore fields. ×16000.
Fig. 14.   External view of frustule showing striation, linking spines and bands. ×6700.
Fig. 15.   Enlargement of valve pole with small apical pore field. ×10700.
Fig. 16.   Valvocopula. ×6700.

Localities

Figs 4, 8.   Fossil deposit, Kanomura (Wamura), Ueda, Nagano.
Figs 5, 6, 9-16.   Akan-ko (Lake Akan), Hokkaido.
Fig. 7.   Yuno-ko (Yu Pond), Tochigi.

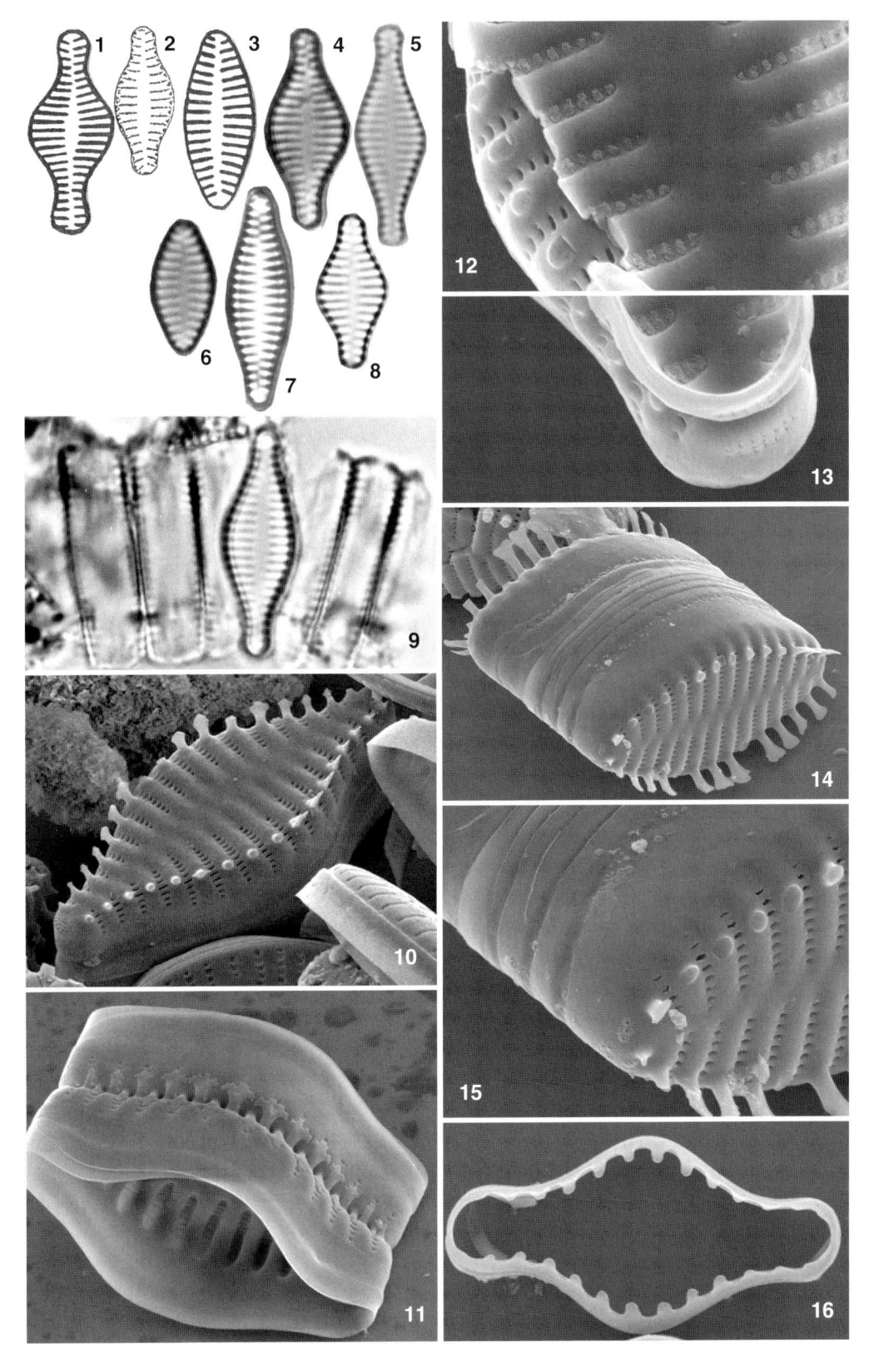

Plate 94

Plate 95 *Staurosira venter* (Ehrenberg) var. *binodis* H. Kobayasi

Fig. 1.   Drawing.
Figs 2-8.   LM. ×2000.
Figs 9-15.   SEM.

Figs 1-7.   Valve view.
Fig. 8.   Girdle view of the filamentous colony.
Fig. 9.   External view of valve showing striation. ×4000.
Fig. 10.   External oblique view of valve showing striae and linking spines. ×5000.
Fig. 11.   External view of valve pole showing striae and apical pore field. ×13000.
Fig. 12.   Internal view of valve showing striation and axial area.
Fig. 13.   Internal view of valve showing inner occlusions of areolae and linking spines. ×16000.
Fig. 14.   Enlargement of valve pole showing apical pore fields. ×17000.
Fig. 15.   Girdle view of sibling thecae. ×6000.

Localities
Fig. 2.   Kinu-gawa (Kinu River), Tochigi.
Figs 3-8.   Fossil deposit Kanomura (Wamura), Ueda, Nagano.
Figs 9-15.   Akan-ko (Lake Akan), Hokkaido.

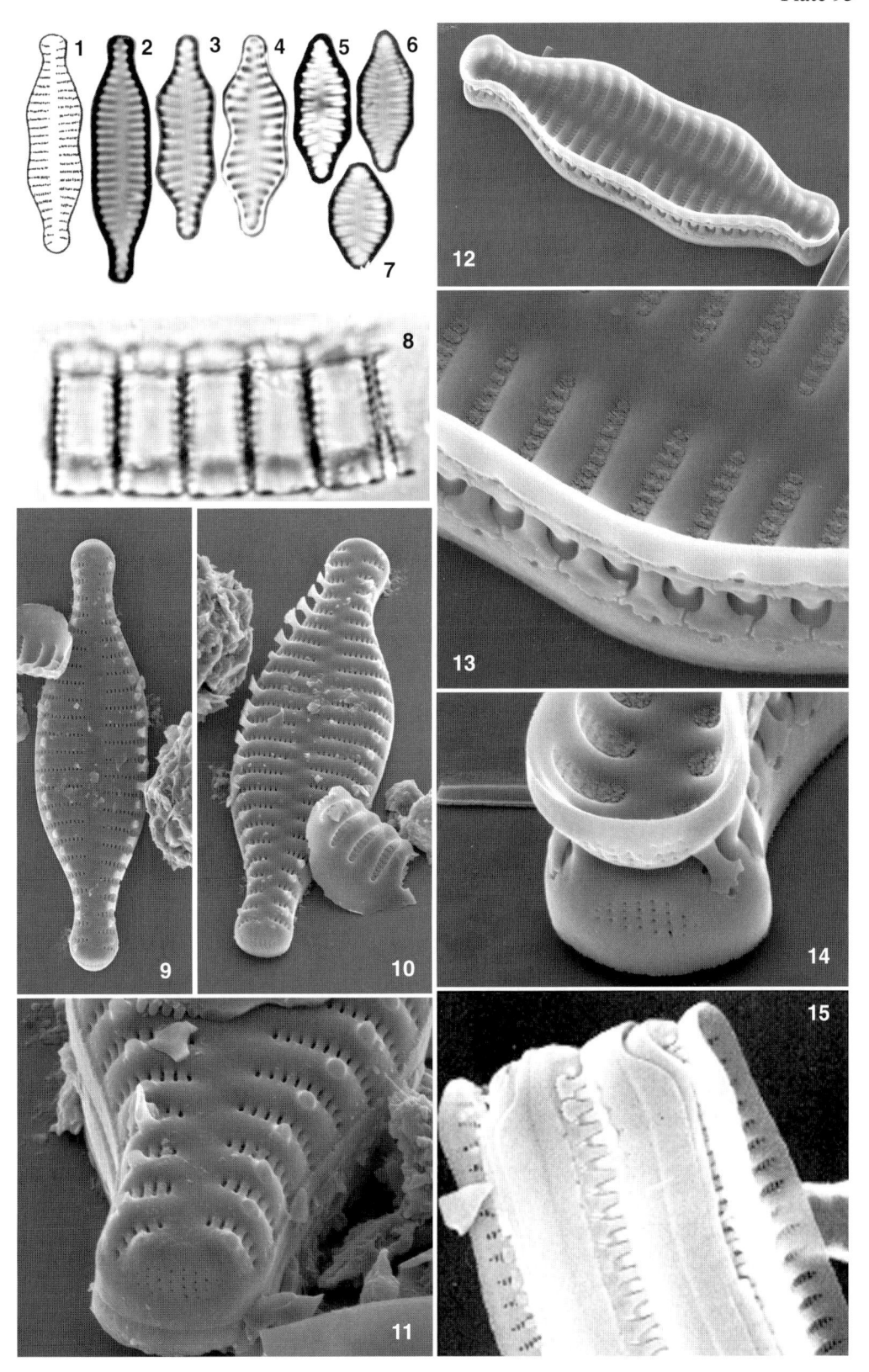

Plate 95

Plate 96 *Staurosirella lapponica* (Grunow) D. M. Williams & Round

Figs 1-4, 6. LM. ×2000.
Fig. 5. Drawing. ×2000.
Figs 7-9. SEM.

Fig. 6. Colony in girdle view.
Fig. 7. Internal view of the valve showing the striae and the interstriae. ×6000.
Fig. 8. Internal oblique view of valve with a valvocopula. ×5000.
Fig. 9. Internal view of the striae with slitlike areolae. ×10000.

Localities
Figs 1, 6. Aoki-ko (Lake Aoki), Nagano.
Figs 2, 4. Yamanaka-ko (Lake Yamanaka), Yamanashi.
Fig. 3. Oze-numa (Oze Pond), Gunma.
Figs 7-9. Akan-ko (Lake Akan), Hokkaido.

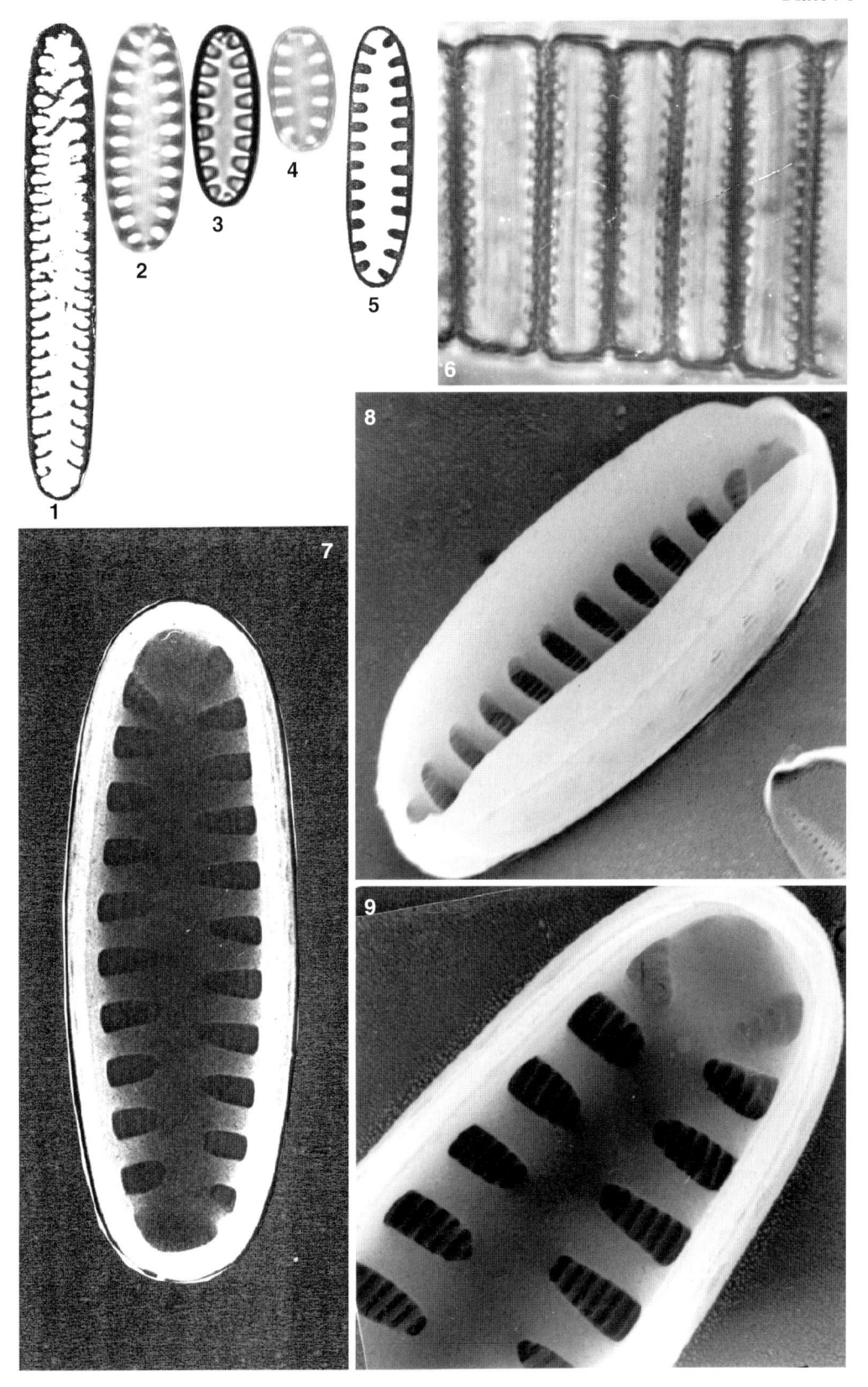

Plate 96

Plate 97 *Staurosirella leptostauron* (Ehrenberg) D. M. Williams & Round

Fig. 1.   Drawing. ×2000.
Figs 2-4.   LM. ×2000.
Figs 5-12.   SEM.

Figs 2-4.   Valve view.
Fig. 5.   External view of valve. ×3000.
Fig. 6.   Internal view of valve. ×3000.
Figs 7, 8.   External view of poles of a valve showing slitlike areolae, linking spines and apical pore field with slits. ×10500.
Fig. 9.   External view of valve showing striation with slitlike areolae. ×16000.
Fig. 10.   Internal view of valve showing inner occlusion of areolae. ×14400.
Fig. 11.   Internal view of valve showing a valvocopula with crenate edge. ×2100.
Fig. 12.   Girdle view of frustules showing linking spines and cingula. ×7000.

Localities
Figs 2, 3.   Akan-ko (Lake Akan), Hokkaido.
Fig. 4.   Naeba-san shitsugen (Naeba Moor), Niigata.
Figs 5-12.   Zigoku-zawa (Zigoku Stream), Nikko, Tochigi.

Plate 97

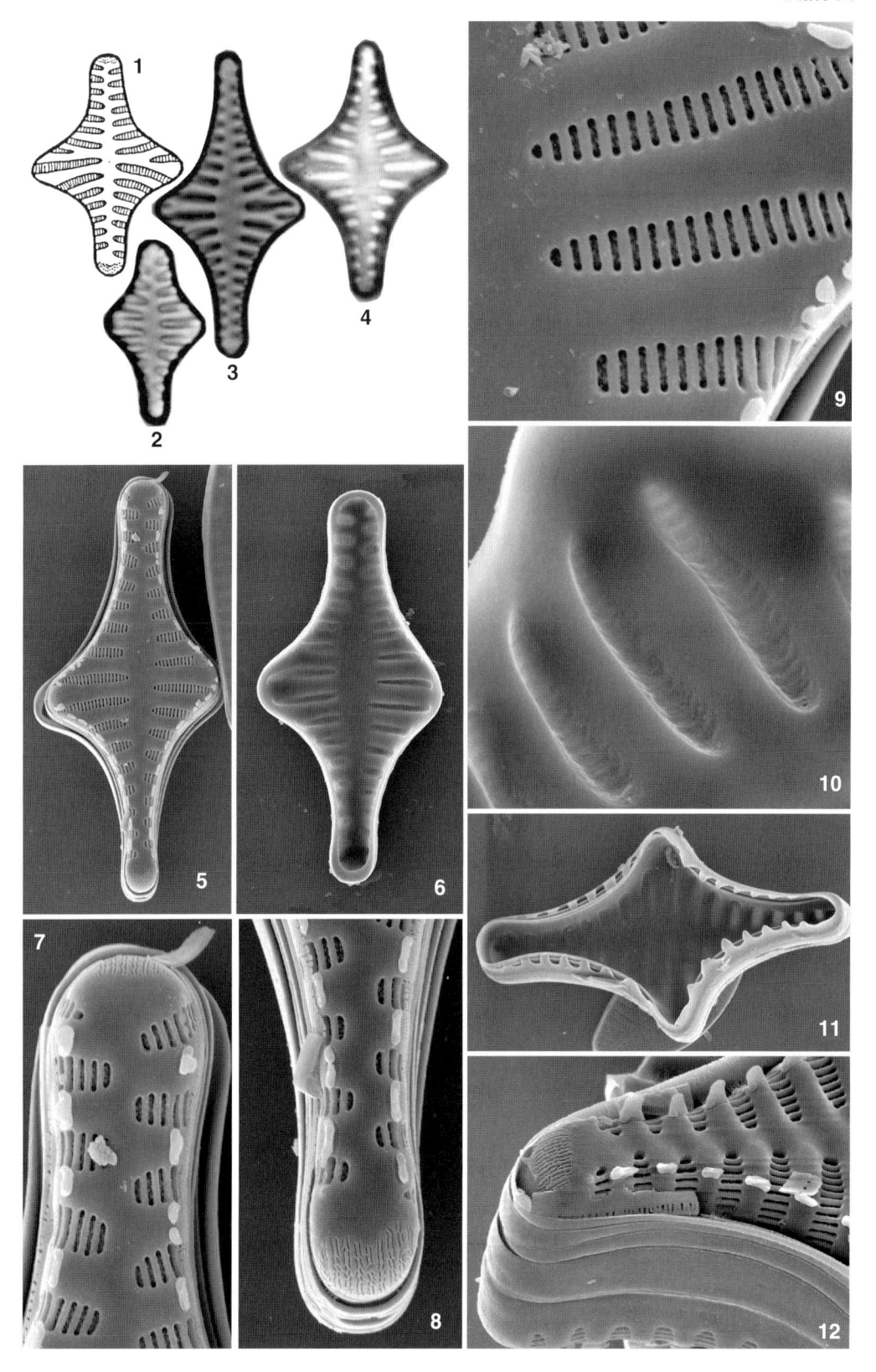

Plate 98 *Staurosirella pinnata* (Ehrenberg) D. M. Williams & Round

Fig. 1.　Drawing. ×2000.

Figs 2-7.　LM. ×2000.

Figs 8-14.　SEM.

Figs 2-7.　Valve view.

Fig. 8.　External view of frustule showing linking spines, slitlike areolae and cingula. ×6400.

Fig. 9.　External view of frustule. ×7000.

Fig. 10.　Internal view of valve. ×10000.

Fig. 11.　External view of valve showing apical pore field at right pole. ×10000.

Fig. 12.　External view of valve showing striation with slitlike areolae and round marks of linking spines. ×20000.

Fig. 13.　Internal view of valve showing inner occlusion of areolae. ×20000.

Fig. 14.　Oblique view of frustule showing linking spines, areolation and cingulum. ×7000.

Localities

Figs 2, 6.　Jyo-numa (Jyo Pond), Gunma.

Figs 3, 4, 8, 11, 12.　Akan-ko (Lake Akan), Hokkaido.

Fig. 5.　Yamadaoh-numa (Yamadaoh Pond), Musashikyuryo-shinrin Park, Saitama.

Fig. 7.　Yamanaka-ko (Lake Yamanaka), Yamanashi.

Figs 9, 10, 13, 14.　Tsuta-numa (Tsuta Pond), Aomori.

Plate 98

Plate 99 *Synedrella parasitica* (W. Smith) Round & Maidana

Figs 1-3.   LM. ×2000.
Fig. 4.   Drawing. ×2000.
Fig. 5.   Drawing. ×5000.
Figs 6-10.   SEM.

Figs 1-4.   Valve view.
Fig. 5.   Oblique view of frustule.
Fig. 6.   External view of frustules attached to *Surirella*. ×2000.
Fig. 7.   External view of frustules showing mucilage from apical pore field. ×5000.
Fig. 8.   External view of valve showing striation and apical pore field. ×5000.
Fig. 9.   External view of valve showing external openings and inner occlusions of areolae. ×20000.
Fig. 10.   External oblique view of frustule. ×10000.

Localities

Fig. 1.   Tatara-numa (Tatara Pond), Gunma.
Figs 2, 8, 9.   Yamagata-jo Hori (The moat of Yamagata Castle), Yamagata.
Fig. 3.   Ida-ike (Ida Pond), Heda, Shizuoka.
Figs 6, 7.   A small stream, Nara.
Fig. 10.   Akan-ko (Lake Akan), Hokkaido.

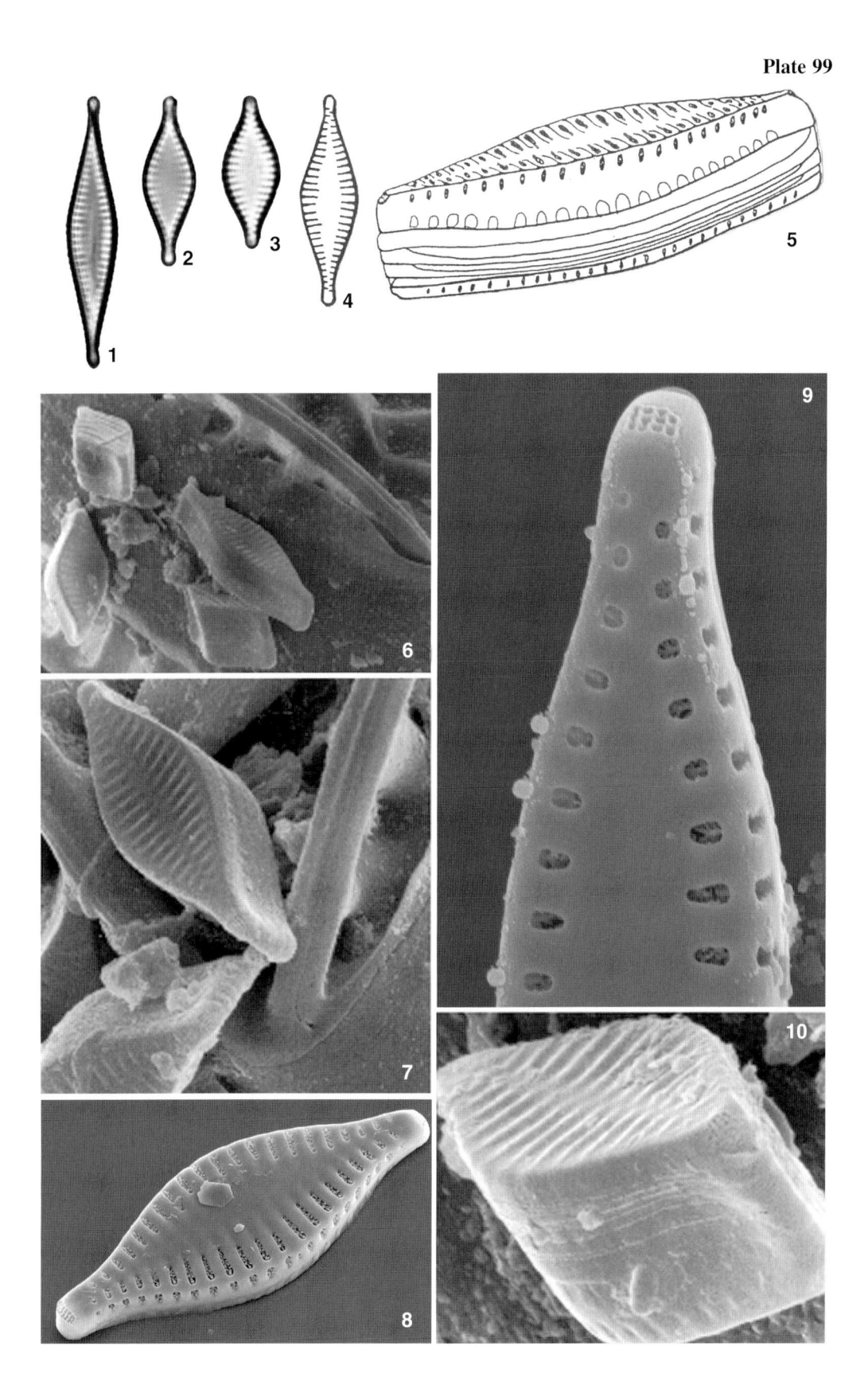

Plate 99

Plate 100 *Tabularia affinis* (Kützing) Snoeijs

Figs 1, 2, 4-9.   LM.   ×2000.
Fig. 3.   Drawing.   ×2000.
Figs 10-14.   SEM.

Figs 1, 2.   Holotype, Triest 169, Coll. Kützing, B.M. 18340.
Fig. 10.   Internal view of broken valve with alveolate striae.
Fig. 11.   External view of valve pole showing biseriate striae, opening of labiate process and ocellulimbus.   ×10000.
Fig. 12.   Internal view of valve pole with labiate process.   ×10000.
Fig. 13.   Internal view of the opposite valve pole without labiate process.   ×10000.

Localities
Figs 1, 2.   Holotype, Triest 169, Coll. Kützing, B.M. 18340.
Figs 4-13.   Seto-naikai (Seto Inland Sea), Hyogo.

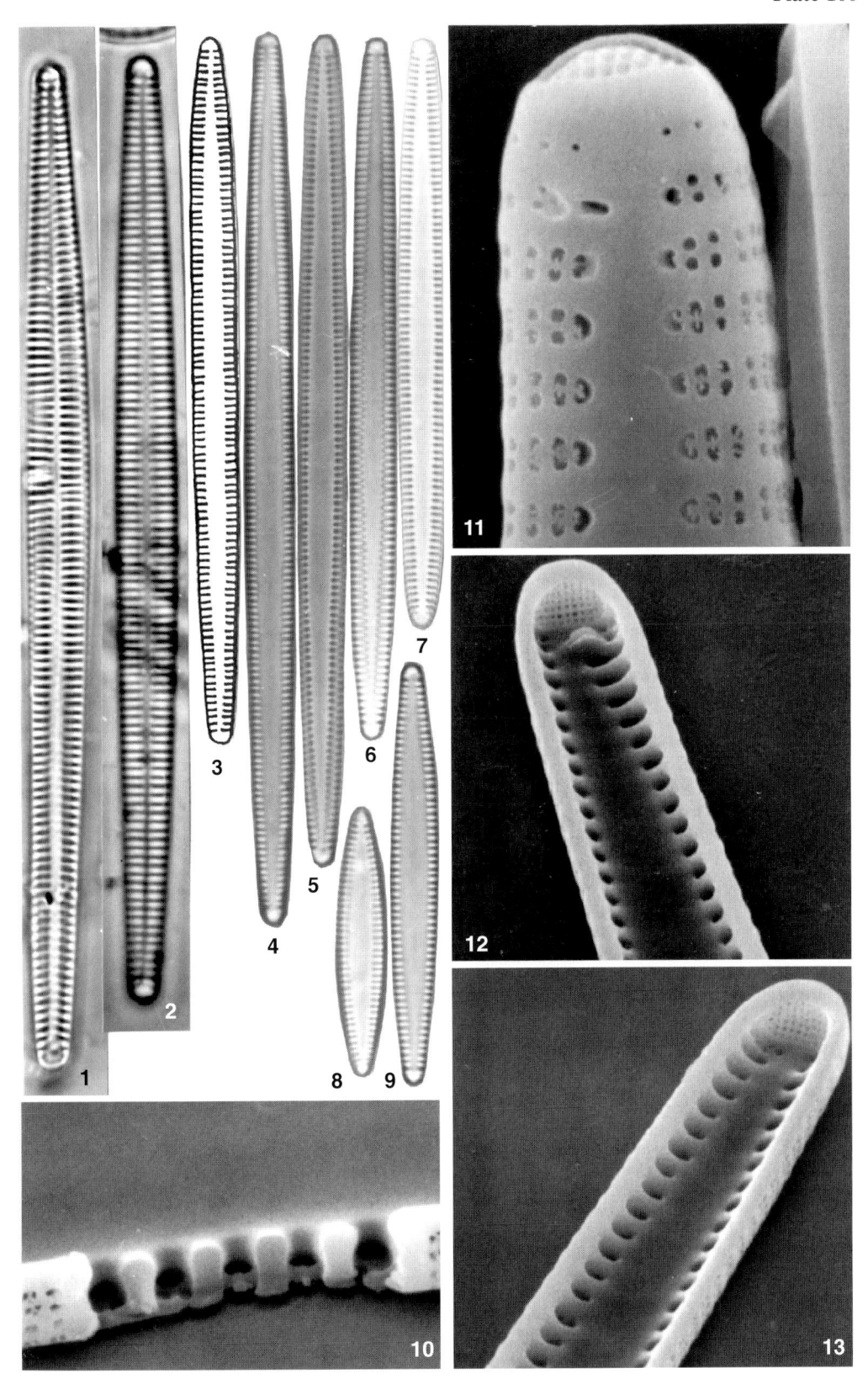

Plate 100

347

Plate 101 *Ulnaria acus* (Kützing) M. Aboal

Fig. 1.   Drawing. ×2000.
Figs 2-4.   LM. ×2000.
Fig. 5.   LM. ×1000.
Figs 6-9.   SEM.

Figs 1-4.   Valve view.
Fig. 5.   Living cells attached to substrate.
Figs 6, 7.   Internal view of both valve poles with a labiate process. ×7000.
Fig. 8.   External polar view showing an opening of labiate process (small arrow) and ocellulimbus (large arrow). ×15000.
Fig. 9.   External view of striae. ×15000.

Localities
Fig. 2.   Sendai-gawa (Sendai River), Kagoshima.
Fig. 3.   Tatara-numa (Tatara Pond), Gunma.
Figs 4, 5.   A paddy field in Shiozawa, Niigata.
Figs 6-9.   Akan-ko (Lake Akan), Hokkaido.

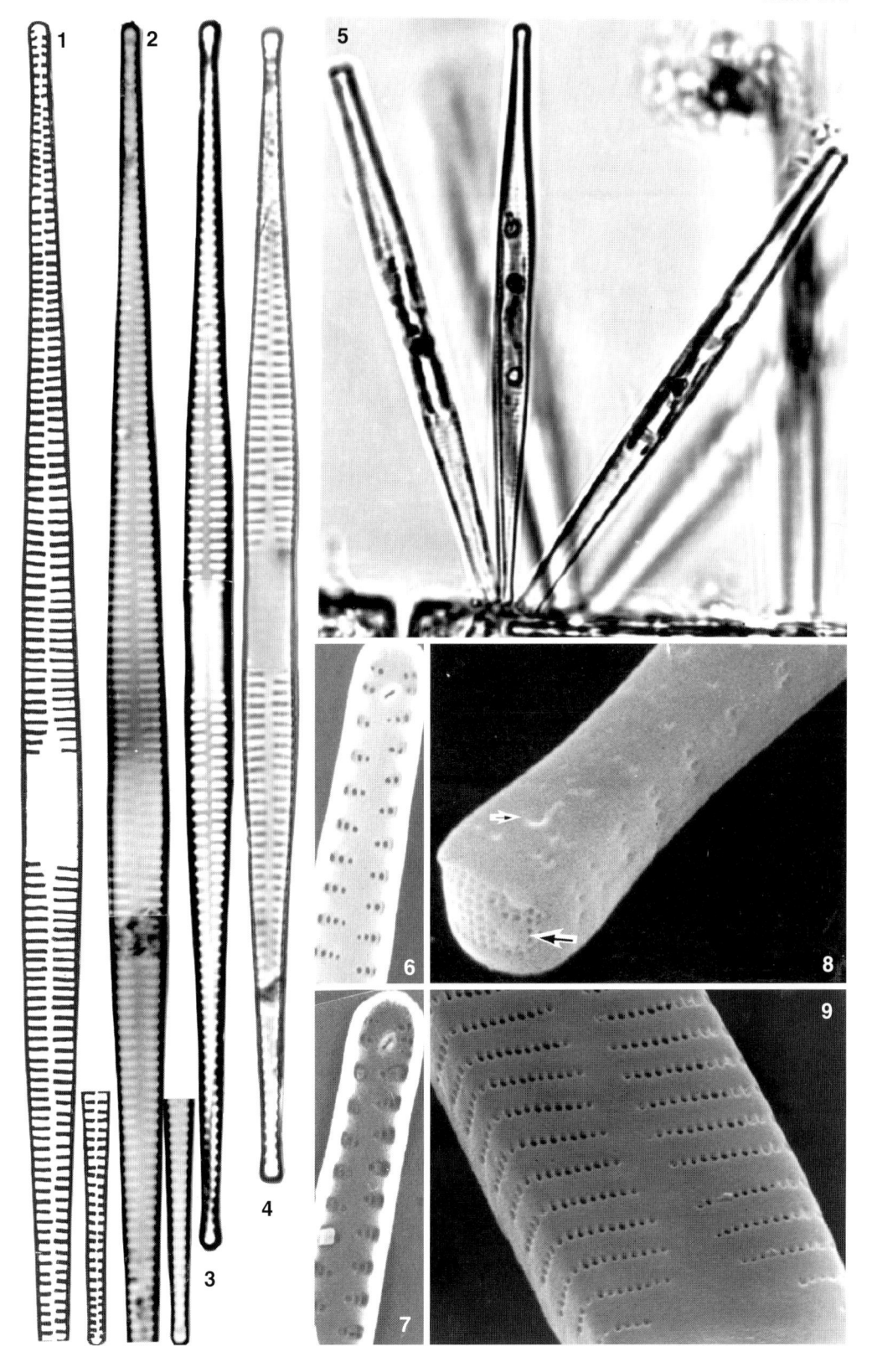

Plate 101

Plate 102 *Ulnaria biceps* (Kützing) Compère

Fig. 1.   Drawing. ×2000.

Figs 2, 3.   LM. ×2000.

Figs 4-6.   SEM.

Fig. 4.   External polar view showing an opening of labiate process (arrow) and ocellulimbi. ×7000.

Fig. 5.   External view of central area.

Fig. 6.   External view of striae. ×7000.

Localities

Fig. 2.   Haruna-ko (Lake Haruna), Gunma.

Figs 3-6.   Akan-ko (Lake Akan), Hokkaido.

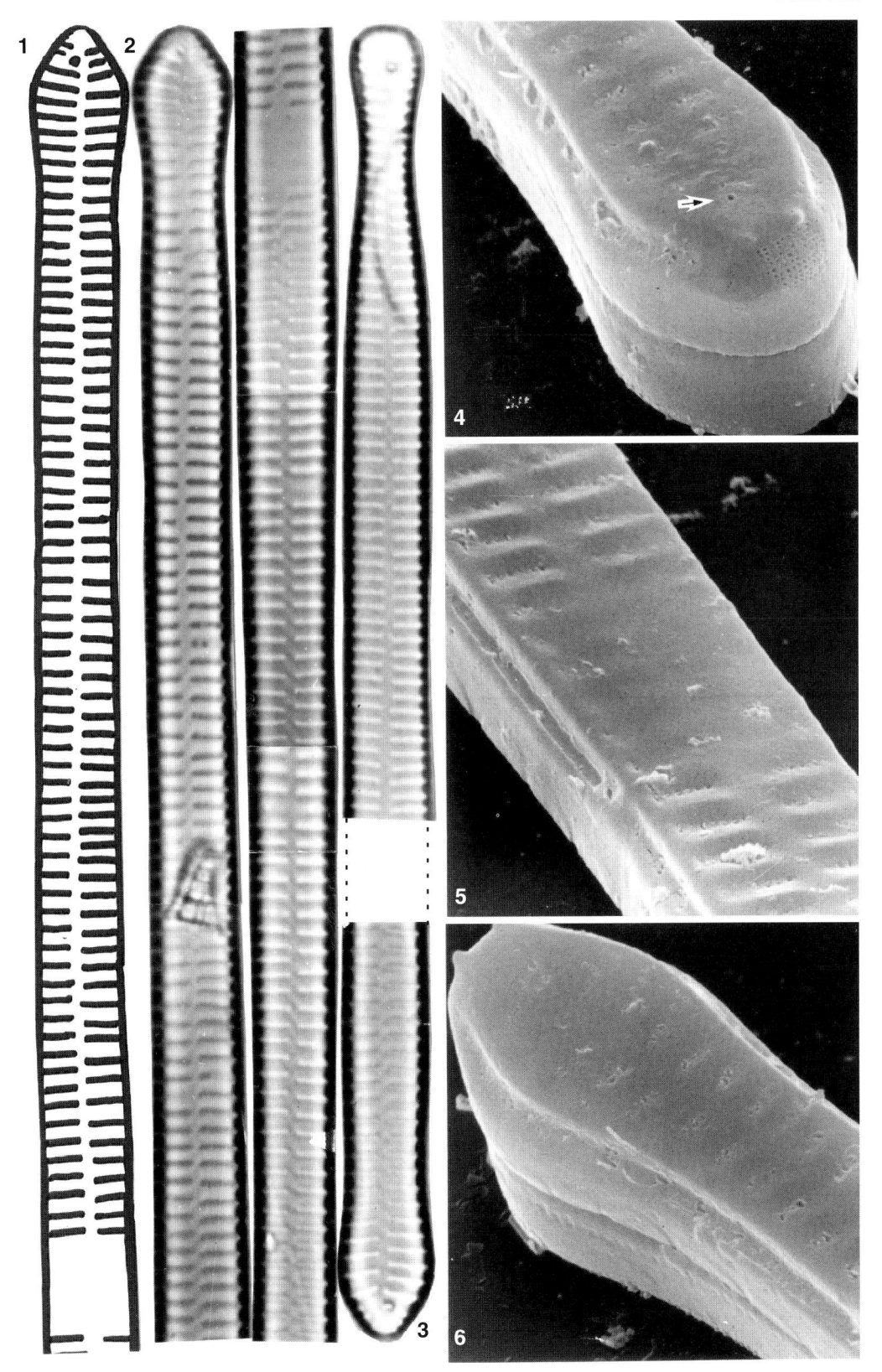

Plate 102

Plate 103 *Ulnaria capitata* (Ehrenberg) Compère

Fig. 1.   Drawing. ×2000.

Fig. 2.   LM. ×2000.

Fig. 3.   LM. ×1000.

Figs 4-6.   SEM.

Fig. 4.   External polar view showing an opening of labiate process and ocellulimbus. ×4000.

Fig. 5.   Internal view of striae. ×8000.

Fig. 6.   Internal polar view showing an opening of labiate process and ocellulimbus. ×7000.

Locality

Figs 2-6.   Akan-ko (Lake Akan), Hokkaido.

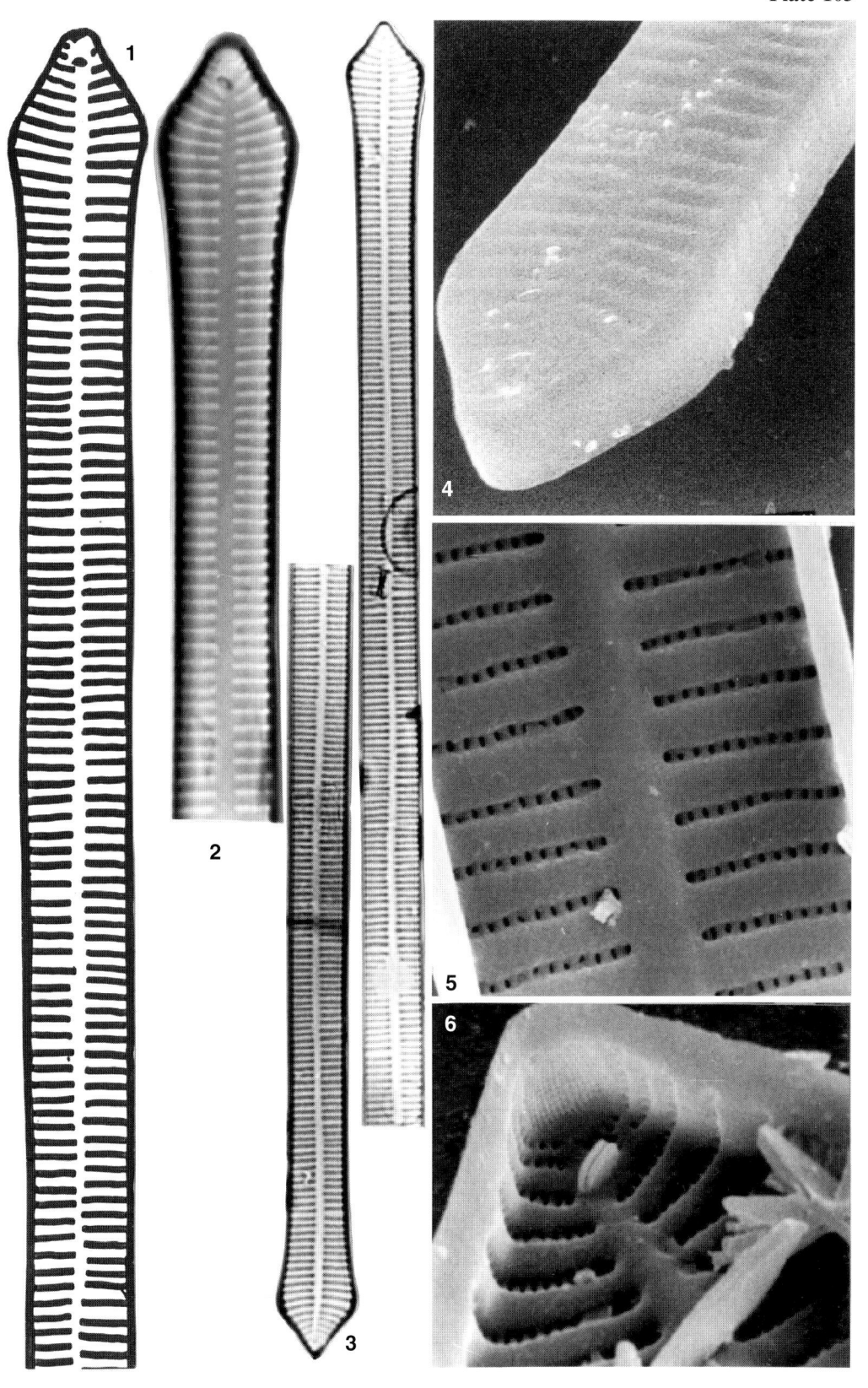

Plate 103

Plate 104 *Ulnaria inaequalis* (H. Kobayasi) M. Idei comb. nov.

Figs 1-3.　LM.　×2000.
Fig. 4.　Drawing.　×2000.
Figs 5-8.　SEM.

Figs 1-4.　Valve view showing irregularly curved subcentral sternum, irregular and delicate striae and ghost striae of central area.
Fig. 5.　Internal polar view showing a labiate process. ×10000.
Fig. 6.　External polar view showing ocellulimbus and an opening of labiate process. ×20000.
Fig. 7.　Internal view of striae. ×10000.
Fig. 8.　External view of areolae with rota occlusion. ×20000.

Localities
Figs 1-3.　Ara-kawa (Ara River), Saitama.
Figs 5-8.　Kanna-gawa (Kanna River), Gunma.

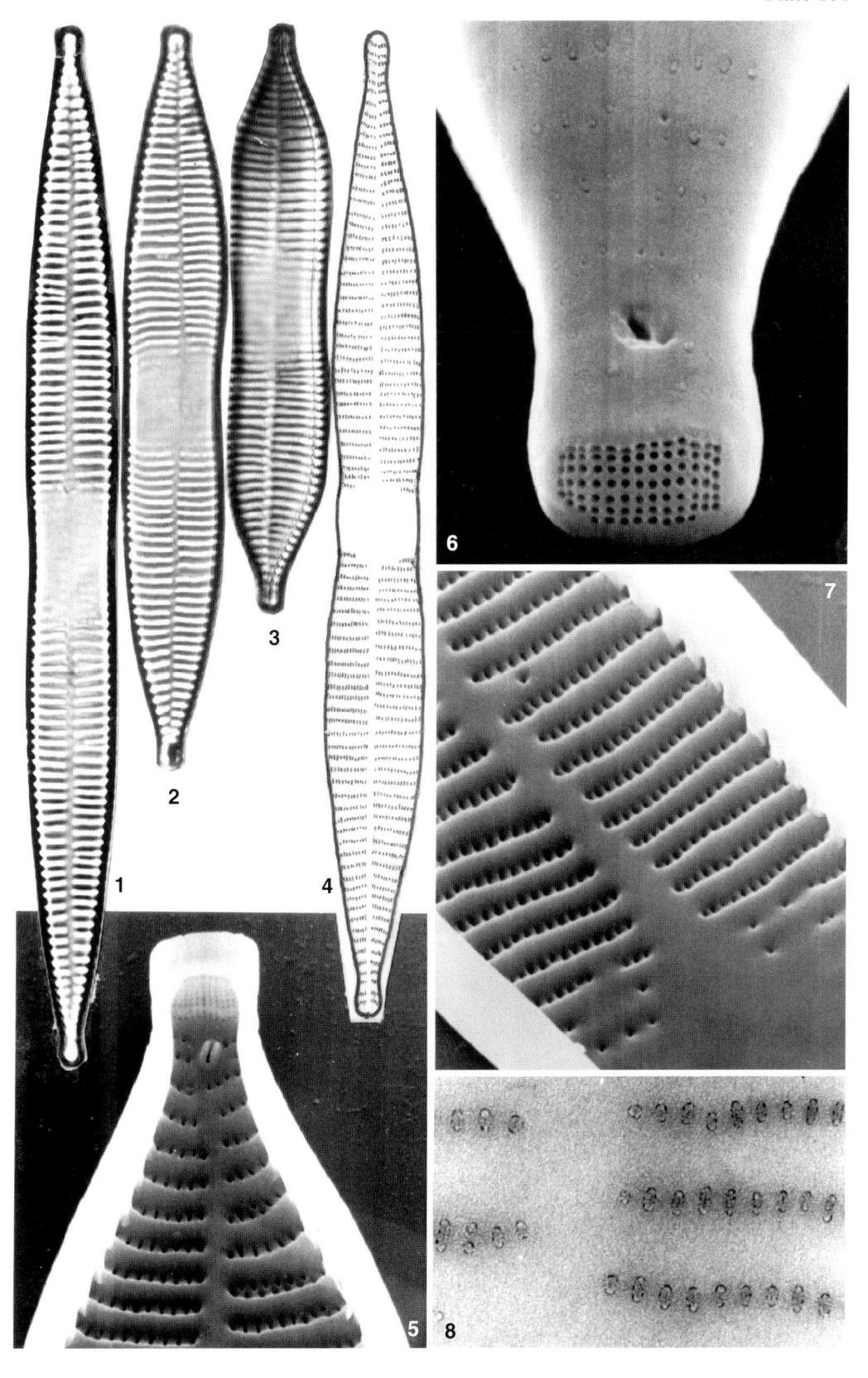

Plate 104

Plate 105 *Ulnaria lanceolata* (Kützing) Compère

Figs 1-3. LM. ×2000.
Fig. 4. Drawing. ×2000.
Figs 5-10. SEM.

Figs 1-4. Valve view.
Fig. 5. External polar view of frustule showing an opening of labiate process and ocellulimbi. ×10000.
Fig. 6. Internal valve pole with a labiate process. ×5400.
Fig. 7. Internal polar view and valvocopula. ×5400.
Fig. 8. External view of striae with double rows of areolae. ×15000.
Fig. 9. Internal view of central area. ×5800.
Fig. 10. Internal view of striae. ×5800.

Locality
Figs 1-3, 5-10. Kino-kawa (Kino River), Wakayama.

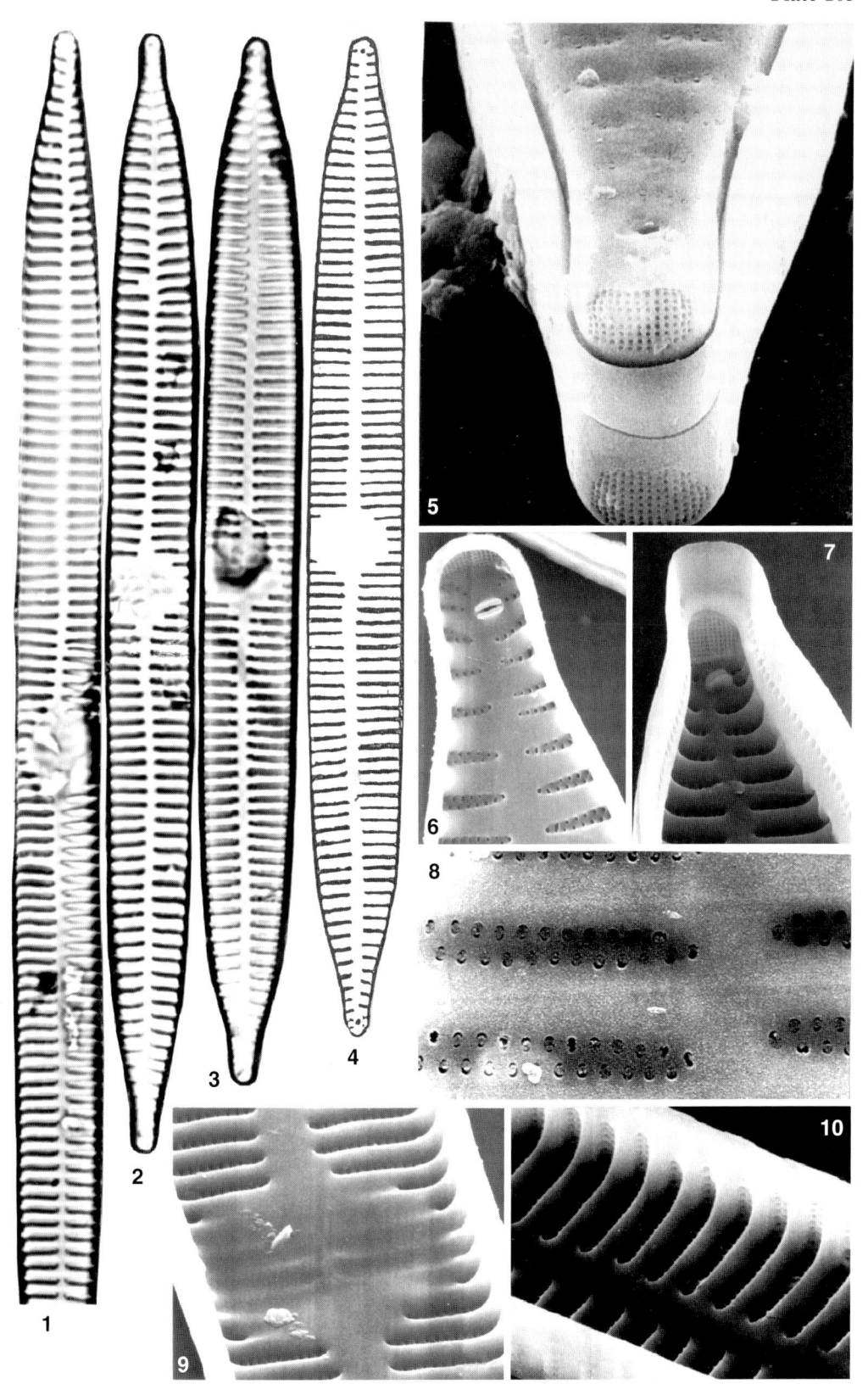

Plate 105

Plate 106 *Ulnaria pseudogaillonii* (H. Kobayasi & M. Idei) M. Idei comb. nov.

Fig. 1.　Drawing. ×700.
Fig. 2.　LM. ×700.
Fig. 3.　LM. ×2000.
Fig. 4.　LM. ×260.
Figs 5-7.　SEM.

Figs 1-3.　Valve view showing parallel striation.
Fig. 4.　Colony of living cells with two plate-like chloroplasts.
Fig. 5.　External polar view showing ocellulimbus, an opening of labiate process, two spines of valve end and areolation. ×6000.
Fig. 6.　Broken valve view showing linking spines. ×7000.
Fig. 7.　Colony. ×1000.

Locality
Figs 2-7.　Chikugo-gawa (Chikugo River), Fukuoka.

Plate 106

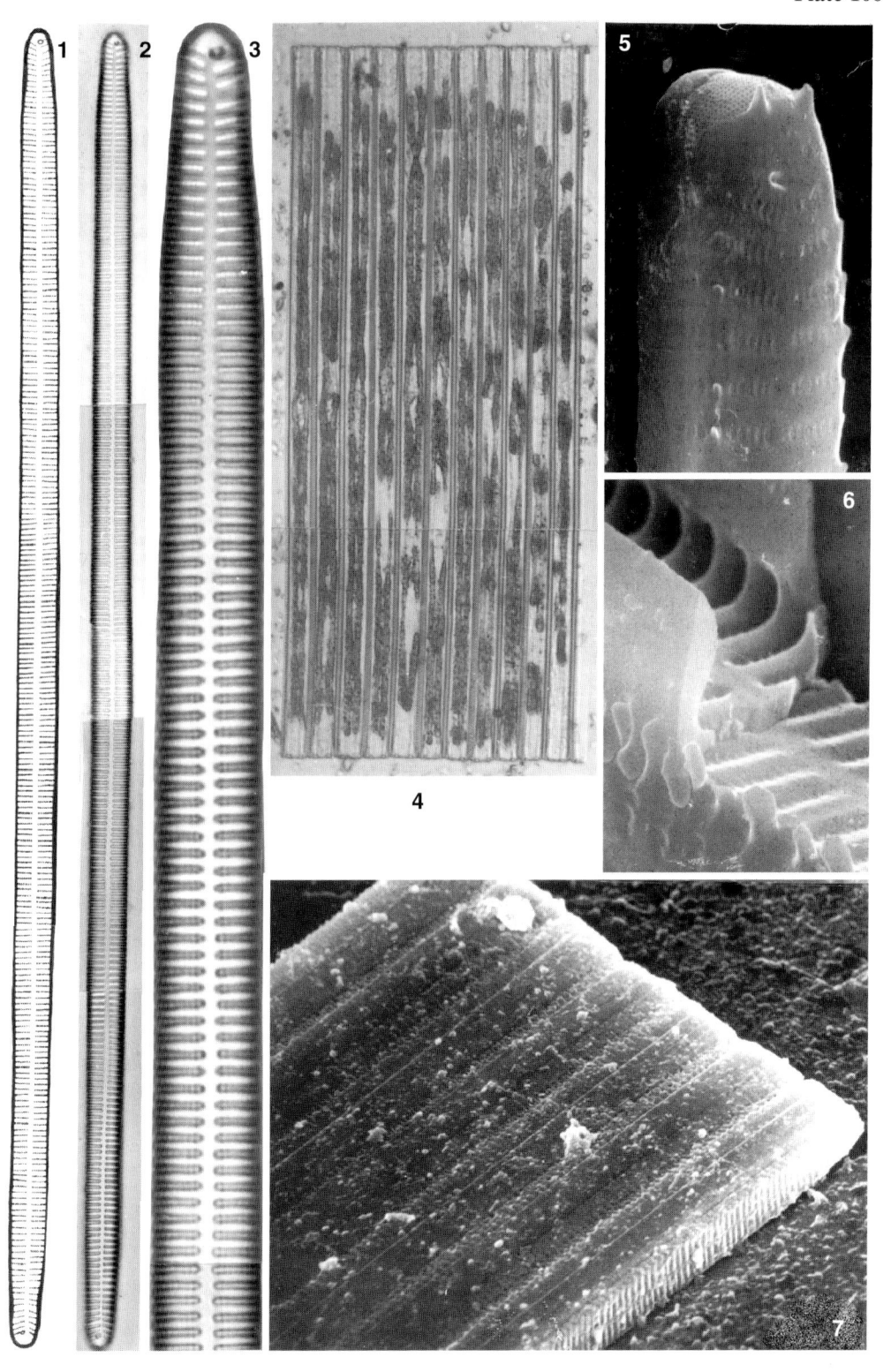

Plate 107 *Ulnaria ulna* (Nitzsch) Compère

Figs 1, 2. LM. ×2000.
Figs 3-7. SEM.

Figs 1, 2. Valve view.
Fig. 3. External view of valve. ×2000.
Fig. 4. Internal view of valve. ×2000.
Fig. 5. Internal view of valve pole showing ocellulimbus and labiate process. ×10000.
Fig. 6. External view of valve pole showing ocellulimbus, an opening of labiate process and areolation. ×10000.
Fig. 7. External view of areolae with rota occlusion. ×10000.

Localities
Figs 1, 2. Kinu-gawa (Kinu River), Tochigi.
Figs 3-7. Kanna-gawa (Kanna River), Gunma.

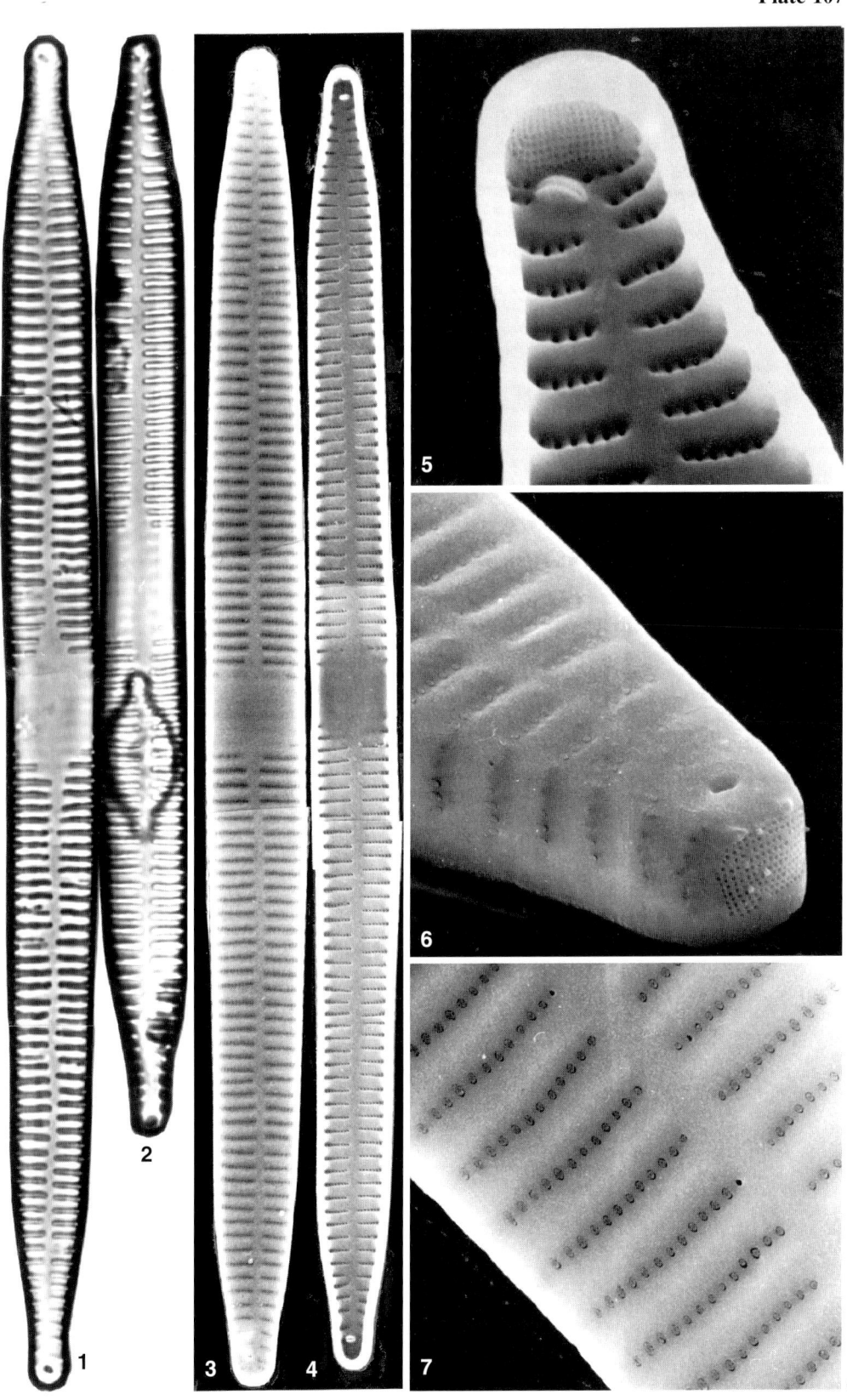

Plate 107

361

Plate 108 *Tabellaria fenestrata* (Lyngbye) Kützing

Fig. 1. LM. ×1000.
Figs 2-4. LM. ×2000.
Fig. 5. Drawing. ×2000.
Fig. 6. Drawing. ×500.
Figs 7-11. SEM.

Fig. 1. Girdle view of dried cells colonized by the connecting mucilage.
Fig. 2. Girdle view of a cleaned specimen.
Figs 3, 5. Valve face.
Fig. 4. Band with septum.
Fig. 6. Girdle view of a chain colony.
Fig. 7. Band with septum. ×800.
Fig. 8. Band without septum. ×800.
Fig. 9. Girdle view of the colony. ×200.
Fig. 10. External view of the enlarged valve end showing apical pore field. ×8000.
Fig. 11. External view of the enlarged valve center showing poroid areolae, outer opening of labiate process (arrow) and two spinules on the margin of the valve face (left margin). ×10000.

Locality
Figs 1-4, 7-11. An irrigation pond in Kamikomatsuki, Shiga.

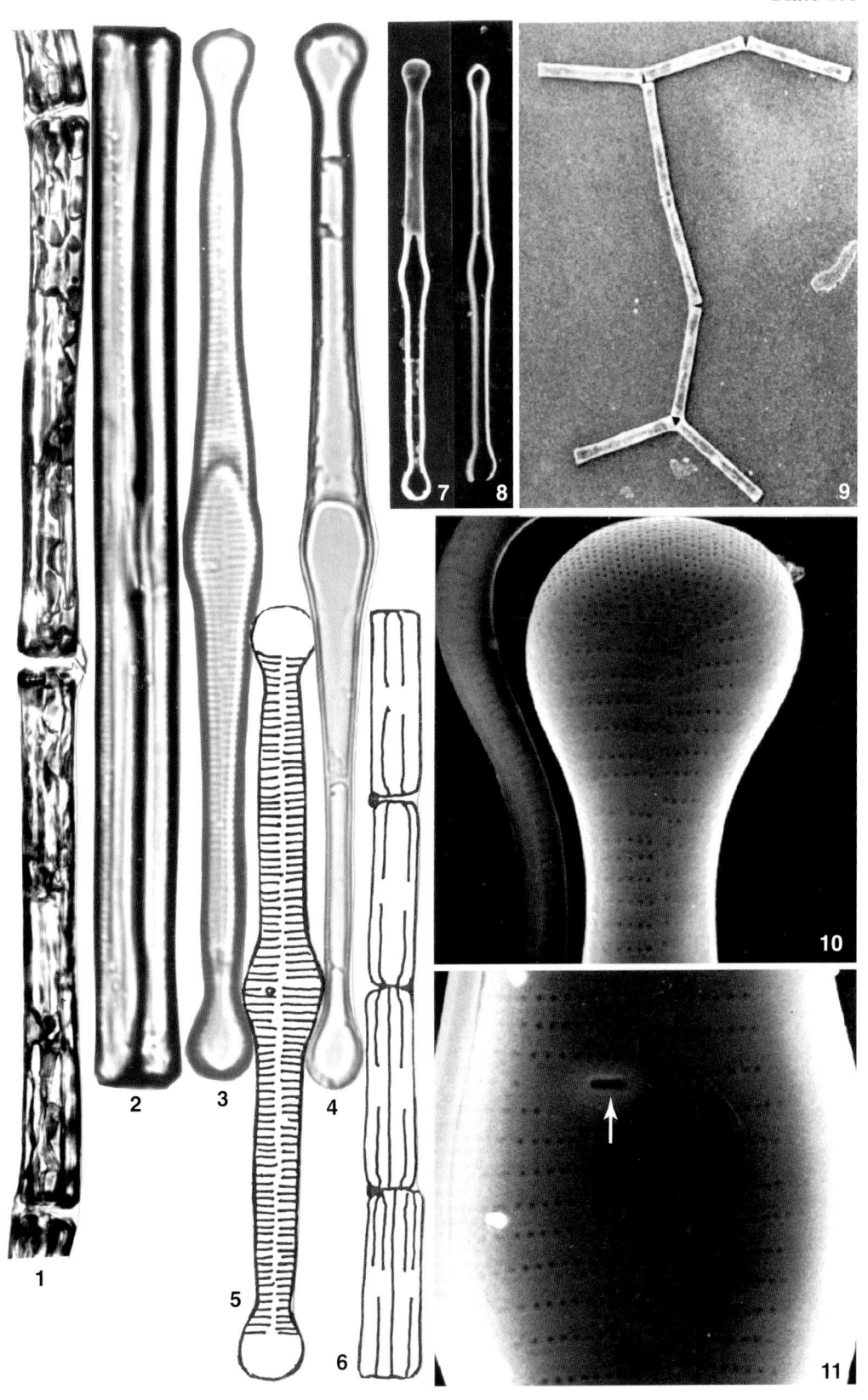

Plate 108

Plate 109 *Tabellaria fenestrata* (Lyngbye) Kützing

Figs 12-17. SEM.

Fig. 12. Oblique view of the valvocopula with septum (s) and the valve (v). ×10000.

Fig. 13. External oblique view of one frustule end showing an apical pore field at the apex of each valve and the girdle band arrangement. ×8000.

Fig. 14. The other frustule end of the same specimen as in Fig. 13. All bands have open ends in either frustule end. ×8000.

Fig. 15. Open ends of the band. ×8000.

Fig. 16. Internal view of the valve (v) associated with two bands, each of which has a septum, showing a labiate process through a window made by these septa. ×5000.

Fig. 17. Valve (v) associated with two bands. On the bottom edge of the second band, which has a single row of areolae, an expanded septum is visible. ×8000.

Locality

Figs 12-17. An irrigation pond in Kamikomatsuki, Shiga.

Plate 109

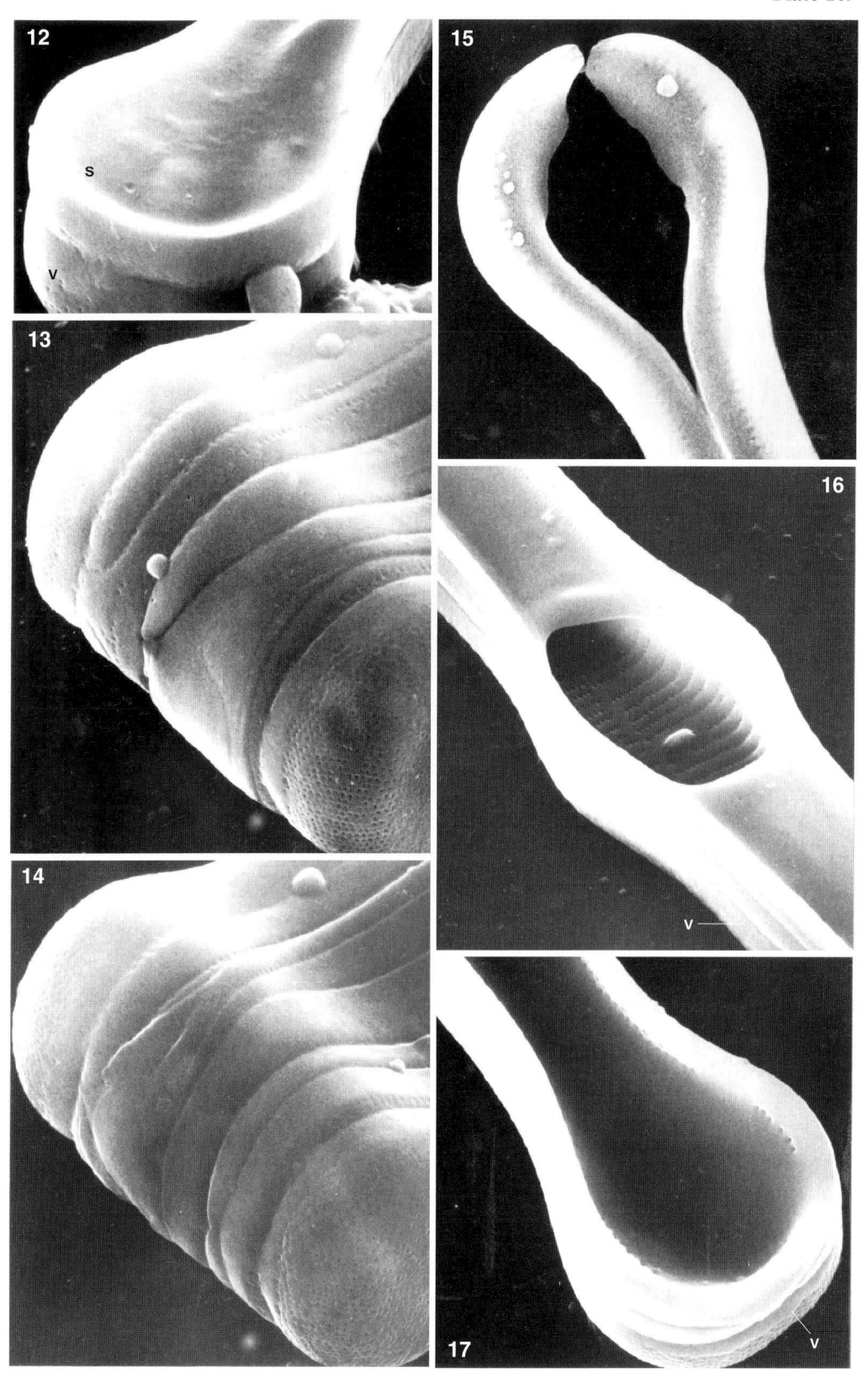

Plate 110 *Tabellaria flocculosa* (Roth) Kützing

Figs 1, 2.   Drawings. ×2000.
Figs 3-9.   LM. ×2000.
Figs 12, 13.   LM. ×1000.
Figs 10, 11, 14.   SEM.

Figs 1, 3-6.   Valve face.
Figs 2, 8, 9.   Band with septum and subseptum.
Fig. 7.   Band with septum.
Fig. 10.   Girdle view of zigzag colony. ×480.
Fig. 11.   External view of the valve face. ×3000.
Fig. 12.   Girdle view of zigzag colony. Dried specimen same as Fig. 10.
Fig. 13.   Girdle view of zigzag colony. Living specimen.
Fig. 14.   Band with septum (s) and rudimentary septum (rs). ×2000.

Locality
Figs 3-14.   Lake Päijänne, Finland.

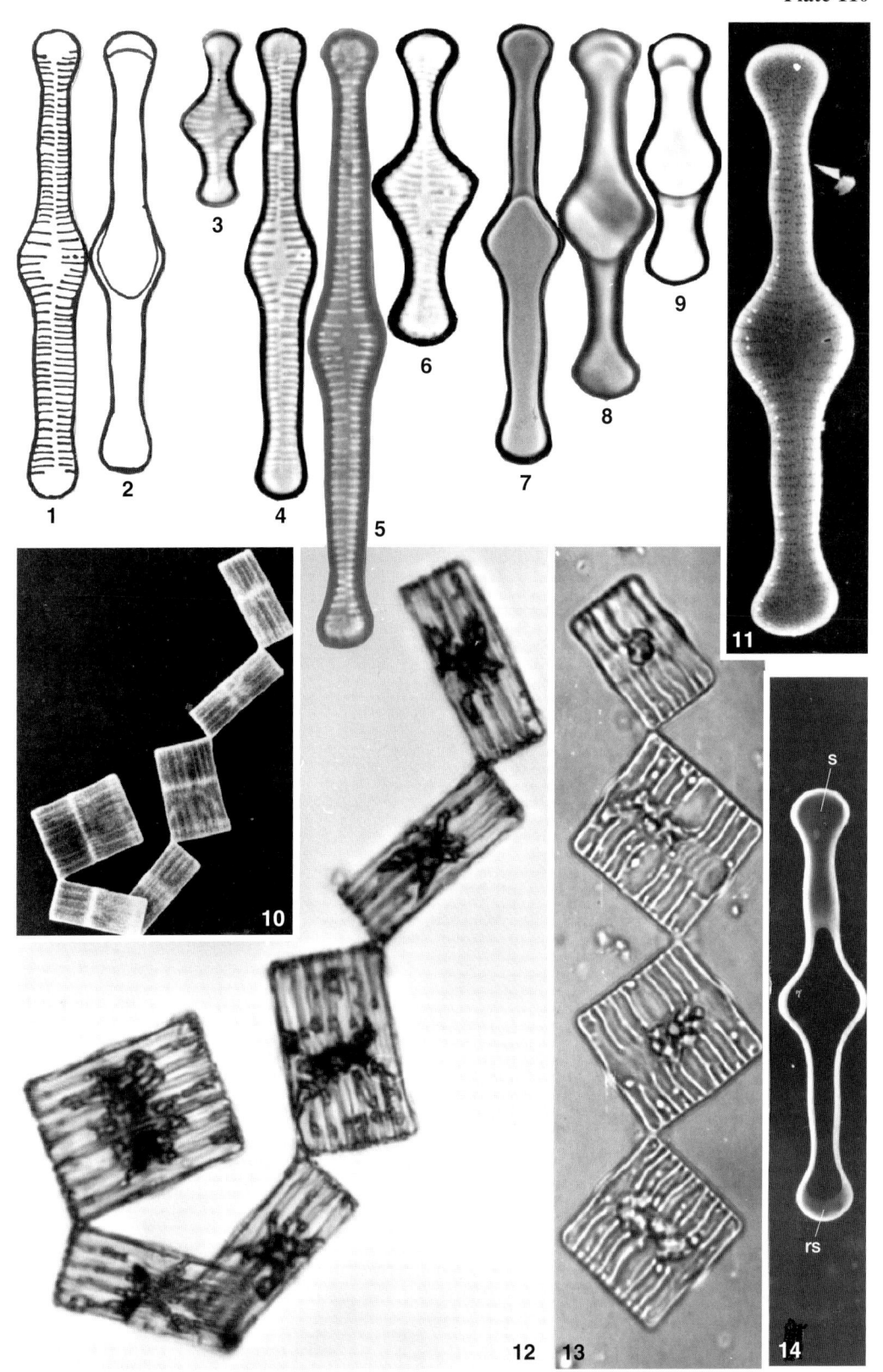

Plate 110

Plate 111 *Tabellaria flocculosa* (Roth) Kützing

Figs 15, 17, 18.   SEM.

Fig. 16.   Drawing of the frustule end.

Fig. 15.   External oblique view of a whole frustule showing spinules on the valve shoulder. ×3000.

Fig. 17.   Enlarged frustules connected by mucilage secreted from apical pore field showing girdle bands with a single row (partly double at apex) of areolae. ×8000.

Fig. 18.   Details of the band with rudimentary septum. ×10000.

Locality

Figs 15, 17, 18.   Lake Päijänne, Finland.

Plate 111

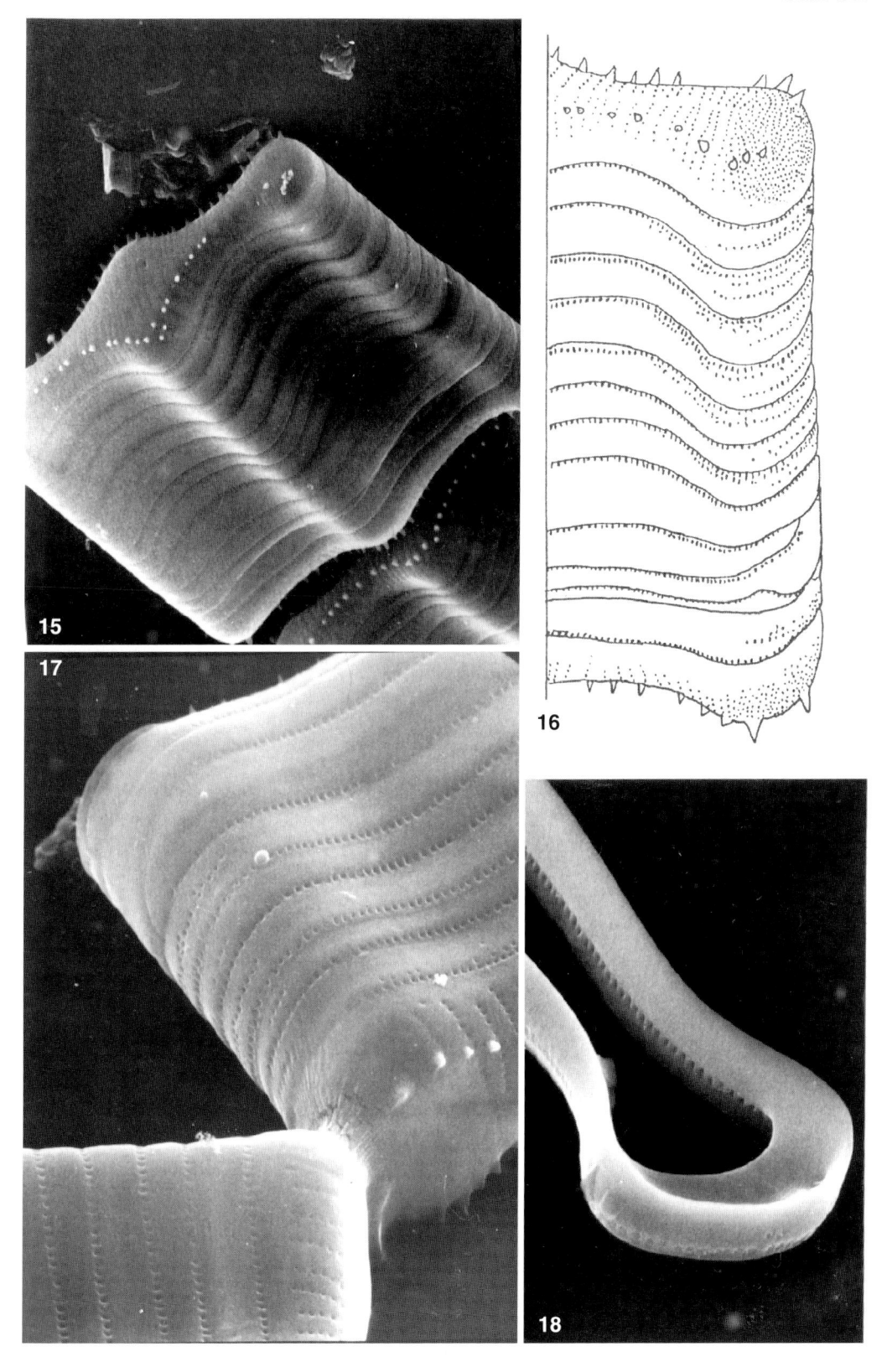

369

Plate 112 *Tabellaria pseudoflocculosa* H. Kobayasi ex Mayama sp. nov.

Figs 1, 2.   Drawings. ×2000.
Fig. 8.   Drawing. ×500.
Figs 3-6.   LM. ×2000.
Fig. 7.   LM. ×1000.
Fig. 9.   LM. ×300.
Figs 10-12.   SEM.

Figs 1, 4, 5.   Valve face.
Figs 2, 6.   Bands with septum.
Fig. 3.   Girdle view.
Fig. 7.   Girdle view of the dried specimen. Note connecting mucilage between cells.
Figs 8, 9.   Girdle view of the zigzag colonies.
Fig. 10.   External oblique view of the valve end with spinules. ×10000.
Fig. 11.   External oblique view of the mid-portion of the valve. ×10000.
Fig. 12.   Enlargement of the frustule end showing band arrangement. ×8000.

Locality

Figs 3-7, 9-12.   An irrigation pond in Kamikomatsuki, Shiga.

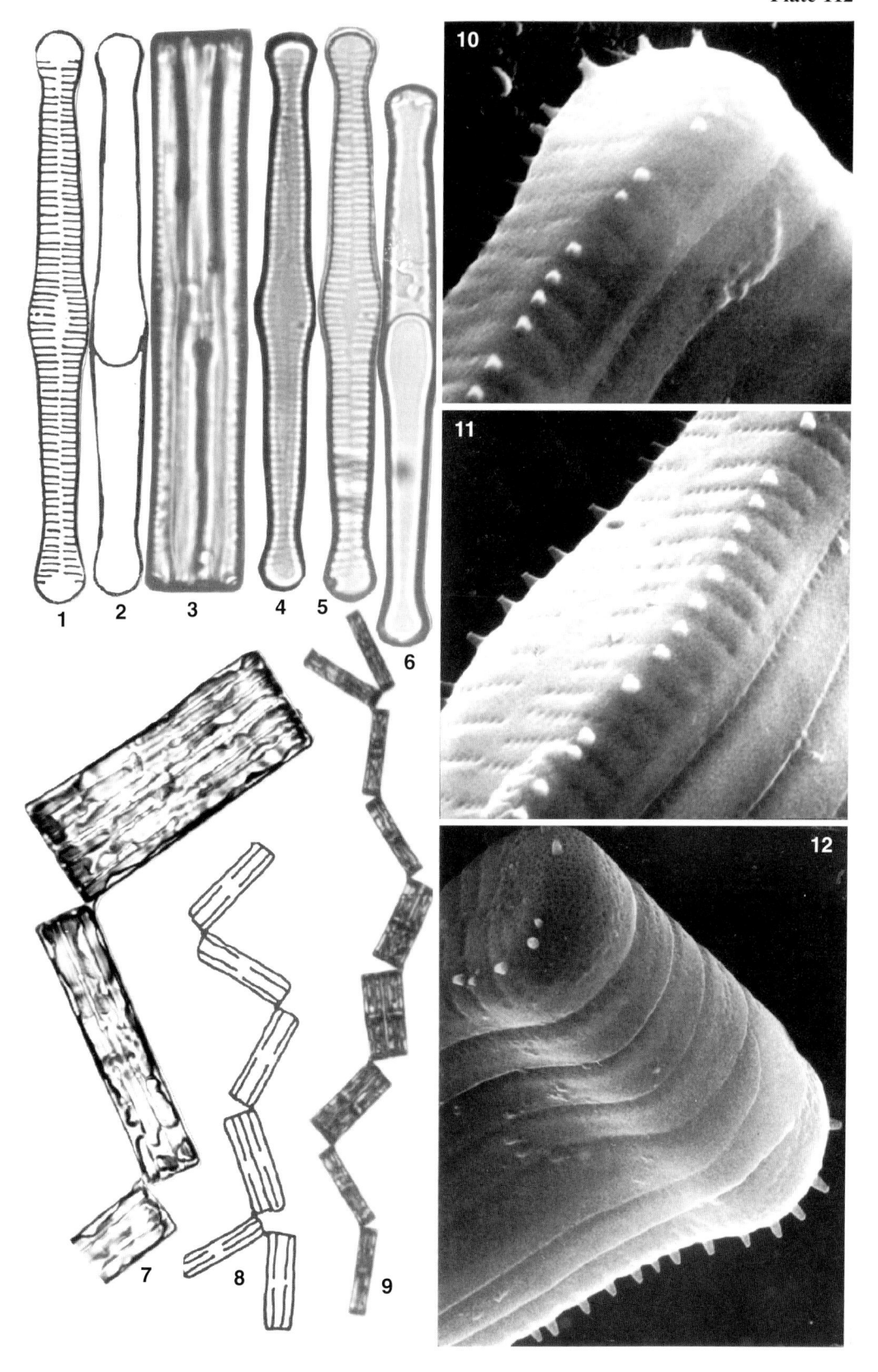

Plate 112

Plate 113 *Actinella brasiliensis* Grunow

Figs 1, 3, 4. LM. ×2000.
Fig. 2. Drawing. ×2000.
Figs 5-7. SEM.

Fig. 5. Oblique view of a whole frustule. ×1000.
Fig. 6. Internal view of the valve on the side of the head pole showing one raphe branch on the mantle. ×5000.
Fig. 7. Internal view of the valve on the side of the foot pole showing the other raphe branch on the mantle. ×5000.

Localities
Figs 1, 5-7. Koridon-no-ike (Koridon Pond), Ojiya, Niigata.
Fig. 3. Jizoin-numa (Jizoin Pond), Hanyu, Saitama.
Fig. 4. A pond in Musashi Kyuryo Shinrin Park, Higashimatsuyama, Saitama.

Plate 113

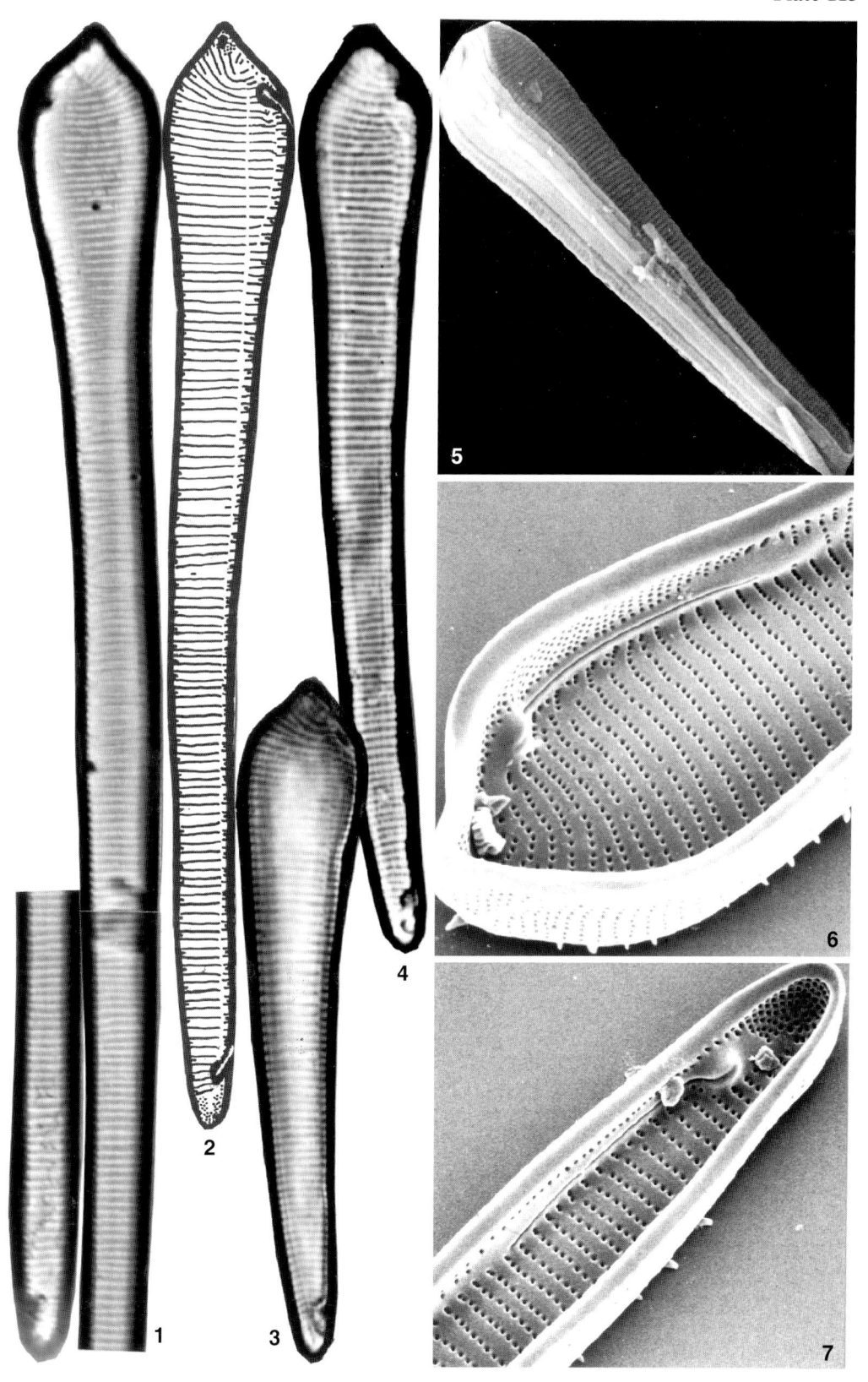

373

Plate 114 *Peronia fibula* (Brébisson ex Kützing) R. Ross

Figs 1, 2.   Drawings. ×2000.
Figs 3-10.   LM. ×2000.
Figs 11-13.   SEM.

Figs 1, 5-7.   Raphe valves.
Figs 2, 8-10.   Rudimentary raphe valves.
Figs 3, 4.   Girdle view of the frustules.
Fig. 11.   External view of the valve on the side of the headpole showing one branch of the raphe. ×4000.
Fig. 12.   External view of the valve on the side of the footpole showing the other branch of the raphe. ×4000.
Fig. 13.   Internal view of the area around the footpole showing a fissure of the rudimentary raphe. ×10000.

Localities
Figs 3, 5-10.   Yashimaga-ike (Yashima Pond), Kirigamine, Nagano.
Figs 4, 11-13.   Koridon-no-ike (Koridon Pond), Ojiya, Niigata.

Plate 114

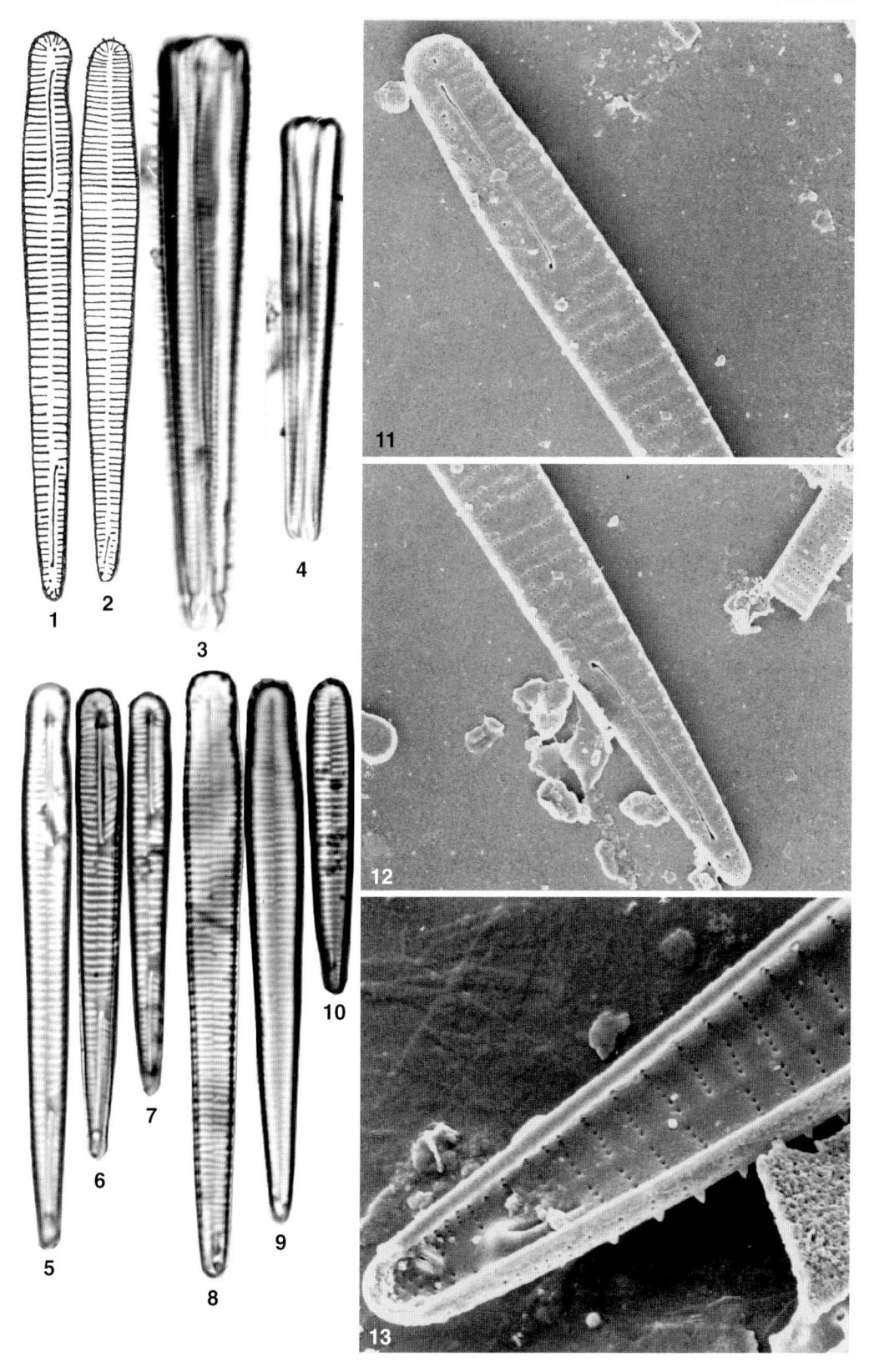

Plate 115 *Rhoicosphenia abbreviata* (C. Agardh) Lange-Bertalot

Figs 1, 7.　Drawings.　×2000.
Figs 2-6, 8-16.　LM.　×2000.
Figs 17, 18.　SEM.

Figs 1-6.　Concave valves with complete raphe.
Figs 7-13.　Convex valves with rudimentary raphe.
Figs 14-16.　Girdle view.
Fig. 17.　Oblique view of a frustule.　×3000.
Fig. 18.　The valve with rudimentary raphe showing an apical pore field in the footpole. ×5000.

Localities

Figs 2, 4, 8, 11.　Kinu-gawa (Kinu River), Tochigi.
Figs 3, 5, 6, 10, 12, 13-15.　Chikuma-gawa (Chikuma River), Kawakami, Nagano.
Fig. 9.　Teich See, Germany.
Fig. 16.　A gutter in Dractin, the Netherlands.
Figs 17, 18.　Matsumoto-jo Hori (The moat of Matsumoto Castle), Nagano.

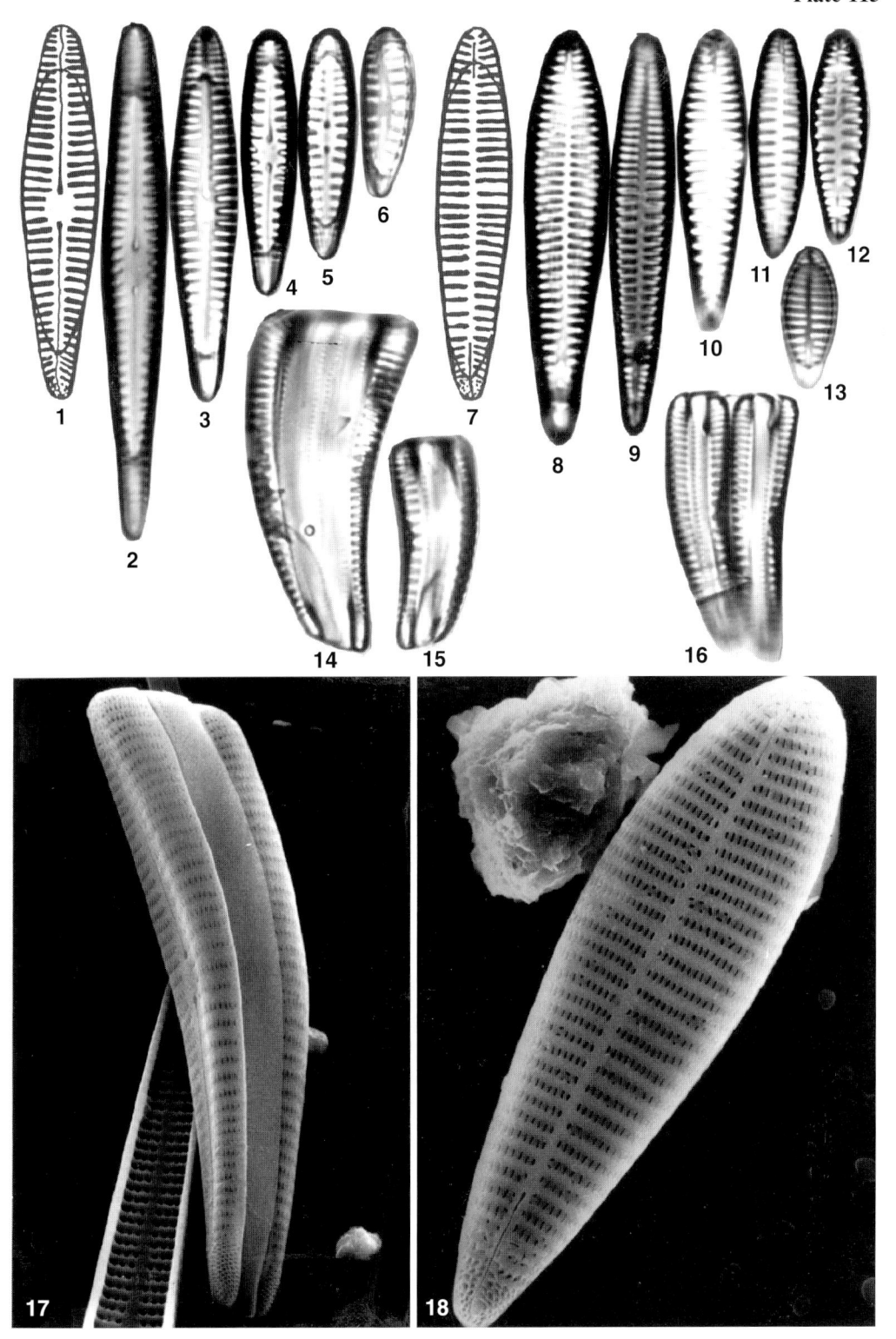

Plate 115

Plate 116 *Rhoicosphenia abbreviata* (C. Agardh) Lange-Bertalot

Figs 19-24. SEM.

Fig. 19. Enlargement of the headpole of the complete raphe valve. ×20000.
Fig. 20. Enlarged central area of the complete raphe valve showing the external raphe branches which end in central pores. ×10000.
Fig. 21. Internal view of the rudimentary raphe valve with the first band showing a septum in the headpole. ×8000.
Fig. 22. External view of the footpole showing the complete raphe fissure and the apical pore field. ×10000.
Fig. 23. Internal view of the rudimentary raphe valve with the first band showing the septum in the footpole. ×8000.
Fig. 24. Internal whole view of the rudimental raphe valve with the first band. ×3000.

Locality

Figs 19-24. Ryugaeshino-taki (Ryugaeshi Falls), Karuizawa, Nagano.

Plate 116

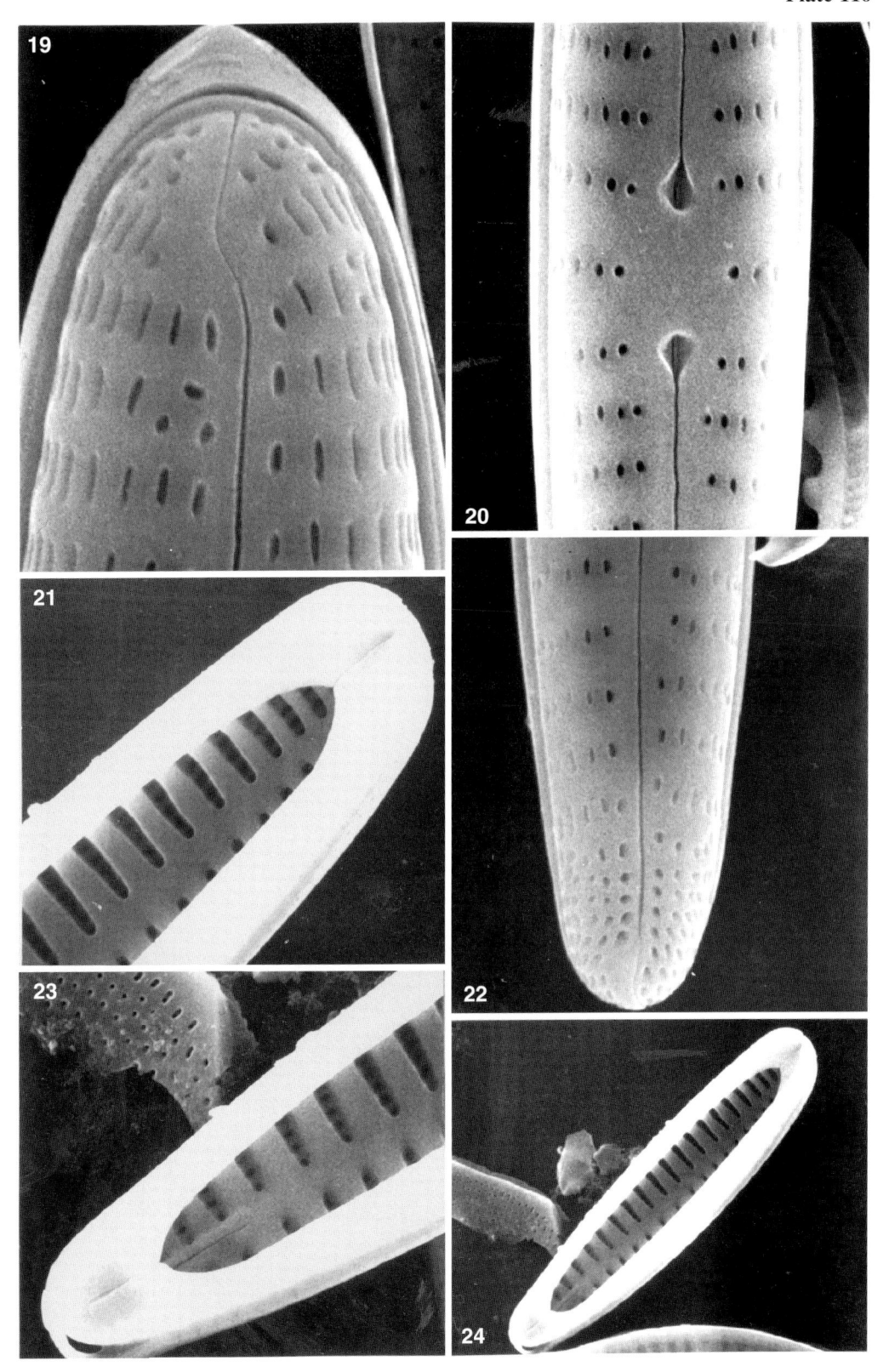

Plate 117 *Rhoicosphenia abbreviata* (C. Agardh) Lange-Bertalot

Fig. 25. Drawing.

Fig. 25. Diagram of the elements of the frustule. EV : epivalve, EB : epibands, HB : hypoband, HV : hypovalve.

Plate 117

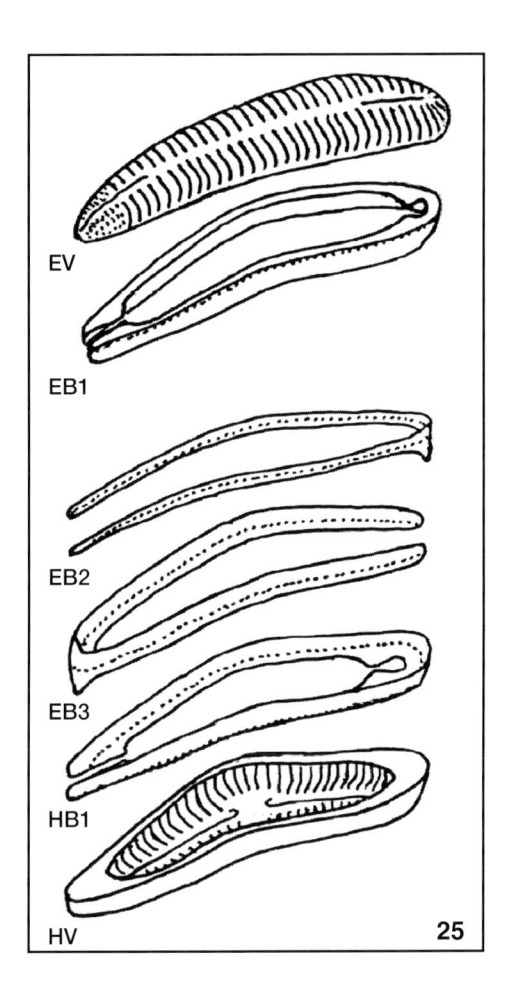

Plate 118 *Gomphoneis heterominuta* **Mayama & Kawashima**

Figs 1-3. LM. ×2000.
Fig. 4. Drawing. ×2000.
Figs 5-9. SEM.

Figs 1-4. Valve view.
Fig. 5. Oblique internal view of the valve showing a pseudoseptum at each apex, nearly straight inner raphe fissures, the large openings of the alveolate striae, a row of large areolae in the valve margin, and unilaterally raised central nodule with a stigma and central raphe endings. ×6000.
Fig. 6. Enlarged external view of the headpole showing a slightly undulated external raphe fissure, the terminal fissure curved to the side bearing the stigma, and the reniform openings of the areolae in each biseriate stria. ×15000.
Fig. 7. Enlarged external view of the valve center showing the central raphe endings terminating in expanded central pores, the relatively large, circular opening of the stigma, and the reniform (or round) openings of the areolae. ×15000.
Fig. 8. Internal view of the central nodule to the headpole, showing a pseudoseptum at the headpole, the terminal raphe ending terminating in a small helictoglossa, a straight inner raphe fissure, hooked central raphe endings, and a stigma opening within an elliptical hollow. ×10000.
Fig. 9. Enlarged external view of the footpole showing the apical pore field composed of round pores, and the terminal fissure curved towards the side bearing the stigma. ×15000.

Localities
Fig. 1. Yamanaka-ko (Lake Yamanaka), Yamanashi.
Figs 2, 3. Kano-gawa (Kano River), Shizuoka.
Figs 5-9. Inabe-gawa (Inabe River), Mie.

Plate 118

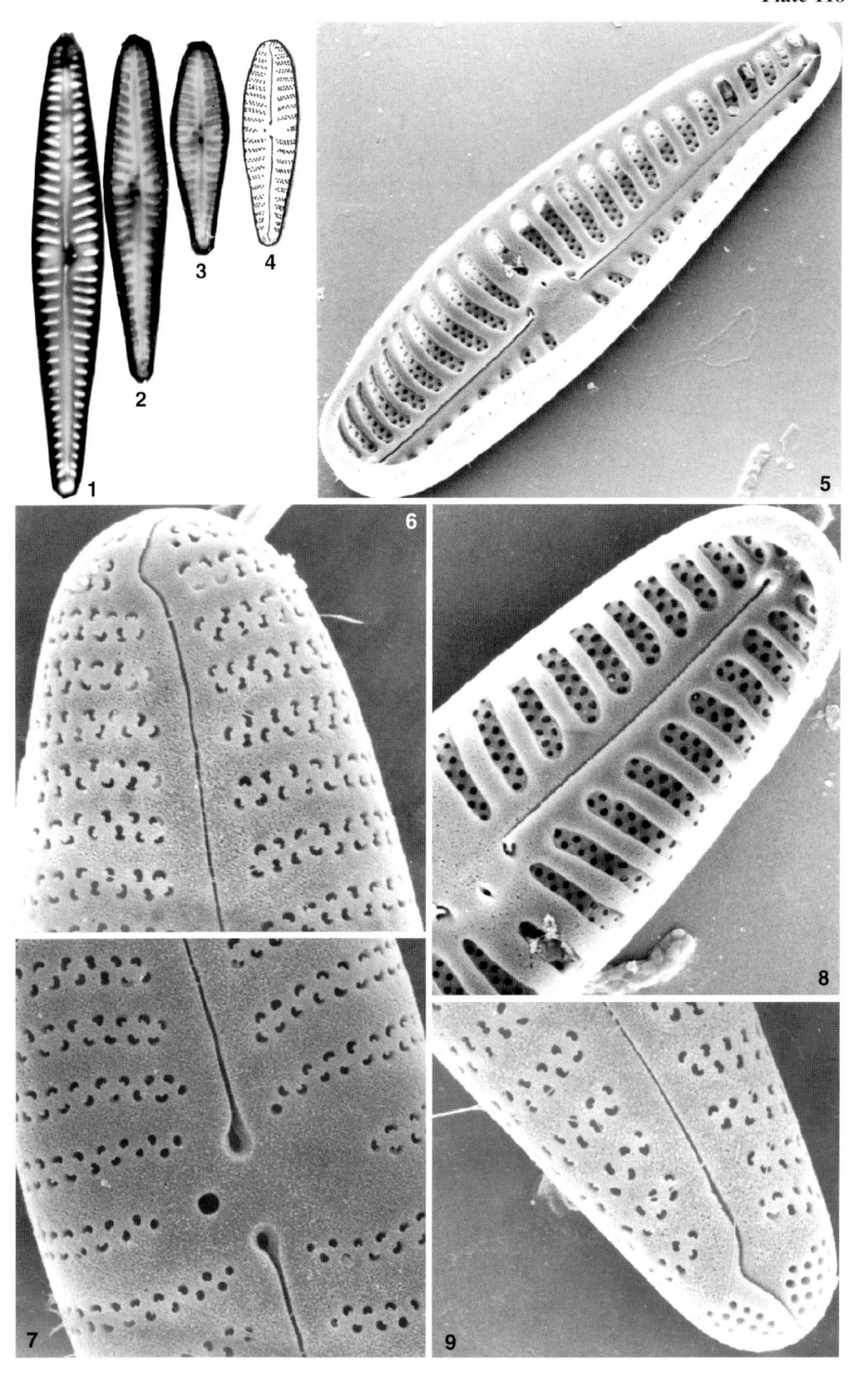

Plate 119 *Gomphoneis rhombica* **(Fricke) V. Merino** *et al.*

Figs 1-6, 9-11. LM. ×2000.
Figs 7, 8. Drawings. ×2000.
Figs 12-14. SEM.

Figs 1-7, 10, 11. Valve view.
Fig. 8. Girdle view of a theca.
Fig. 9. Girdle view of a frustule.
Fig. 12. External view of the headpole showing a narrow external raphe fissure, the terminal fissure slightly curved to the side bearing the stigma, and the areolae alternately arranged in each biseriate stria. ×10000.
Fig. 13. Enlarged external view of the valve centre showing the smooth surface of the axial area, the round opening of the stigma, slightly expanded central raphe endings, and double rows of the areolae. ×10000.
Fig. 14. External view of the footpole showing a slightly curved terminal fissure, the apical pore field composed of the pores similar to the areolae in size, and a single row of the areolae in a few striae at the valve end. ×8000.

Localities

Figs 1, 2, 4-6, 9-11. Ara-kawa (Ara River), Saitama.
Fig. 3. Chigonosawa (Chigonosawa Creek), Kisofukushima, Nagano.
Figs 12-14. Nippara-gawa (Nippara River), Tokyo.

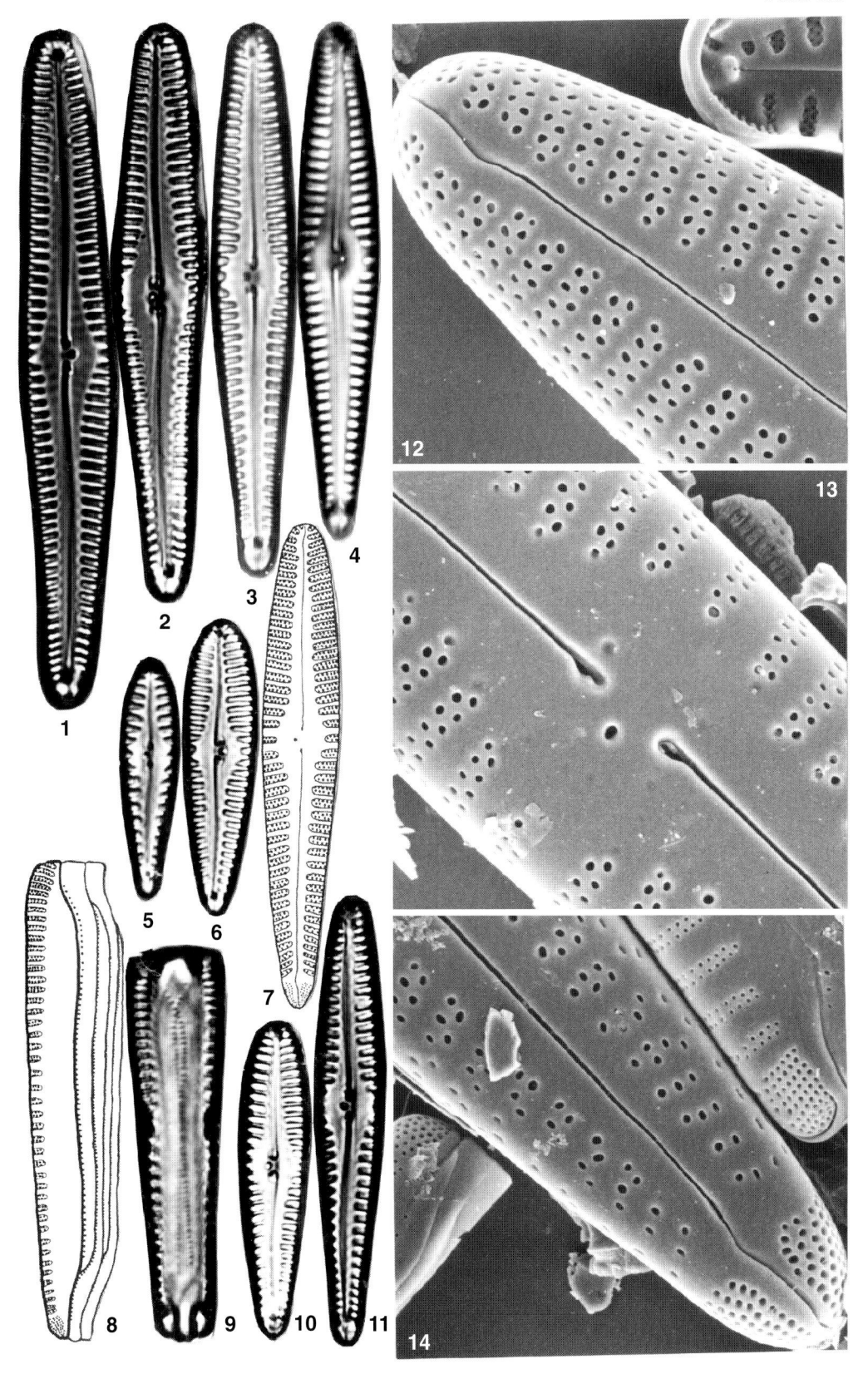

Plate 119

Plate 120 *Gomphoneis rhombica* **(Fricke) V. Merino** *et al.*

Figs 15-20. SEM.

Fig. 15. Internal view of the headpole showing a relatively large pseudoseptum, the helictoglossa slightly deflected towards the non-stigma side, and the large foramina of the alveoli. ×10000.

Fig. 16. Enlarged internal view of the valve centre showing a unilaterally raised central nodule, the slit-like opening of the stigma, the central raphe endings bent towards the side bearing the stigma and then turned back. ×10000.

Fig. 17. Internal view of the footpole showing a developed pseudoseptum, the helictoglossa slightly deflected towards the non-stigma side, and the elliptical foramina of the alveoli. ×10000.

Fig. 18. Enlarged girdle view of the headpole of the frustule, showing the bands each with a row of the areolae lying along the suture. ×8000.

Fig. 19. Enlarged girdle view of the footpole of the same frustule as Fig. 18, showing the apical pore field, and the bands with a row of areolae. ×8000.

Fig. 20. Enlarged oblique view of the broken valve showing the foramina barely occluded in the margin by the very narrow projections of the inner wall. ×15000.

Locality

Figs 15-20. Nippara-gawa (Nippara River), Tokyo.

Plate 120

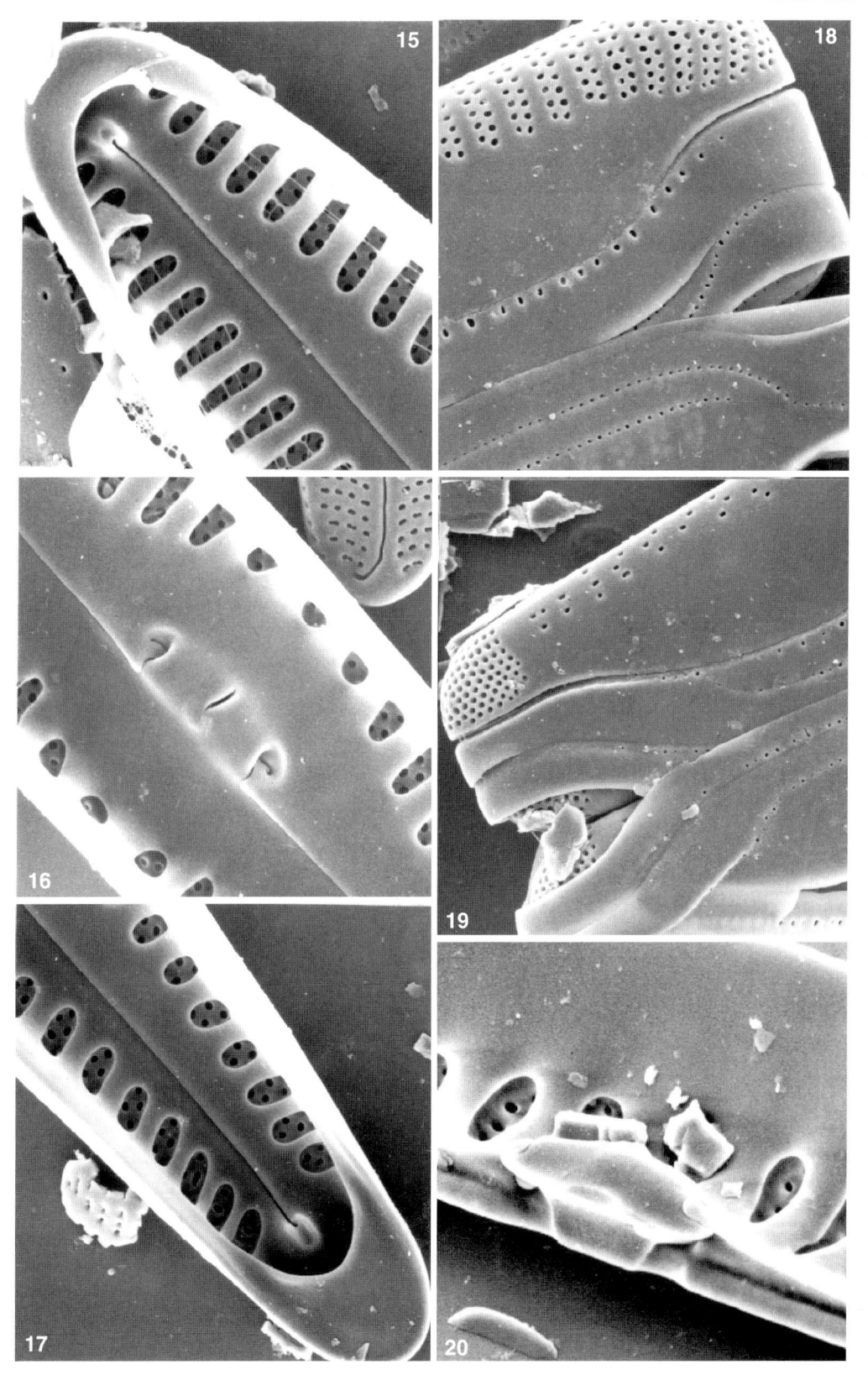

Plate 121　*Gomphonema clevei* Fricke

Figs 1-3.　Drawings.　×2000.
Figs 4-6.　LM.　×2000.
Figs 7-11.　SEM.

Figs 1-6.　Valve view.
Fig. 7.　External oblique view of a whole valve. ×5000.
Fig. 8.　Internal oblique view of a whole valve showing nearly straight inner raphe fissures, the large openings of the alveoli, pseudosepta (arrows) at both apices, straight inner raphe fissures, the terminal raphe endings terminated in small helictoglossae, and a central nodule with hooked central raphe endings and the elliptical hollow of the stigma. ×8000.
Fig. 9.　Girdle view of the headpole of the frustule, showing the closed ends of the first (B1, valvocopula) and the second (B2) bands, and the open end of the third band (B3). ×10000.
Fig. 10.　Details of the headpole to the valve centre in Fig. 7 showing an undulate raphe fissure, the terminal fissure curved towards the side bearing the stigma, the central raphe endings terminating in the central pores slightly expanded in lateral, a round stigma opening with a thickened margin, and the areolae occluded by raised flaps. ×10000.
Fig. 11.　Enlarged internal view of the valve centre showing the central raphe endings bent first towards the side bearing the stigma and then hooked, the stigma opening within a laterally elongated hollow. ×20000.

Locality
Figs 4-11.　Kalimabenge River, the east of Congo.

Plate 121

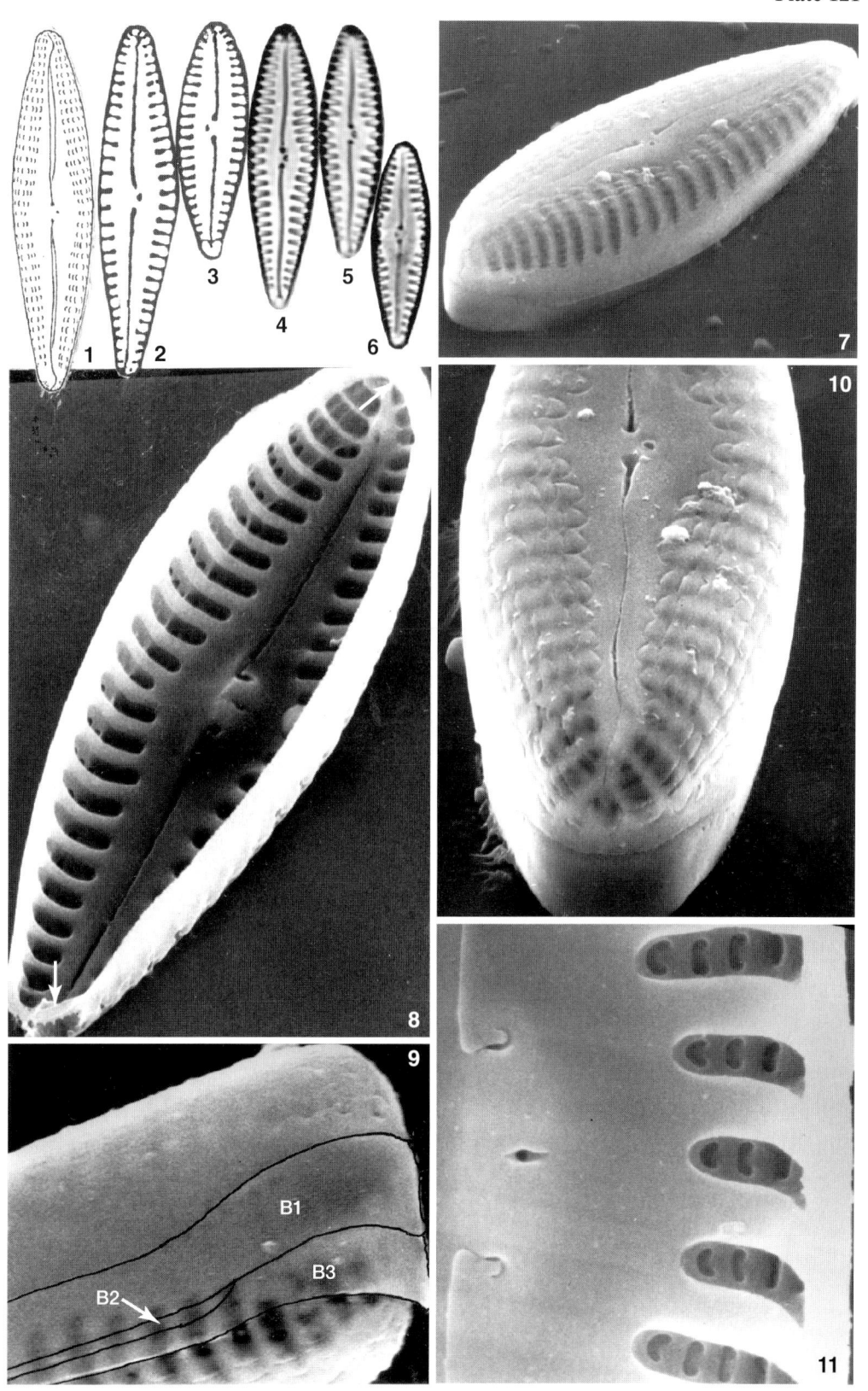

Plate 122 *Gomphonema curvipedatum* H. Kobayasi ex K. Osada sp. nov.

Figs 1, 7. Drawings. ×2000.
Figs 2-6. LM. ×2000.
Figs 8-13. SEM.

Figs 1-7. Valve view.
Fig. 8. External view of a whole valve. ×5000.
Fig. 9. Girdle view of the headpole of the frustule showing the closed ends of the first band (B1, valvocopula) and the third band (B3). ×10000.
Fig. 10. Girdle view of the footpole of the frustule showing the apical pore fields of the epivalve and the hypovalve, the open end of the valvocopula (B1), and the closed end of the second band (B2). ×10000.
Fig. 11. Internal view of the footpole showing straight inner fissure, the terminal raphe ending terminating in the helictoglossa facing the side bearing the stigma, the large inner openings of the alveoli, and a slightly extended pseudoseptum at the valve apex. ×10000.
Fig. 12. Enlarged external view of the valve center showing many roundish depressions in the axial area, slightly expanded central raphe endings, the small, round opening of the stigma, and uniseriate striae with the C-, I-or 3-shaped slits of the areolae. ×10000.
Fig. 13. Enlarged internal view of the valve center showing a unilaterally raised central nodule, the central raphe endings bent first towards the side bearing the stigma and then hooked, and the small, roundish opening of the stigma. ×20000.

Localities
Fig. 2. Ippeki-ko (Lake Ippeki), Shizuoka.
Figs 3-6, 8-13. A small concrete tank in Tokyo Gakugei University, Tokyo.

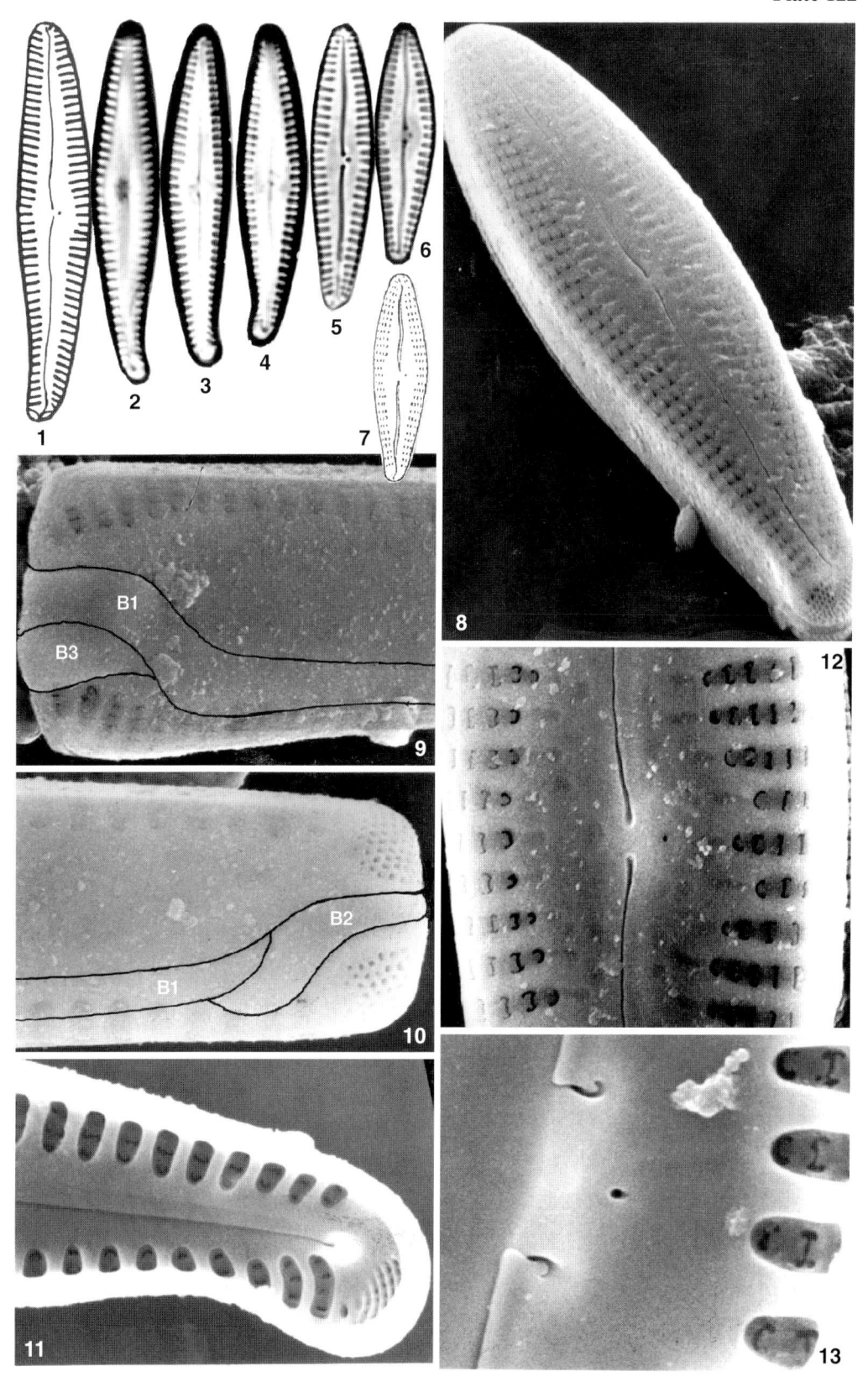

Plate 122

Plate 123 *Gomphonema gracile* Ehrenberg

Figs 1, 3, 4.   LM.   ×2000.
Fig. 2.   Drawing.   ×2000.
Figs 5-9.   SEM.

Figs 1-4.   Valve view.
Fig. 5.   External view of the headpole showing a narrow external raphe fissure, the terminal fissure strongly curved first to the side bearing the stigma and then back to the opposite side, and the C-shaped slits of the areolae.   ×10000.
Fig. 6.   Enlarged external view of the valve center showing the central raphe endings terminating straight in round central pores, the large, circular opening of the stigma, and the crescent and the C-shaped slits of the areolae.   ×10000.
Fig. 7.   Enlarged internal view of the valve center showing the inner central raphe endings hooked to the side bearing the stigma, the stigma opening within an elongated depression, and the stubs (or struts) lying on the vimines within each alveolus.   ×10000.
Fig. 8.   Enlarged external oblique view of the broken valve showing the plicate slit of the raphe branch and the flaps of the areolae.   ×20000.
Fig. 9.   External oblique view of the footpole showing the apical pore field composed of round pores, the terminal fissure curved to the stigma side and then terminating close to the valve margin of the opposite side.   ×10000.

Localities
Figs 1, 3, 5-9.   Small concrete tanks in Tokyo Gakugei University, Tokyo.
Fig. 4.   A paddy field in Ojiya, Niigata.

Plate 123

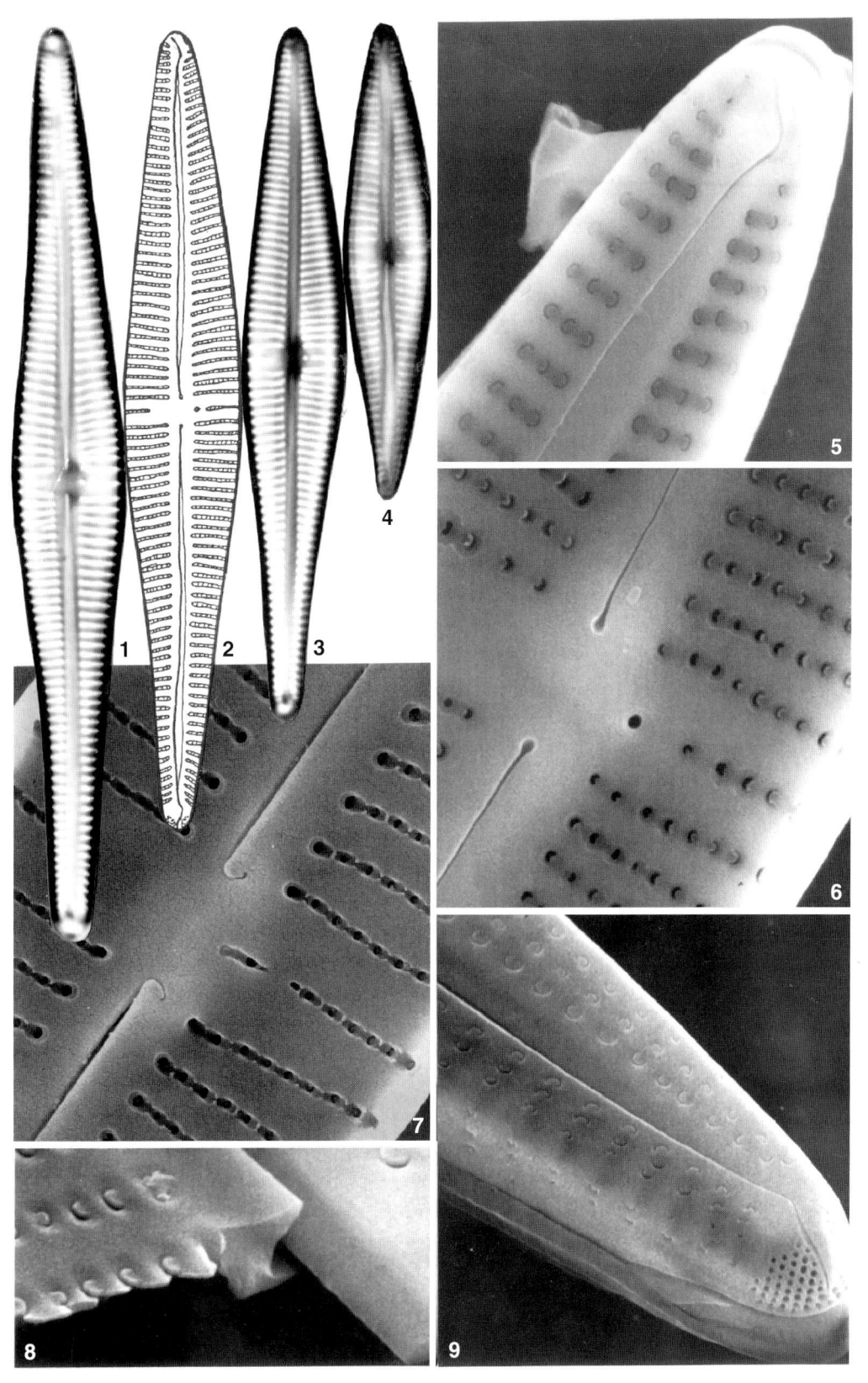

393

Plate 124 *Gomphonema inaequilongum* (H. Kobayasi) H. Kobayasi

Figs 1, 6.   Drawings. ×2000.
Figs 2-5.   LM. ×2000.
Figs 7-11.   SEM.

Figs 1-6. Valve view.
Fig. 7.   Oblique view of the headpole of the frustule showing the terminal fissure curved first to the side bearing the stigma and then backed to the opposite side, and the uniseriate striae arranged radially around the terminal nodule. ×10000.
Fig. 8.   Girdle view of the headpole of the frustule showing the epicingulum composed of four open bands (B1, B2, B3, B4), and a row of areolae in the first band. ×10000.
Fig. 9.   Enlarged external view of the valve centre showing narrow raphe fissures, slightly expanded central raphe endings, the small, round opening of the stigma, and the "C"-, "S"-, or "3"-shaped slits of the areolae. ×8000.
Fig. 10.   Enlarged internal view of the valve centre showing the elliptic foramina of the alveoli, the slightly elongated opening of the stigma, the hooked central endings with spine-like structures. ×20000.
Fig. 11.   External view of the footpole showing an apical pore field, the terminal fissure curved slightly to the side bearing the stigma, and a rough surfaced axial area. ×10000.

Locality
Figs 2-5, 7-11. Hagure-gawa (Hagure River), Saitama.

Plate 124

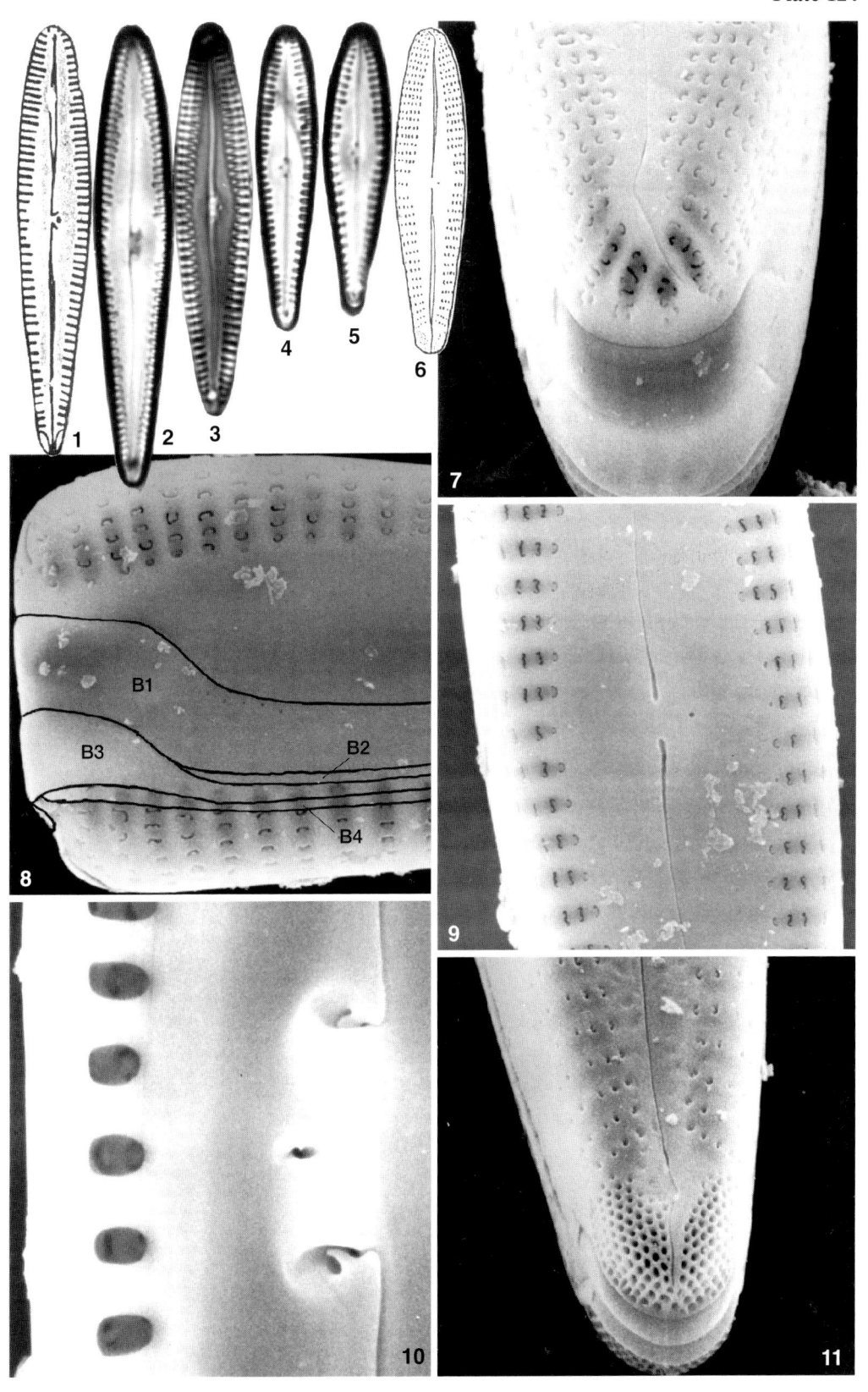

Plate 125 *Gomphonema kinokawaensis* H. Kobayasi ex K. Osada sp. nov.

Figs 1, 2, 4.   LM. ×2000.
Fig. 3.   Drawing. ×2000.
Figs 5-9.   SEM.

Figs 1-4.   Valve view.
Fig. 5.   External oblique view of a whole valve showing elongated depressions in the axial area, undulated raphe fissures, the terminal fissures curved to the side bearing the stigma, and uniseriate striae. ×5000.
Fig. 6.   Girdle view of the headpole of the frustule showing the closed ends of the first (B1, valvocopula) and the third bands (B3), and the open ends of the second (B2) and the fourth bands (B4). ×10000.
Fig. 7.   Girdle view of the footpole of the frustule showing the apical pore fields of the epi- and the hypo valves, the open ends of the valvocopula (B1) the third bands (B3), and the closed ends of the second (B2) and the fourth bands (B4). ×10000.
Fig. 8.   Enlarged external view of valve center showing the central raphe endings terminating in slightly expanded central pores, the small opening of the stigma, the C-shaped slits of the areolae adjacent to the axial area, and the straight slits of the other areolae. ×20000.
Fig. 9.   Enlarged internal view of valve center showing the central raphe fissures hooked towards the side bearing the stigma, a stigma opening within an elongated short depression, and the foramina of the alveoli. ×20000.

Localities
Figs 1, 2, 5-9.   Kino-kawa (Kino River), Wakayama.
Fig. 4.   Ippeki-ko (Lake Ippeki), Shizuoka.

Plate 125

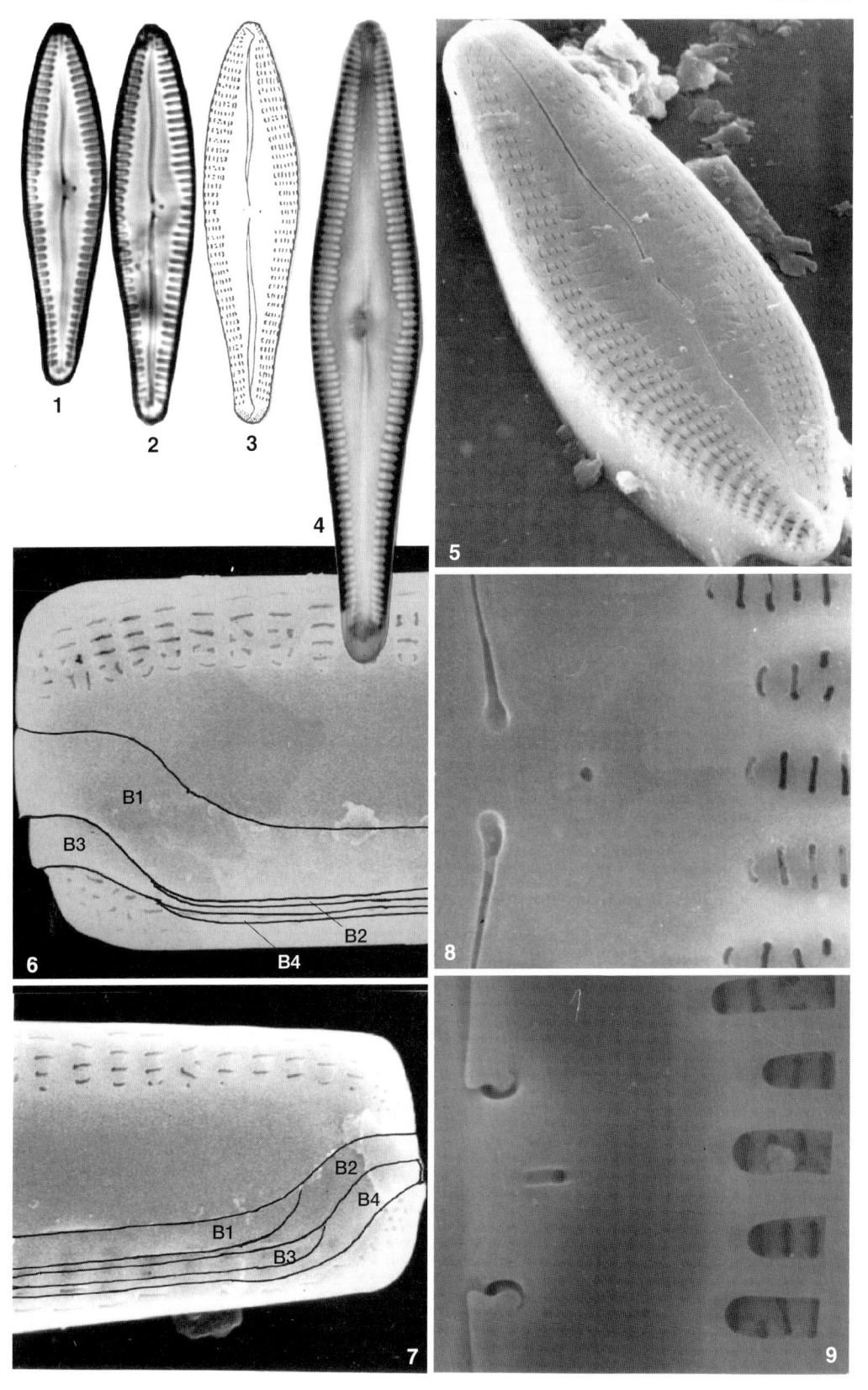

Plate 126 *Gomphonema micropus* Kützing

Figs 1-8.   LM. ×2000.
Figs 9, 10, 12.   SEM.
Fig. 11.   TEM.

Figs 1-8.   Valve view.
Fig. 9.   External view of the headpole to the valve centre showing the terminal fissure deflected towards the side lacking the stigma, a slightly undulated external raphe fissure, the uniseriate striae, and the short striae arranged radially at the valve apex. ×10000.
Fig. 10.   External view of the footpole to the valve centre showing the apical pore field composed of small, roundish pores and the terminal fissure slightly deflected towards the side lacking the stigma. ×10000.
Fig. 11.   Enlarged areolae showing the tiny flaps protruding from the areolar wall within each areola. ×100000.
Fig. 12.   Enlarged external view of the valve centre showing the elliptical opening of the stigma near the apex of the median stria, the roundish openings of the areolae, and round central raphe pores. ×10000.

Localities

Figs 1-3.   Neotype slide, BM 18607. Coll. Kützing, Falaise 370.
Fig. 4.   Ohguri-gawa (Ohguri River), Tokyo.
Fig. 5.   Taba-gawa (Taba River), Yamanashi.
Fig. 6.   Tama-gawa (Tama River), Yanokuchi, Tokyo.
Fig. 7.   Tama-gawa (Tama River), Rokugo-bashi (Rokugo Bridge), Tokyo.
Fig. 8.   Tama-gawa (Tama River), Maruko-bashi (Maruko Bridge), Tokyo.
Figs 9-12.   Headwater of Minamiasa-kawa (Minamiasa River), Tokyo.

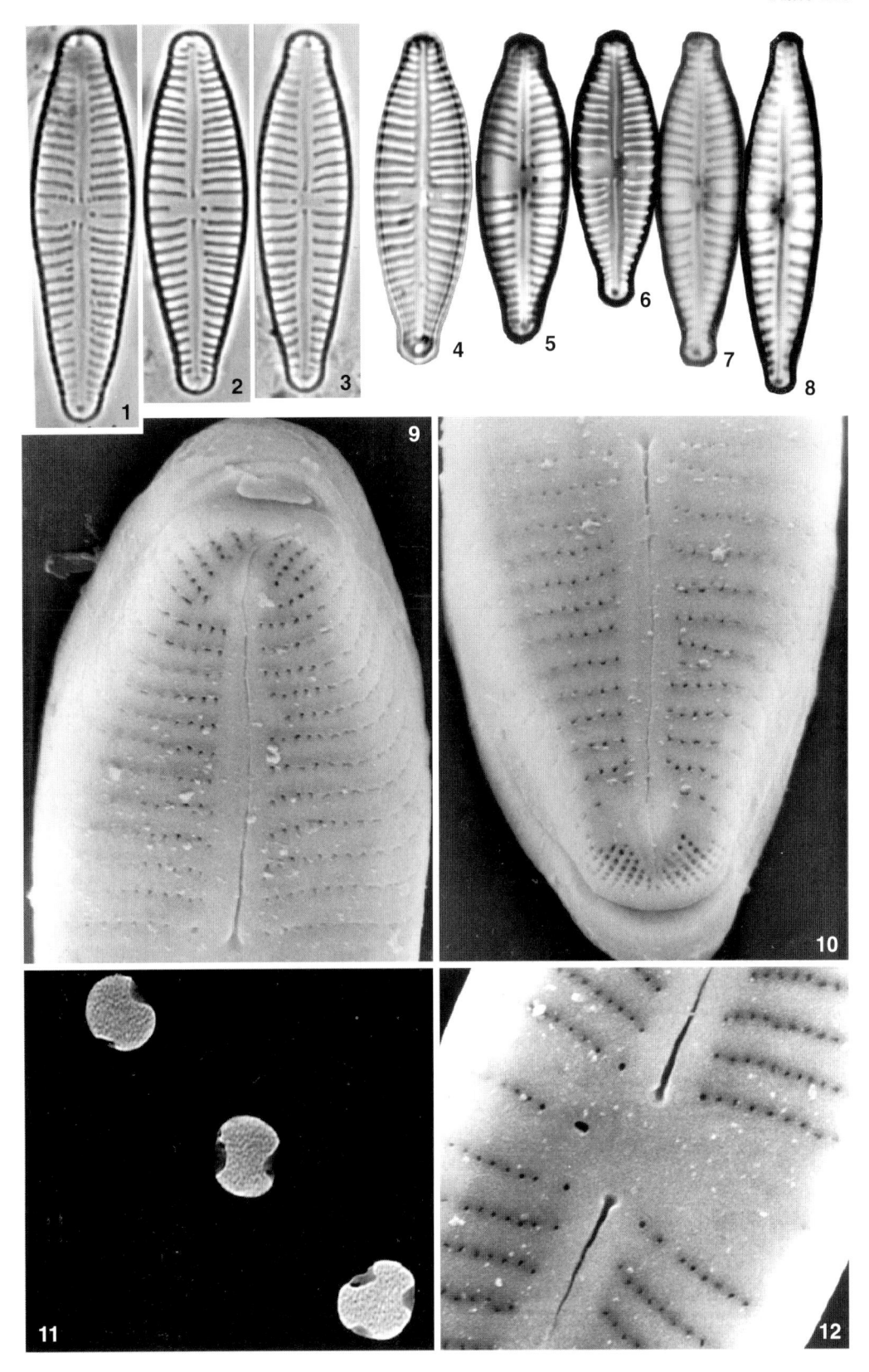

Plate 126

Plate 127 *Gomphonema nipponicum* Skvortsov

Figs 1-3. LM. ×2000.
Fig. 4.   Drawing. ×2000.
Figs 5-7. SEM.

Figs 1-4. Valve view.
Fig. 5.   External view of the headpole showing a narrow external raphe fissure, the terminal fissure curved to the side bearing the stigma, the crescent depressions of the areolae in the uniseriate striae, and radially arranged striae at the valve end. ×8000.
Fig. 6.   External view of the footpole showing the apical pore field composed of roundish pores, the terminal fissure curved to the stigma side, and the crescent and the C-shaped depressions of the areolae. ×8000.
Fig. 7.   Enlarged external view of the valve centre showing straightly terminated central raphe endings, the small, roundish opening of the stigma at the apex of the longer median stria, and the narrow slits within the crescent depressions of the areolae. ×8000.

Localities
Figs 1, 3. Tatara-numa (Tatara Pond), Gunma.
Fig. 2.   Kondo-numa (Kondo Pond), Gunma.
Figs 5-7. Hozoji-numa (Hozoji Pond), Saitama.

Plate 127

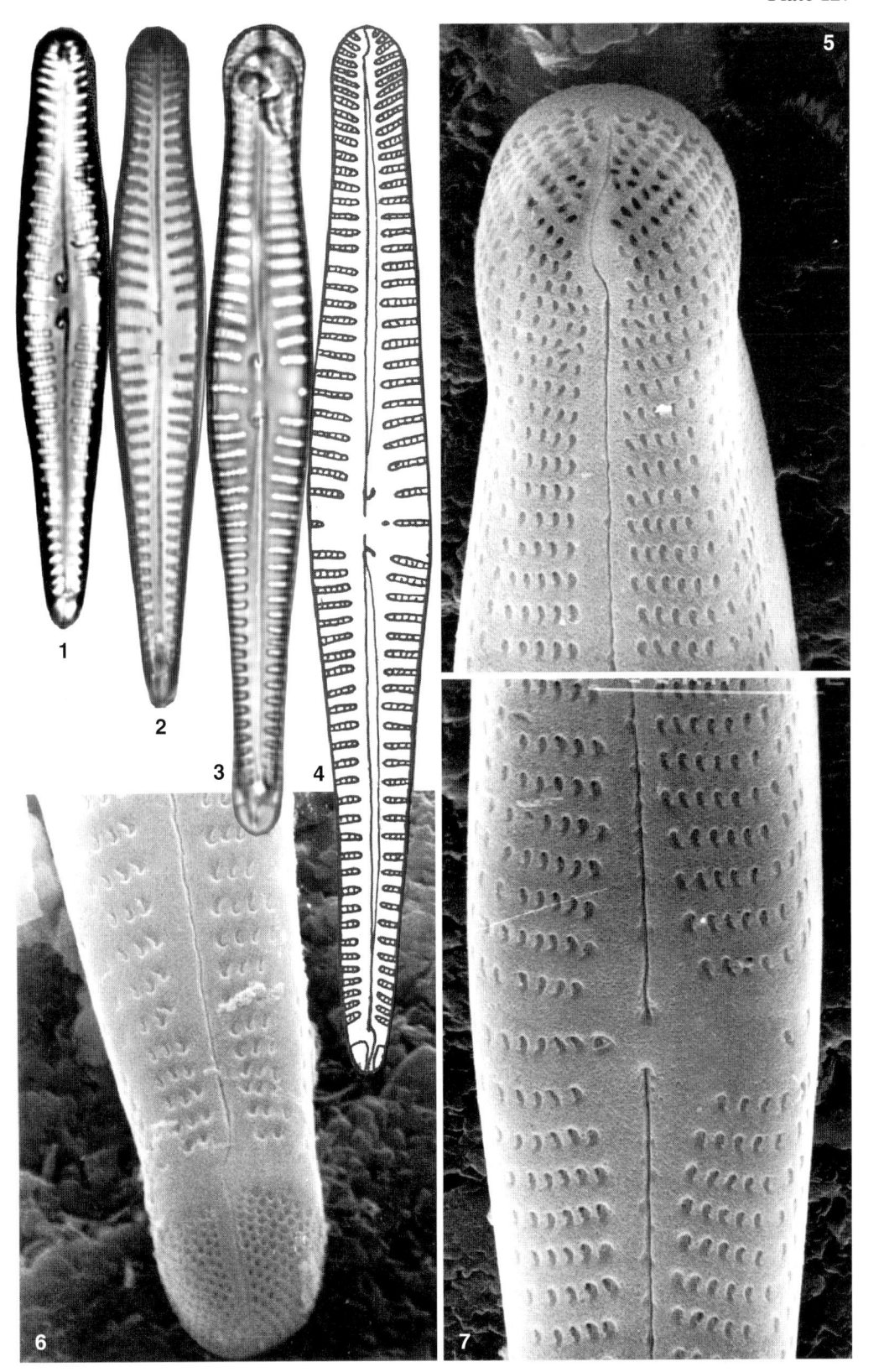

401

Plate 128 *Gomphonema parvulum* var. *lagenula* (Kützing) Frenguelli

Figs 1-4. LM. ×2000.
Fig. 5. Drawing. ×2000.
Figs 6-10. SEM.

Figs 1-5. Valve view.
Fig. 6. Oblique view of a whole frustule, showing the external raphe fissures undulated slightly and the areolar rows extending from the sternum edge near to the valve margin. ×5000.
Fig. 7. Enlarged internal view of the footpole of the same valve as in Fig. 8 showing the terminal raphe ending terminating in a small helictoglossa, a narrow pseudoseptum at the valve apex, and a terminal pore field. ×20000.
Fig. 8. Internal oblique view of the valve showing narrow, nearly straight internal raphe fissures, the slitlike opening of the stigma (arrow), the central raphe endings bent towards the side bearing the stigma, the alveoli continuing from the margin of the raphe sternum to the valve shoulder, and the areola rows of the uniseriate striae in the valve mantle. ×10000.
Fig. 9. Oblique view of the headpole of the frustule showing the terminal fissure slightly curved first to the side bearing the stigma and then extending to the opposite mantle, the C-and S-shaped slits of the areolae, and the closed ends of both the first (B1) and the third bands (B3). ×20000.
Fig. 10. Oblique view of the footpole of the frustule showing the apical pore field composed of roundish pores, the terminal fissure curved to the stigma side, and the closed ends of both the second (B2) and the fourth bands (B4). ×20000.

Localities
Figs 1, 2. Type slide, BM 18638. Coll. Kützing, Tacarigua 389.
Figs 3, 4. An oxbow of Kano-gawa (Kano River), Shizuoka.
Fig. 6. No-gawa (No River), Setagaya, Tokyo.
Figs 8-10. Ohguri-gawa (Ohguri River), Tokyo.

Plate 128

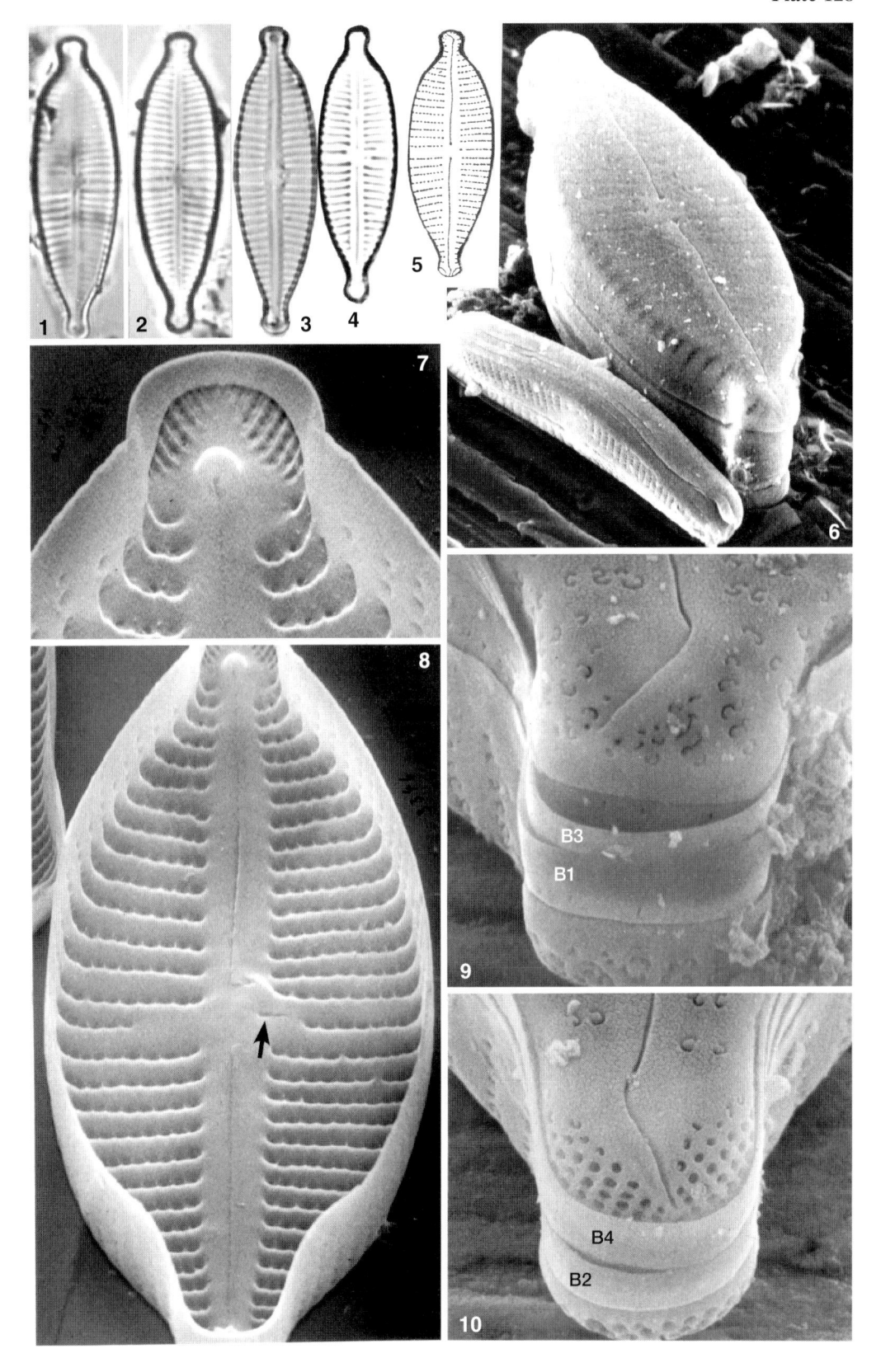

Plate 129 *Gomphonema parvulum* var. *neosaprophilum* H. Kobayasi ex K. Osada var. nov.

Figs 1-4, 6, 7. LM. ×2000.
Fig. 5. Drawing. ×2000.
Figs 8-12. SEM.

Figs 1-7. Valve view.
Fig. 8. External view of the whole valve showing the external raphe fissures undulated slightly, uniseriate striae, and both curved terminal fissures. ×4000.
Fig. 9. Enlargement of the valve centre in Fig. 8 showing the central raphe endings expanded slightly, the variously shaped slits of the areolae, and the round stigma opening at the apex of the median stria. ×10000.
Fig. 10. External view of the headpole showing the terminal fissure slightly curved to the side bearing the stigma and then extending to the opposite mantle, and the C- and 3-shaped slits of the areolae. ×15000.
Fig. 11. Enlarged external view of the valve centre showing the round opening of the stigma and slightly expanded central raphe endings. ×15000.
Fig. 12. External view of the footpole showing a curved terminal fissure and the apical pore field composed of radiate rows of roundish pores. ×15000.

Locality
Figs 1-4, 6-12. Sakai-gawa (Sakai River), Tokyo.

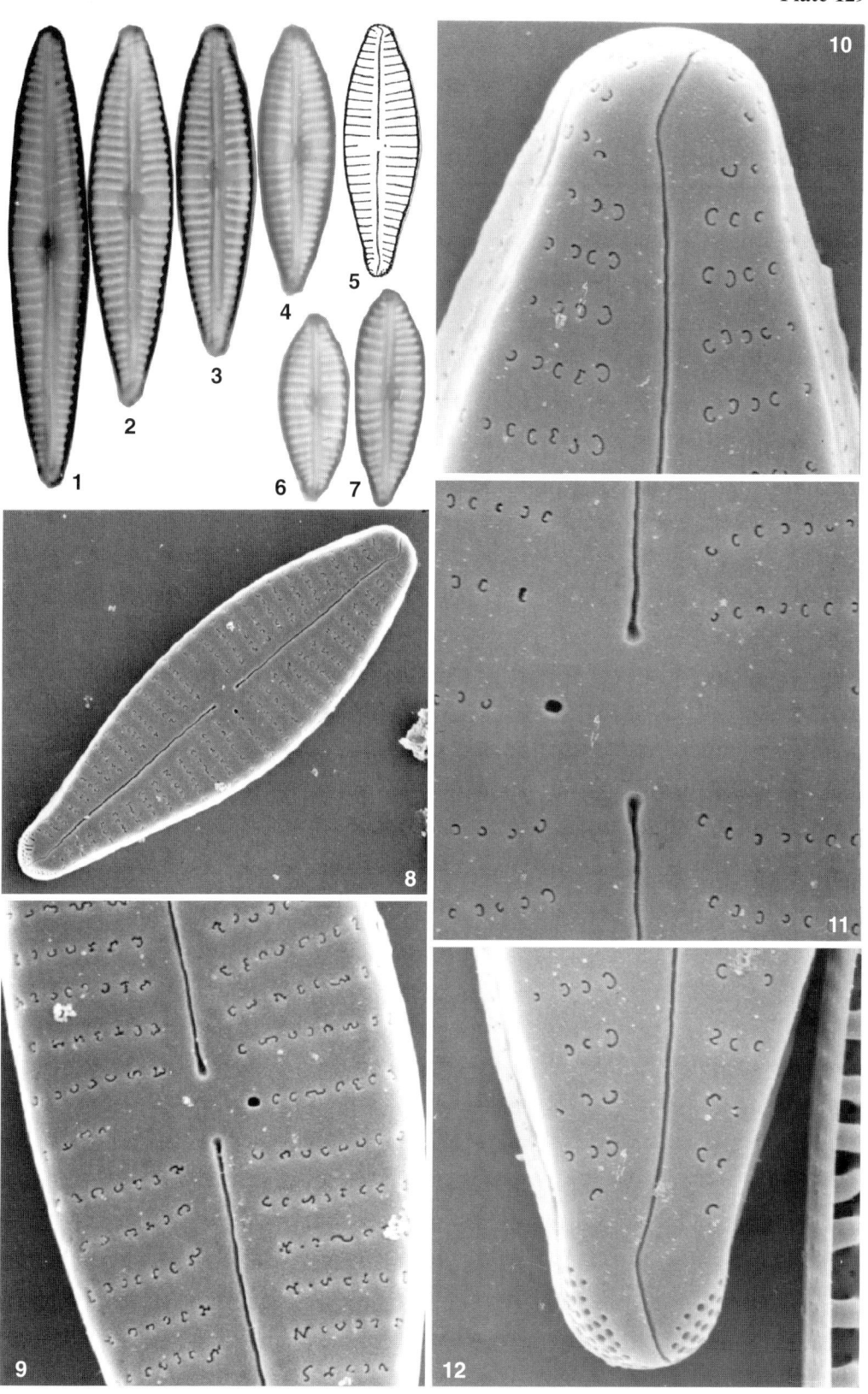

Plate 129

Plate 130 *Gomphonema parvulum* var. *neosaprophilum* H. Kobayasi ex K. Osada var. nov.

Figs 13-18. SEM.

Fig. 13. Internal view of the whole valve showing straight internal raphe fissures, the alveoli with struts, and both of the helictoglossae deflected just a little towards the valve side lacking a stigma. ×5000.

Fig. 14. Enlarged oblique view of the footpole of the frustule showing the slits of the valve areolae, a curved terminal fissure, an apical pore field, and a row of areolae in the band. ×15000.

Fig. 15. Enlarged internal view of the alveoli showing the struts lying over the vimines, and the flap occluding each areola. ×40000.

Fig. 16. Internal oblique view of the headpole showing the terminal raphe ending terminating in a helictoglossa, a very narrow pseudoseptum at the valve apex, and the struts in the alveoli extending to the valve mantle. ×15000.

Fig. 17. Enlarged internal view of the valve center showing the elongated opening of the stigma, hooked central raphe endings, and the struts of the alveoli. ×15000.

Fig. 18. Internal view of the footpole showing the raphe fissure terminating in a small helictoglossa and the alveoli accompanied by areolae and struts. ×15000.

Locality

Figs 13-18. Sakai-gawa (Sakai River), Tokyo.

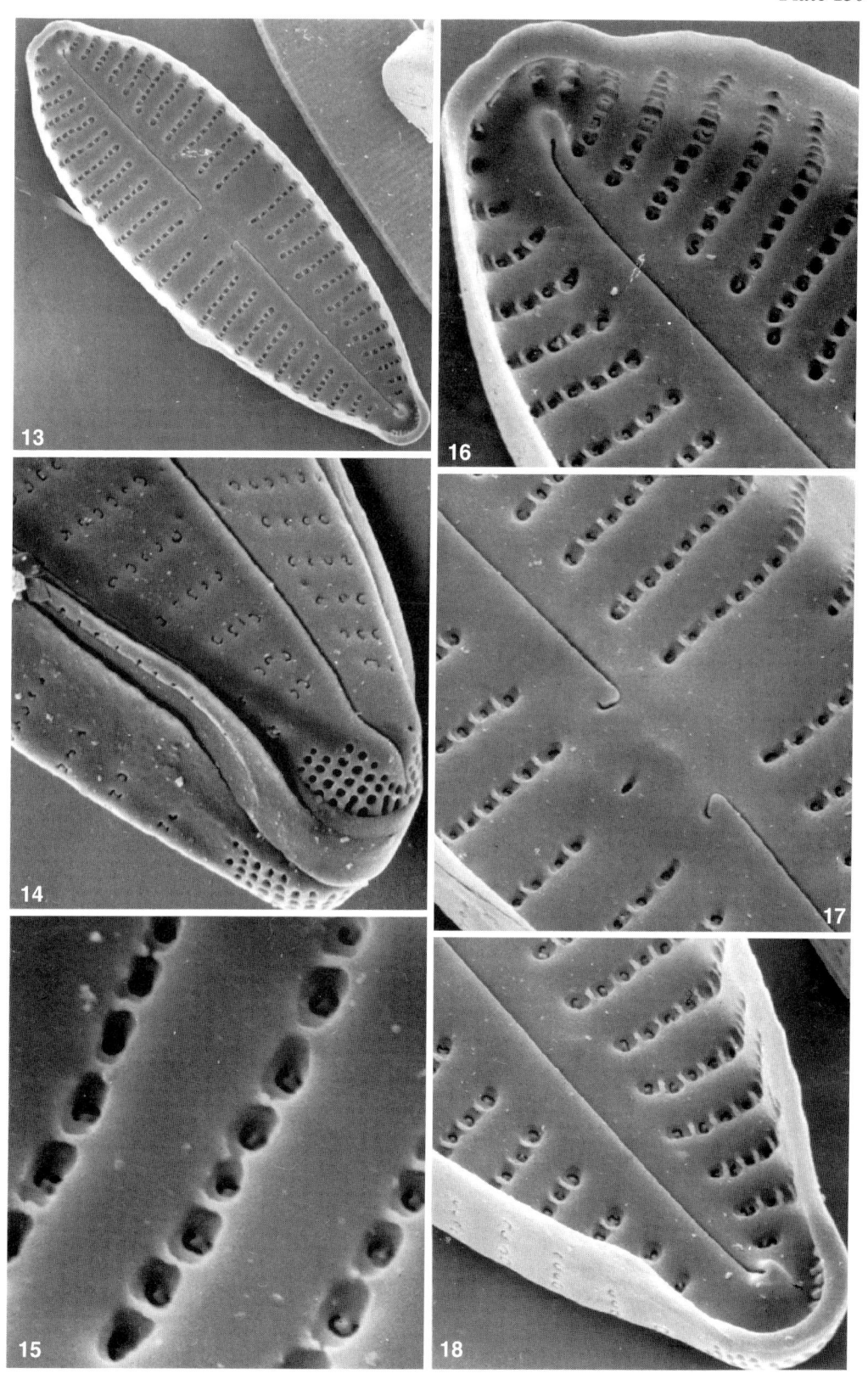

Plate 130

Plate 131 *Gomphonema parvulum* (Kützing) Kützing var. *parvulum*

Figs 1-20. LM. ×2000.
Fig. 21. Drawing. ×2000.
Figs 22, 23. SEM.

Figs 1-21. Valve view.
Fig. 22. Oblique view of a whole frustule. ×5000.
Fig. 23. Enlarged external view of the valve centre showing narrow external raphe fissures, the central raphe endings terminating straightly, the round opening of the stigma lying at a distance from the apex of the median stria, and uniseriate striae. ×10000.

Localities
Figs 1-5. Type slide, BM 18696. Coll. Kützing, Falaise 1260.
Fig. 6. Kondo-numa (Kondo Pond), Gunma.
Figs 7-9. Kano-gawa (Kano River), Shizuoka.
Fig. 10. Ara-kawa (Ara River), Saitama.
Figs 11, 12. Ippeki-ko (Lake Ippeki), Shizuoka.
Fig. 13. Chigonosawa, a stream in Kisofukushima, Nagano.
Figs 14, 15. Oh-numa (Lake Oh), Hokkaido.
Figs 16-18. Culture.
Fig. 19. Kasumiga-ura (Lake Kasumigaura), Ibaraki.
Fig. 20. Jizoin-numa (Jizoin Pond), Saitama.

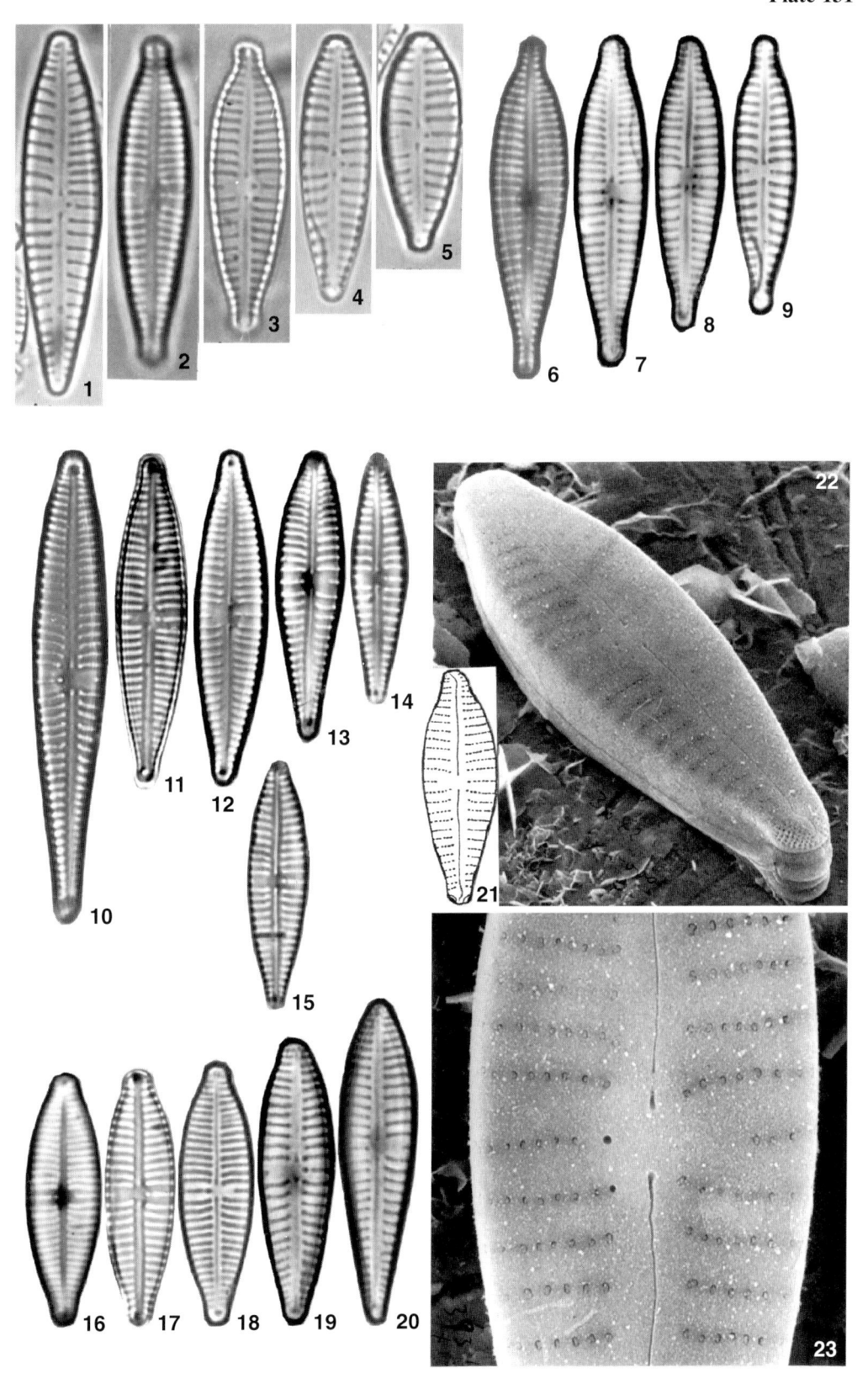

Plate 131

Plate 132 *Gomphonema parvulum* (Kützing) Kützing var. *parvulum*

Figs 24-29.　SEM.

Fig. 24.　Oblique view of the headpole of the frustule, showing the terminal fissure curved to the side bearing the stigma and then back towards the opposite side, and the bands connected with the epivalve (bottom). ×20000.

Fig. 25.　Enlarged external view of the valve centre showing the round opening of the stigma, the C- and 3-shaped slits of the areolae, and slightly expanded central raphe endings. ×30000.

Fig. 26.　Oblique view of the footpole of the same frustule as in Fig. 24, showing the apical pore field composed of small, roundish pores, and a curved terminal fissure. ×20000.

Fig. 27.　Internal view of the headpole showing a narrow pseudoseptum at the valve apex, a small helictoglossa, and alveoli with many stubs. ×20000.

Fig. 28.　Enlarged internal view of the valve centre showing hooked central raphe endings, the slitlike depression of the stigma, the round openings of the areolae, and the stubs lying over inner surface of the vimina within each alveolus. ×30000.

Fig. 29.　Internal view of the footpole of the same valve as in Fig. 27, showing a helictoglossa and a terminal pore field. ×20000.

Localities

Figs 24-26.　No-gawa (No River), Setagaya, Tokyo.

Figs 27, 29.　Ippeki-ko (Lake Ippeki), Shizuoka.

Fig. 28.　Kabe-gawa (Kabe River), a tributary of Kuma-gawa (Kuma River), Kumamoto.

Plate 132

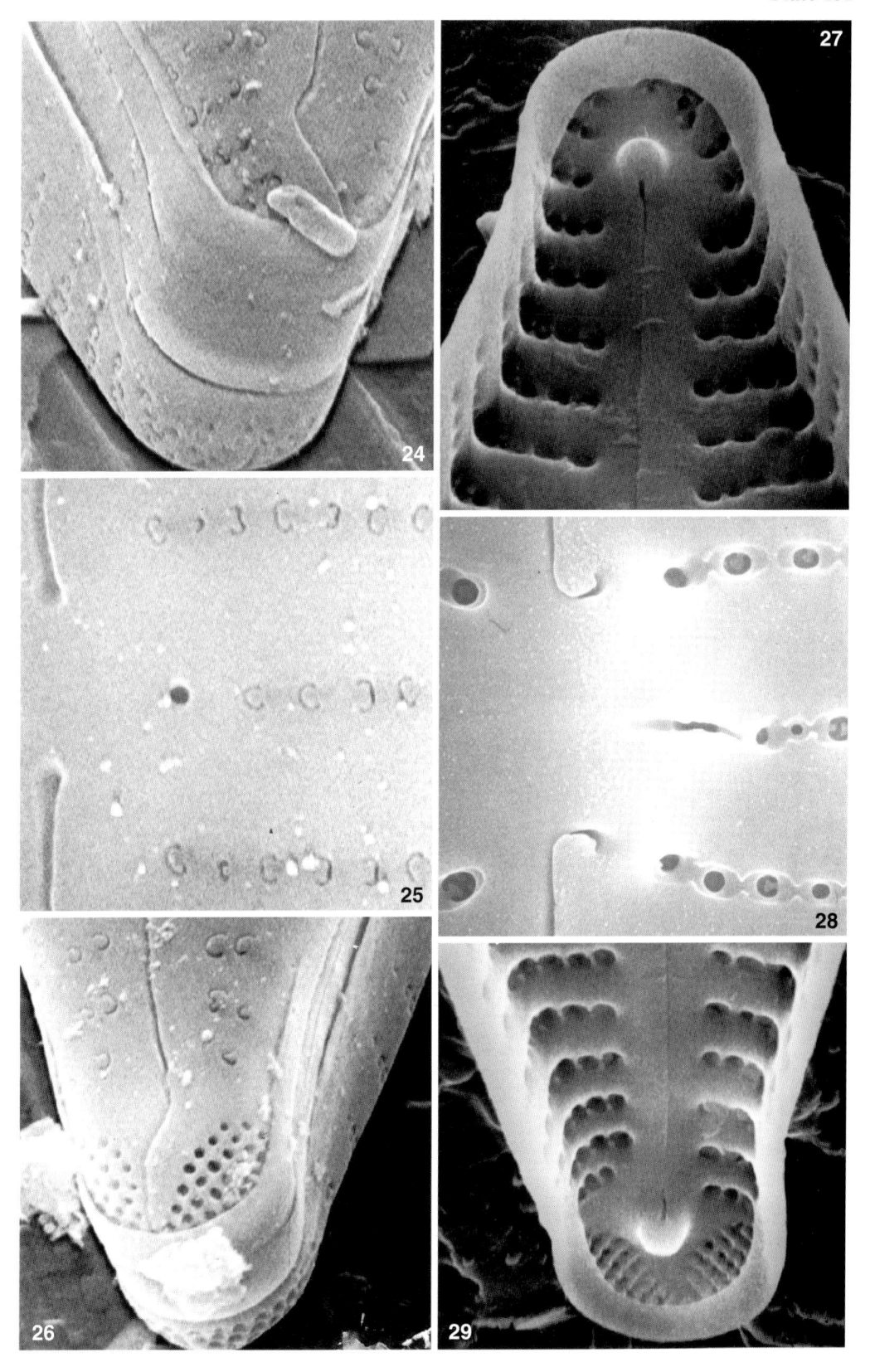

Plate 133 *Gomphonema pseudoaugur* Lange-Bertalot

Fig. 1.   Drawing. ×2000.
Figs 2-5.   LM. ×2000.
Figs 6-9.   SEM.

Figs 1-5.   Valve view.
Fig. 6.   Oblique view of a whole frustule showing narrow, more-or-less undulated external raphe fissures, the terminal fissures curved to the side bearing the stigma, and uniseriate striae. ×3000.
Fig. 7.   Enlarged internal oblique view of the broken valve showing the plicate slit of the raphe branch, a central nodule, and the inner openings of the alveoli. ×5000.
Fig. 8.   Enlarged external view of the valve center showing the central raphe endings terminating straight in round central pores, the laterally elongated opening of the stigma, the narrow, variously shaped slits of the areolae. ×10000.
Fig. 9.   Enlarged internal view of the valve center showing the inner central raphe endings curved roundly to the side bearing the stigma and then hooked, the slitlike opening of the stigma, and the vimines in the alveoli lacking obvious struts and/or stubs. ×10000.

Localities

Figs 2-4.   Tama-gawa (Tama River), Maruko-bashi (Maruko Bridge), Tokyo.
Fig. 5.   Slide Zu 2/9, Main bei Muhlheim.
Figs 6, 8.   Sumida-gawa (Sumida River), Odai-bashi (Odai Bridge), Tokyo.
Fig. 7.   Tama-gawa (Tama River), Yanokuchi, Tokyo.
Fig. 9.   Sakai-gawa (Sakai River), Tokyo.

Plate 133

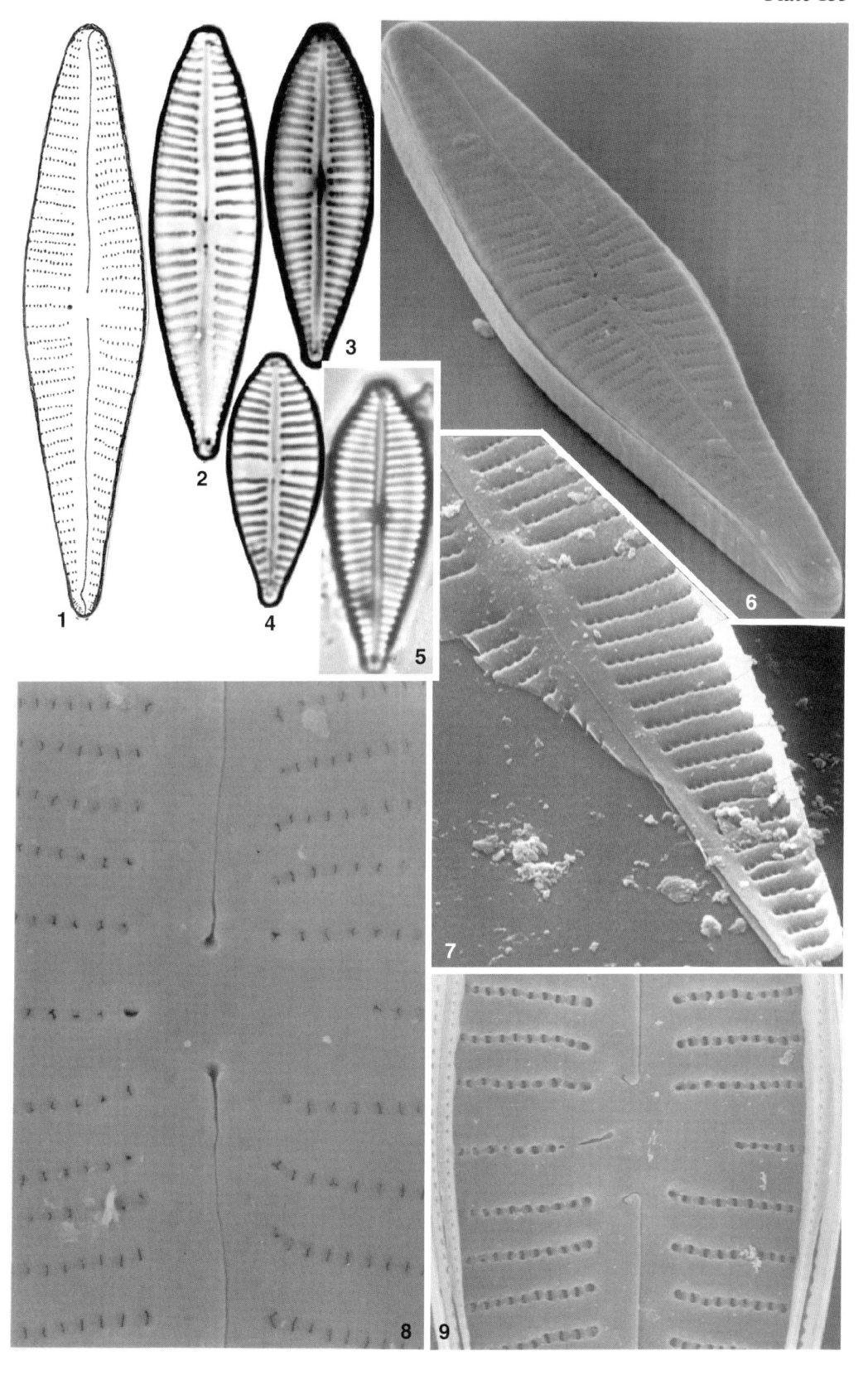

Plate 134 *Gomphonema truncatum* Ehrenberg

Figs 1, 3, 7. LM. ×2000.
Fig. 2. Drawing. ×2000.
Figs 4-6, 8. SEM.

Figs 1-3, 7. Valve view.
Fig. 4. Enlarged external view of the headpole showing a narrow, undulated external raphe fissure, the terminal fissure curved to the side baring the stigma, and the C-shaped slits of the areolae. ×8000.
Fig. 5. Enlarged external view of the valve centre showing slightly expanded central raphe endings, the small, the round opening (arrow) of the stigma at a distance from the apex of the median stria, and uniseriate striae. ×8000.
Fig. 6. Internal view of the same valve as Fig. 7, showing the elongate, large openings of the alveoli, nearly straight inner raphe fissures, and the small opening of the stigma. ×5000.
Fig. 8. External view of the footpole showing the apical pore field composed of small, round pores, and the terminal fissure curved to the side baring the stigma. ×8000.

Localities
Fig. 1. Koridon-no-ike (Koridon Pond), Ojiya, Niigata.
Figs 3-8. A small concrete tank in Tokyo Gakugei University, Tokyo.

Plate 134

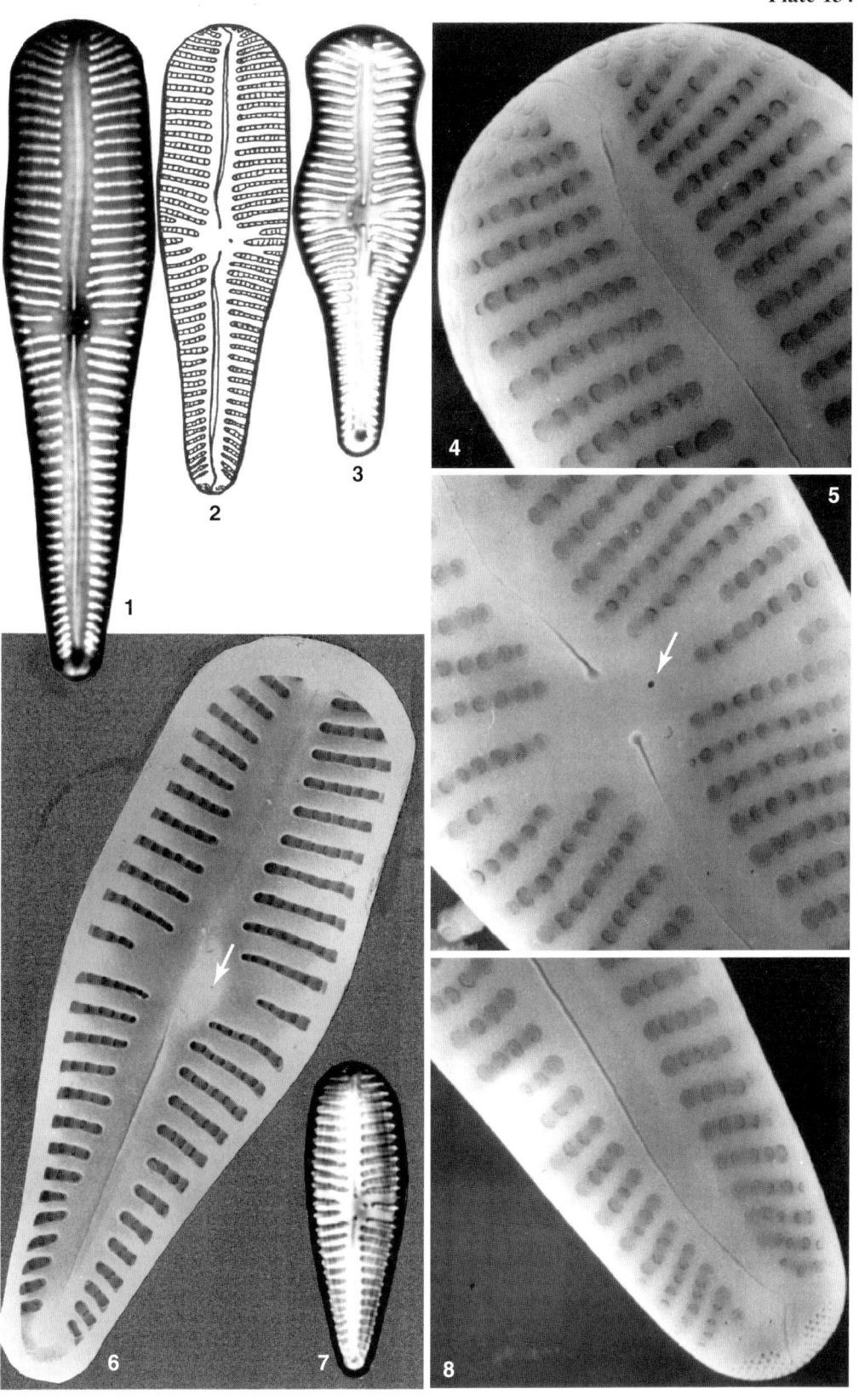

Plate 135 *Gomphonema yamatoensis* H. Kobayasi ex K. Osada sp. nov.

Figs 1-6.　LM.　×2000.
Figs 7-9.　Drawings.　×2000.
Figs 10-14.　SEM.

Figs 1-8.　Valve view.
Fig. 9.　Oblique view of a frustule.
Fig. 10.　External oblique view of the footpole showing the apical pore field composed of round pores, the terminal fissure curved to the side bearing the stigma, and the C-shaped slits of the areolae. ×15000.
Fig. 11.　Enlarged internal view of the headpole showing a pseudoseptum at the valve apex, the terminal raphe ending terminating in a helictoglossa, the elliptical openings of the areolae in the alveoli. ×20000.
Fig. 12.　Enlarged internal view of the valve center showing a raised central nodule, the elliptical opening of the stigma, the round openings of the areolae in the foramina of the alveoli, and the central raphe endings bent towards the side bearing the stigma and then turned back acutely. ×20000.
Fig. 13.　Enlarged external view of the valve center showing the central raphe endings terminating in round central pores, a circular stigma opening with a thickened margin, and the areolar slits curved to the axial area in the valve face and reversed in the valve mantle. ×15000.
Fig. 14.　Enlarged external oblique view of the headpole showing the terminal fissure slightly curved to the side bearing the stigma, and the C-shaped and the S-shaped slits of the areolae. ×15000.

Localities
Figs 1-6, 11-14.　Inabe-gawa (Inabe River), Mie.
Fig. 10.　Hashino-gawa (Hashino River), Kamikurihashi, Iwate.

Plate 135

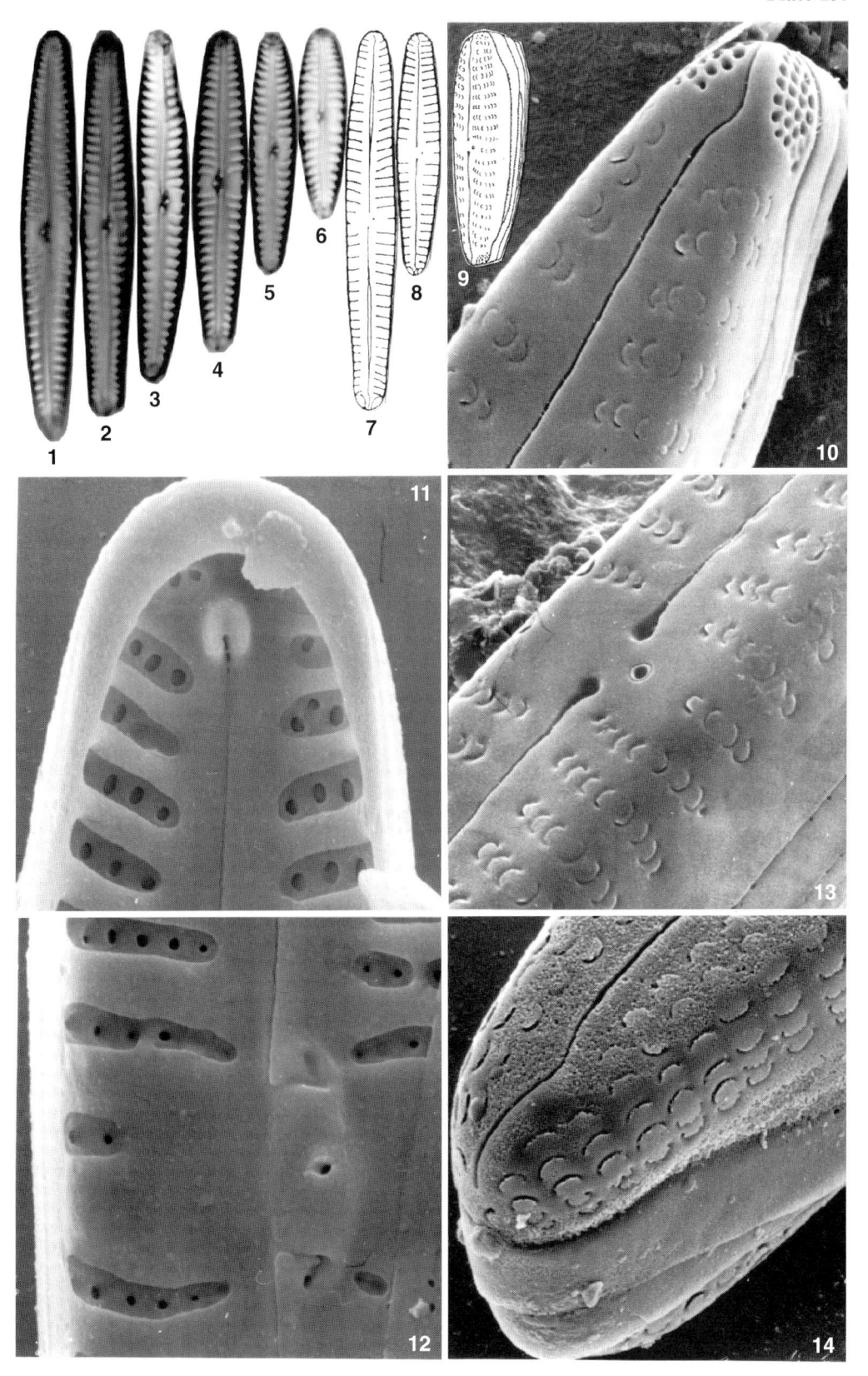

Plate 136 *Achnanthes coarctata* (Brébisson) Grunow

Figs 1-3, 6-9.　LM.　×2000.
Figs 4, 5.　Drawings.　×2000.
Figs 10-13.　SEM.

Figs 1, 5, 7, 8, 10-13.　Raphid valves.
Figs 2, 4, 6, 9.　Araphid valves.
Fig. 3.　Girdle view.
Fig. 10.　Oblique external view of the partly broken valve showing raphe branches with inflated central endings.　×10000.
Fig. 11.　Enlarged section of the valve showing a plicate raphe system and areolae occluded outwardly by volae.　×20000.
Fig. 12.　Details of the volae on the raphid valve.　×30000.
Fig. 13.　Oblique external view of the valve end showing a weakly bent polar fissure.　×10000.

Localities

Figs 1, 2.　Towada-ko (Lake Towada), Towada, Aomori.
Figs 3, 7-13.　Kanna-gawa (Kanna River), Gunma.
Fig. 6.　Ara-kawa (Ara River), Nagatoro, Saitama.

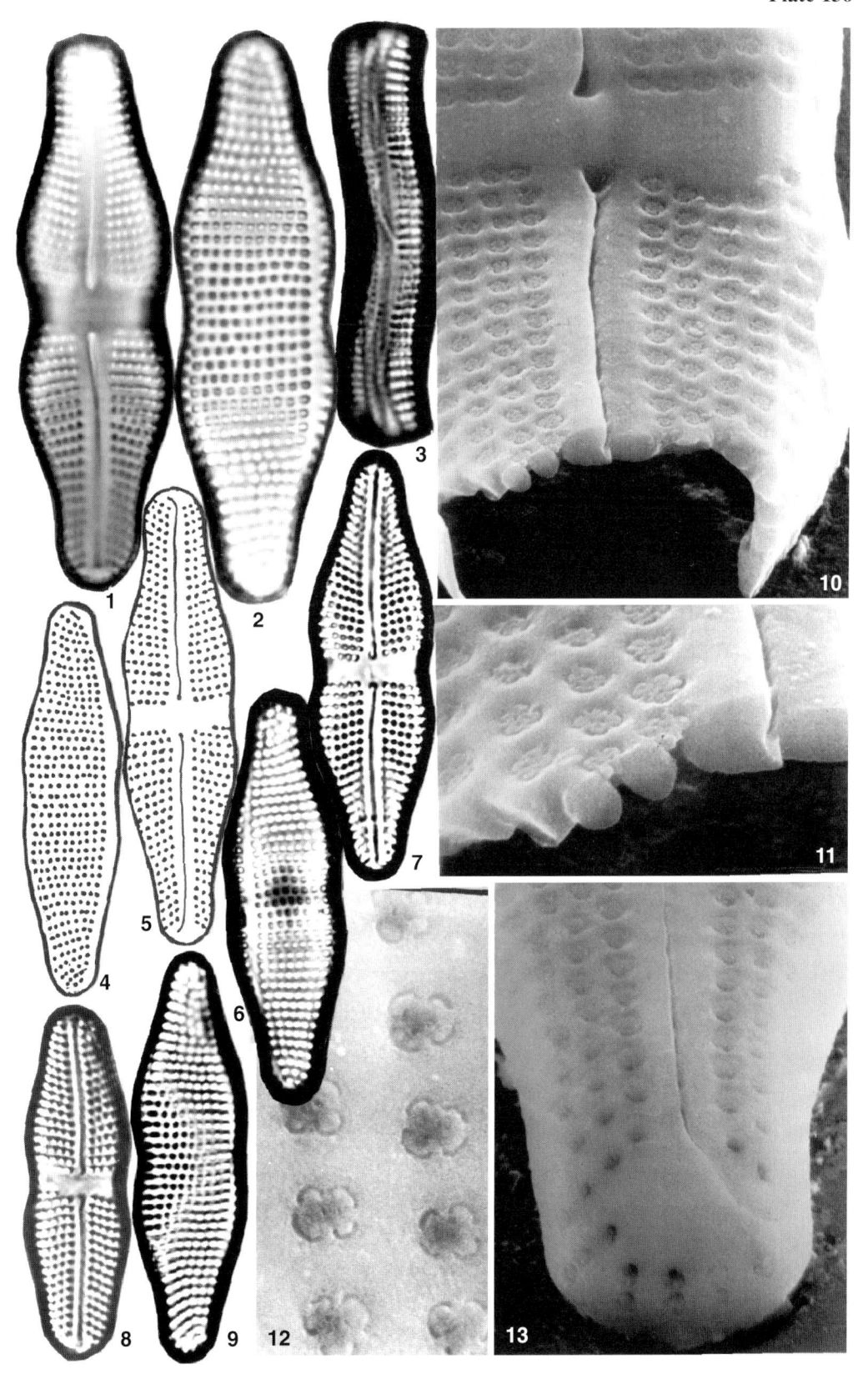

Plate 136

Plate 137 *Achnanthes coarctata* (Brébisson) Grunow

Figs 14-20. SEM.

Fig. 14. Internal view of the raphid valve center showing a fascia and central raphe endings, which strongly hook to the same side. ×10000.

Fig. 15. External view of the araphid valve end showing a transapically concave valve face and a well developed ridge on the valve shoulder. ×8000.

Fig. 16. Details of the frustule end showing alternating arrangement of open or closed ends of the bands in epi- and hypocingula. ×8000.

Fig. 17. Broken araphid valve showing areolar occlusions by volae on the surface of the valve. ×20000.

Fig. 18. Girdle view of the two connected frustules. ×2000.

Fig. 19. Details of volae on the araphid valve. ×50000.

Fig. 20. Oblique view of the araphid theca showing developed interstria and a sternum located on the left margin of the valve face. ×5000.

Locality

Figs 14-20. Kanna-gawa (Kanna River), Gunma.

Plate 137

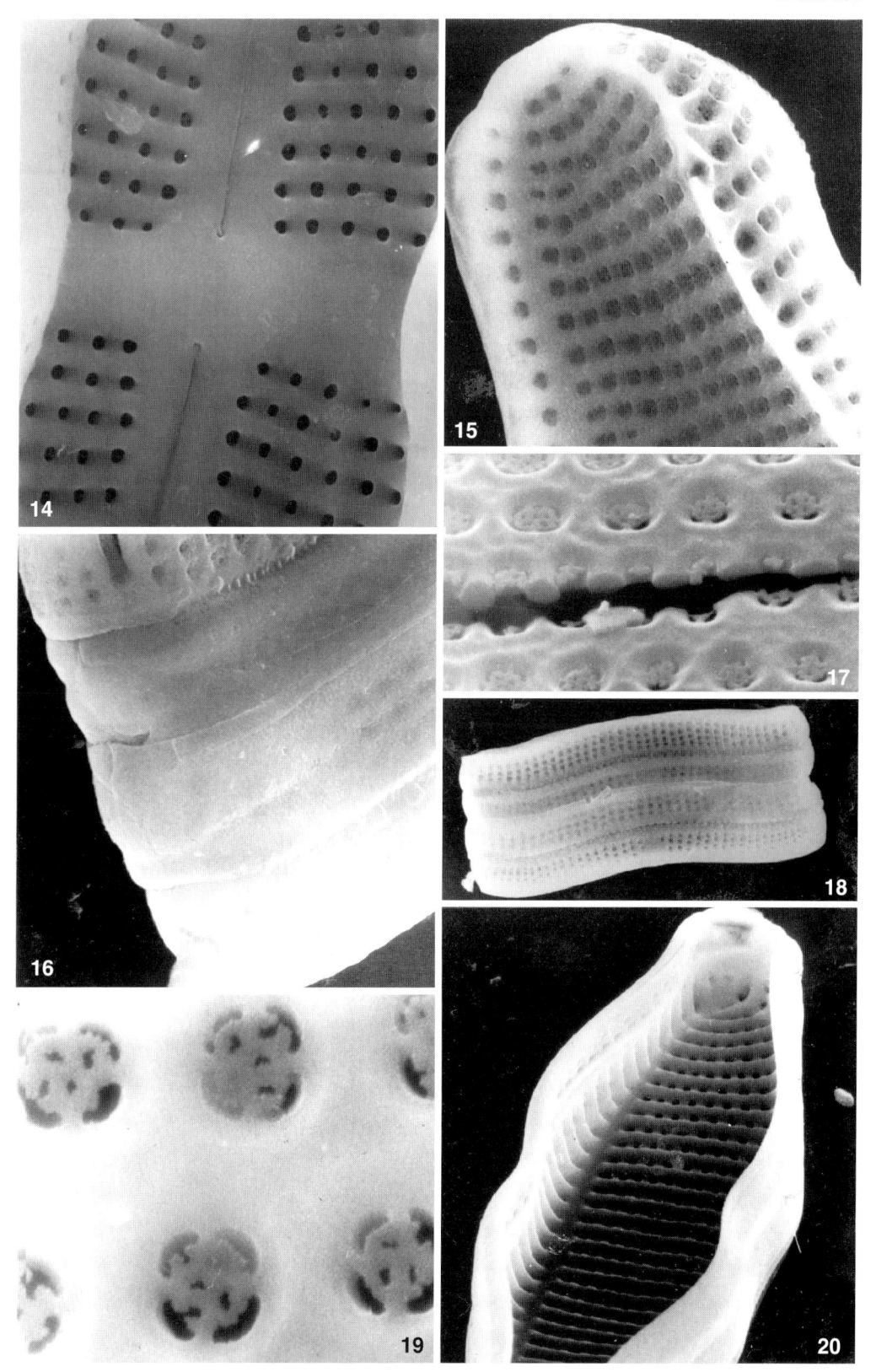

Plate 138 *Achnanthes crenulata* Grunow

Figs 1, 2, 5. LM. ×2000.
Figs 3, 4. Drawings. ×2000.
Figs 6-9. SEM.

Figs 1, 3, 6-9. Raphid valves.
Figs 2, 4. Araphid valves.
Fig. 5. Girdle view.
Fig. 6. Whole view of a frustule. ×2000.
Fig. 7. Internal view of a broken valve showing a plicate raphe, a strongly hooked central ending and sectioned areolar chambers. ×10000.
Fig. 8. Valve fragment showing occlusions of the outer openings of areolae by volae. ×20000.
Fig. 9. Details of the volae in external view. ×20000.

Locality
Figs 1, 2, 5-9. Kano-gawa (Kano River), Shimofunabara, Shizuoka.

Plate 138

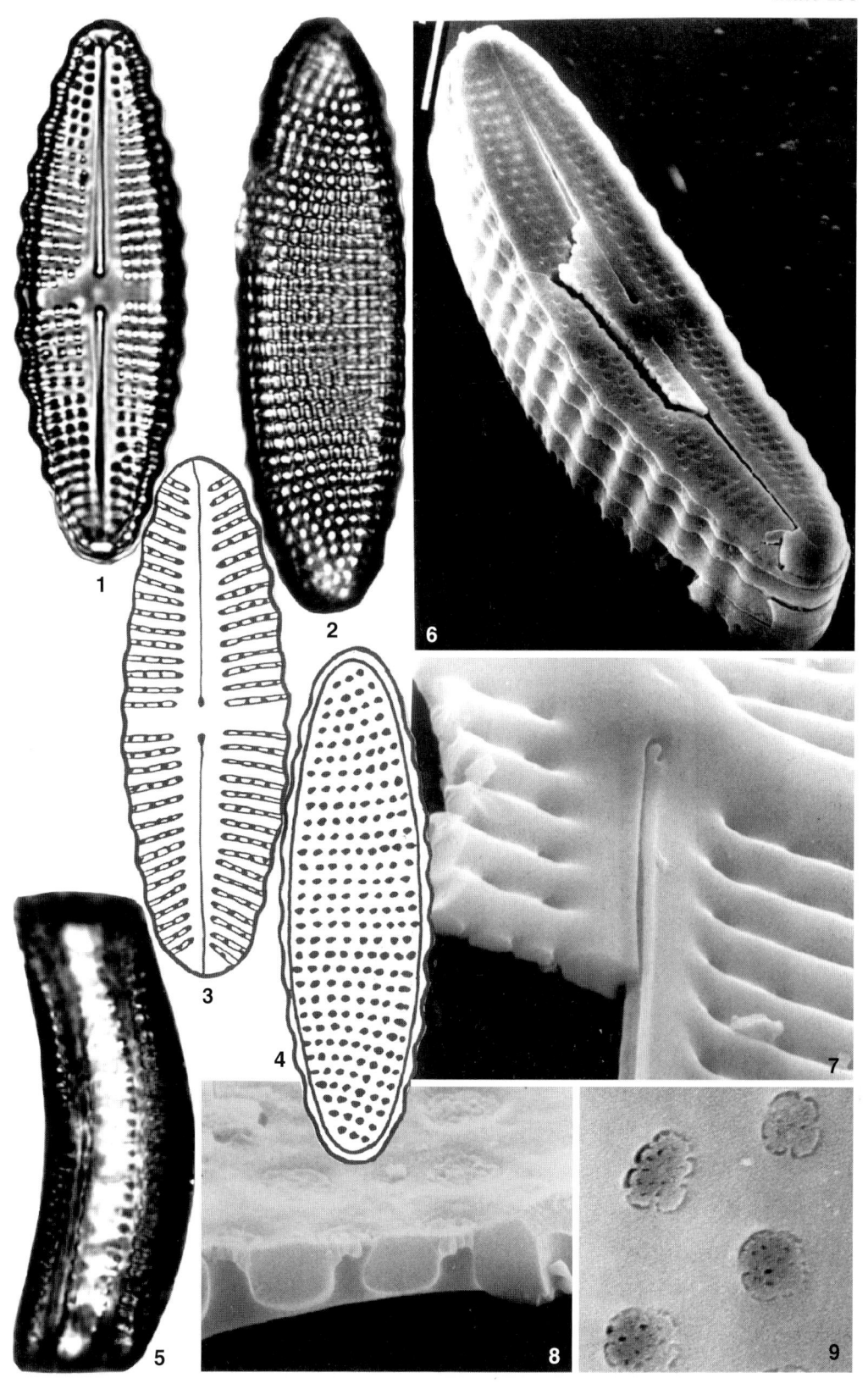

423

Plate 139 *Achnanthes crenulata* Grunow

Figs 10-15. SEM.

Fig. 10. End of the frustule showing a hypo-araphid valve, three epibands and an epivalve mantle. ×5000.

Fig. 11. Partly broken frustule showing the configuration of its elements. ×5000.

Fig. 12. Internal oblique view of an araphid valve showing well developed interstria. ×8000.

Fig. 13. Details of volae in external araphid valve view. ×20000.

Fig. 14. Internal view of an araphid valve showing rows of areolae aligned between the well developed interstria. ×20000.

Fig. 15. Internal view of the raphid valve showing circular openings of areolae. ×20000.

Locality

Figs 10-15. Kano-gawa (Kano River), Shimofunabara, Shizuoka.

Plate 139

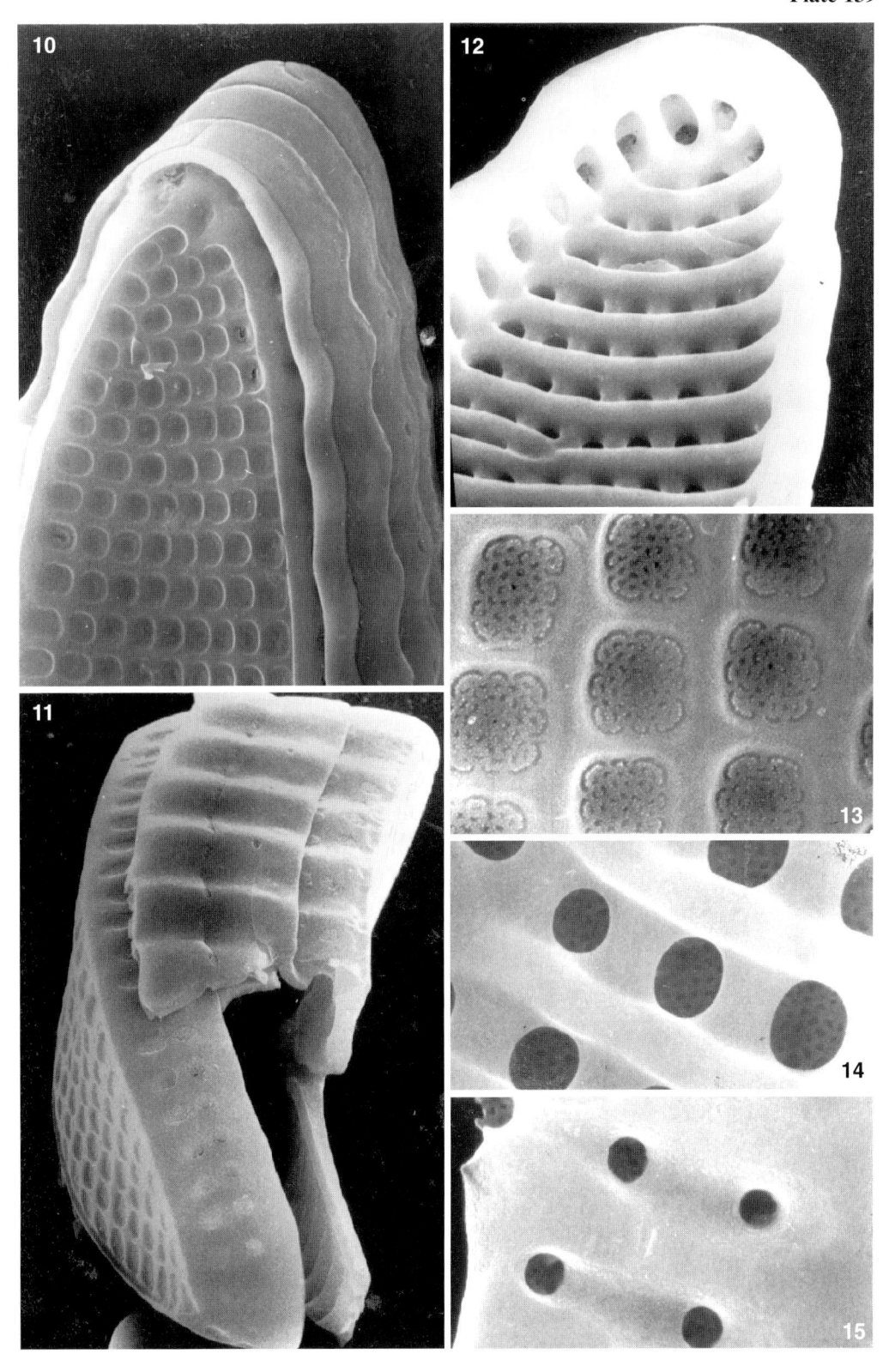

Plate 140 *Achnanthes inflata* (Kützing) Grunow

Figs 1-5. LM. ×2000.
Figs 6, 7. Drawings. ×2000.
Figs 8-10. SEM.

Figs 1, 3, 5, 9, 10. Raphid valves.
Figs 2, 4, 7. Araphid valves.
Fig. 5. Girdle view.
Fig. 8. Oblique view of connected cells (two frustules and one theca). ×1000.
Fig. 9. External view of the enlarged valve center showing inflated central raphe endings. ×8000.
Fig. 10. External view of the enlarged valve apex showing a bent polar fissure. ×10000.

Localities

Figs 1-5, 8. A small stream in Sohrohsenen, Koganei, Tokyo.
Figs 9, 10. Kano-gawa (Kano River), Shimofunabara, Shizuoka.

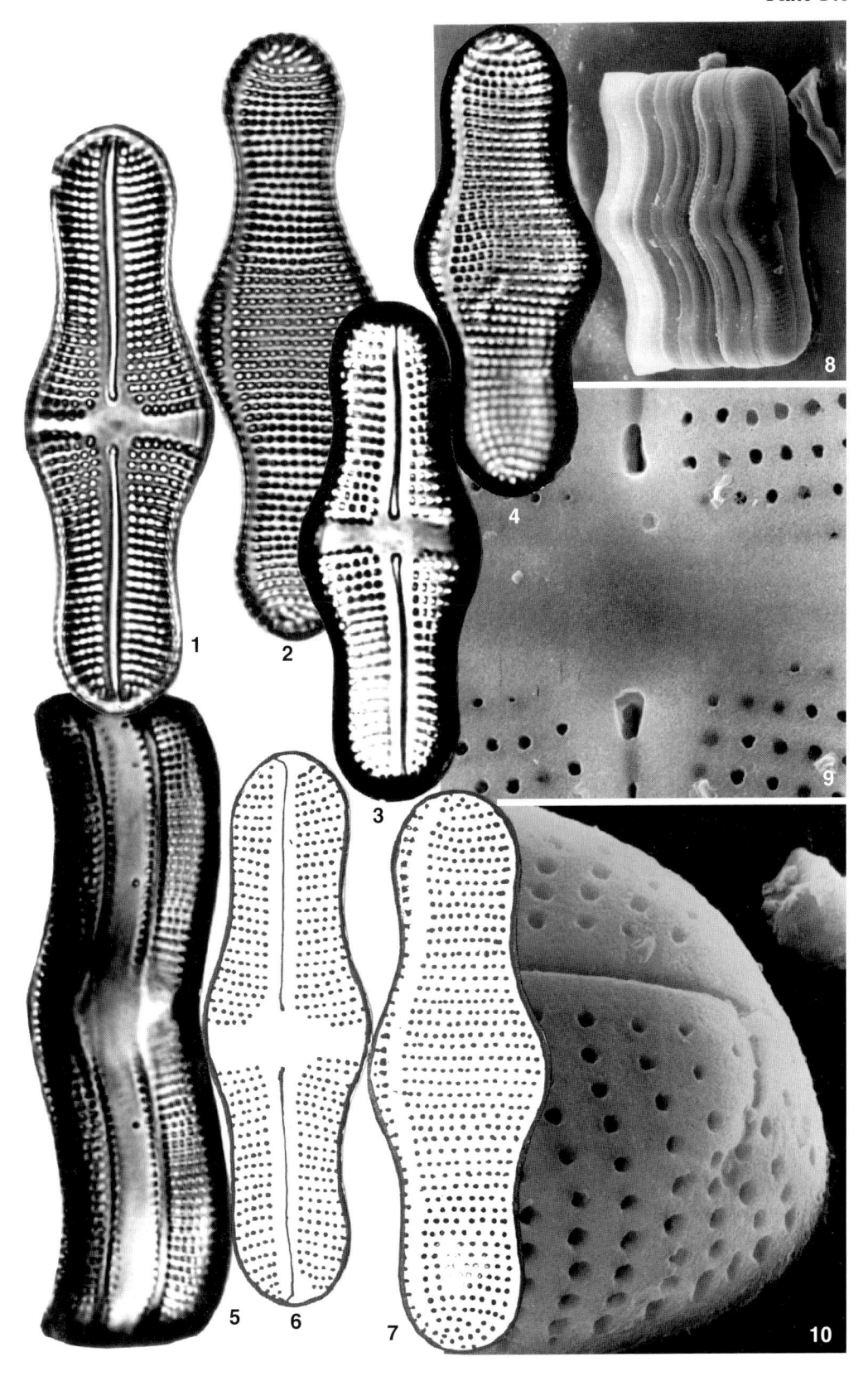

Plate 140

Plate 141 *Achnanthes inflata* (Kützing) Grunow

Figs 11-19. SEM.

Fig. 11. Details of a central raphe ending and volae occluding the outer openings of areolae. ×20000.

Fig. 12. Internal view of raphid valve center showing a developed fascia and the central raphe endings. Note both endings curving to the same side. ×10000.

Fig. 13. Fractured valve showing the plicate raphe system and the volate areolar occlusions. ×20000.

Fig. 14. Internal view of a raphid valve showing the circular inner-openings of areolae. ×30000.

Fig. 15. Internal oblique view of an araphid valve showing a sternum running along the left juncture of the valve face and mantle. ×5000.

Fig. 16. External oblique view of an araphid valve apex showing a transversely concave valve face and a lightly elevated valve shoulder. ×8000.

Fig. 17. Section of araphid valve showing outward occlusions of areolae by volae. ×10000.

Fig. 18. Details of the volae on an araphid valve. ×20000.

Fig. 19. Internal view showing the circular openings of areolae between well developed interstria. ×20000.

Localities

Figs 11, 14, 16.   A stream in Sohrohsenen, Koganei, Tokyo.

Figs 12, 13, 15, 17-19.   Myojin-ike (Myojin Pond), Heda, Shizuoka.

Plate 141

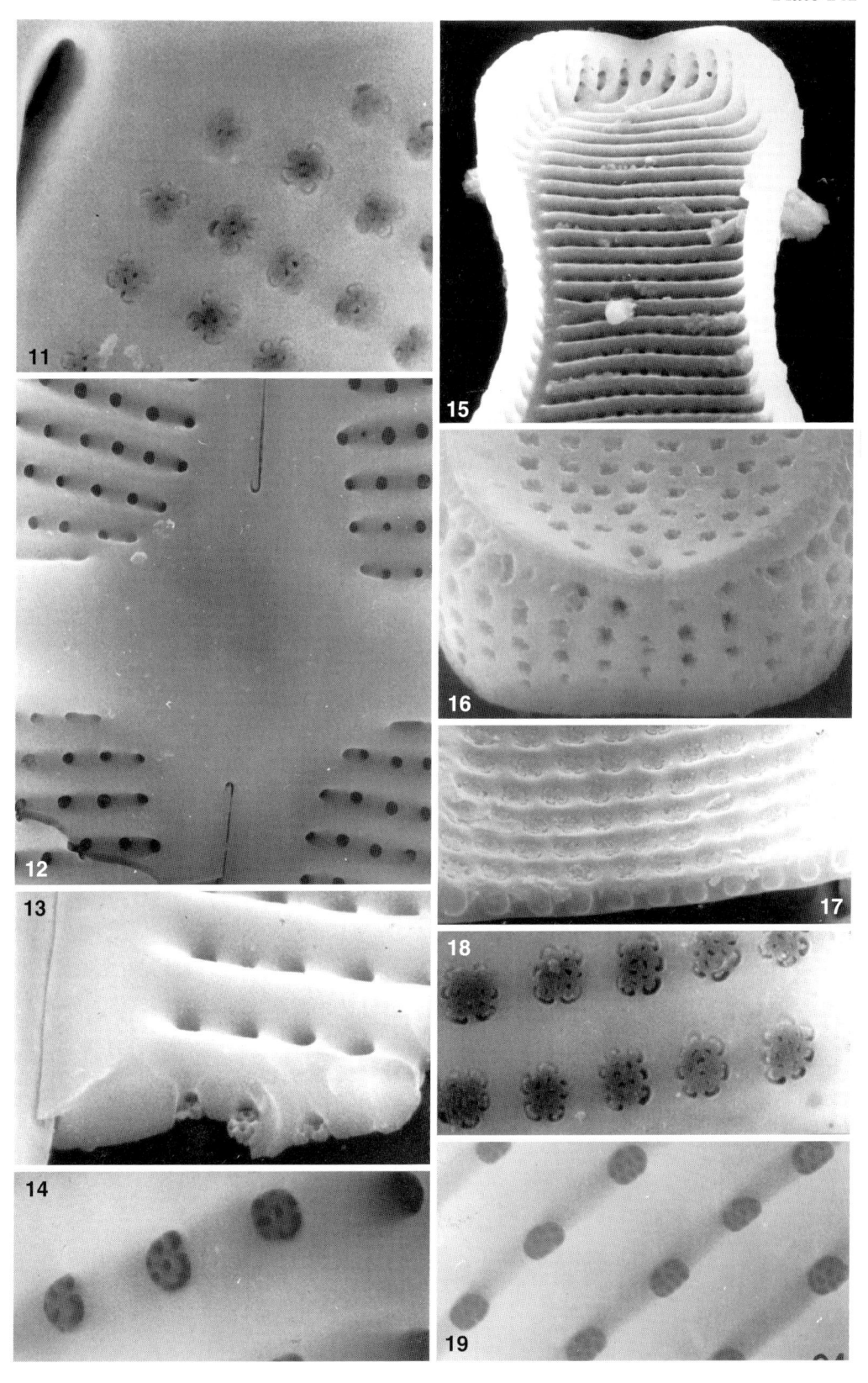

Plate 142 *Achnanthes kuwaitensis* Hendey

Figs 1, 2.   Drawings. ×2000.
Figs 3, 4, 7.   LM. ×2000.
Figs 5, 6, 8-11.   SEM.

Figs 1, 3, 5, 8, 9, 10.   Raphid valves.
Figs 2, 4, 6, 11.   Araphid valves.
Fig. 5.   External view of a raphid valve. ×2000.
Fig. 6.   Internal view of an araphid valve. ×2000.
Fig. 7.   Girdle view of two connected frustules.
Fig. 8.   Details of the external view of inflated central raphe endings. ×8000.
Fig. 9.   Enlarged internal valve showing a fascia and both central endings. ×10000.
Fig. 10.   External view of the raphid valve end. ×8000.
Fig. 11.   Internal view of the araphid valve end showing a large polar vola. ×10000.

Localities
Figs 3-7, 10, 11.   Aono-gawa (Aono River), river mouth, Koine, Izu, Shizuoka.
Figs 8, 9.   Seashore in Kujiranami, Niigata.

Plate 142

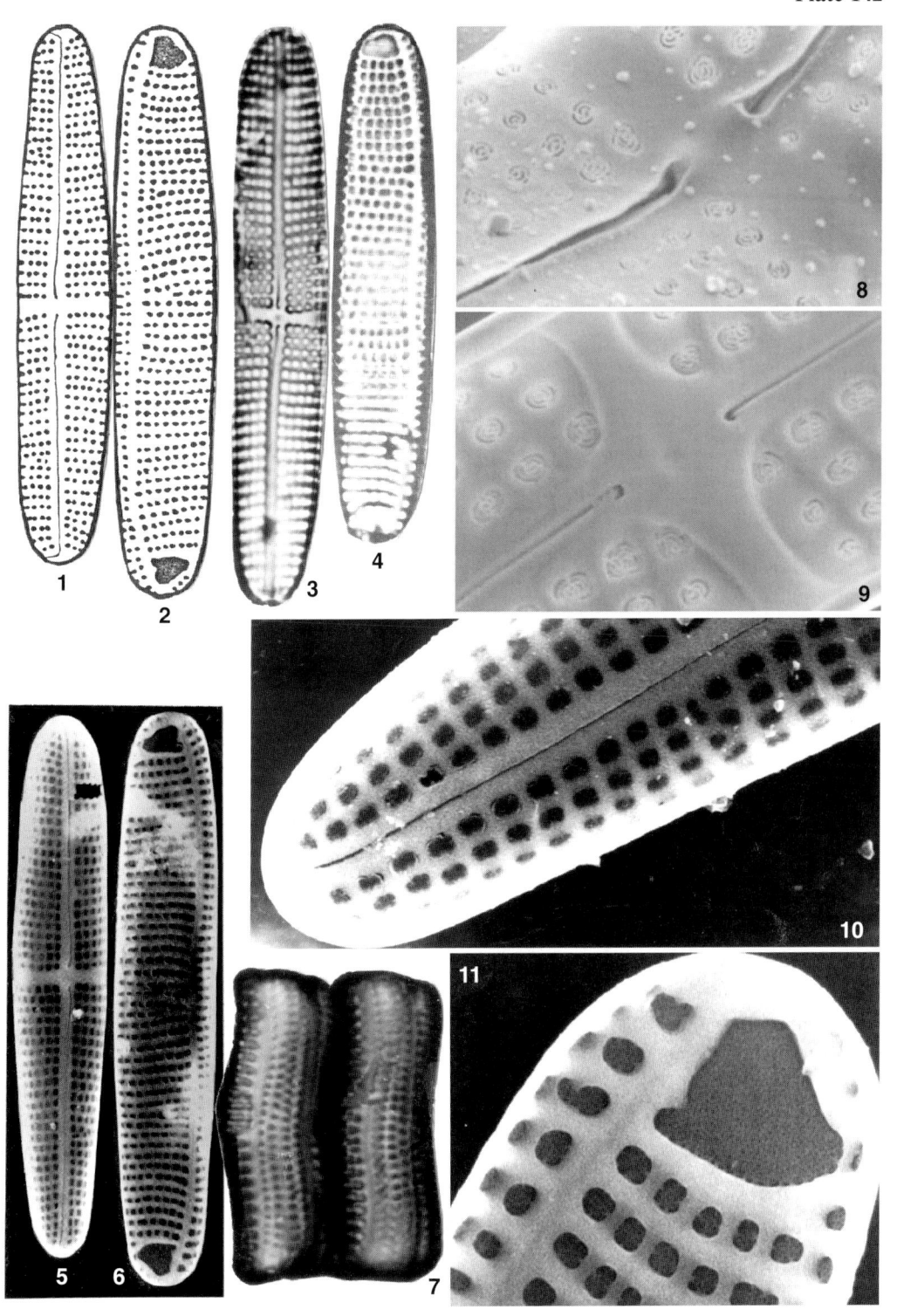

Plate 143 *Cocconeis diminuta* Pantocsek

Figs 1, 2.   Drawings. ×2000.
Figs 3-7.   LM. ×2000.
Figs 8-10.   SEM.

Figs 1, 3, 4.   Valve view of RV.
Figs 2, 5-7.   Valve view of ARV.
Fig. 8.   Internal view of RV with a valvocopula, showing a ring-like marginal ridge (arrow). ×8000.
Fig. 9.   External view of RV showing round areolae and external raphe fissures. ×8000.
Fig. 10.   External view of ARV showing elongated areolae and depressed axial area. ×5000.

Locality
Figs 3-10.   Otome-numa (Otome Pond), Fukushima.

Plate 143

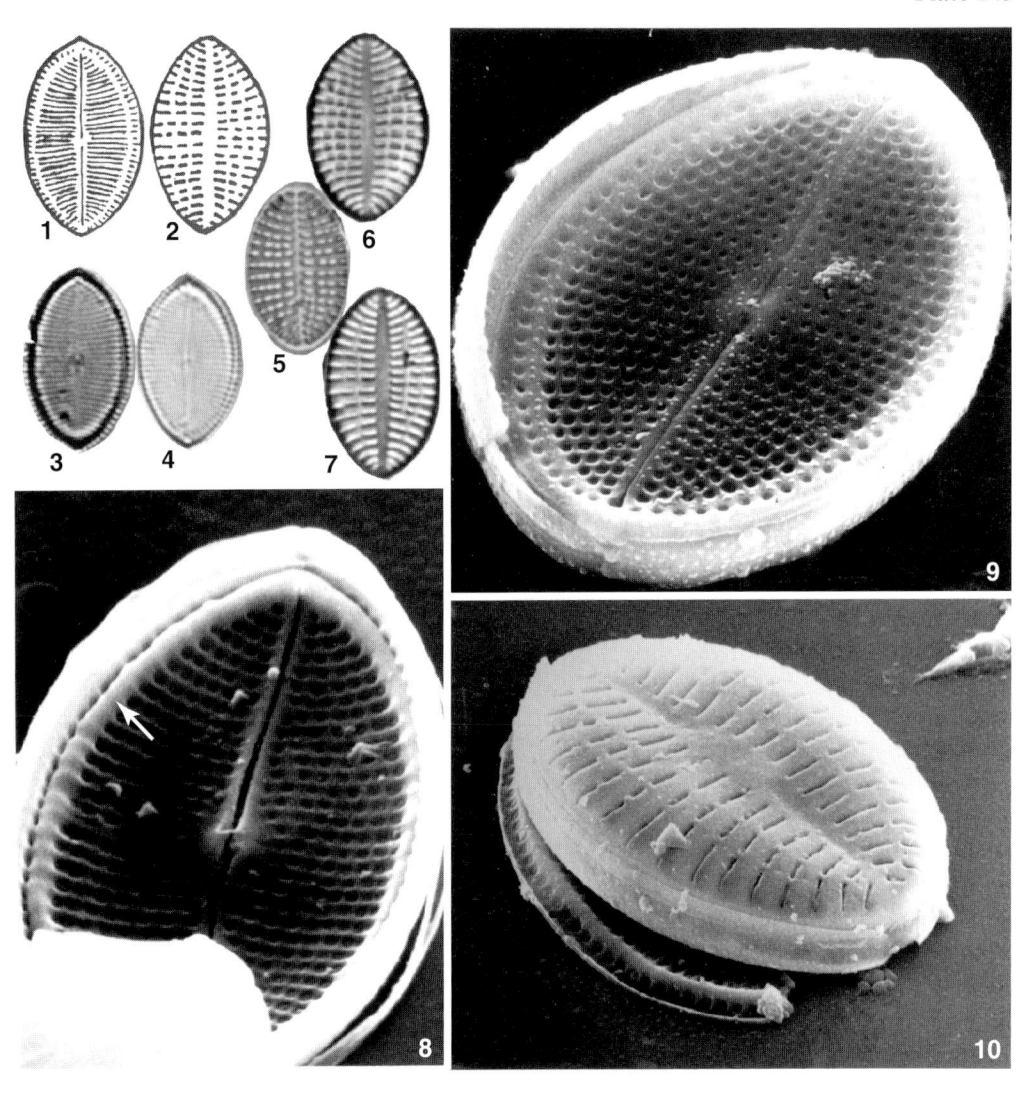

Plate 144 *Cocconeis pediculus* Ehrenberg

Figs 1, 2. Drawings. ×2000.

Figs 3, 4. LM. ×2000.

Figs 5-9. SEM.

Figs 1, 3. Valve view of RV.

Figs 2, 4. Valve view of ARV.

Fig. 5. External view of a RV. ×2500.

Fig. 6. Internal view of an ARV and the lobed fimbriae of the valvocopula detached from the RV. ×3000.

Fig. 7. Enlargement of the lobed fimbriae of the valvocopula. ×10000.

Fig. 8. Internal view of a broken ARV showing the foramina and the longitudinal sections of the areolae. ×20000.

Fig. 9. Short fimbriae of the valvocopula of ARV. ×20000.

Localities

Figs 3, 4. Tama-gawa (Tama River) Tokyo.

Figs 5, 7-9. Biwa-ko (Lake Biwa), Ogoto, Shiga.

Fig. 6. Funada-gawa (Funada River), Tokyo.

Plate 144

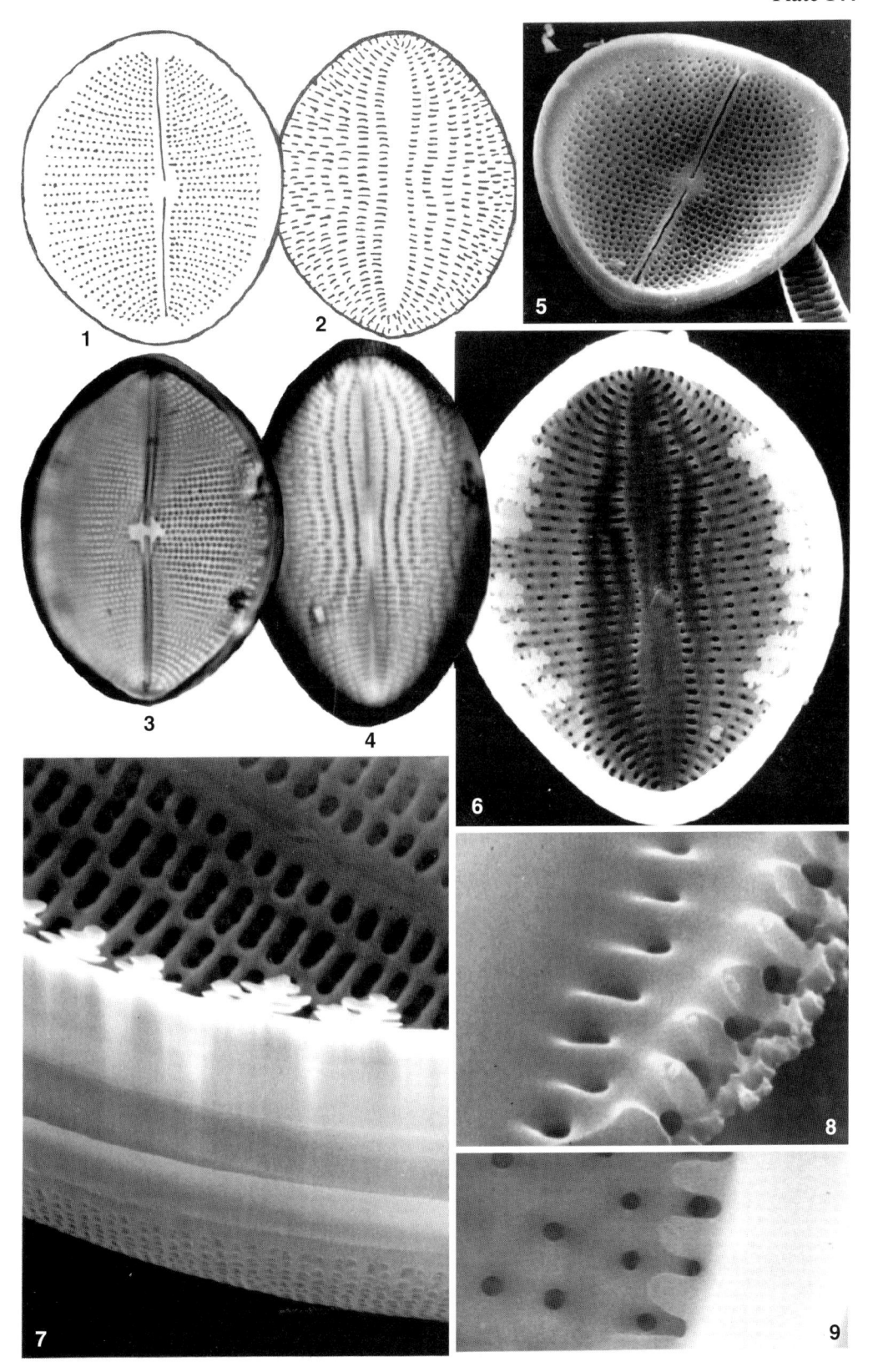

Plate 145 *Cocconeis pediculus* Ehrenberg

Figs 10, 11. SEM.

Fig. 10. Cells attached on a flamentous alga. ×600.
Fig. 11. Enlargement of Fig.10, showing the exterior of the ARV. ×2000.

Locality

Figs 10, 11. Kawaguchi-ko (Lake Kawaguchi), Yamanashi.

Plate 145

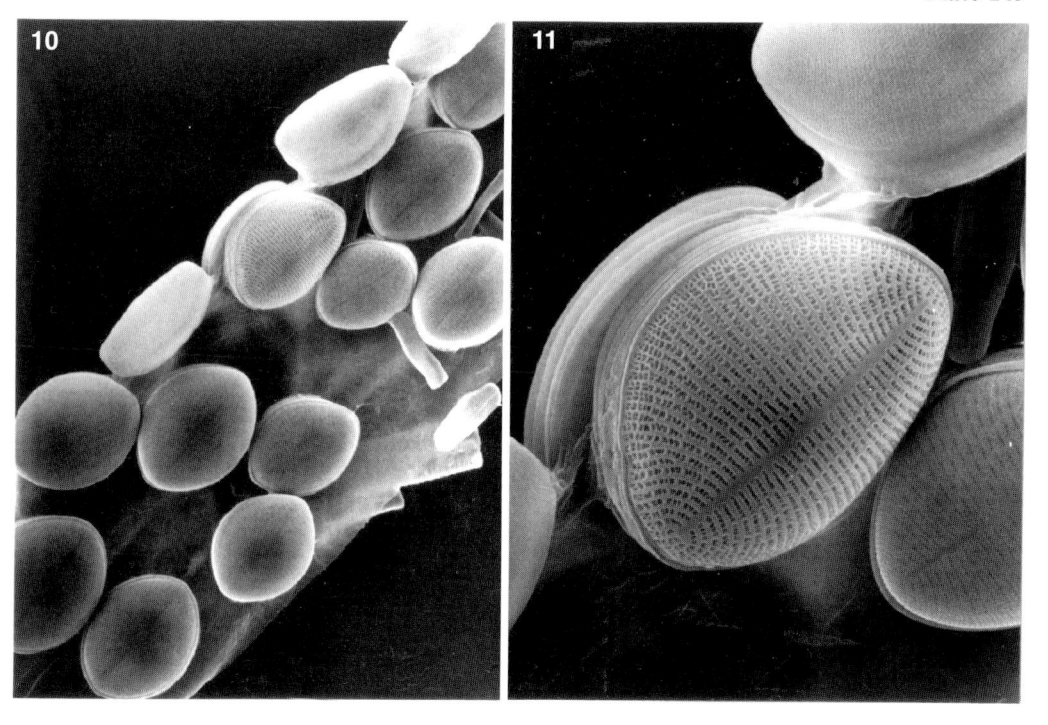

Plate 146 *Cocconeis placentula* **Ehrenberg var.** *placentula*

Figs 1, 4-7.   LM. ×2000.
Figs 2, 3.   Drawings. ×2000.
Fig. 8.   SEM.

Figs 1, 2.   Valve view of RV.
Figs 3-5, 7.   Valve view of ARV.
Fig. 6.   Valvocopula of RV.
Fig. 8.   Community on substratum. ×300.

Localities
Figs 1, 4, 7.   Ohchi-gawa (Ohchi River), Saitama.
Figs 5, 6, 8.   Minamiasa-kawa (Minamiasa River), Tokyo.

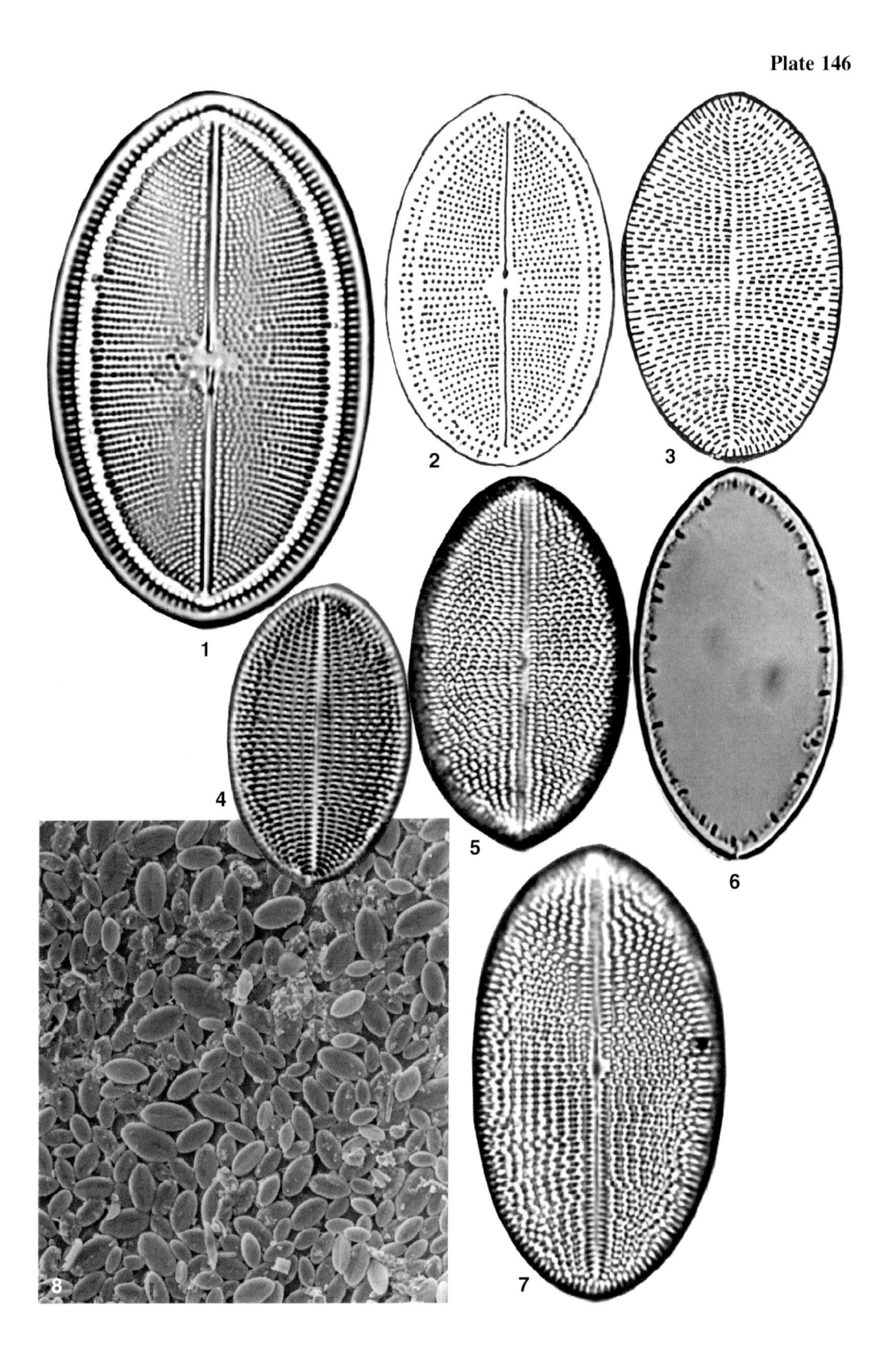

Plate 146

Plate 147 *Cocconeis placentula* Ehrenberg var. *placentula*

Figs 9-15.   SEM.

Fig. 9.   External view of the RV showing central raphe endings and round areolae. ×7000.

Fig. 10. External view of the AV showing laterally elongated areolae. ×7000.

Fig. 11. Internal view of the RV with a valvocopula, showing a ring-like marginal ridge (arrow). ×2000.

Fig. 12. Valvocopula of RV. ×2000.

Fig. 13. Enlargement of the end of RV valvocopula, showing open end and two kinds of projections. ×10000.

Fig. 14. ARV valvocopula with serrate fimbriae. ×2000

Fig. 15. Enlargement of the RV valvocopula attached to RV, showing the thick projections; they extend onto the ling-like marginal ridge of RV, and the small projections of the pars interior. ×30000.

Localities

Figs 9, 15.   Ohchi-gawa (Ohchi River), Saitama.

Figs 10-14. Minamiasa-kawa (Minamiasa River), Tokyo.

Plate 147

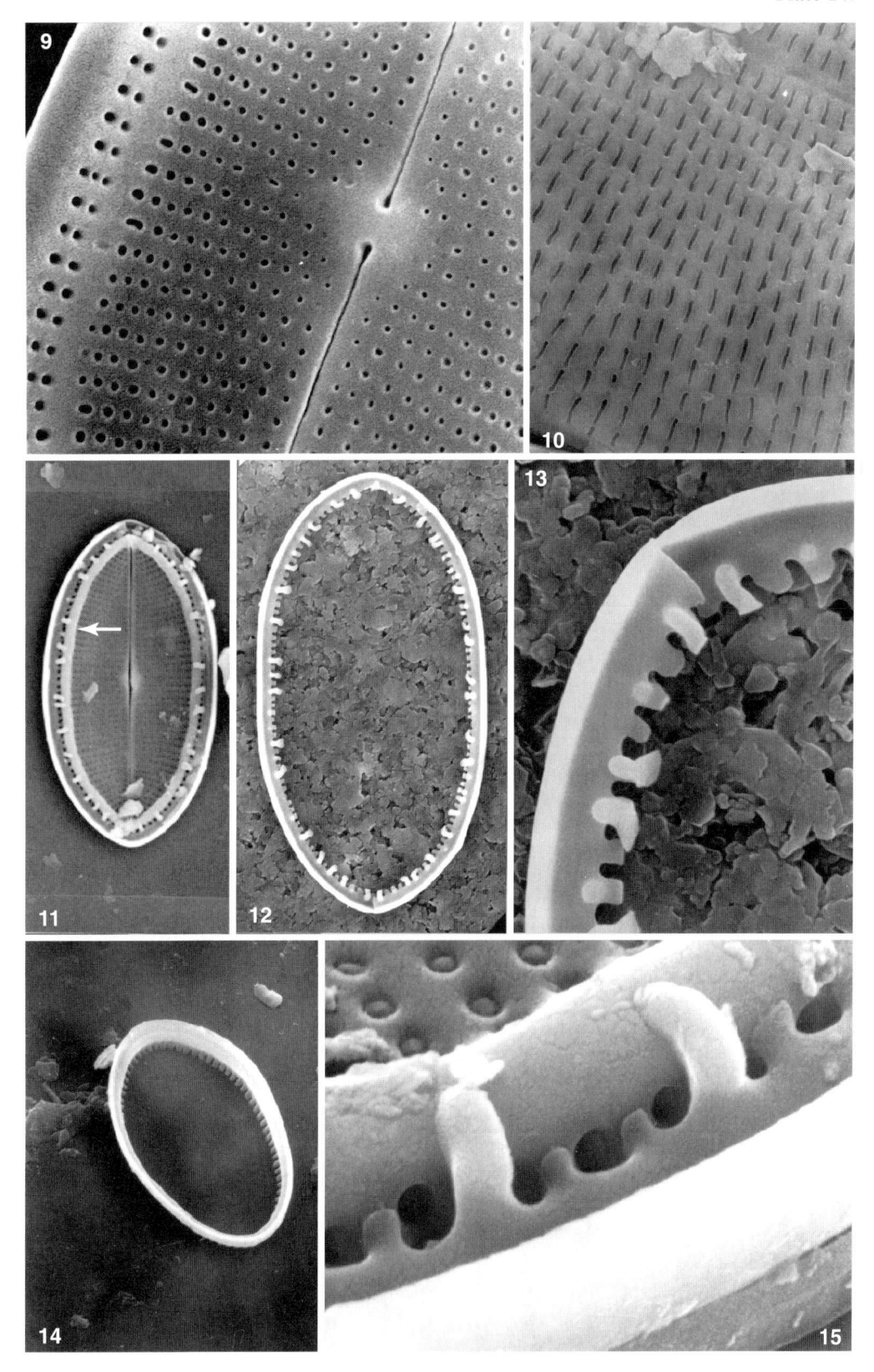

Plate 148 *Cocconeis placentula* var. *lineata* (Ehrenberg) Van Heurck

Figs 1-5. LM. ×2000.
Figs 6, 7. SEM.
Figs 8, 9. TEM.

Fig. 1. Valve view of RV.
Figs 2-5. Valve view of ARV.
Fig. 6. External view of a frustule showing the RV with a ring-like hyaline area (arrows). ×4000.
Fig. 7. Oblique view of ARV. ×2000.
Fig. 8. Hymens of areolae in RV, having the marginal elongated perforations arranged in radial and the central ones arranged in parallel. ×40000.
Fig. 9. Hymens of areolae in ARV, having marginal elongated perforations and central ones. ×40000.

Localities
Figs 1, 3-9. Ohchi-gawa (Ohchi River), Saitama.
Fig. 2. Minamiasa-kawa (Minamiasa River), Kobotoke-tohge (Kobotoke Pass), Tokyo.

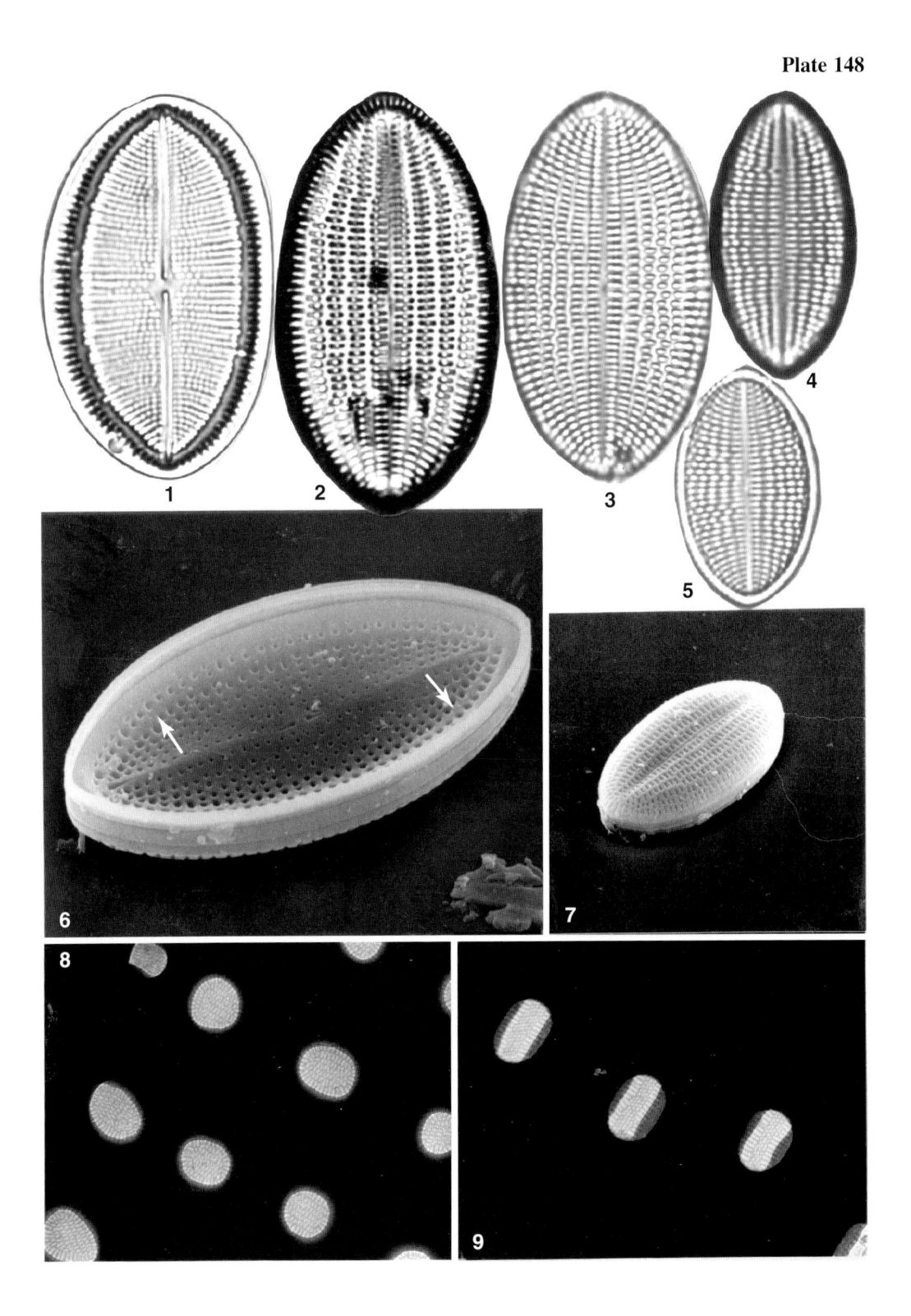

Plate 148

Plate 149 *Cocconeis placentula* var. *lineata* (Ehrenberg) Van Heurck

Figs 10-15. SEM.

Fig. 10. A separated frustule showing the exterior of the ARV, the serrate fimbriae of the valvocopula of ARV, and the interior of the RV with a ring-like marginal ridge (arrow). ×4500.

Fig. 11. Enlargement of the broken ARV showing the elongated areolae accompanied externally by flaps (arrowheads) and a hymen, and internally by a small foramen (arrow). ×50000.

Fig. 12. External view of the ARV showing the elongated areolae arranged longitudinally and shallow grooves (arrow) on the sternum. ×10000.

Fig. 13. External view of the RV centre showing round areolae and the central raphe endings expanded slightly. ×10000.

Fig. 14. Internal view of the RV centre showing the round areolae occluded by a domed hymen, and the central raphe endings bent slightly in the opposite direction. ×30000.

Fig. 15. Internal view of ARV showing the small, round foramina of the areolae. ×10000.

Locality

Figs 10-15. Ohchi-gawa (Ohchi River), Saitama.

Plate 149

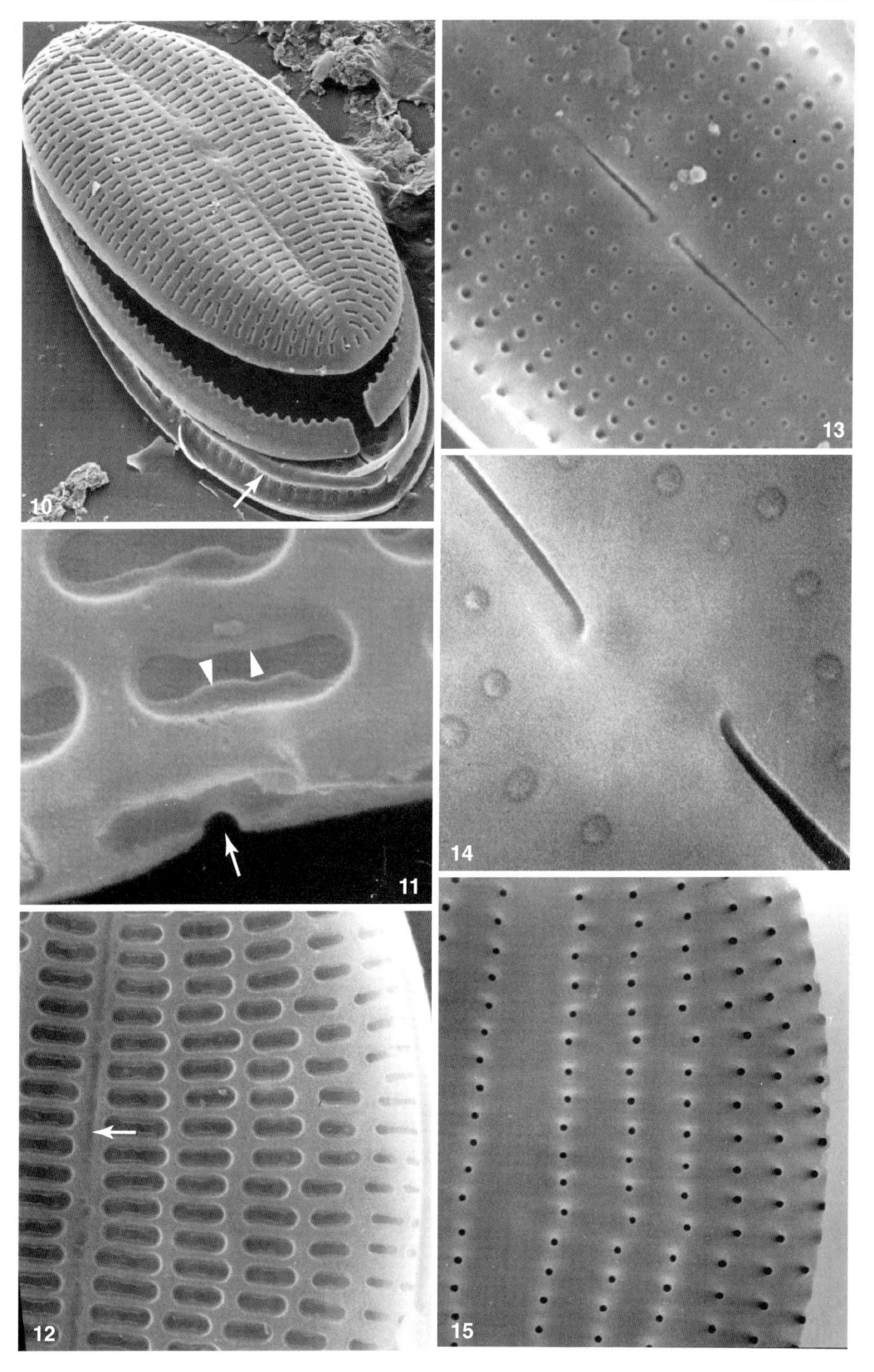

Plate 150 *Cocconeis scutellum* Ehrenberg

Figs 1, 5.   Drawings. ×2000.
Figs 2-4.   LM. ×2000.
Figs 6, 7.   SEM.

Figs 1, 4.   Valve view of RV.
Figs 2, 3, 5.   Valve view of ARV.
Fig. 6.   External view of RV. ×2000.
Fig. 7.   Broken frustule showing the ARV, the valvocopula (AVC) detached from the ARV, and RV with a ring-like marginal ridge (arrow) and a valvocopula (RVC). ×3000.

Localities
Figs 2-4, 6.   Tomari, Akita.
Fig. 7.   Take-shima (Take Island) in Mikawa-wan (Mikawa Bay), Aichi.

Plate 150

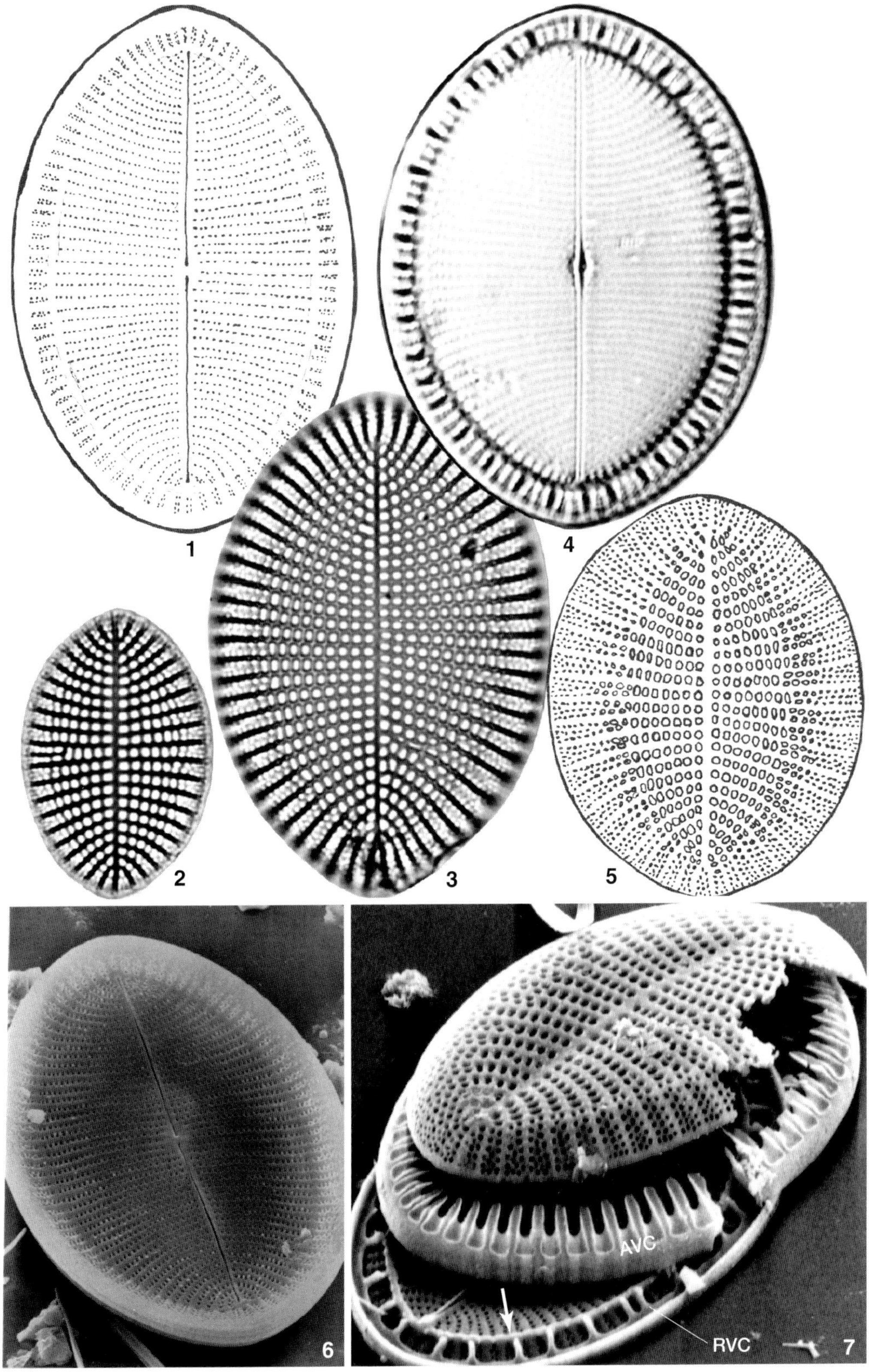

Plate 151 *Cocconeis stauroneiformis* (Rabenhorst) Okuno

Figs 1, 2.   Drawings. ×2000.
Figs 3, 4.   LM. ×2000.
Figs 5-8.   SEM.

Figs 1, 3.   Valve view of RV.
Figs 2, 4.   Valve view of ARV.
Fig. 5.   Internal view of RV showing a ring-like marginal ridge (arrow) of the valvocopula. ×4000.
Fig. 6.   Internal view of ARV showing rows of areolae and the projections (arrowheads) of the valvocopula. ×12000.
Fig. 7.   Enlargement of the areolae bearing a reticulate velum. ×36000.
Fig. 8.   External view of eroded ARV. ×4000.

Locality
Figs 1-8.   Take-shima (Take Island) in Mikawa-wan (Mikawa Bay), Aichi.

Plate 151

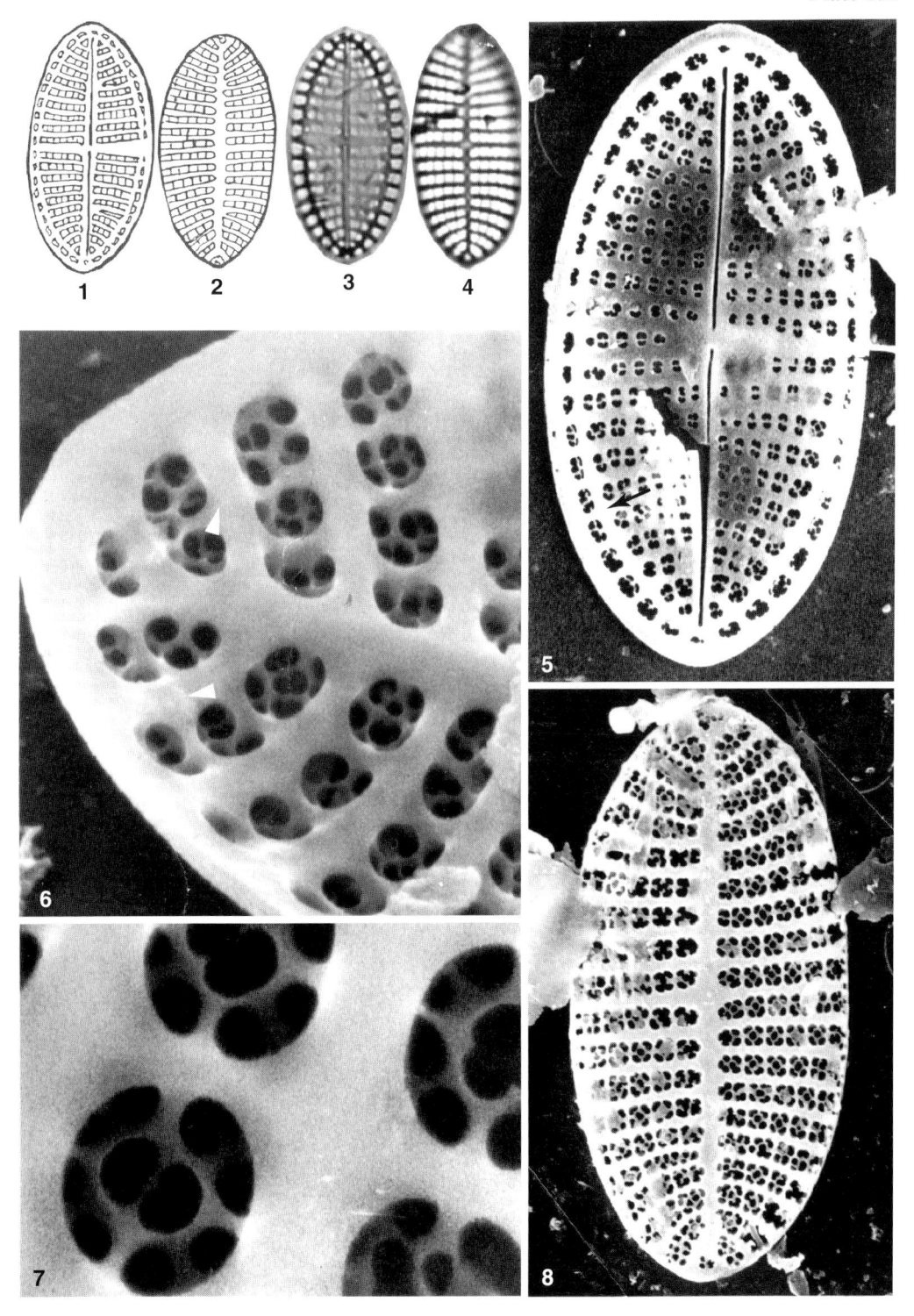

Plate 152 *Achnanthidium convergens* (H. Kobayasi) H. Kobayasi

Figs 1, 2, 5-8. LM. ×2000.
Figs 3, 4. Drawings. ×2000.
Figs 9, 10. TEM. ×4000.
Figs 11-17. SEM.
Fig. 18. TEM. ×6000.

Figs 1, 3, 5, 7, 9. Raphid valves.
Figs 2, 4, 6, 8, 10. Araphid valves.
Fig. 11. External oblique view of a whole frustule showing a raphid valve. ×8000.
Fig. 12. Internal oblique view of the valve. ×8000.
Fig. 13. Enlargement of the frustule end showing a hooked terminal fissure and the open ends of a split band (*). ×17000.
Fig. 14. Details of the external oblique view of the valve showing hyaline rim along the valve shoulder and the outer openings of areolae. Note the linear to elliptic areolar openings with constrictions (arrows). ×30000.
Fig. 15. External view of the araphid valve. ×600.
Fig. 16. External oblique view of the araphid valve. ×700.
Fig. 17. Enlargement of the valve interior. The occlusions of the areolae, which align to form stria, are not clearly separated from each other. ×30000.
Fig. 18. Details of the hymenate pore occlusion (parallel array type) and the foramina with constrictions (arrows) of the raphid valve areolae.

Locality
Figs 1, 2, 5-18. Ara-kawa (Ara River), Saitama.

Plate 152

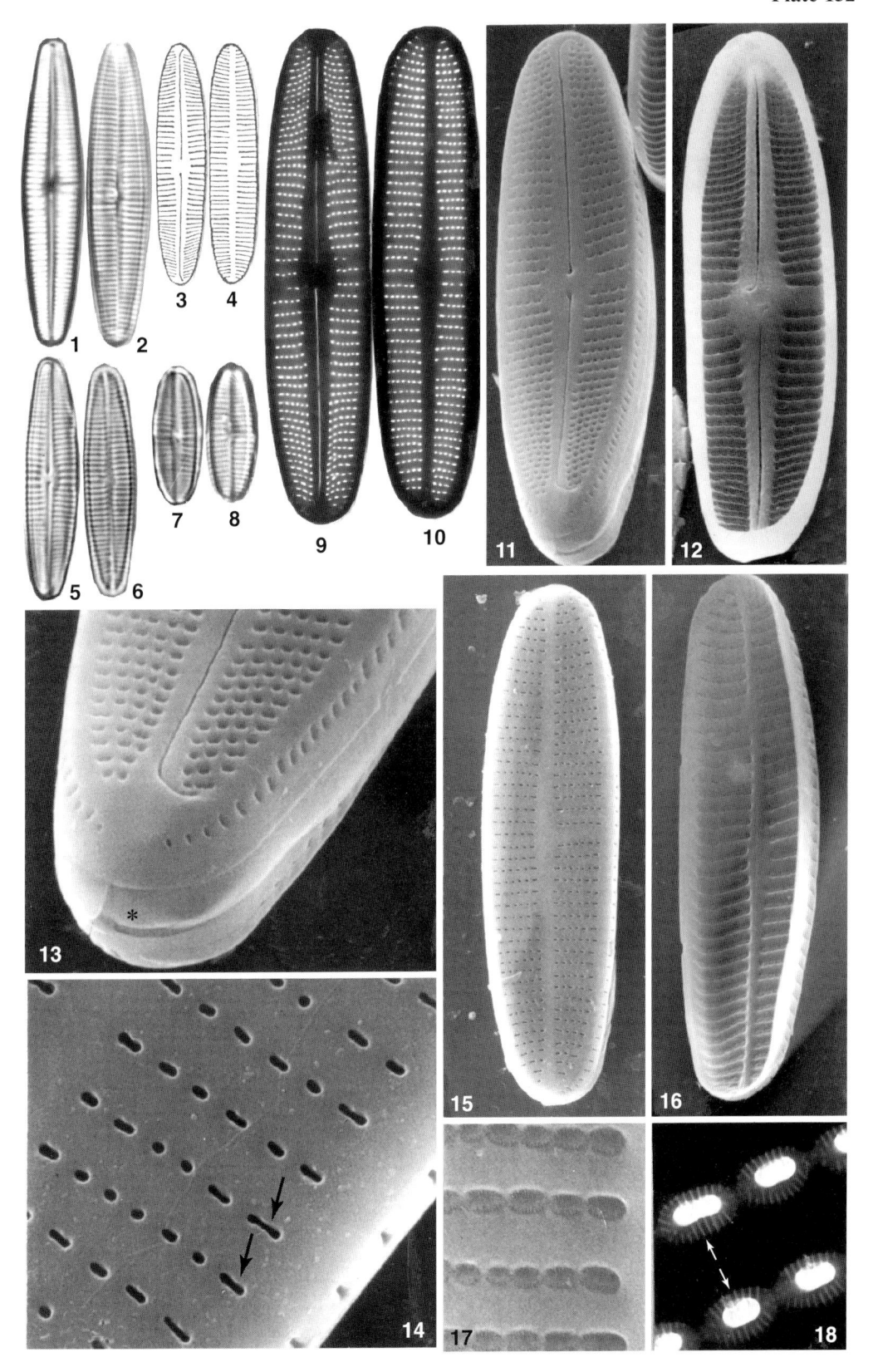

Plate 153 *Achnanthidium exiguum* (Grunow) Czarnecki

Figs 1-6.   LM. ×2000.
Figs 7, 8.   Drawings. ×2000.
Figs 9-14. SEM.

Figs 1, 3, 5, 7.   Raphid valves.
Figs 2, 4, 6, 8.   Araphid valves.
Fig. 9.   External view of the valve center showing the central endings of raphe branches, which are located in shallow grooves. ×20000.
Fig. 10.   External oblique view of the raphid valve showing areolation. Note both Voigt faults (arrows) are located on the opposite side of the valve. ×8000.
Fig. 11.   Oblique view of a whole frustule. ×5000.
Fig. 12.   Oblique external view of the araphid valve. ×5000.
Fig. 13.   Enlarged external view of the valve end showing deflected terminal fissure. ×20000.
Fig. 14.   Enlarged internal view of the valve end showing the areolation with hymenate pore occlusions and a trace of the raphe (arrow), which was filled in by silica. ×20000.

Locality

Figs 1-6, 9-14.   A small concrete tank in Tokyo Gakugei University, Tokyo.

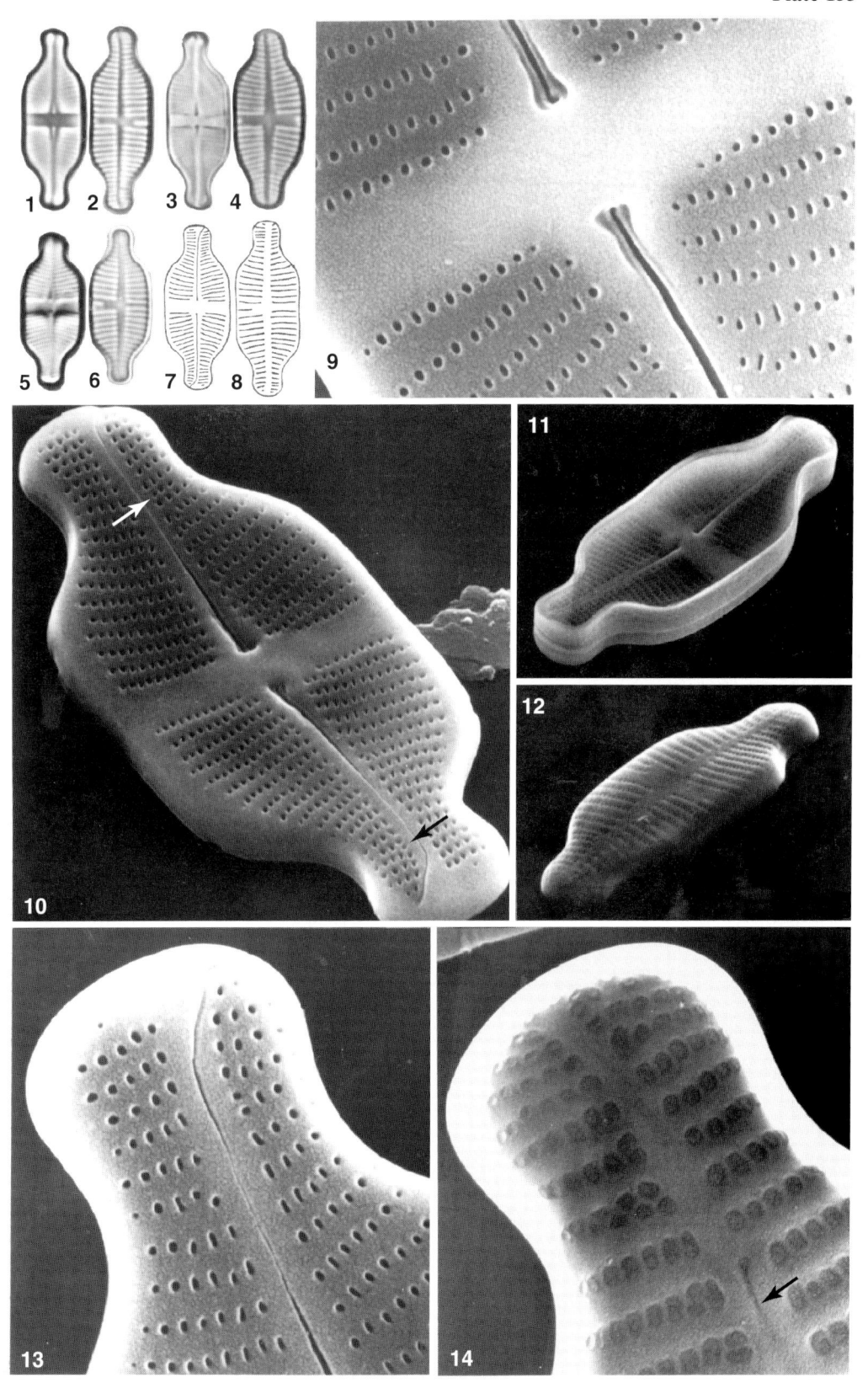

Plate 153

453

Plate 154 *Achnanthidium gracillimum* (Meister) Mayama comb. nov.

Figs 1-6, 9.   LM. ×2000.
Figs 7, 8.   Drawings. ×2000.
Figs 10-14.   SEM.

Figs 1, 3-5, 7.   Raphid valves.
Figs 2, 6, 8.   Araphid valves.
Fig. 9.   Girdle view of a frustule.
Fig. 10.   External view of the raphid valve end showing a bent polar fissure. ×20000.
Fig. 11.   Enlarged external view of the raphid valve showing the central endings of the raphe branches and the slit-like openings of areolae. ×20000.
Fig. 12.   Oblique view of a frustule end composed of an araphid epi-valve, split bands and a raphid hypo-valve. ×20000.
Fig. 13.   Center of the araphid valve exterior. Note the transversely elongated openings of the areolae. ×20000.
Fig. 14.   Internal oblique view of the raphid valve showing the areolae with occlusions and the central endings of the raphe fissures, which slightly deflect to opposite sides. ×20000.

Localities
Figs 1-6, 9, 12.   Sho-kawa (Sho River), Naride, Gifu.
Figs 10, 11, 13, 14.   Chikuma-gawa (Chikuma River), Kawakami, Nagano.

Plate 154

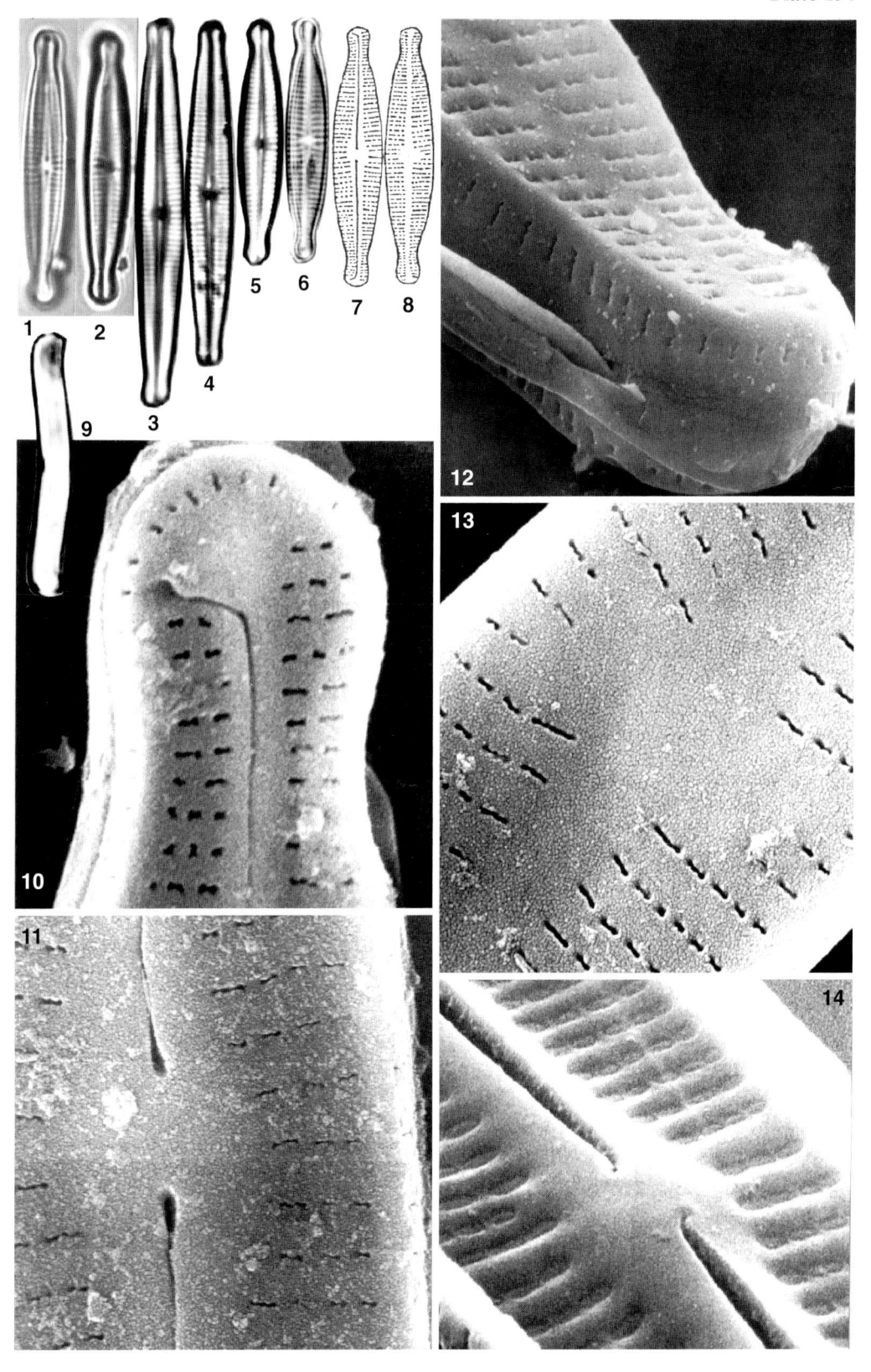

Plate 155 *Achnanthidium japonicum* (H. Kobayasi) H. Kobayasi

Figs 1, 2.   Drawings. ×2000.
Figs 3-8.   LM. ×2000.
Figs 9, 10.   TEM. ×4000.
Figs 11-17.   SEM.
Fig. 18.   TEM. ×60000.

Figs 1, 3, 5, 7, 9.   Raphid valves.
Figs 2, 4, 6, 8, 10.   Araphid valves.
Fig. 11.   External view of the raphid valve. ×7000.
Fig. 12.   Internal view of the raphid valve. ×7000.
Fig. 13.   Enlargement of the raphid valve end showing weakly hooked terminal fissure and hyaline rim along the valve shoulder, i.e., the valve face and mantle junction. ×20000.
Fig. 14.   External view of the araphid valve. ×7000.
Fig. 15.   Internal view of the araphid valve with three open bands. ×700.
Fig. 16.   Details of the external araphid valve showing continuous rows of areolae on the valve shoulder. ×30000.
Fig. 17.   Details of the areolar occlusion (hymenate type) on valve inside. ×30000.
Fig. 18.   Hymens with perforations of the centric array type and foramina with median constriction (arrows) of the areolae in raphid valve.

Locality
Figs 3-18.   Ara-kawa (Ara River), Saitama.

Plate 155

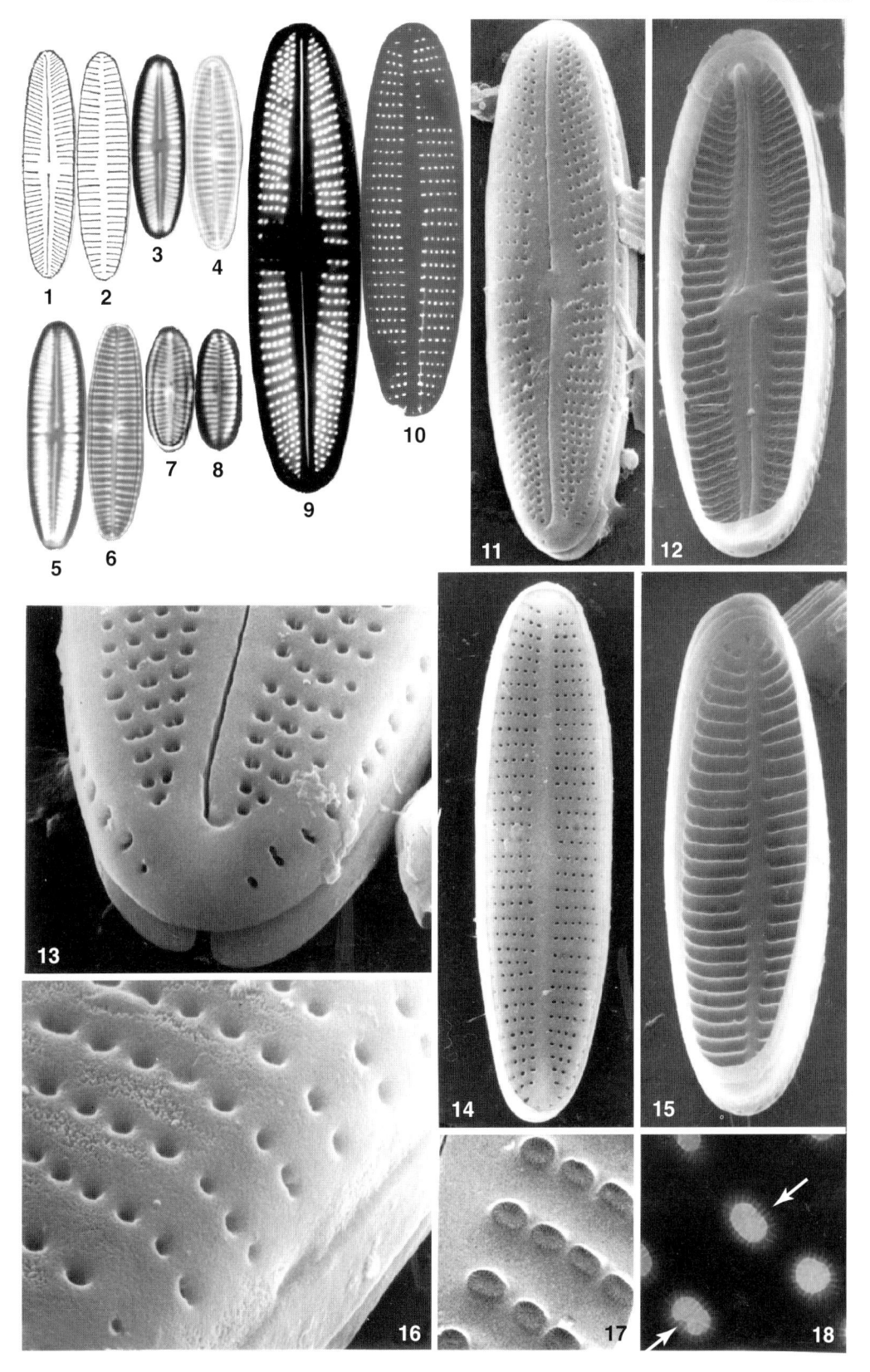

Plate 156 *Achnanthidium minutissimum* (Kützing) Czarnecki

Figs 1-12, 15, 16, 19-27. LM. ×2000.
Figs 13, 14, 17, 18. Drawings. ×2000.
Figs 28-34. TEM. ×5000.
Fig. 35. SEM.

Figs 1-5, 7-34. Valve view (Fig. 9 in part).
Figs 6, 9, 35. Girdle view (Fig. 9 in part).
Figs 28-34. Morphological variation in specimens occurring in a small concrete tank (4 m × 4 m × 0.5 m). Figs 28-31. Raphid valve. Figs 32-34. Araphid valve.
Fig. 35. External oblique view of a theca with the raphid valve. ×6000.

Localities

Figs 1-4. Type slide. BM 77949, Aschersleben. Kütz. Alg. Ex. Dec. VIII. No.75.
Figs 5, 6. Lectotype slide of "*Achnanthidium microcephalum* Kützing." BM 18434. Coll. Kützing. Triest, Biassoleto 289. Desig. by Lange-Bertalot & Ruppel (1980).
Figs 7, 8. Type slide of "*Achnanthes minutissima* var. *macrocephala* Hustedt." Coll. Hustedt Mo2/46. Toba See, Sumatra (BRM).
Fig. 9. Coll. Hustedt. Ma4/54. Berhmerhaven (BRM).
Figs 10-12, 35. Tama-gawa (Tama River), Oiranbuchi, Yamanashi.
Figs 15, 16. Ara-kawa (Ara River), Saitama.
Figs 19-23. "*Achnanthes lineariformis* Kobayasi (1965)," Ara-kawa (Ara River), Saitama.
Figs 24, 27. A spring in Kawagoe, Saitama.
Figs 25, 26. Sanpoji-ike (Sanpoji Pond), Tokyo.
Figs 28-34. A small concrete tank in Tokyo Gakugei University, Tokyo.

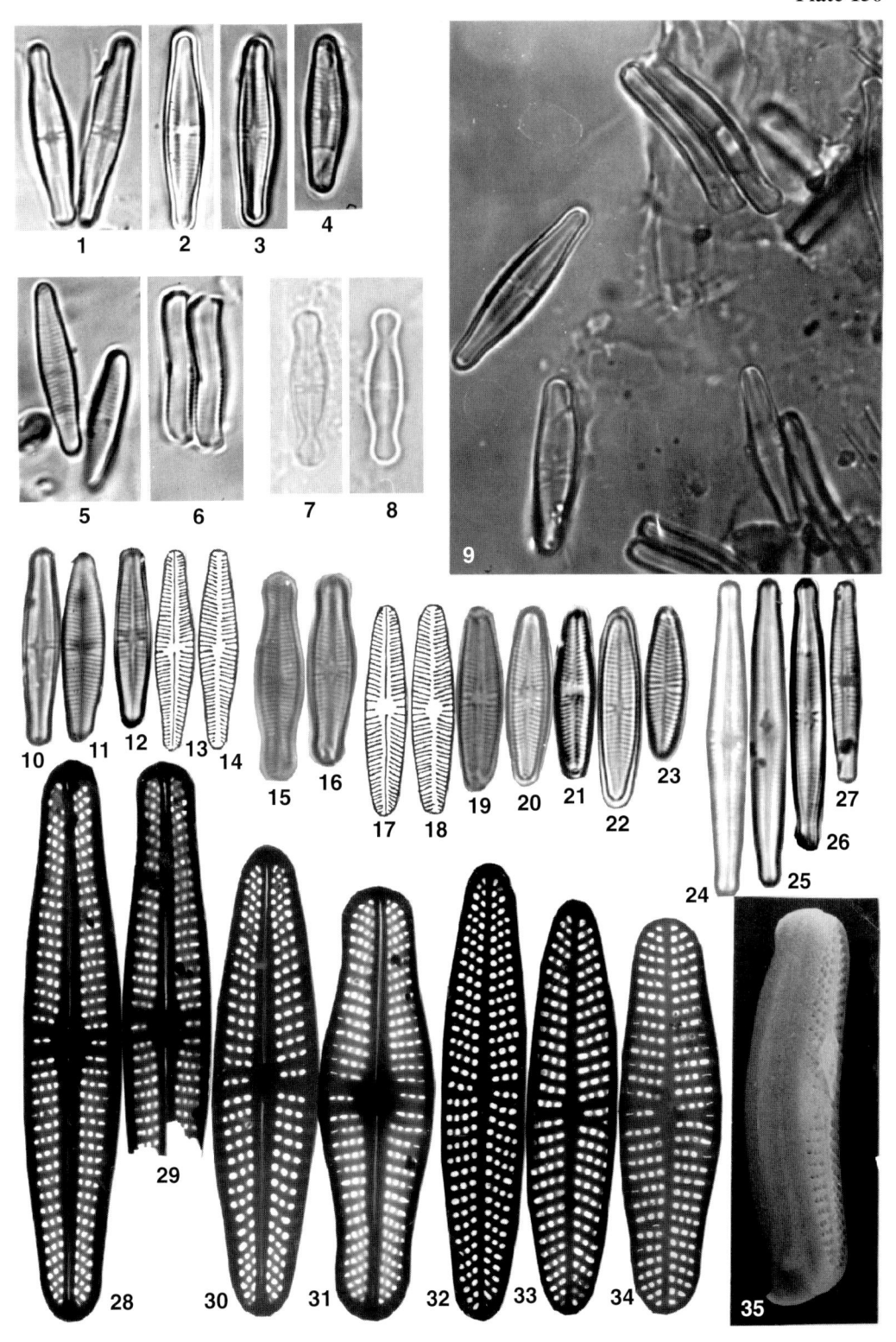

Plate 156

Plate 157 *Achnanthidium minutissimum* **(Kützing) Czarnecki**

Figs 36-42. SEM.

Fig. 36. External view of the raphid valve with both the central and the terminal deflected raphe endings. ×8000.

Figs 37, 38. Internal oblique view of the raphid valve showing occluded areolae and the central raphe endings which deflect in opposite directions to each other. ×8000.

Fig. 39. External oblique view of the raphid valve showing a longitudinal row of slit-like areolae. ×8000.

Fig. 40. Internal oblique view of the araphid valve showing occluded areolae and linear axial area. ×8000.

Fig. 41. Details of the bottom end of the valve in Fig. 36 showing larger areolae, which make striae with stronger radiation, and the deflected terminal ending of the raphe. ×20000.

Fig. 42. The frustule end showing a hypovalve (top), an epicingulum composed of three bands and an epivalve (bottom). ×20000.

Localities

Figs 36, 41. Tama-gawa (Tama River), Oiranbuchi, Yamanashi.

Figs 37-39. A small concrete tank in Tokyo Gakugei University, Tokyo.

Figs 40, 42. Chikuma-gawa (Chikuma River), Kamiyamada, Nagano.

Plate 157

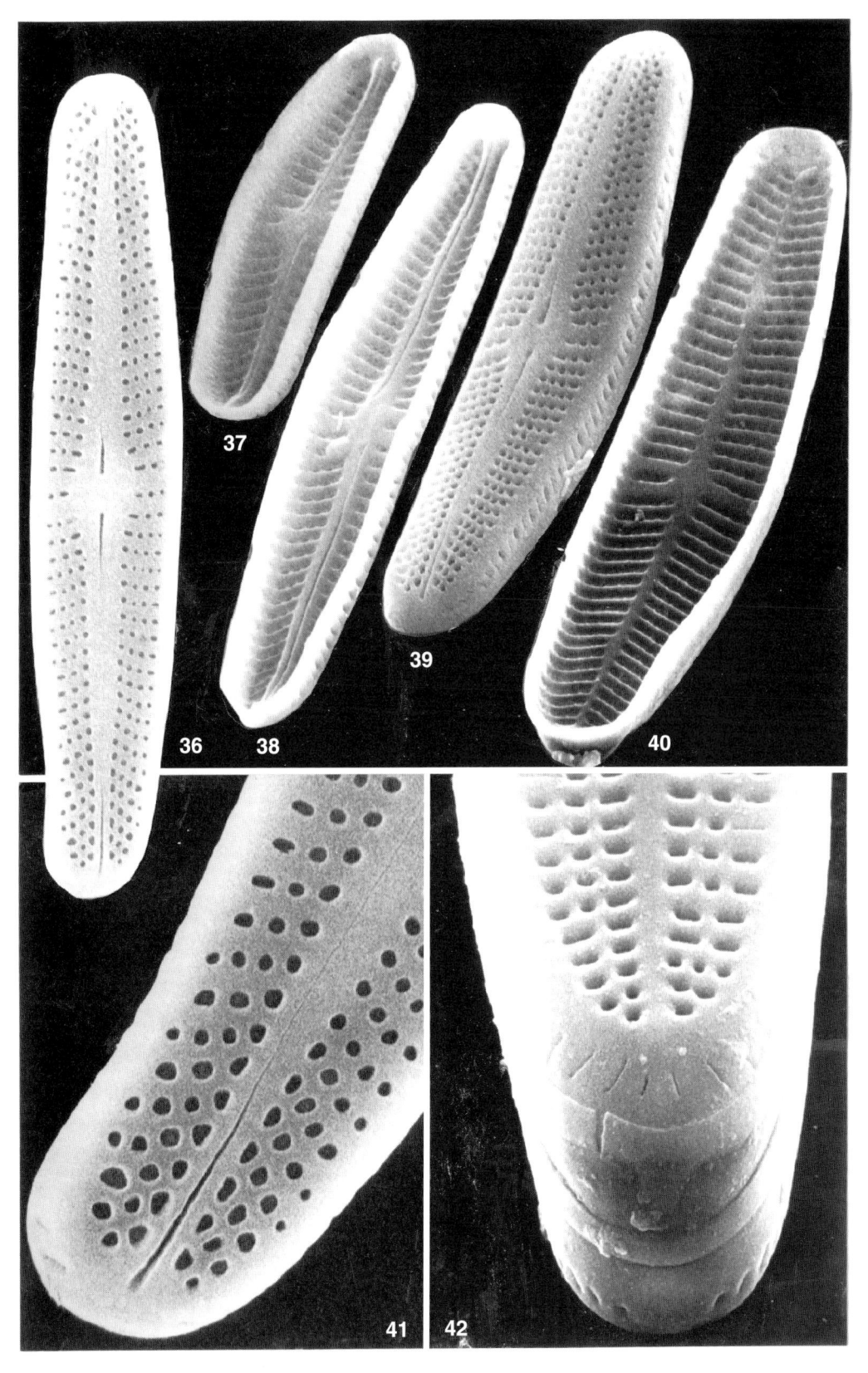

Plate 158 *Achnanthidium pusillum* **(Grunow) Czarnecki**

Figs 1-5.   LM. ×2000.
Figs 6, 7.   Drawings. ×2000.
Figs 8, 9.   TEM. ×5000.
Figs 10-13. SEM.
Fig. 14.   TEM. ×4000.

Figs 1, 3, 5, 6, 8.   Raphid valves.
Figs 2, 4, 7, 9.   Araphid valves.
Fig. 10.   External view of the raphid valve showing raphe branches with the swollen terminal endings. ×8000.
Fig. 11.   External view of the araphid valve with hyaline rim along the valve shoulder. ×8000.
Fig. 12.   Internal view of the raphid valve showing the striae, each of which consists of a series of areolae covered by hymenate pore occlusions or ricae. Both of the central raphe endings are deflected in opposite directions to each other. ×20000.
Fig. 13.   Enlarged external valve center showing outer openings of the areolae and central pores of the raphe endings. ×20000.
Fig. 14.   Details of the hymenate pore occlusions.

Locality

Figs 1-5, 8-14.   On moss at Ryugaeshino-taki (Ryugaeshi Falls), Karuizawa, Nagano.

Plate 158

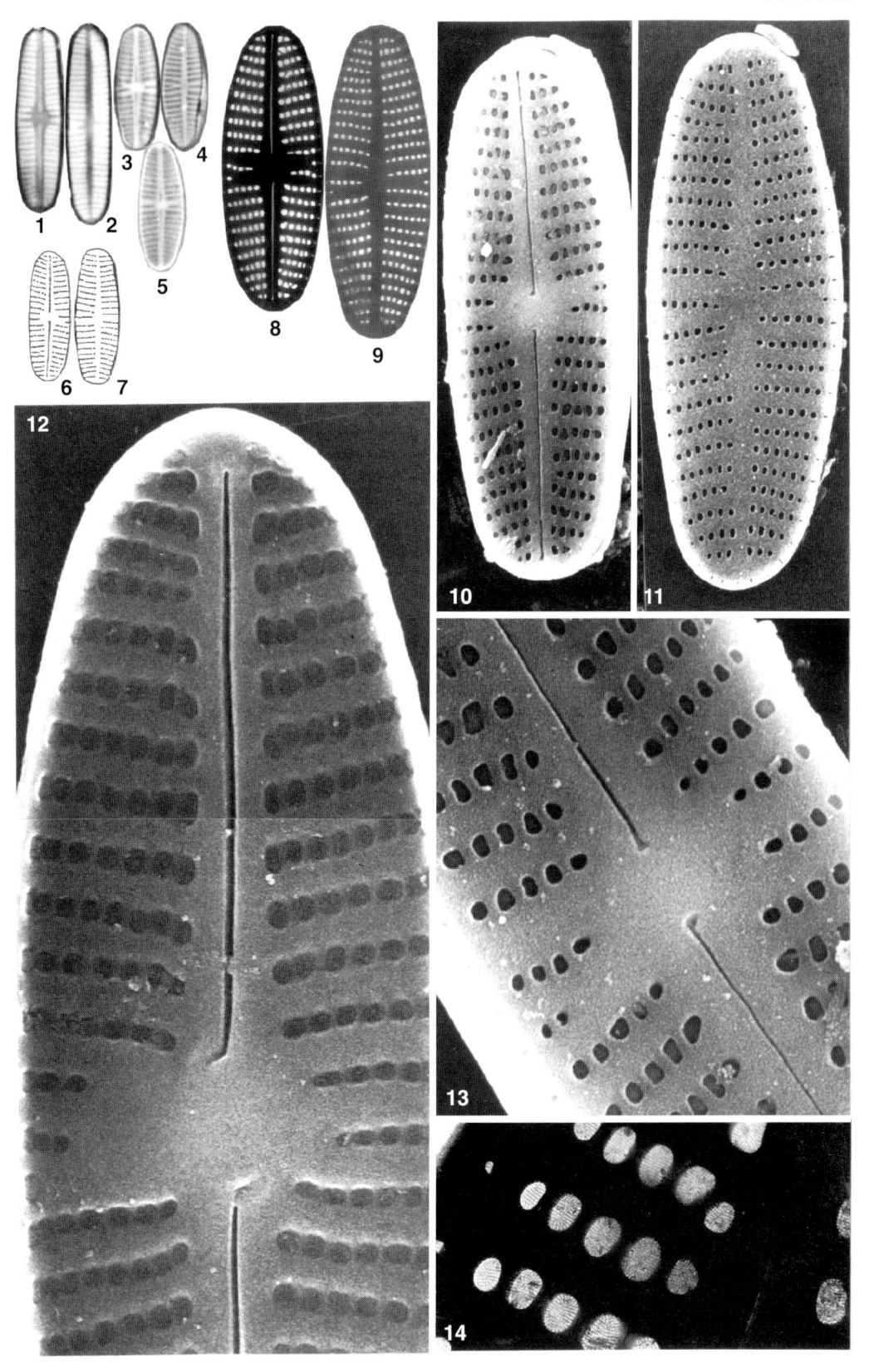

Plate 159 *Achnanthidium pyrenaicum* (Hustedt) H. Kobayasi

Figs 1, 2.   Drawings. ×2000.
Figs 3-7.   LM. ×2000.
Figs 8-12.   SEM.

Figs 1, 3-6, 8-11.   Raphid valves.
Fig. 12.   Araphid valve.
Fig. 8.   Whole view of a raphid valve. ×5000.
Fig. 9.   Enlargement of one valve end showing a bent terminal fissure. ×20000.
Fig. 10.   Valve center showing the central pores of raphe endings. ×20000.
Fig. 11.   Enlargement of the other end opposite that in Fig. 9. ×20000.
Fig. 12.   Loose frustule showing a pair of raphid (internal view) and araphid (external view) valves. ×5000.

Locality

Figs 3-12.   Inabe-gawa (Inabe River), Hokusei, Mie.

Plate 159

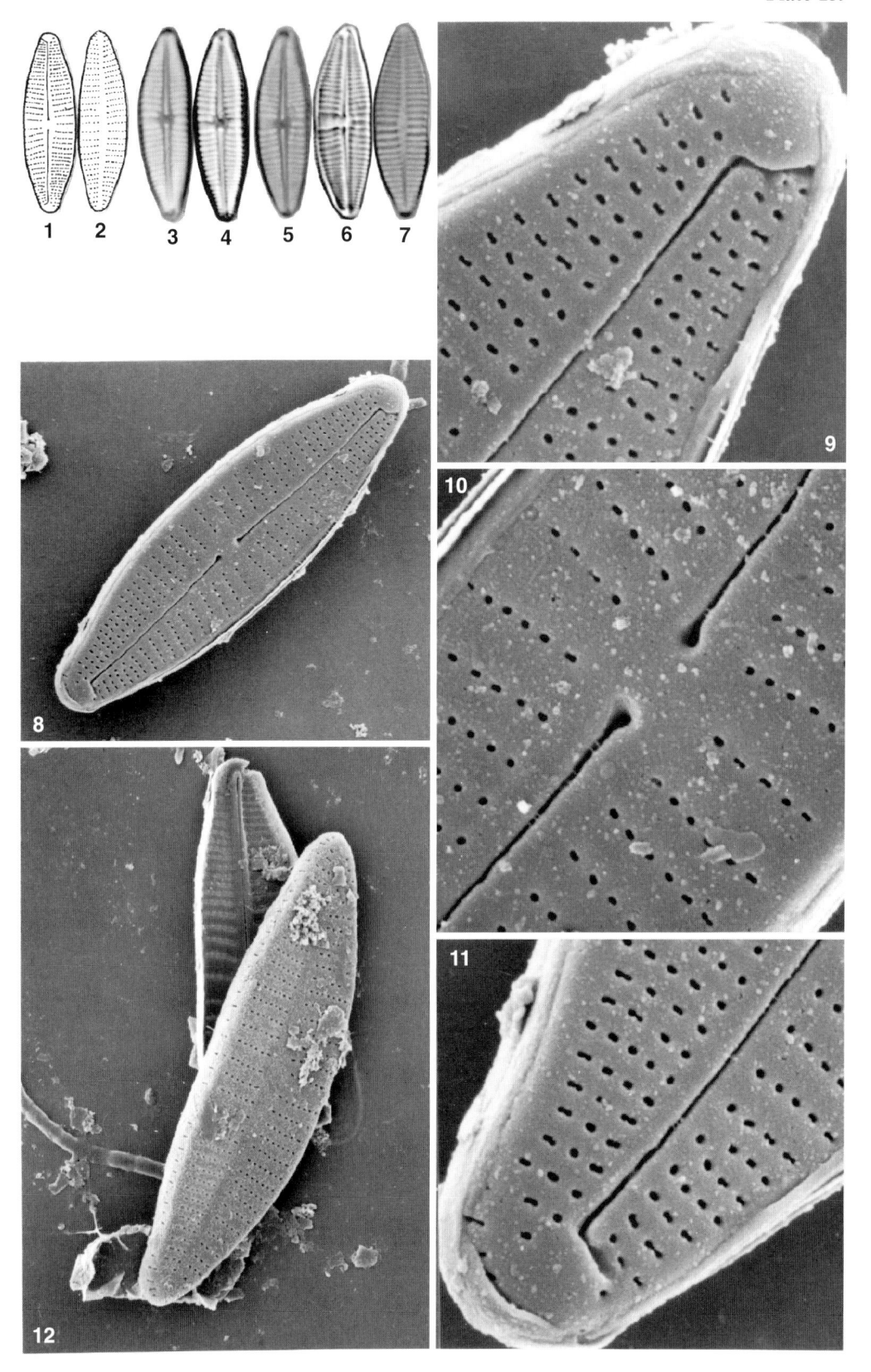

Plate 160 *Achnanthidium pyrenaicum* (Hustedt) H. Kobayasi

Figs 13-18.　SEM.

Fig. 13.　External view of an araphid valve with a hyaline valve shoulder.　×7000.

Fig. 14.　Internal view of an araphid valve showing eroded areolae.　×5000.

Fig. 15.　Whole view of a internal raphid valve.

Fig. 16.　Internal view of the raphid valve showing a developed axial area and a straight raphe slit. Note the majority of adjacent ricae fusing with each other.　×15000.

Fig. 17.　Internally, both central endings of the raphe curve to opposite sides of each other.　×15000.

Fig. 18.　The other valve end opposite that in Fig. 16.　×15000.

Locality

Figs 13-18.　Inabe-gawa (Inabe River), Hokusei, Mie.

Plate 160

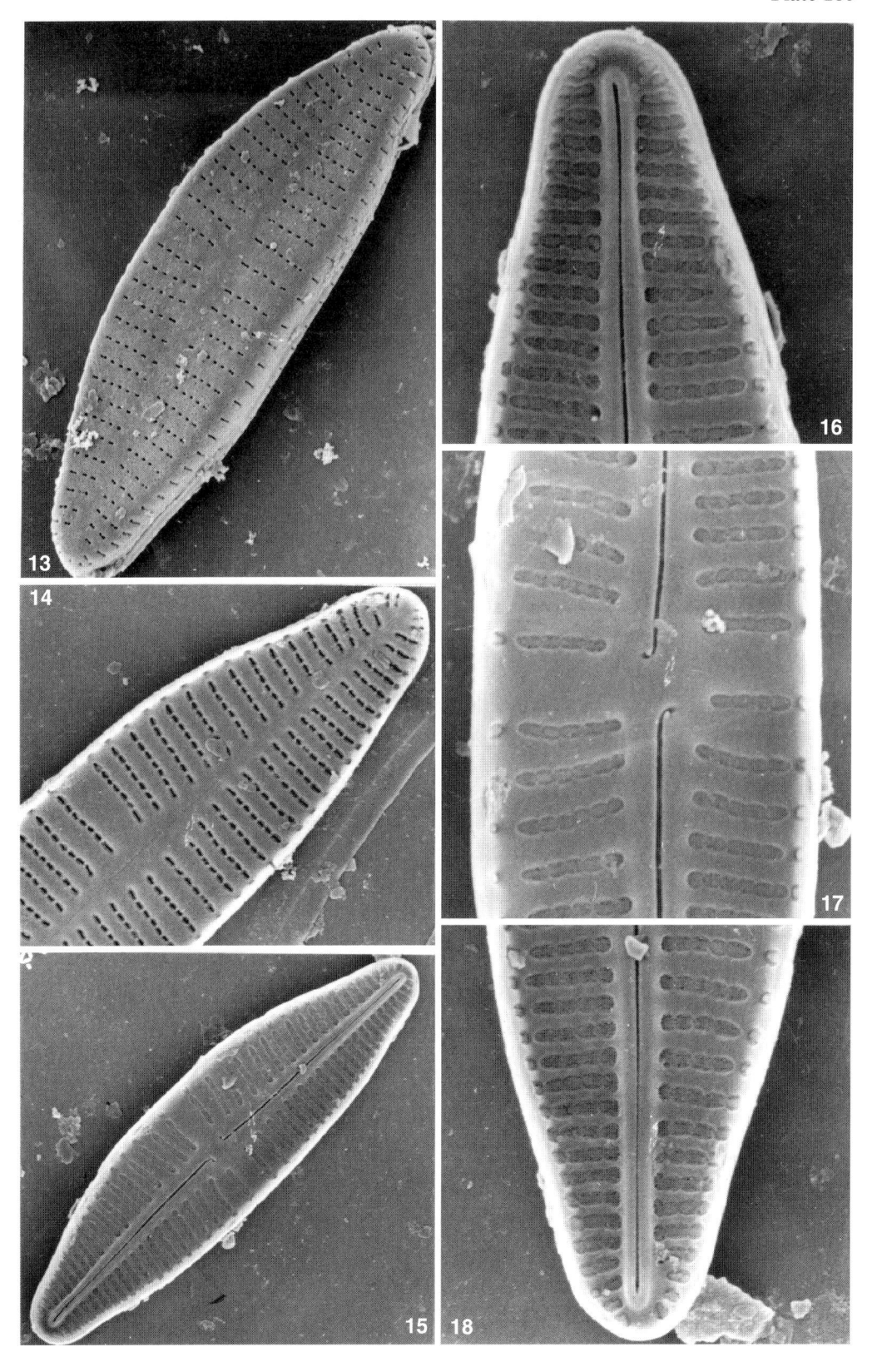

Plate 161 *Achnanthidium saplophilum* (H. Kobayasi & Mayama) Round & Bukhtiyarova

Figs 1, 2, 5-7.   LM. ×2000.

Figs 3, 4.   Drawings. ×2000.

Figs 8-13.   TEM. ×6000.

Figs 14-16.   SEM.

Fig. 17.   TEM. ×39000.

Figs 1, 3, 5, 7-10.   Raphid valves.

Figs 2, 4, 6, 11-13.   Araphid valves.

Fig. 14.   External view of the raphid valve showing the areolation and the swollen central endings of the raphe branches. ×9000.

Fig. 15.   External view of the araphid valve showing the areolation and the narrow lanceolate axial area. ×9000.

Fig. 16.   Enlarged external view of the raphid valve end showing almost straight terminal ending of the raphe. ×20000.

Fig. 17.   Enlargement of the areolar occlusions (hymenate type) with perforations arrayed decussately.

Localities

Figs 1, 2, 6-8, 11.   Minamiasa-kawa (Minamiasa River), Tokyo. Figs 1, 2. Holotype. Figs 3, 4. Iconotype.

Fig. 5.   Sen-gawa (Sen River), Setagaya, Tokyo.

Figs 9, 12, 17.   Zanbori-gawa (Zanbori River), Tachikawa, Tokyo.

Figs 10, 13.   Yanase-gawa (Yanase River), Kiyose, Tokyo.

Fig. 14.   Funada-gawa (Funada River), Hachioji, Tokyo.

Fig. 15.   Hatsusawa-gawa (Hatsusawa River), Hachioji, Tokyo.

Fig. 16.   No-gawa (No River), Setagaya, Tokyo.

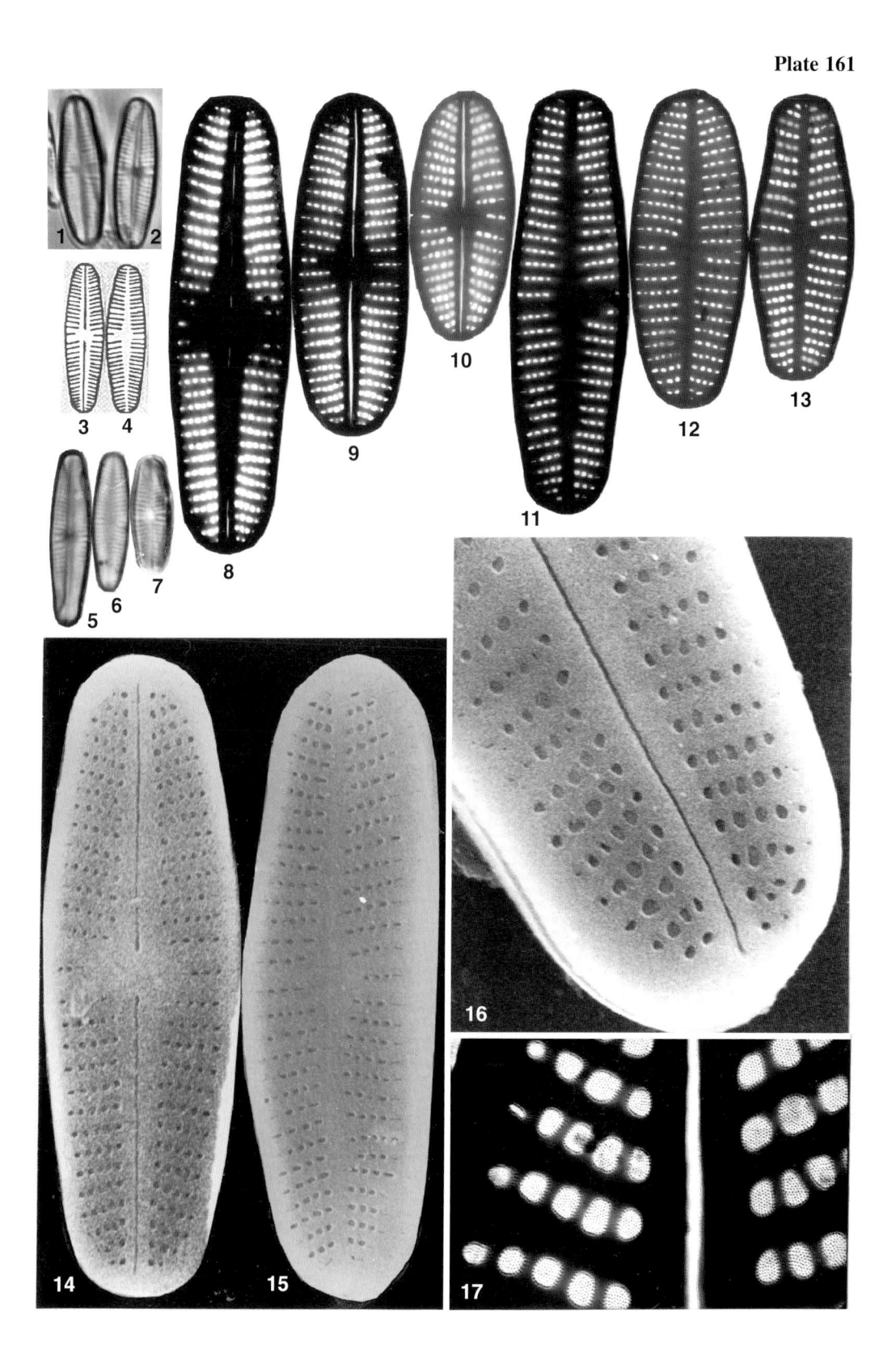

Plate 161

Plate 162 *Achnanthidium subhudsonis* (Hustedt) H. Kobayasi comb. nov.

Figs 1, 2.   Drawings. ×2000.
Figs 3-12.   LM. ×2000.
Figs 13-16.   SEM.

Figs 1, 3, 5, 7, 9, 11, 13, 15, 16.   Raphid valves.
Figs 2, 4, 6, 8, 10, 12.   Araphid valves.
Fig. 13.   External view showing a valve with terminal raphe fissures, which hook in the same direction. ×8000.
Fig. 14.   Partly broken frustule showing the deep mantle of the epivalve. ×10000.
Fig. 15.   Fractured internal valve showing a section of the valve mantle. Note each areola bearing a rica (arrow) near the outer opening of the chamber. ×20000.
Fig. 16.   Internal view of the valve center showing that both raphe endings deflect to the same side. ×20000.

Localities
Figs 3, 4.   Yuno-ko (Yu Pond), Nikko, Tochigi.
Figs 5, 6.   Chigonosawa (Chigosawa Creek), Kisofukushima, Nagano.
Figs 7-12, 15, 16.   Nakazawa (a tributary of the Sakura-gawa), Kowada, Tsukuba, Ibaraki.
Fig. 14.   Chikuma-gawa (Chikuma River), Kawakami, Nagano.

Plate 162

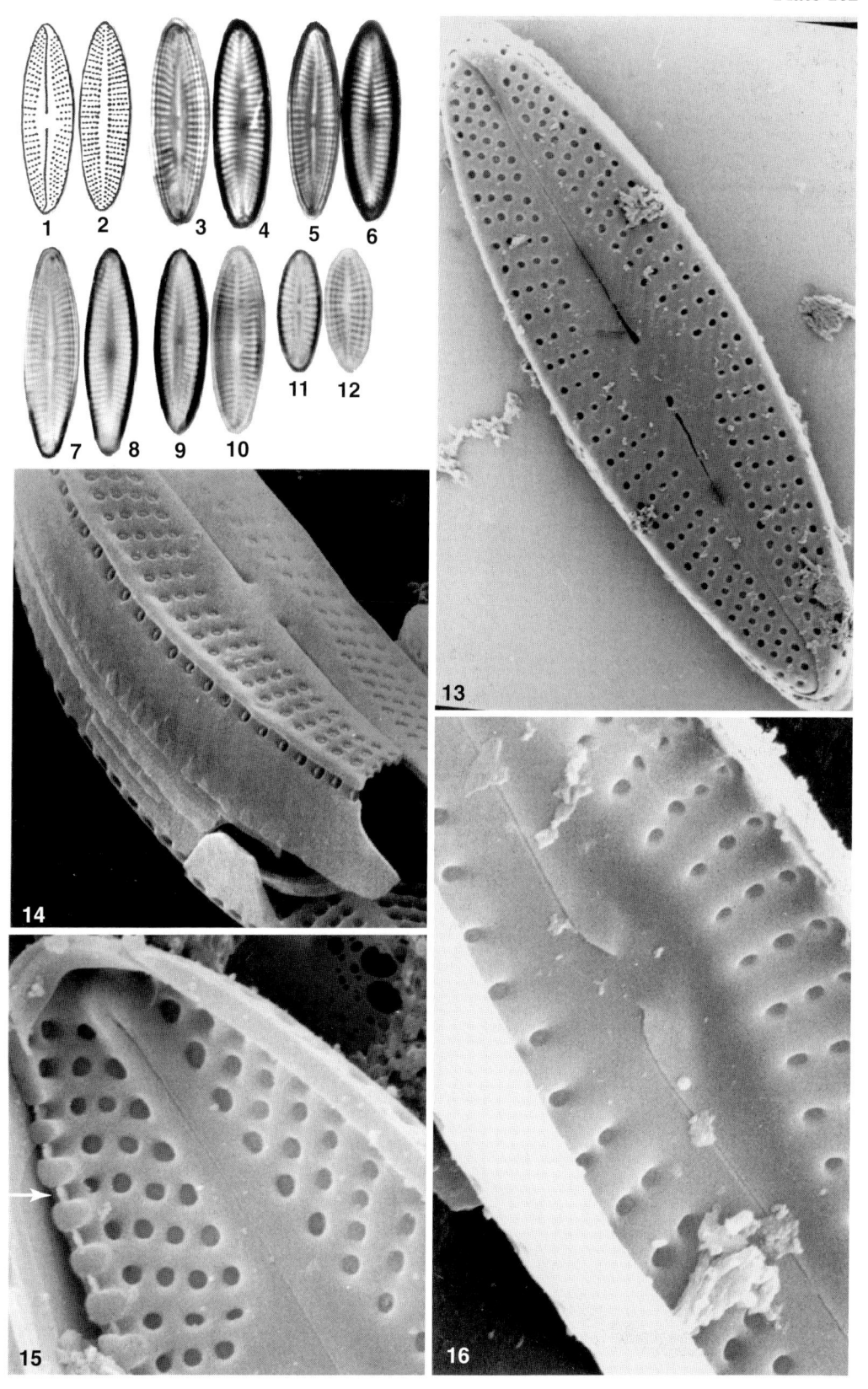

Plate 163 *Achnanthidium subhudsonis* (Hustedt) H. Kobayasi comb. nov.

Figs 17, 18.   Drawings.
Figs 19-22.   SEM.

Fig. 19.   Oblique view of the frustule showing a hypovalve (araphid) and an epitheca with the deep valve mantle accompanied by four split bands. ×7000.
Fig. 20.   Details of the frustule apex. The 1st and the 3rd bands show their open ends, and alternatly, the 2nd and the 4th closed. ×15000.
Fig. 21.   External view of the araphid valve showing areolae with ricae, which are domed and project from the inward side. ×10000.
Fig. 22.   External view of the araphid valve showing an obvious ridge on the valve shoulder and a vestigial raphe. ×9000.

Locality

Figs 19-22.   Nakazawa (a tributary of the Sakura-gawa), Kowada, Tsukuba, Ibaraki.

Plate 163

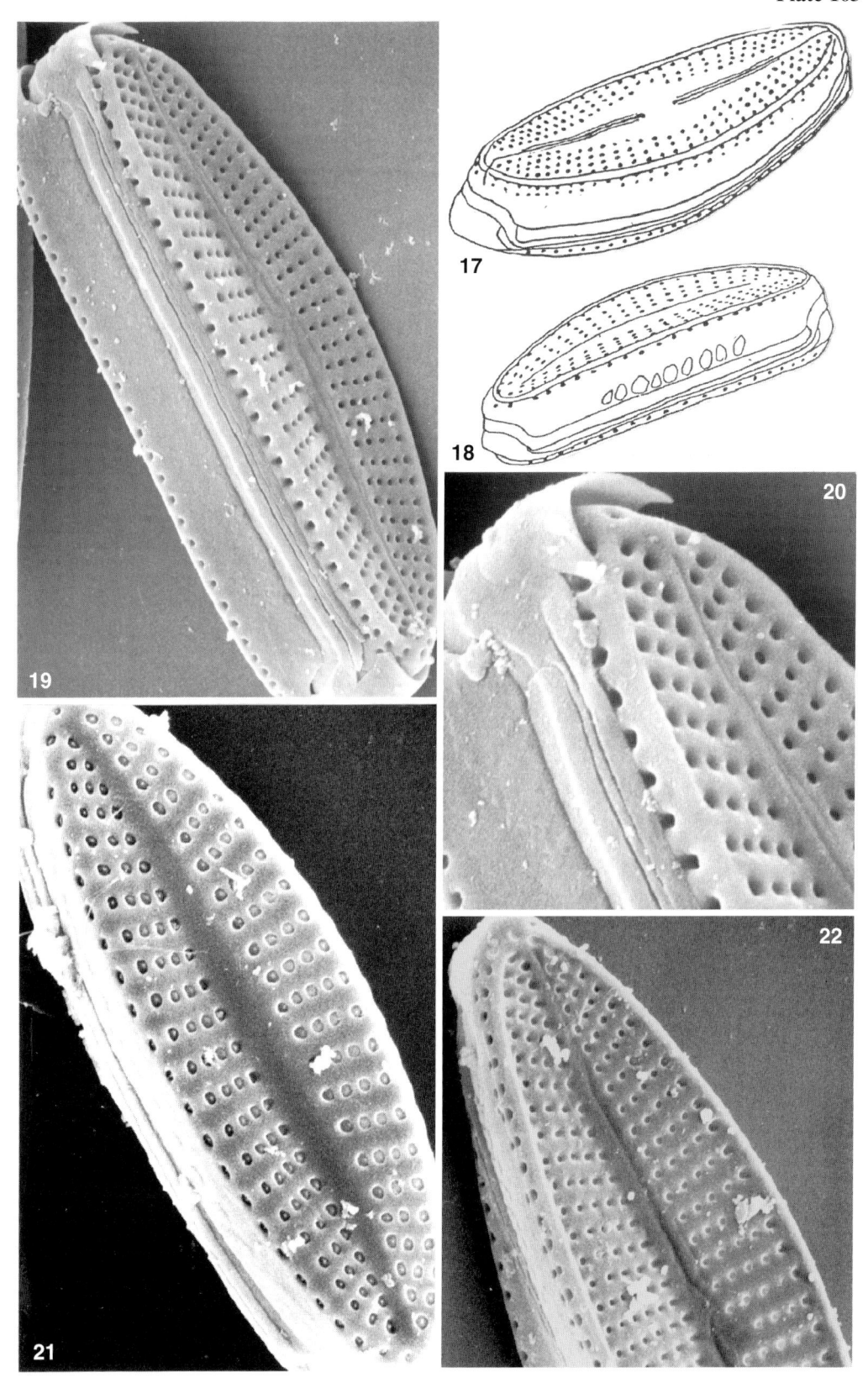

Plate 164 *Lemnicola hungarica* **(Grunow) Round & Basson**

Figs 1, 2, 5-12.   LM.   ×2000.
Figs 3, 4.   Drawings.   ×2000.
Figs 13-15.   SEM.

Figs 1, 3, 5, 7, 8, 13, 15.   Raphid valves.
Figs 2, 4, 6, 9, 14.   Araphid valves.
Figs 10-12.   Living cells. Each has a single plate-like plastid.
Fig. 13.   External view of the raphid valve. Note the terminal fissures curving in opposite directions.   ×3000.
Fig. 14.   Internal view of the araphid valve showing a narrower central area than in the raphid valve.   ×3000.
Fig. 15.   Internal view of the raphid valve showing thickened axial and central areas.   ×3000.

Locality
Figs 1, 2, 5-7, 10-15.   A glass tank for fish at the H. Kobayasi home, Koganei, Tokyo.

Plate 164

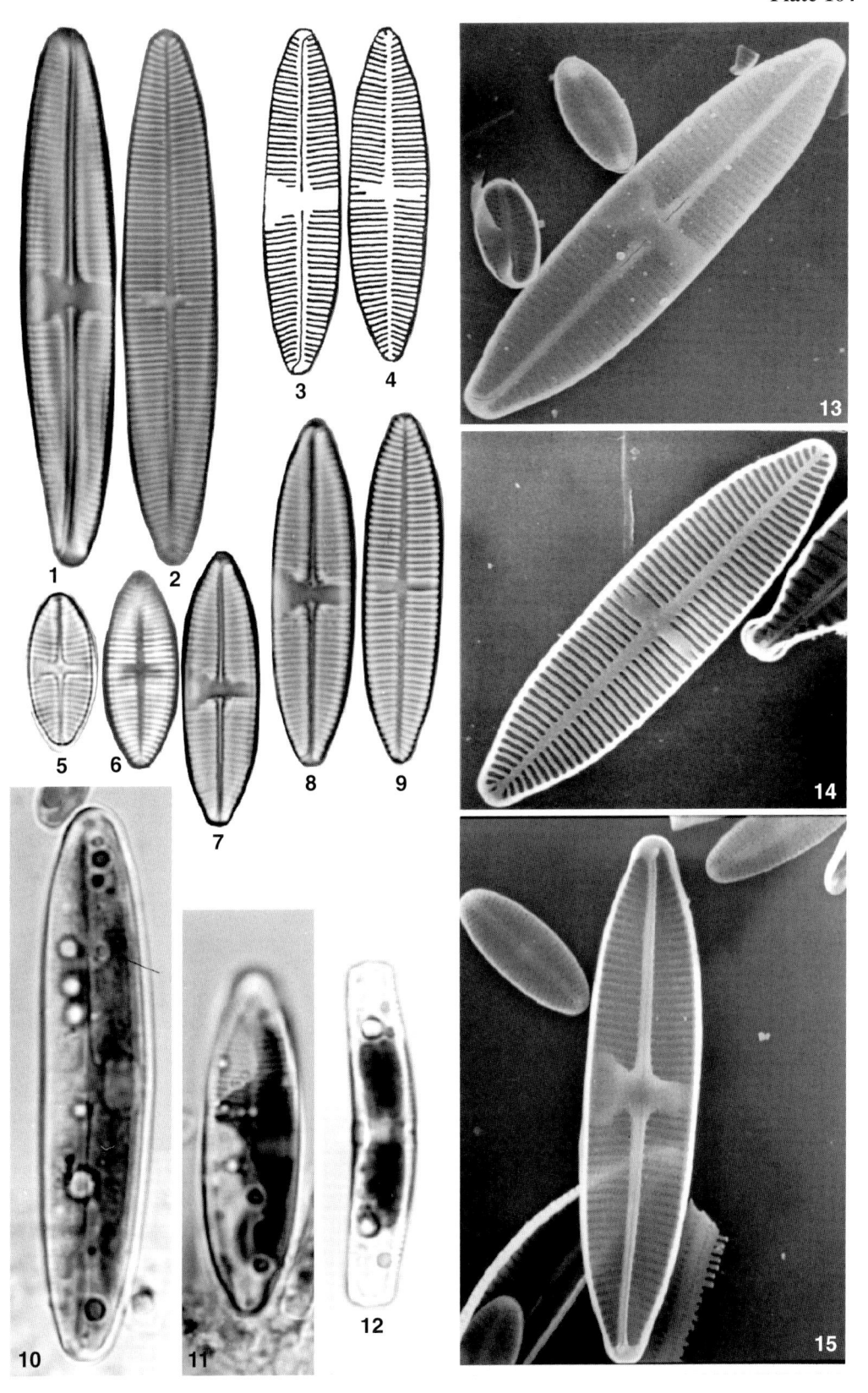

Plate 165 *Lemnicola hungarica* **(Grunow) Round & Basson**

Figs 16-22.   SEM.

Figs 16-20, 22.   Raphid valves.
Fig. 21.   Araphid valve.
Fig. 16.   External view of the valve end showing striae, each of which is composed of a double row (partly single row) of areolae. ×8000.
Fig. 17.   Enlarged central area showing central raphe endings on small, depressed areas. ×20000.
Fig. 18.   External view of the valve center with an asymmetric central area. ×8000.
Fig. 19.   Internal view of the valve end showing the raphe fissure, which terminates in a helictoglossa. ×10000.
Fig. 20.   Inner central raphe endings deflected to opposite side. ×20000.
Fig. 21.   Internal view of the araphid valve showing well developed interstriae. ×50000.
Fig. 22.   Enlarged internal surface of the raphid valve showing occluded areolae. ×50000.

Locality
Figs 16-22.   A glass tank for fish at the H. Kobayasi home, Koganei, Tokyo.

Plate 165

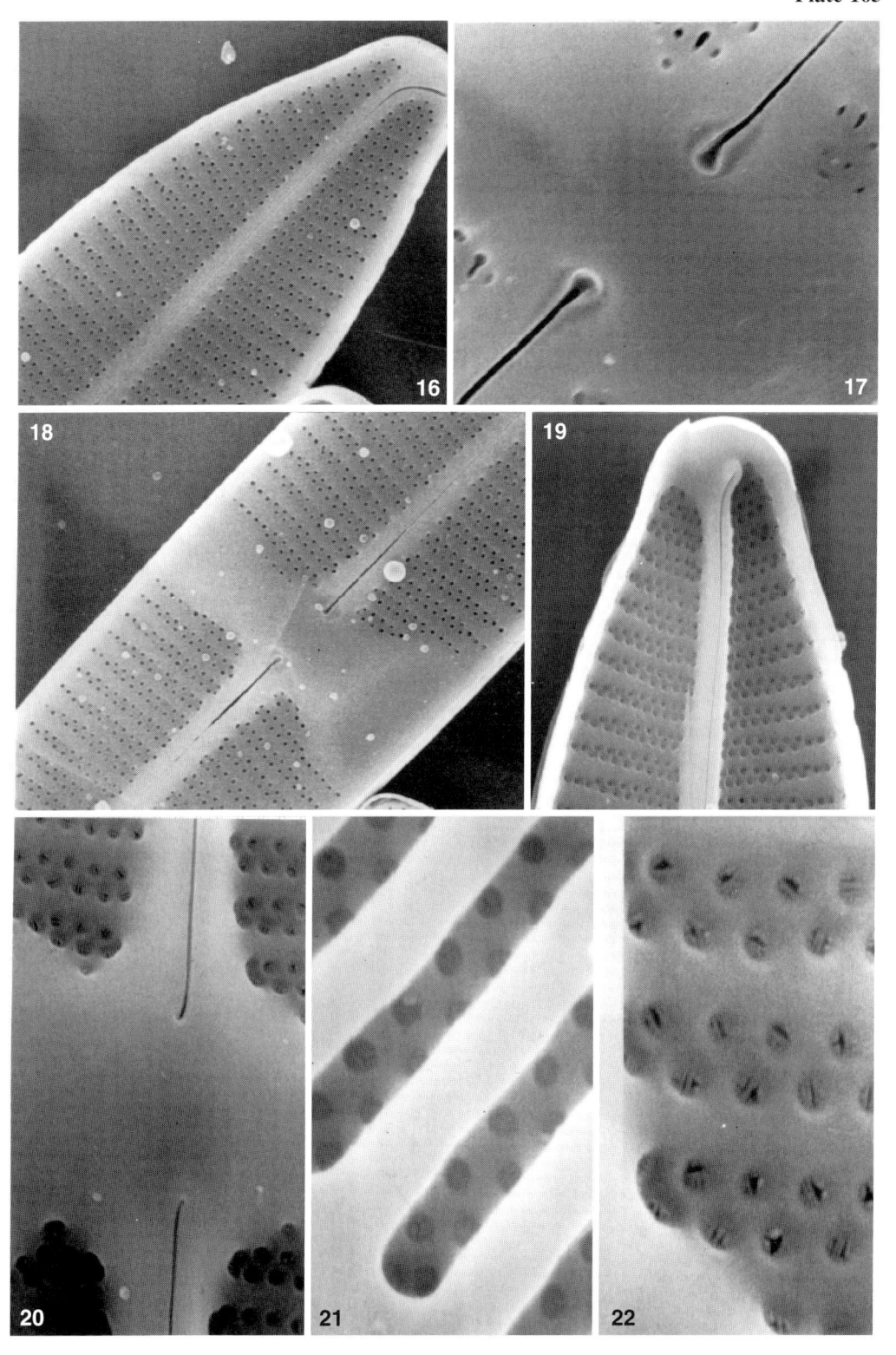

Plate 166 *Planothidium frequentissimum* (Lange-Bertalot) Lange-Bertalot

Figs 1-8.　LM.　×2000.
Figs 9-13.　SEM.

Figs 1, 3, 5, 7, 9, 11.　Raphid valves.
Figs 2, 4, 6, 8, 10, 12, 13.　Araphid valves.
Fig. 9.　Internal view of a raphid valve showing alveolate striae and raphe branches, which end almost straight centrally.　×8000.
Fig. 10.　Internal oblique view of the araphid valve showing alveoli and a hood developed in the horseshoe-shaped area.　×8000.
Fig. 11.　External view of the raphid valve showing swollen the central endings of the raphe branches and each alveolate stria composed of 3-5 rows of areolae.　×10000.
Fig. 12.　External view of the araphid valve showing the horseshoe-shaped area.　×8000.
Fig. 13.　Enlargement of the inner valve center with the well developed hood.　×20000.

Localities
Figs 1-4, 7-13.　A river in Taroko Gorge, Taroko National Park, Taiwan.
Figs 5, 6.　Nonomi mire, Iiyama/Sakae, Nagano.

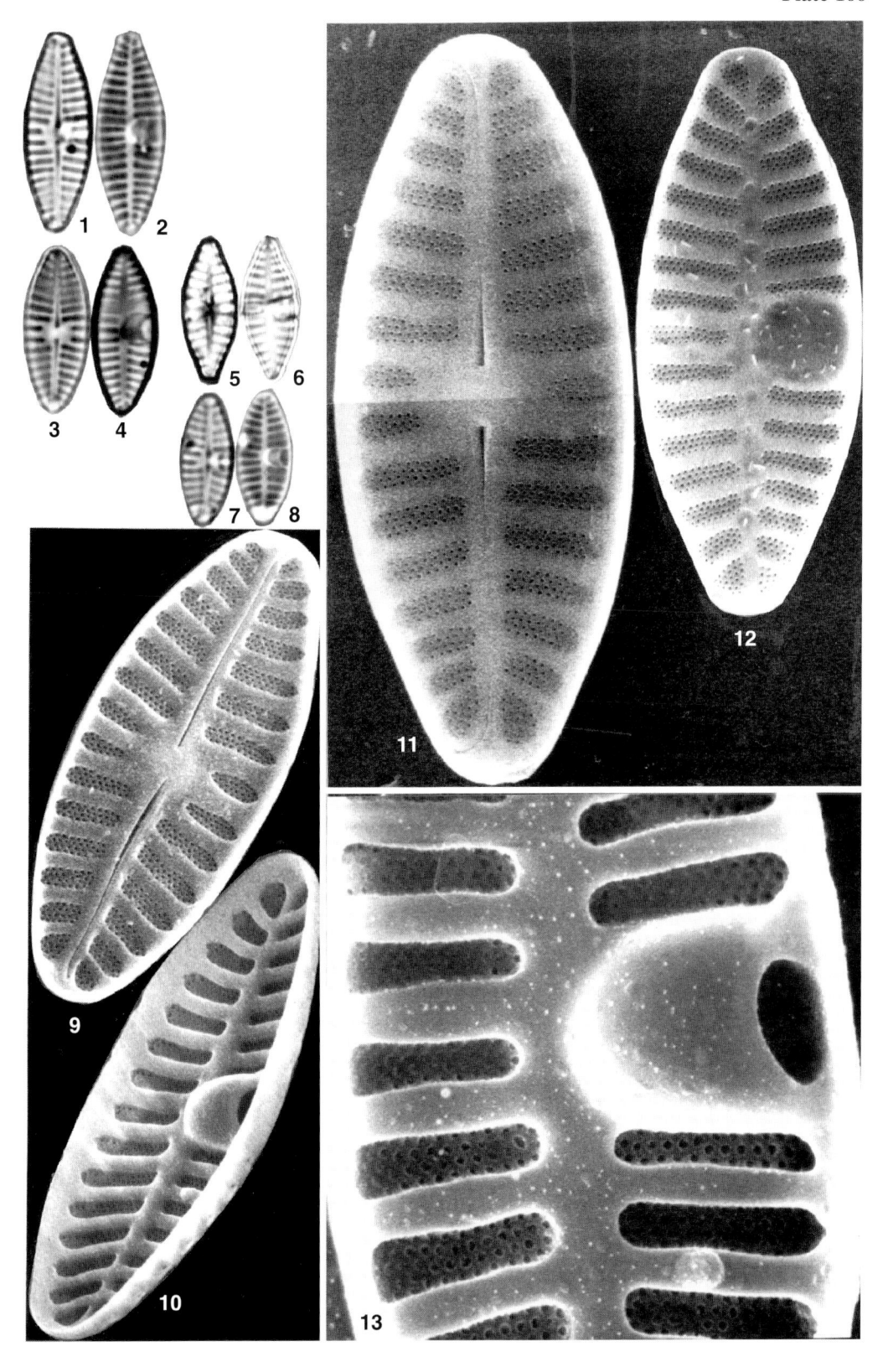

Plate 166

479

Plate 167 *Planothidium lanceolatum* (Brébisson ex Kützing) Lange-Bertalot

Figs 1, 2.   Drawings. ×2000.
Figs 3-8.   LM. ×2000.
Figs 9-14.   SEM.

Figs 1, 3, 5, 6.   Raphid valves.
Figs 2, 4, 7, 8.   Araphid valves.
Fig. 9.   External view of an araphid valve. ×3000.
Fig. 10.   Internal view of an araphid valve. ×2000.
Fig. 11.   External view of the raphid valve showing the raphe with central pores and polar bends and the striae composed of multiple rows of areolae. ×10000.
Fig. 12.   Enlargement of the same specimen as that in Fig. 10. showing alveoli, in which double rows of areolae with occlusions are aligned. ×10000.
Fig. 13.   External view of the araphid valve showing stria composed of triple rows of areolae. ×5000.
Fig. 14.   Internal oblique view of the araphid valve showing a depressed horseshoe-shaped area. ×5000.

Localities

Figs 3, 4, 6, 7.   Sakawa-gawa (Sakawa River), Ayusawa, Kanagawa.
Fig. 5.   Chigonosawa (Chigonosawa Creek), Nagano.
Fig. 8.   A spring in Kawagoe, Saitama.
Fig. 9.   Ara-kawa (Ara River), Saitama.
Figs 10, 12, 14.   Kari-gawa (Kari River), Kanagawa.
Fig. 11.   Hi-numa (Lake Hi), Ibaraki.
Fig. 13.   Chikuma-gawa (Chikuma River), Kawakami, Nagano.

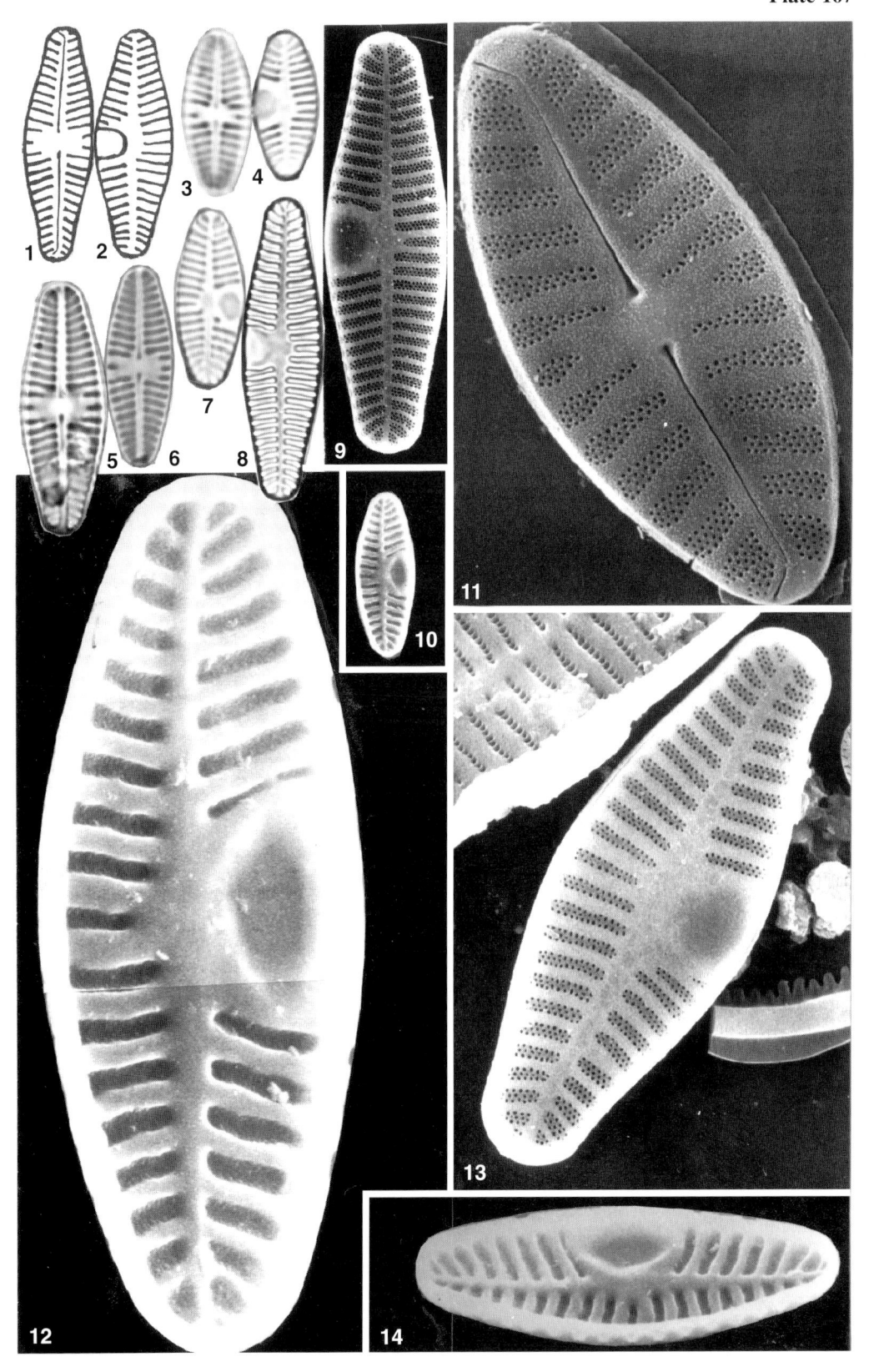

Plate 167

Plate 168 *Planothidium septentrionale* (Østrup) Round & Bukhtiyarova ex U. Rumrich *et al.*

Figs 1, 2. *Planothidium delicatulum* (Kützing) Round & Bukhtiyarova. Lectotype of *Achnanthidium delicatulum* Kützing, desig. by Lange-Bertalot.

Figs 3-12. *P. septentrionale*.

Figs 1, 2, 5-8. LM. ×2000.
Figs 3, 4. Drawings. ×2000.
Figs 9-12. SEM.

Figs 1, 3, 5, 6. Raphid valves.
Figs 2, 4, 7, 8. Araphid valves.
Fig. 9. Enlarged internal view of the raphid valve showing the alveoli with triple or quadruple rows of areolae. ×10000.
Fig. 10. Internal oblique view showing a bend in the raphid valve. ×5000.
Fig. 11. External view of the araphid valve showing the concave araphid valve. ×8000.
Fig. 12. Enlargement of the external araphid valve face showing the outer openings of areolae distributed entirely except on the axial area. ×20000.

Localities

Figs 1, 2. BM 26545, V. H. Type No.234, Angleterre (BM).
Figs 5-12. U-kawa (U River), Kyoto.

Plate 168

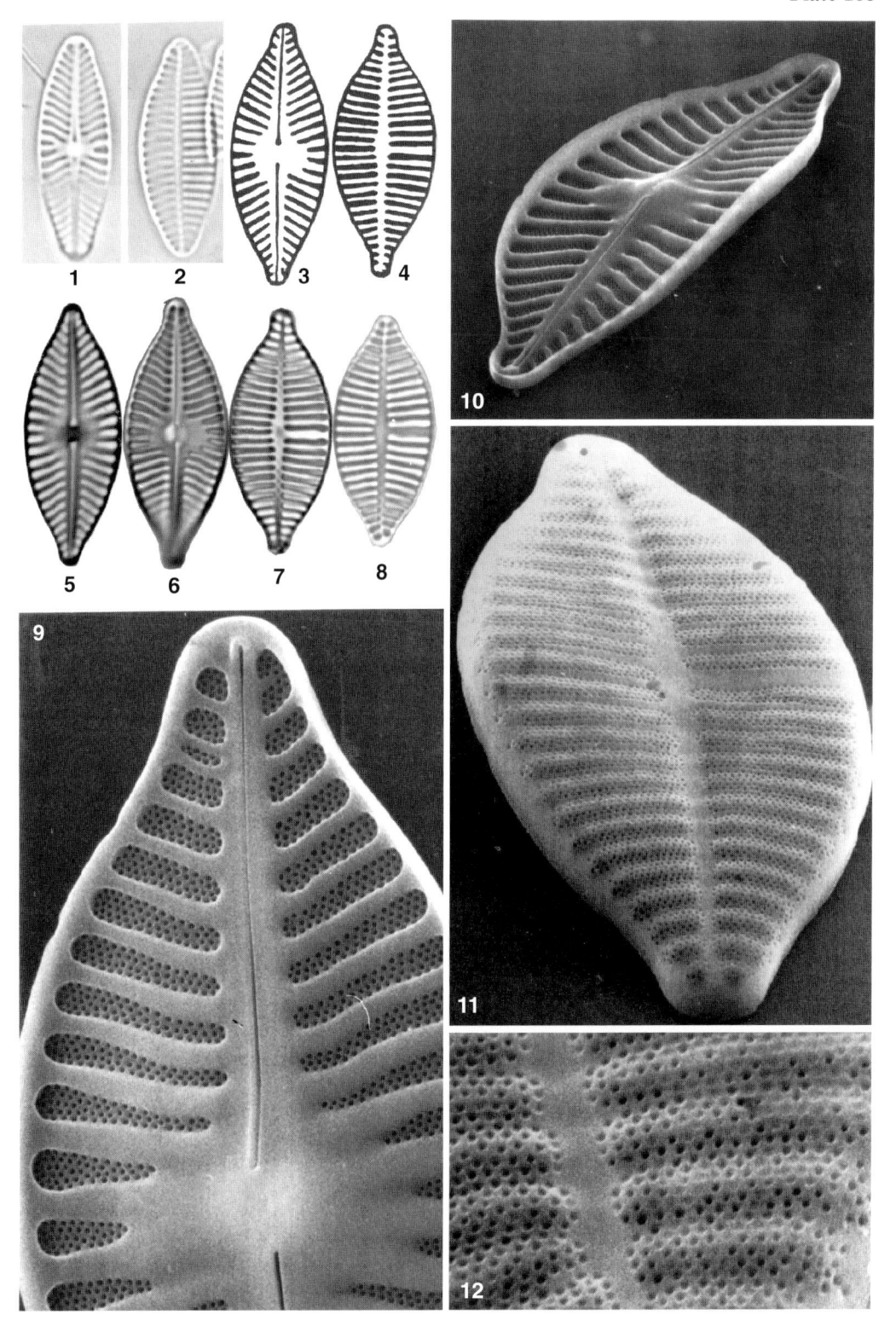

Plate 169 *Psammothidium helveticum* (Hustedt) Bukhtiyarova & Round

Figs 1-4, 7-12.   LM. ×2000.
Figs 5, 6.   Drawings. ×2000.
Figs 13-17.   SEM.

Figs 1, 3, 5, 7, 9, 11.   Raphid valves.
Figs 2, 4, 6, 8, 10, 12.   Araphid valves.
Fig. 13.   External oblique view of the raphid valve showing the striation and the raphe fissure, which opens to the shallow ditch. ×5000.
Fig. 14.   Internal oblique view of the araphid valve showing lightly thickened axial area. ×5000.
Fig. 15.   External oblique view of the frustule with the concave araphid valve. ×5000.
Fig. 16.   External oblique view of the frustule with the slightly convex raphid valve showing terminal fissures, which bent to opposite sides of each other. ×10000.
Fig. 17.   Enlargement of the valve end interior. ×30000.

Localities

Figs 1-4.   Holotype slide. Coll. Hustedt, Ma1/12, Kl Flüelasee 135, Grund, Davos, Schweitz (BRM).
Figs 7, 8.   A paddy field in Ojiya, Niigata.
Figs 9-12.   Kamega-ike (Kame Pond), Mt. Norikura, Gifu.
Figs 13, 15.   Kamaga-ike (Kama Pond), Kirigamine, Nagano.
Figs 14, 17.   Chikuma-gawa (Chikuma River), Kawakami, Nagano.
Fig. 16.   Kamega-ike (Kame Pond), Mt. Norikura, Gifu.

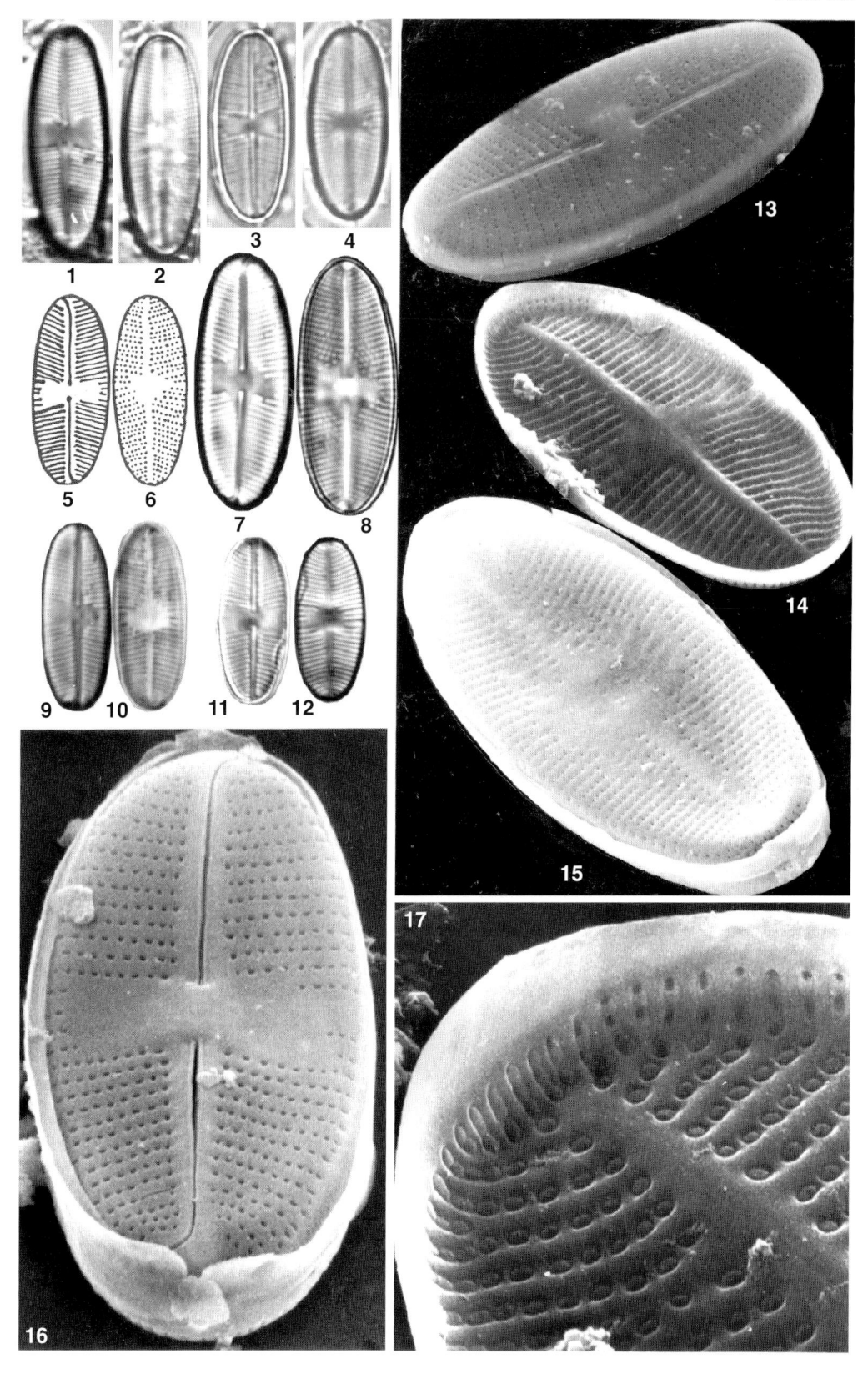

Plate 169

Plate 170 *Psammothidium hustedtii* (Krasske) Mayama

Figs 1-4.   LM. ×2000.
Figs 5, 6.   Drawings. ×2000.
Figs 7-13.   SEM.

Figs 1, 3, 5, 7, 9, 11, 12.   Raphid valves.
Figs 2, 4, 6, 8, 10, 13.   Araphid valves.
Fig. 7.   External oblique view of a raphid valve. ×6000.
Fig. 8.   External oblique view of an araphid valve showing reticulate relief in the axial area. ×6000.
Fig. 9.   Internal oblique view of the raphid valve showing areolae aligned in shallow troughs. ×6000.
Fig. 10.   Internal oblique view of the araphid valve showing the smooth surface of the axial area. ×6000.
Fig. 11.   Enlarged inner valve center. The central endings of the raphe hook in directions opposite to each other. ×13000.
Fig. 12.   External view of the raphid valve showing striae composed of one or two rows of areolae and outer raphe fissures, which open to the ditches near the valve center. ×9000.
Fig. 13.   External view of the araphid valve showing striae composed of one or two rows of areolae. ×9000.

Locality
Figs 1-4, 7-13.   Ohchi-gawa (Ohchi River), Ohtaki, Saitama.

Plate 170

Plate 171 *Psammothidium marginulata* (Grunow) Bukhtiyarova & Round

Figs 1-8.   LM. ×2000.
Figs 9, 10.   TEM. ×4000.
Figs 11, 12.   Drawings. ×2000.
Figs 13-17.   SEM.

Figs 1, 2, 5, 7, 9, 11.   Raphid valves.
Figs 3, 4, 6, 8, 10, 12.   Araphid valves.
Fig. 13.   External oblique view of a frustule showing the striation of the raphid valve. Note both terminal fissures end straight. ×8000.
Fig. 14.   Girdle view of the frustule with the convex raphid hypo-valve. ×8000.
Fig. 15.   Internal view of the raphid valve showing the central endings of raphe fissures and the occluded areolae. ×20000.
Fig. 16.   Internal oblique view of the raphid valve. The valve plane is not flat but concave. ×8000.
Fig. 17.   External oblique view of the araphid valve with a rhomboid axial area. ×8000.

Localities
Figs 1-4.   Holotype slide. Coll. Grunow, 2024. Memerutungen, Norwegen (W).
Figs 5, 6.   Gongen-ike (Gongen Pond), Mt. Norikura, Gifu.
Figs 7, 8.   Gono-ike (Go Pond), Mt. Norikura, Gifu.
Figs 9, 10.   Kamega-ike (Kame Pond), Mt. Norikura, Gifu.
Figs 13-17.   Kamaga-ike (Kama Pond), Kirigamine, Nagano.

Plate 171

Plate 172 *Psammothidium montanum* (Krasske) Mayama

Figs 1-4. LM. ×2000.
Figs 5-11, 13. SEM.
Fig. 12. TEM.

Figs 1, 3, 5, 7, 9-11, 13. Raphid valves.
Figs 2, 4, 6, 8, 12. Araphid valves.
Fig. 5. Whole valve of a frustule showing the striation and raphe of the raphid valve, which has a convex profile. ×5000.
Fig. 6. Whole view of a frustule showing the araphid valve with reticulated relief on the surface. ×8000.
Fig. 7. Whole view of a frustule showing straight outer fissures, which open to shallow ditch near the valve center.
Fig. 8. Internal oblique view of the araphid valve showing a smooth, rhombic central area. ×8000.
Fig. 9. Internal oblique view of the raphid valve showing a somewhat thickened central area and the inner fissures of the raphe. ×8000.
Fig. 10. Details of the outer openings of areolae. ×80000.
Fig. 11. Details of the inner openings of areolae occluded by ricae. ×80000.
Fig. 12. Ricae with regularly scattered perforations in an araphid valve. ×80000.
Fig. 13. Details of the central endings of the inner fissures; they hook in opposite directions. ×20000.

Localitiy
Figs 1-13. Moss at the bank of the Ohchi-gawa (Ohchi River), Ohtaki, Saitama.

Plate 172

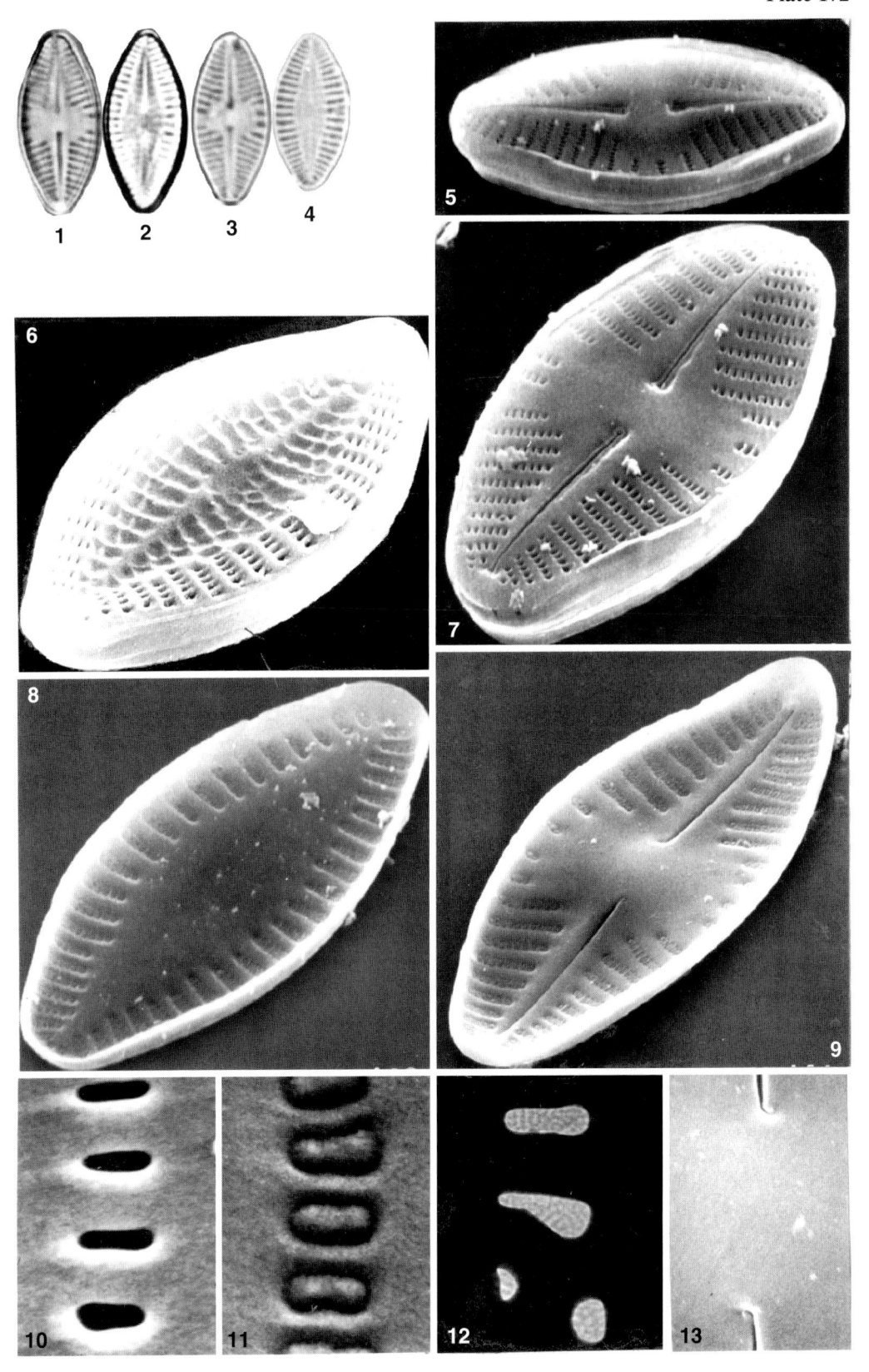

491

Plate 173 *Psammothidium subatomoides* (Hustedt) Bukhtiyarova & Round

Figs 1-4. LM. ×2000.
Figs 5-11. SEM.
Fig. 12. TEM.

Figs 1, 3, 5-7, 10, 11. Raphid valves.
Figs 2, 4, 8, 9. Araphid valves.
Fig. 5. Oblique view of a whole frustule with a partly broken epivalve mantle showing straight ends of the raphe both centrally and terminally. ×10000.
Fig. 6. Internal oblique view of the raphid valve showing both central raphe endings deflected to opposite sides and areolae occluded by ricae. ×10000.
Fig. 7. External oblique view of the valve showing a convex profile. ×10000.
Fig. 8. External oblique view of the araphid theca showing simple openings of areolae in the face but elongate openings in the mantle. ×10000.
Fig. 9. Internal oblique view of the araphid valve showing vestigial raphe branches near both ends (arrows). ×10000.
Fig. 10. Outer openings of the areolae of a raphid valve. ×80000.
Fig. 11. Inner openings of the areolae of a raphid valve occluded by ricae. ×80000.
Fig. 12. Ricae with regularly scattered pores centrally and radiating slits peripherally. ×80000.

Localities

Figs 1, 2, 5-12. Oze-numa (Oze Pond), Oze, Gunma.
Figs 3, 4. Oh-numa (Oh Pond), Shiobara, Tochigi.

Plate 173

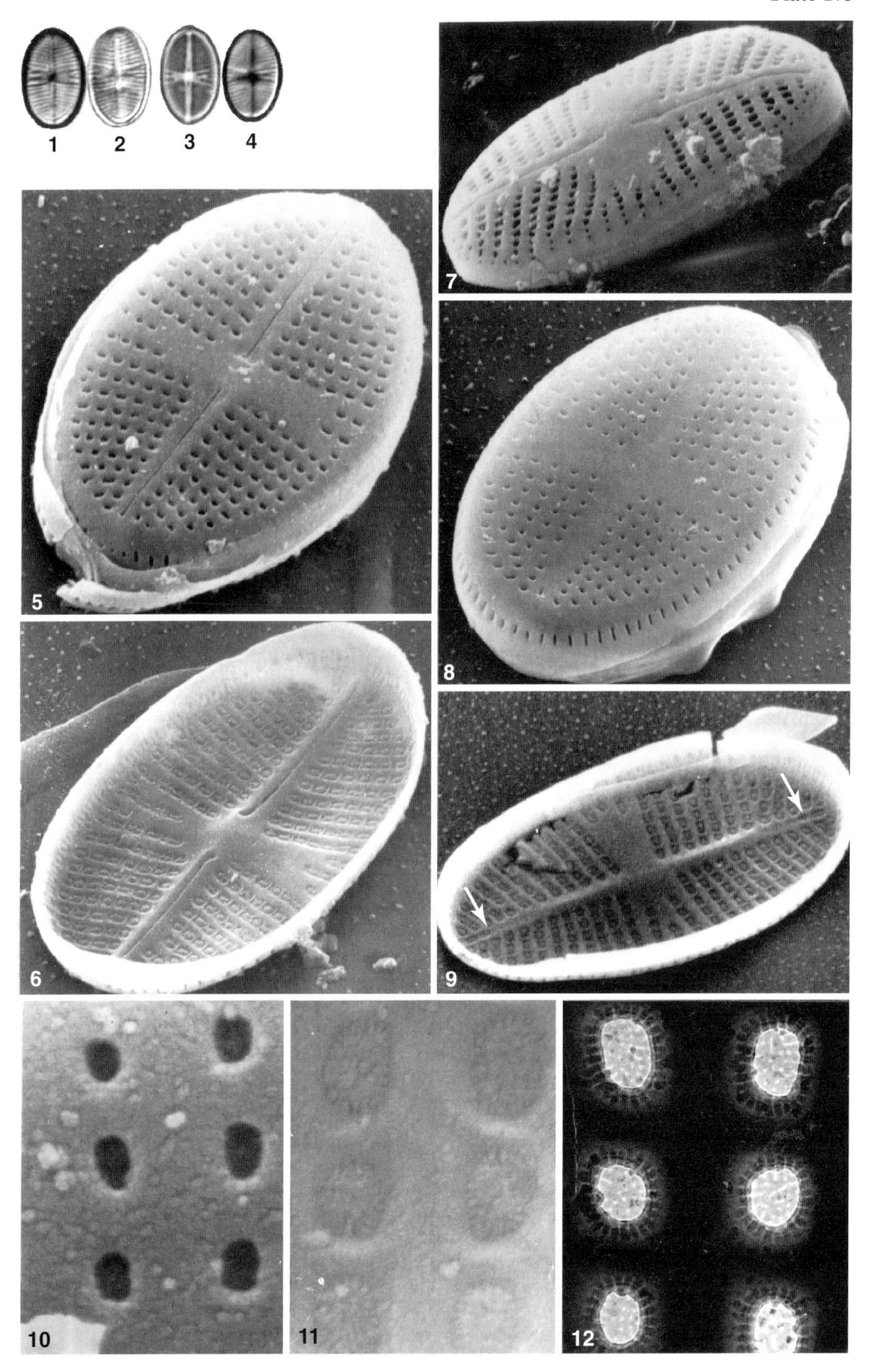

Plate 174 *Nupela lapidosum* (Krasske) Lange-Bertalot

Figs 1-6. Drawings. ×2000.
Figs 7-13. LM. ×2000.
Figs 14-18. SEM.

Figs 1, 3, 5, 7, 9, 16-18. Raphid valves.
Figs 2, 4, 6, 8, 10-15. Araphid valves.
Fig. 14. Oblique internal araphid valve showing a narrow lanceolate axial area, in which vestigial raphe branches (arrows) are recognizable. ×5000.
Fig. 15. Enlarged end of the araphid valve showing simple openings of the areolae. ×20000.
Fig. 16. Inner oblique view of the raphid valve with a partly broken valvo copula. ×10000.
Fig. 17. Enlarged valve end showing the raphe which terminates in a helictoglossa. ×20000.
Fig. 18. Enlarged valve center showing the raphe branches which end centrally and T-shaped. ×20000.

Localities
Figs 7, 8. Ara-kawa (Ara River), Chichibu, Saitama.
Figs 9, 10. A small stream in Sohrohsenen, Koganei, Tokyo.
Figs 11, 12. Kosuge-gawa (Kosuge River), Kosuge, Yamanashi.
Figs 13-18. Ryugaeshino-taki (Ryugaeshi Falls), Karuizawa, Nagano.

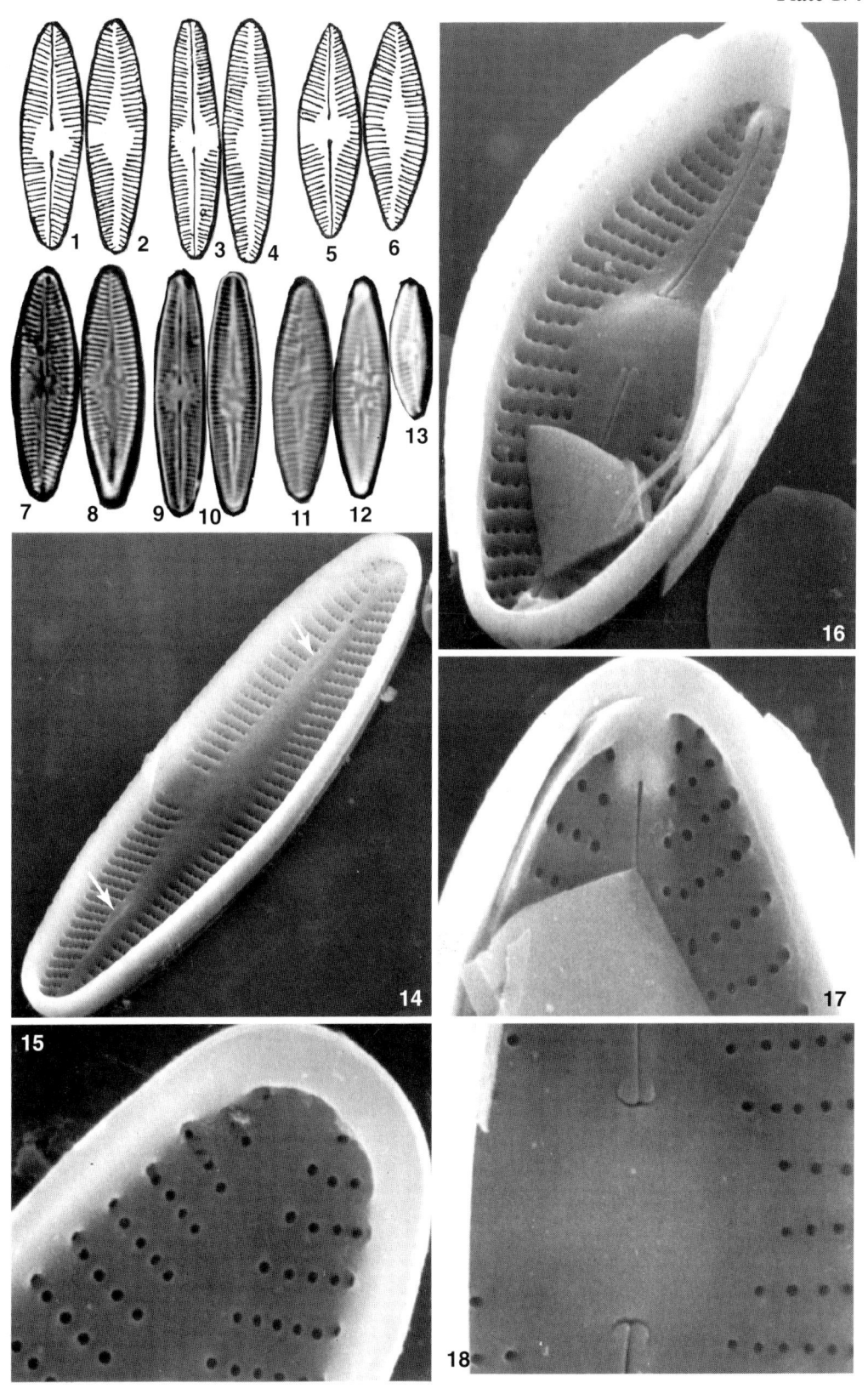

Plate 174

Plate 175 *Diatomella balfouriana* Greville

Figs 1-4.   LM. ×2000.
Fig. 5.   Drawing. ×2000.
Figs 6-9.   SEM.

Figs 1, 2.   Valve face.
Fig. 3.   Image focused on a septum.
Fig. 4.   Girdle view.
Fig. 6.   Oblique view of an epitheca end showing four bands. ×10000.
Fig. 7.   Oblique view of the epitheca showing less areolated valve face. ×10000.
Fig. 8.   Internal oblique view of the theca showing the deep mantle and cingulum, as well as the zigzag connection of both sides of the septum. ×10000.
Fig. 9.   Internal view of the theca showing three holes in the septa and the inner fissure of the raphe. ×8000.

Locality

Figs 1-4, 6-9.   On moss at Ryugaeshino-taki (Ryugaeshi Falls), Karuizawa, Nagano.

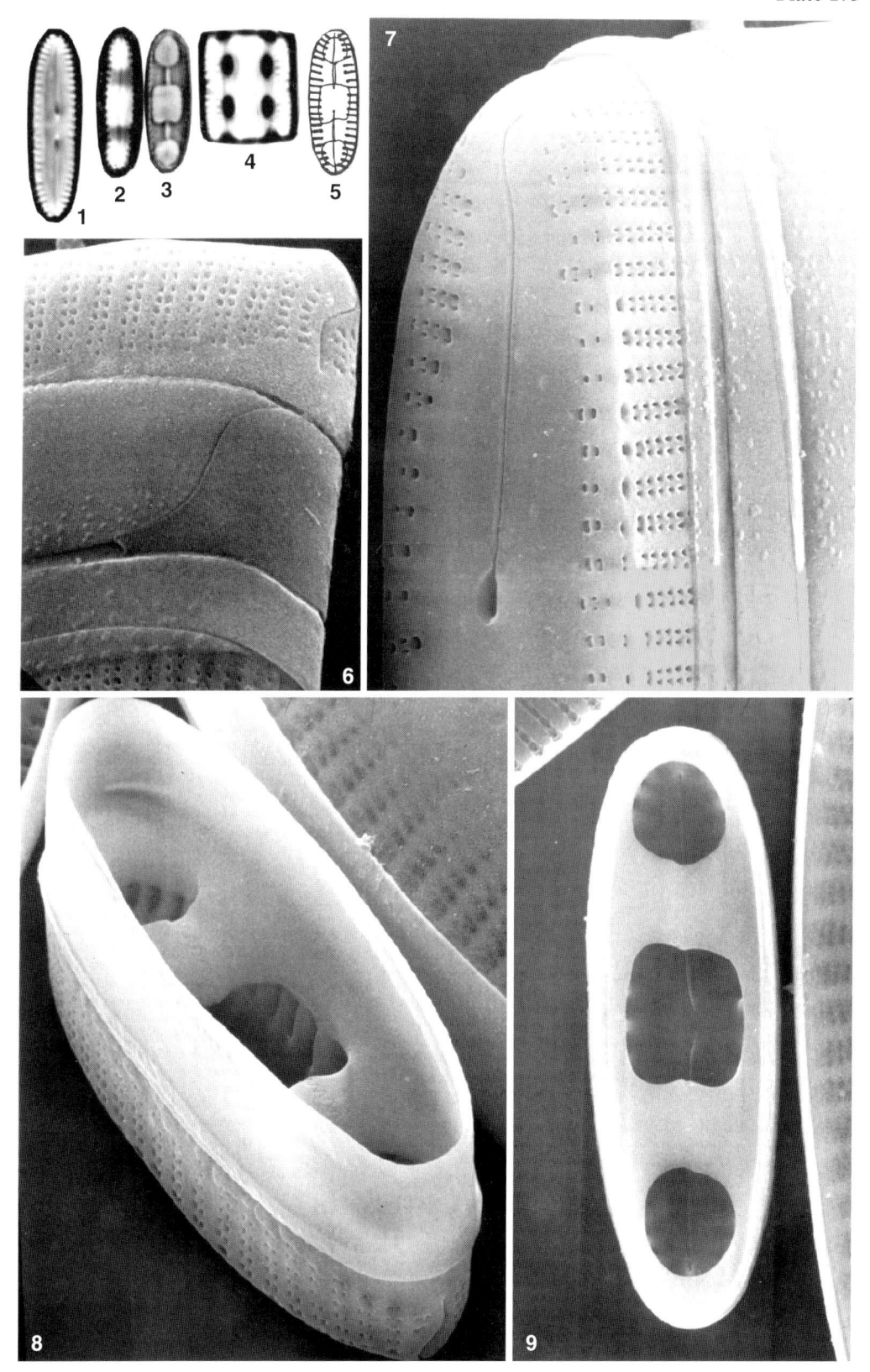

Plate 175

Plate 176　*Diploneis elliptica* (Kützing) Cleve

Fig. 1.　Drawing. ×2000.
Figs 2-4.　LM. ×2000.
Figs 5-7.　SEM.

Figs 1-4.　Valve view.
Fig. 5.　External view of whole valve. ×2000.
Fig. 6.　External view of areola with depressed cribra. ×30000.
Fig. 7.　Internal view of whole valve. ×2000.

Locality
Figs 2-7.　Moss in a small stream, Hannou, Saitama.

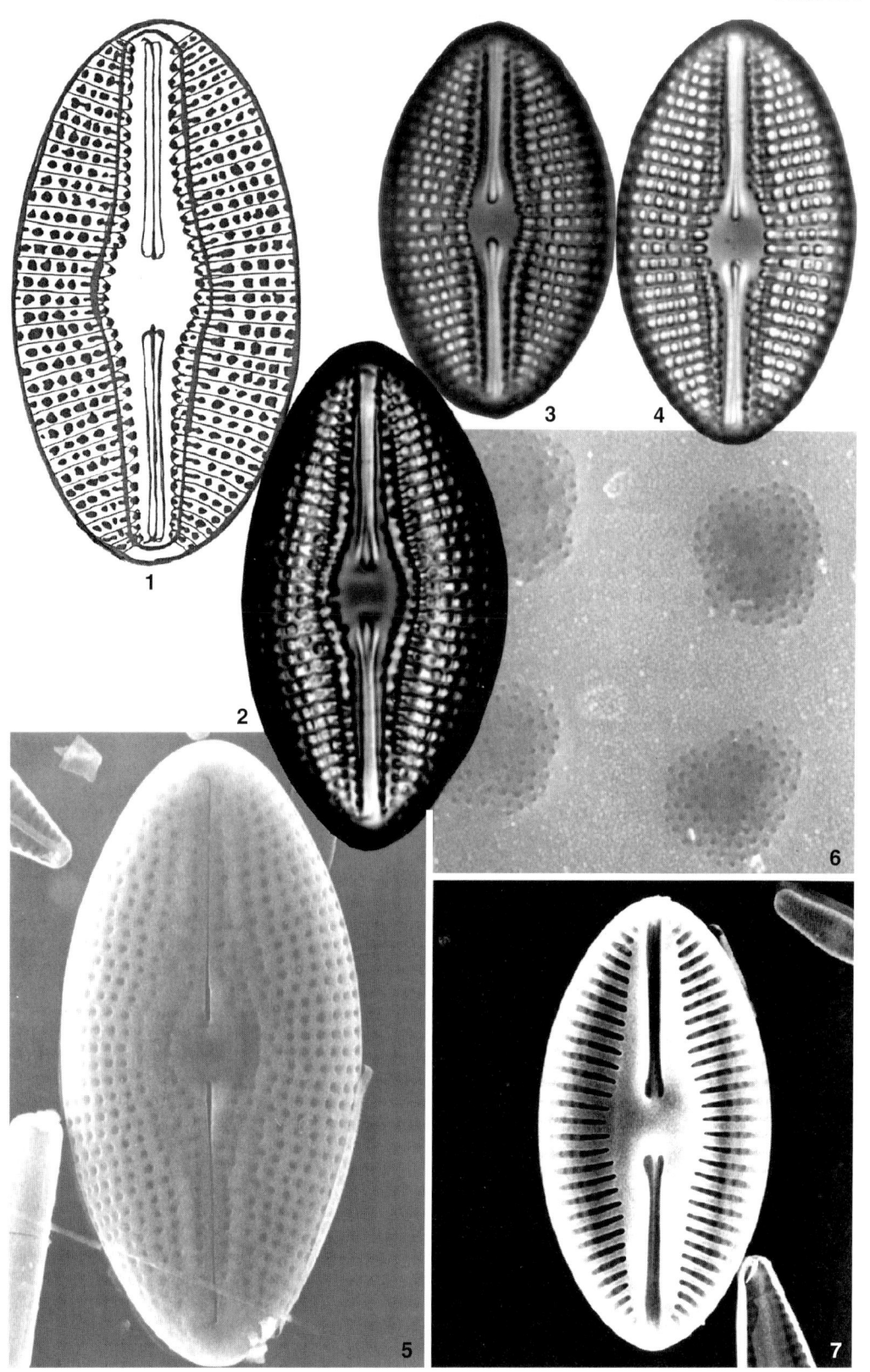

Plate 176

499

Plate 177 *Diploneis ovalis* (Hilse) Cleve

Figs 1-2. LM. ×2000.
Fig. 3. Drawing. ×2000.
Figs 4-8. SEM.

Figs 1-3. Valve view.
Fig. 4. External polar view of whole frustule showing terminal fissure and open end of valvocopula. ×10000.
Fig. 5. External view of valve showing areolae with cribra and filamentous ornamentation on valvocopula. ×20000.
Fig. 6. Internal view of whole valve. ×3000.
Fig. 7. External view of central area showing central fissures, areolae of longitudinal canal and striae. ×10000.
Fig. 8. Internal oblique view of broken valve showing transapical costae and inner walls of alveoli. ×20000.

Locality
Figs 1, 2, 4-8. Odanai-numa (Odanai Pond), Aomori.

Plate 177

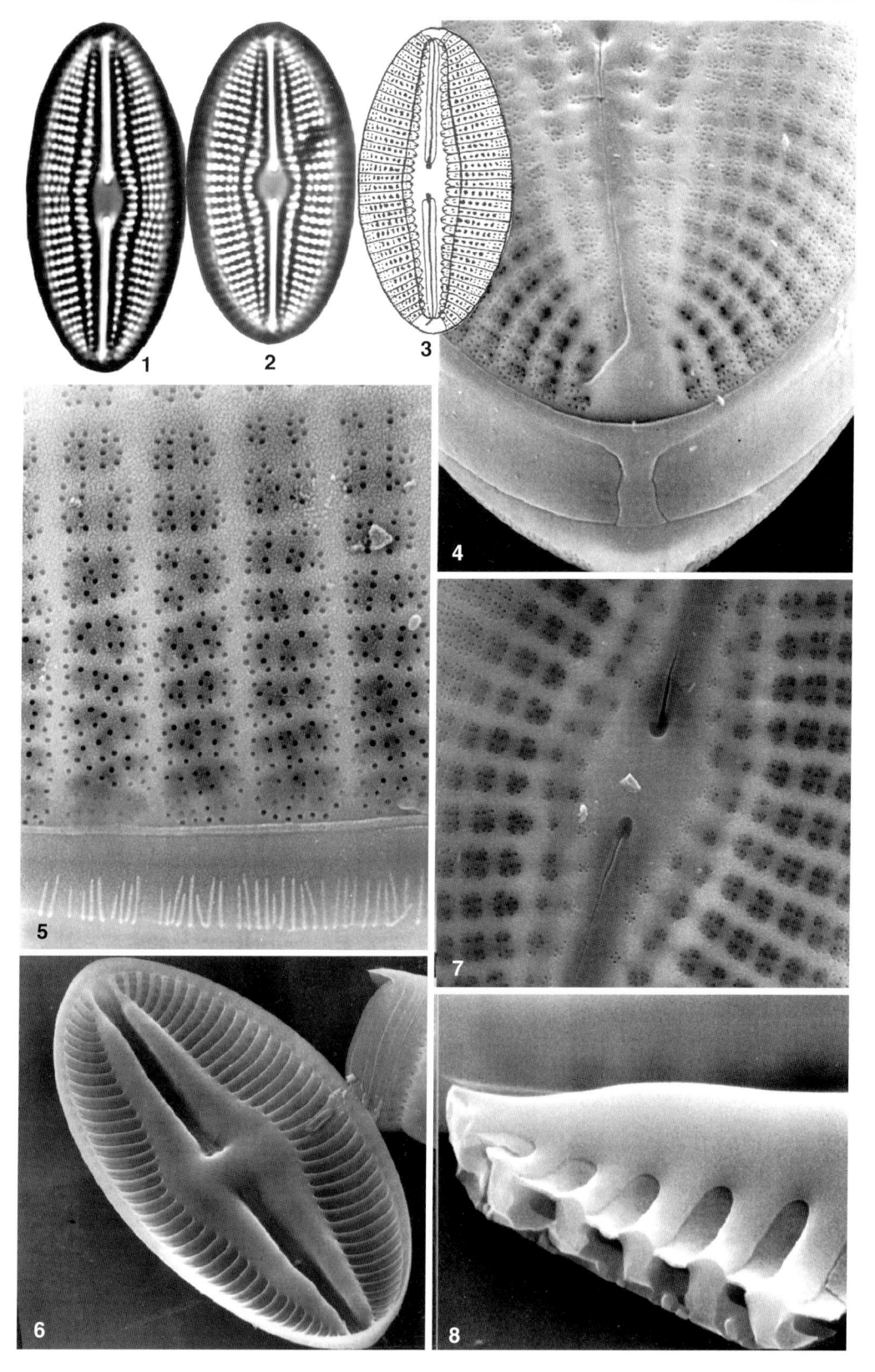

Plate 178 *Diploneis smithii* (Brébisson ex W. Smith) Cleve

Fig. 1. Drawing from paratype specimens designated by Ross. BM 35963. ×2000.
Figs 2-3. LM. ×2000.
Figs 4-7. SEM.

Fig. 2. BM 23457. *Navicula elliptica* W. Smith. Dorset (Pool Bay).
Fig. 3. Valve view.
Fig. 4. External view of valve center showing central fissures and striae with double rows of areolae. ×10000.
Fig. 5. External view of areolae with cribra. ×20000.
Fig. 6. External view of whole valve. ×5000.
Fig. 7. External polar view showing terminal fissure. ×8000.

Localities
Fig. 2. Dorset (Pool Bay)
Figs 3-7. Jyusan-ko (Lake Jyusan), Aomori.

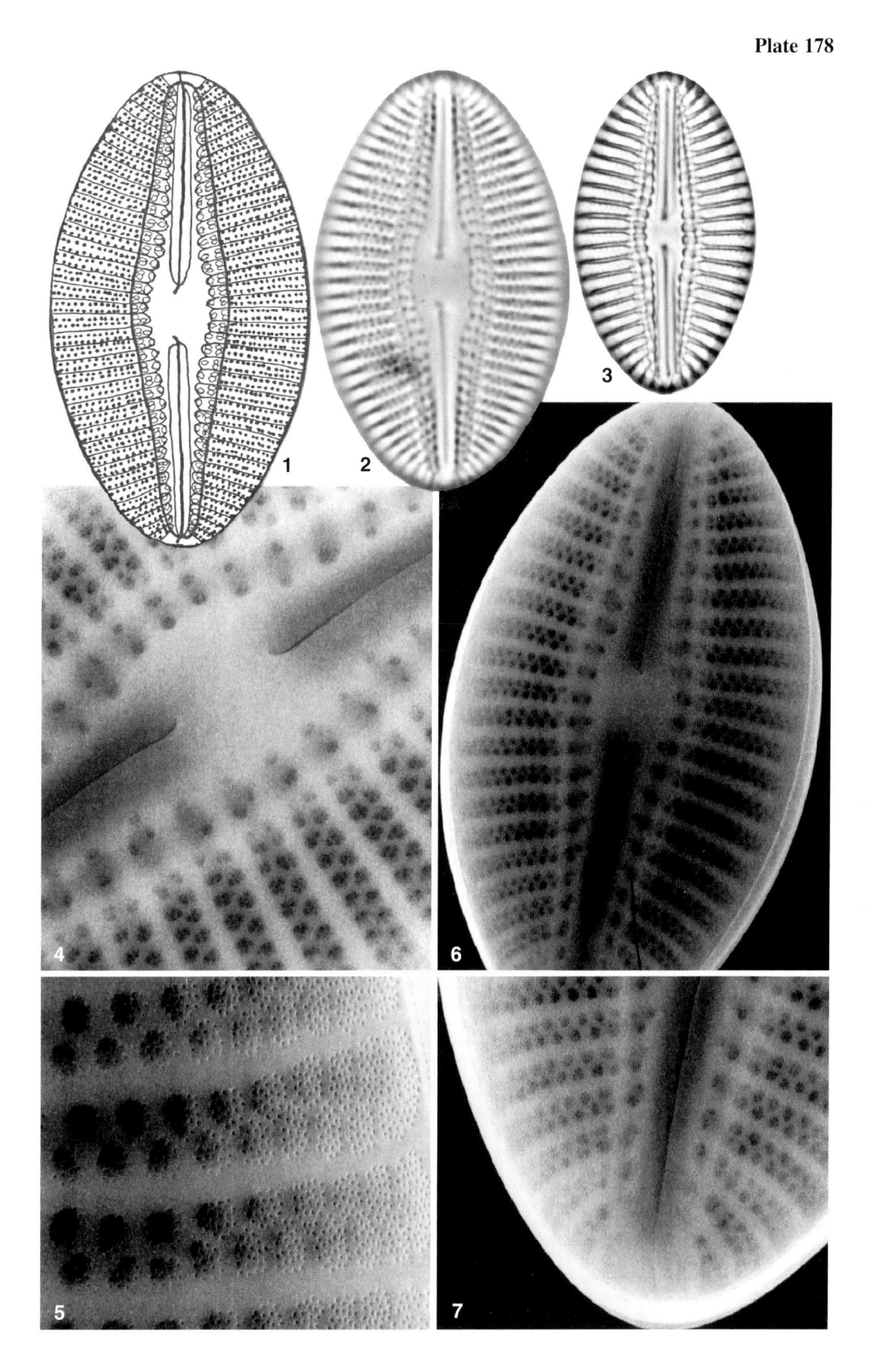

Plate 178

Plate 179 *Entomoneis japonica* (Cleve) K. Osada

Figs 1-4.   LM. ×870.
Figs 5-8.   SEM.

Figs 1, 4.   Girdle view of a frustule.
Fig. 2.   Valve view of a valve.
Fig. 3.   Girdle view of a valve.
Fig. 5.   Valve view of a valve showing an elevated sigmoid keel and marginal scuti (arrows). ×1700.
Fig. 6.   Enlargement of a half valve showing a prominent wing, bisinuosity on the border between wing and valve body and a helictoglossa (arrow). ×1900.
Fig. 7.   Girdle view of a frustule with scuti (arrows) in the valve margin. ×800.
Fig. 8.   Transapical section of wing showing three kinds of fibulae linking opposite wing costae arranged in parallel. ×4600.
   bf ; basal fibula. if ; intermediate fibula. rc ; raphe canal. rf ; raphe fibula.

Localities

Figs 1, 3, 5-8.   Tama-gawa (Tama River), Kanagawa.
Figs 2, 4.   Culture (Fig. 2: KE-343, Fig. 4: KE-213) isolated from Tama-gawa (Tama River), Kanagawa.

Plate 179

Plate 180 *Entomoneis paludosa* (W. Smith) Reimer

Figs 1-3.　LM.　×1600.
Figs 4, 6.　SEM.
Fig. 5.　TEM.

Figs 1, 2.　Girdle view of valve.
Fig. 3.　Valve view of valve.
Fig. 4.　Enlargement of valve exterior showing wing bars, the hymenate strips between adjacent wing bars, and the hymenes of areolae on the valve body and the double canal of the keel. ×12000.
Fig. 5.　Perforations of hymenate strips and the hymenes of the areolae in the double canal. ×30000.
Fig. 6.　Transapical section of valve showing the wing composed of wing bars, raphe fibulae, a raphe canal and an additional canal. ×7200.
　ac ; additional canal. rc ; raphe canal. rf ; raphe fibula. wb ; wing bar.

Localities
Fig. 1.　Lectotype, BM 23406.
Figs 2, 3, 6.　Culture KE-984 isolated from Ara-kawa (Ara River), Tokyo.
Fig. 4.　Mookoppe-gawa (Mookoppe River), Hokkaido.
Fig. 5.　Hi-numa (Lake Hi), Ibaraki.

Plate 180

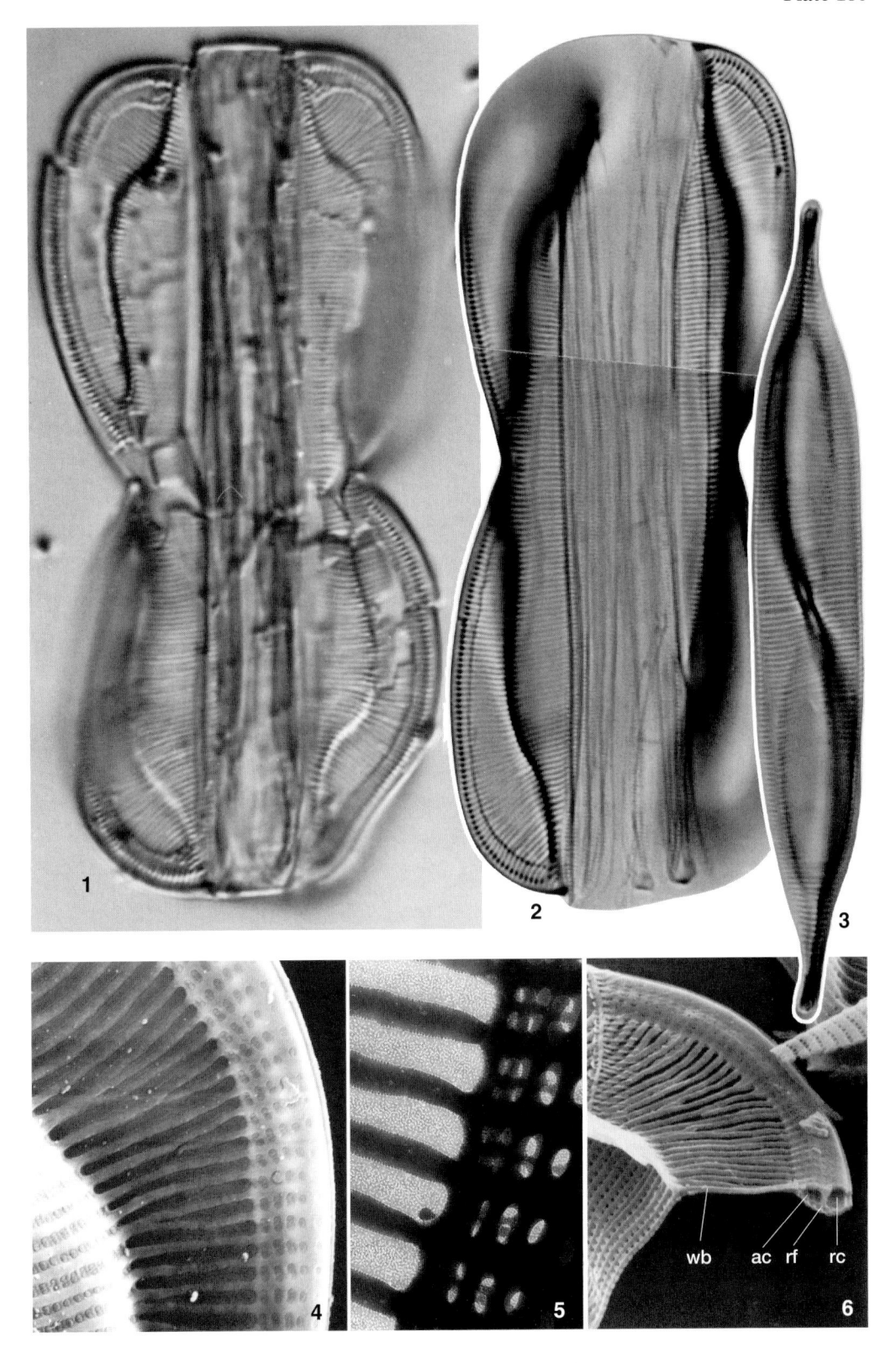

属の学名 - 和名対照表

学名 → 和名

A

Abas	ネジリヒモケイソウ属
Acanthoceras	ジャバラケイソウ属
Achnanthes	ツメケイソウ属
Achnanthidium	ツメワカレケイソウ属
Actinella	ツルギケイソウ属
Actinocyclus	ヒトツメケイソウ属
Actinodictyon	ムカシノユキジルシケイソウ属
Actinoptychus	カザグルマケイソウ属
Adlafia	スカシノマツバケイソウ属
Afrocymbella	アフリカクチビルケイソウ属
Ailuretta	オオメフタヅノケイソウ属
Amicula	フチアナマメケイソウ属
Amphipleura	アミバリケイソウ属
Amphitetras	ヨツカドケイソウ属
Amphora	ニセクチビルケイソウ属
Anaulus	ミズマクラケイソウ属
Andrewsiella	フトスジマメビシケイソウ属
Aneumastus	ゴウリキケイソウ属
Anomoeoneis	サミダレケイソウ属
Anorthoneis	イビツコメツブケイソウ属
Arachnoidiscus	クモノスケイソウ属
Archaegladiopsis	ハクアノトッキケイソウ属
Arcocellulus	ボウスイガタツメダマシケイソウ属
Ardissonea	デカハリケイソウ属
Astartiella	ホシツキツメケイソウ属
Asterionella	ホシガタケイソウ属
Asterionellopsis	シオホシガタケイソウ属
Asterolampra	クンショウケイソウ属
Asteromphalus	エツキクンショウケイソウ属
Attheya	カクダコケイソウ属
Aulacodiscus	コウロケイソウ属
Aulacoseira	スジタルケイソウ属
Auliscus	ギョロメケイソウ属
Auricula	ミミタブケイソウ属
Austariella	ジクビロダエンケイソウ属
Azpeitia	コアミダマシケイソウ属

B

Bacillaria	クサリケイソウ属
Bacteriastrum	タコアシケイソウ属
Bacteriosira	チョクレツケイソウ属
Baxteriopsis	ヒシノバシケイソウ属
Bellerochea	ヒモケイソウ属
Belonastrum	ヒモガタオニジュウジケイソウ属
Bennettella	クジラツキケイソウ属
Berkeleya	ヒメクダズミケイソウ属
Biddulphia	イトマキケイソウ属
Biddulphiopsis	イトマキモドキケイソウ属
Biremis	シンヨリケイソウ属
Bleakeleya	マメホシガタケイソウ属
Brachysira	サミダレモドキケイソウ属
Brassierea	ボウシケイソウ属
Brebissonia	ミバエフネケイソウ属
Brevisira	テンモンタルケイソウ属
Briggera	ムカシノアナヒモケイソウ属
Brightwellia	アンパンボウケイソウ属
Brockmanniella	ニセハネダマシケイソウ属

C

Caloneis	ニセフネケイソウ属
Campylodiscus	クラガタケイソウ属
Campyloneis	カザリコメツブケイソウ属
Campylosira	クチビルダマシケイソウ属
Campylopyxis	シオマガリケイソウ属
Caponea	フタツマクケイソウ属
Catacombas	オオハリケイソウ属
Catenula	ニセイチモンジケイソウ属
Cavinula	ニセコメツブケイソウ属
Centronella	ミツマタケイソウ属
Cerataulina	アナヌキノヒモケイソウ属
Cerataulus	オオメダマモドキケイソウ属
Ceratoneis	ネジレツツケイソウ属
Ceratophora	オオツノケイソウ属
Chaetoceros	ツノケイソウ属
Chamaepinnularia	ヒメハネケイソウ属
Chrysanthemodiscus	キクノハナケイソウ属
Cistula	クケイケイソウ属
Climacodium	オオアナノヒモケイソウ属
Climaconeis	オオナガケイソウ属
Climacosphenia	オオヘラケイソウ属
Cocconeiopsis	ダエンケイソウ属
Cocconeis	コメツブケイソウ属
Colliculoamphora	ムカシノシオイチモンジケイソウ属
Corbellia	ネジリコノハケイソウ属
Corethron	イガクリケイソウ属
Corona	ヒノヒカリケイソウ属
Coscinodiscus	コアミケイソウ属
Cosmioneis	フルイノメケイソウ属

学名 → 和名

Craspedodiscus	ウズマキケイソウ属	*Druridgea*	カサネモチケイソウ属
Craticula	ガイコツケイソウ属	**E**	
Crucidenticula	ジュウジハナラビケイソウ属	*Ehrenbergia*	ボンガタケイソウ属
		Ellerbeckia	オオタルモドキケイソウ属
Ctenophora	ミバエハリケイソウ属	*Encynopsis*	ニセハラミケイソウ属
Cuneolus	ウミマガリケイソウ属	*Encyonema*	ハラミクチビルケイソウ属
Cyclophora	シンツキケイソウ属	*Endictya*	アミカゴケイソウ属
Cyclostephanos	タイコトゲカサケイソウ属	*Entomoneis*	ヨジレケイソウ属
Cyclotella	タイコケイソウ属	*Entopyla*	ミゾナシツメケイソウ属
Cylndrotheca	ネジレケイソウ属	*Eolimna*	コブケイソウ属
Cymatoneis	ウミガメケイソウ属	*Epipellis*	カワツキケイソウ属
Cymatonitzschia	ヨコスジササノハケイソウ属	*Epiphalaina*	クジラツキダマシケイソウ属
Cymatopleura	ハダナミケイソウ属	*Epithemia*	ハフケイソウ属
Cymatosira	オビダマシケイソウ属	*Ethmodiscus*	オオコアミケイソウ属
Cymbella	クチビルケイソウ属	*Eucampia*	アナヒモケイソウ属
Cymbellonitzschia	クチビルササノハケイソウ属	*Eucocconeis*	ネジレコメツブケイソウ属
		Eunophora	イチモンジノツボケイソウ属
Cymbopleura	フナガタクチビルケイソウ属	*Eunotia*	イチモンジケイソウ属
D		*Eunotogramma*	カイコガタケイソウ属
Dactyliosolen	ナガツノナシツツガタケイソウ属	*Euodiella*	ムカシノイトマキケイソウ属
Decussata	コウサフネケイソウ属	*Eupodiscus*	ヨツメケイソウ属
Delicata	ホソミノクチビルケイソウ属	*Extubocellulus*	イボツキハネダマシケイソウ属
Delphineis	オカメモドキケイソウ属	**F**	
Denticula	ハナラビケイソウ属	*Falcula*	シオマガタマケイソウ属
Denticulopsis	ハナラビモドキケイソウ属	*Fallacia*	マクハリタテゴトケイソウ属
Desmogonium	イチモンジモドキケイソウ属	*Fistulifera*	ウスゴロモケイソウ属
Detonula	ヒメホネツギケイソウ属	*Florella*	ウミノハナケイソウ属
Diadema	チクビレツモドキケイソウ属	*Fogedia*	シグレケイソウ属
		Fontigonium	ツクエノアシケイソウ属
Diadesmis	オビフネケイソウ属	*Fragilaria*	オビケイソウ属
Diatomella	フシアナケイソウ属	*Fragilariforma*	オビモドキケイソウ属
Diatoma	イタケイソウ属	*Fragilariopsis*	オビササノハケイソウ属
Dickensoniaforma	ムカシノオカメケイソウ属	*Frankophila*	ミゾツキオビケイソウ属
Dicladia	フタエダユメケイソウ属	*Frickea*	ハマキガタケイソウ属
Dictyoneis	ニセチクビレツケイソウ属	*Frustulia*	ヒシガタケイソウ属
Didymosphenia	オニクサビケイソウ属	**G**	
Dimeregramma	ハネガタモドキケイソウ属	*Geissleria*	ハシツブフネケイソウ属
Dimeregrammopsis	ニセイタケイソウ属	*Gephyria*	ウマノクラケイソウ属
Dimidiata	ムカシノタテアナケイソウ属	*Glorioptychus*	スカシグルマケイソウ属
		Glyphodesmis	シンマルニセハネケイソウ属
Diplomenora	マルオカメケイソウ属	*Gomphocymbella*	クサビクチビルケイソウ属
Diploneis	マユケイソウ属	*Gomphoneis*	クサビフネケイソウ属
Diprora	ミナミノコブネケイソウ属	*Gomphonema*	クサビケイソウ属
Discostella	ホシノタイコケイソウ属	*Gomphonemopsis*	ウミクサビケイソウ属
Distrionella	ホソイタツナギケイソウ属	*Gomphonitzschia*	クサビササノハケイソウ属
Ditylum	クシノハサンカクチョウチンケイソウ属	*Gomphoseptatum*	シオクサビケイソウ属
Donkinia	ミネエスケイソウ属		

学名 → 和名

Gomphotheca	ナガクサビササノハケイソウ属	**M**	
Gonioceros	ヒゲツキツノケイソウ属	Martyana	クサビノオビケイソウ属
Gossleriella	ホネカサケイソウ属	Mastogloia	チクビレツケイソウ属
Grammatophora	ウミヌサガタケイソウ属	Mastogonia	チチマルケイソウ属
Guinardia	ツノナシツツガタケイソウ属	Mayamaea	ヒメツブケイソウ属
		Medlinia	ハクアノカナエケイソウ属
Gyrosigma	エスジケイソウ属	Melosira	タルケイソウ属
H		Meridion	ヘラケイソウ属
Hannaea	クノジケイソウ属	Mesodictyon	アミナカケイソウ属
Hantzschia	ユミケイソウ属	Microcostatus	マメセンタクイタケイソウ属
Haslea	アオイロケイソウ属	Minidiscus	ミツメコゾウケイソウ属
Hemiaulus	シマヒモケイソウ属	Minutocellus	マメハネダマシケイソウ属
Hemidiscus	ハンマルケイソウ属	Monocladia	ヒトエダユメケイソウ属
Hippodonta	ウマノハケイソウ属	Muelleria	ヨコヒゲフネケイソウ属
Hyalinella	ヨコアナヨウジケイソウ属	**N**	
Hyalodiscus	ドラヤキケイソウ属	Nanofrustulum	マメオビダマシケイソウ属
Hyalosira	ウミヌサガタモドキケイソウ属	Navicula	フネケイソウ属
		Navigiolum	フナガタモドキケイソウ属
Hyalosynedra	フトハリケイソウ属	Neidiopsis	ニセハスフネケイソウ属
Hydrosera	サンカクガサネケイソウ属	Neidium	ハスフネケイソウ属
Hydrosilicon	ジュウモンジケイソウ属	Neodelphineis	ハリオカメケイソウ属
I		Neodenticula	アラテハナラビケイソウ属
Isthmia	ダイガタケイソウ属	Neostreptotheca	ウスカラケイソウ属
K		Neosynedra	アラテハリケイソウ属
Karayevia	ツブスジツメワカレケイソウ属	Nitzschia	ササノハケイソウ属
		Nupela	スジカクレケイソウ属
Kittonia	ヒメツノケイソウ属	**O**	
Kobayasiella	ホソミノマツバケイソウ属	Odontella	ヤッコケイソウ属
Kolbesia	ホソスジツメワカレケイソウ属	Opephora	クサビハリケイソウ属
		Orthoseira	ウスガサネケイソウ属
Kurpiszia	カタアキフネケイソウ属	Östrupia	ウリボウケイソウ属
L		Oxyneis	ワラジガタケイソウ属
Lacunicula	アナミゾケイソウ属	**P**	
Lampriscus	オオミツメケイソウ属	Pachyneis	タテスジフネケイソウ属
Lauderia	ヒメホネツギモドキケイソウ属	Palmeria	フタゴケイソウ属
		Papiliocellulus	マメガタツメダマシケイソウ属
Lemnicola	シマツメワカレケイソウ属	Paralia	タルモドキケイソウ属
Lepidodiscus	カザリグルマケイソウ属	Parlibellus	クダズミケイソウ属
Leptocylindrus	ホソミドロケイソウ属	Pauliella	シオツメワカレケイソウ属
Leyanella	ツメダマシケイソウ属	Perissonoë	カクオカメケイソウ属
Licmophora	オウギケイソウ属	Peronia	ツマヨウジケイソウ属
Licmosoma	フジツボオウギケイソウ属	Perrya	オトヒメノクシケイソウ属
Licmosphenia	ホソオウギケイソウ属	Petrochiscus	ムカシノックエケイソウ属
Lioloma	ウミノアミヒモケイソウ属	Petrodictyon	コバンモドキケイソウ属
Lithodesmioides	サンカクチョウチンモドキケイソウ属	Petroneis	チョウガタイロイタケイソウ属
Lithodesmium	サンカクチョウチンケイソウ属	Phaeodactylum	デキソコナイケイソウ属
Lucohuia	ミズマガリナシケイソウ属	Pinnularia	ハネケイソウ属
Lunella	カキノタネケイソウ属	Placoneis	ダエンフネケイソウ属
Luticola	タマスジケイソウ属	Plagiodiscus	ジンゾウガタケイソウ属
Lyrella	タテゴトモヨウケイソウ属	Plagiogramma	ニセハネケイソウ属

学名 → 和名

学名	和名
Plagiogrammopsis	フナガタダマシケイソウ属
Plagiotropis	イカノフネケイソウ属
Planktoniella	カサケイソウ属
Planothidium	フトスジツメワカレケイソウ属
Pleurocyclus	ニセトゲカサケイソウ属
Pleurosigma	メガネケイソウ属
Pleurosira	ジグザグオオメダマケイソウ属
Pliocaenicus	ムカシノタイコケイソウ属
Podocystis	シャモジケイソウ属
Podosira	ニレンキュウケイソウ属
Pogoneis	アゴツメワカレケイソウ属
Porannulus	アナノワアレイケイソウ属
Poretzkia	ビンノフタケイソウ属
Porosira	オワンケイソウ属
Praetriceratium	チャブダイケイソウ属
Pravifusus	スカショウジケイソウ属
Proboscia	ゾウノハナケイソウ属
Progonoia	イキノビケイソウ属
Proschkinia	ウミジュウジケイソウ属
Protokeelia	ムカシノフネケイソウ属
Protoraphis	ハジメノミゾモドキケイソウ属
Psammococconeis	スナジノコメツブケイソウ属
Psammodictyon	タテミゾケイソウ属
Psammodiscus	スナマルケイソウ属
Psammothidium	スナツメワカレケイソウ属
Pseudauliscus	オオメダマケイソウ属
Pseudogomphonema	ニセクサビケイソウ属
Pseudohimantidium	バナナケイソウ属
Pseudonitzschia	ニセササノハケイソウ属
Pseudopodosira	ニセニレンキュウケイソウ属
Pseudorutilaria	ノコギリケイソウ属
Pseudosolenia	ミミナシツツガタケイソウ属
Pseudostaurosira	オビジュウジモドキケイソウ属
Pseudotriceratium	ミツカドダマシケイソウ属
Pteroncola	シオヌサガタケイソウ属
Punctastriata	ニセオニジュウジケイソウ属
Puncticulata	ハナビタイコケイソウ属
Pyxilla	トンガリボウシケイソウ属

R

学名	和名
Raphidodiscus	エリマキモドキケイソウ属
Reimeria	カイコマメケイソウ属
Rhabdonema	ドウナガケイソウ属
Rhaphidophora	キョクジツケイソウ属
Rhaphoneis	オカメケイソウ属
Rhizosolenia	ツツガタケイソウ属
Rhoikoneis	マガリフネケイソウ属
Rhoicosigma	マガリエスジケイソウ属
Rhoicosphenia	マガリクサビケイソウ属
Rhopalodia	クシガタケイソウ属
Rocella	ハグルマケイソウ属
Roperia	ヒトツメコアミケイソウ属
Rossia	ハスフネモドキケイソウ属
Rutilaria	シンボウツナギケイソウ属

S

学名	和名
Sceptroneis	コウガイケイソウ属
Scolioneis	ネジレフネケイソウ属
Scoliopleura	シンネジケイソウ属
Scoliotropis	シンネジモドキケイソウ属
Sellaphora	エリツキケイソウ属
Seminavis	クチビルマガイケイソウ属
Semiorbis	マガタマケイソウ属
Sextiputeus	ムカシノメノワケイソウ属
Sheshukovia	サンカクケイソウ属
Sieminskia	ネコヒゲケイソウ属
Simonsenia	ヒメヨコスジササノハケイソウ属
Skeletonema	ホネツギケイソウ属
Sphynctolethus	ナガフタツノケイソウ属
Stauroneis	ジュウジケイソウ属
Stauronella	ホソジュウジケイソウ属
Staurophora	シオジュウジケイソウ属
Stauropsis	ジュウジモドキケイソウ属
Staurosira	オビジュウジケイソウ属
Staurosirella	オニジュウジケイソウ属
Stellarima	ホシゾラケイソウ属
Stenoneis	ヒメナガケイソウ属
Stenopterobia	ミミズケイソウ属
Stephanodiscus	トゲカサケイソウ属
Stephanopyxis	クシダンゴケイソウ属
Stictocyclus	ニセヒトツメケイソウ属
Stictodiscus	ハスノミケイソウ属
Strangulata	クビレスジタルケイソウ属
Streptotheca	ネジレオビケイソウ属
Striatella	ハラスジケイソウ属
Subsilicea	ナガヒモケイソウ属
Surirella	コバンケイソウ属
Syndendrium	ボンサイユメケイソウ属
Syndetocystis	オオシンボウツナギケイソウ属
Synedra	ウミハリケイソウ属
Synedrella	ホカツキケイソウ属
Synedropsis	ホソハリモドキケイソウ属
Synedrosphenia	ヘラガタハリケイソウ属

T

学名	和名
Tabellaria	ヌサガタケイソウ属
Tabularia	シオハリケイソウ属
Terpsinoe	オシャブリケイソウ属
Tertiarius	アナタイコケイソウ属

学名 → 和名

学名	和名	学名	和名
Tetracyclus	タテジュウジケイソウ属	*Trinacria*	サンカクノヒモケイソウ属
Thalassioneis	ナンキョクハリケイソウ属	*Trossulus*	キクノモンケイソウ属
Thalassionema	ウミノイトケイソウ属	*Tryblionella*	タテナミケイソウ属
Thalassiophysa	ハンカケイソウ属	*Tryblioptychus*	ウミノハダナミケイソウ属
Thalassiosira	ニセコアミケイソウ属	*Tumulopsis*	ハイザラケイソウ属
Thalassiothrix	オオナガウミハリケイソウ属	*Tursiocola*	ニセクジラツキダマシケイソウ属
Toxarium	アミカケイソウ属	**U**	
Toxonidea	ナポレオンケイソウ属	*Ulnaria*	ハリケイソウ属
Trachyneis	ザラメフネケイソウ属	*Undatella*	キテレツケイソウ属
Trachysphenia	クサビハリモドキケイソウ属	*Urosolenia*	マミズツツガタケイソウ属
		V	
Triceratium	ミツカドケイソウ属	*Veigaludwigia*	ニセマツバケイソウ属
Trichotoxon	ウミノユミケイソウ属	*Vikingea*	カイゾクケイソウ属
Trigonium	ミスミケイソウ属	*Vulcanella*	マルキンツバケイソウ属

和名 ⟶ 学名

ア
和名	学名
アオイロケイソウ属	*Haslea*
アゴツメワカレケイソウ属	*Pogoneis*
アナタイコケイソウ属	*Tertiarius*
アナヌキノヒモケイソウ属	*Cerataulina*
アナノワアレイケイソウ属	*Porannulus*
アナヒモケイソウ属	*Eucampia*
アナミゾケイソウ属	*Lacunicula*
アフリカクチビルケイソウ属	*Afrocymbella*
アミカケケイソウ属	*Toxarium*
アミカゴケイソウ属	*Endictya*
アミナカケイソウ属	*Mesodictyon*
アミバリケイソウ属	*Amphipleura*
アラテハナラビケイソウ属	*Neodenticula*
アラテハリケイソウ属	*Neosynedra*
アンパンボウケイソウ属	*Brightwellia*

イ
和名	学名
イガクリケイソウ属	*Corethron*
イカノフネケイソウ属	*Plagiotropis*
イキノビケイソウ属	*Progonoia*
イタケイソウ属	*Diatoma*
イチモンジケイソウ属	*Eunotia*
イチモンジノツボケイソウ属	*Eunophora*
イチモンジモドキケイソウ属	*Desmogonium*
イトマキケイソウ属	*Biddulphia*
イトマキモドキケイソウ属	*Biddulphiopsis*
イビツコメツブケイソウ属	*Anorthoneis*
イボツキハネダマシケイソウ属	*Extubocellulus*

ウ
和名	学名
ウスガサネケイソウ属	*Orthoseira*
ウスカラケイソウ属	*Neostreptotheca*
ウスゴロモケイソウ属	*Fistulifera*
ウズマキケイソウ属	*Craspedodiscus*
ウマノクラケイソウ属	*Gephyria*
ウマノハケイソウ属	*Hippodonta*
ウミガメケイソウ属	*Cymatoneis*
ウミクサビケイソウ属	*Gomphonemopsis*
ウミジュウジケイソウ属	*Proschkinia*
ウミヌサガタケイソウ属	*Grammatophora*
ウミヌサガタモドキケイソウ属	*Hyalosira*
ウミノアミヒモケイソウ属	*Lioloma*
ウミノイトケイソウ属	*Thalassionema*
ウミノハダナミケイソウ属	*Tryblioptychus*
ウミノハナケイソウ属	*Florella*
ウミノユミケイソウ属	*Trichotoxon*
ウミハリケイソウ属	*Synedra*
ウミマガリケイソウ属	*Cuneolus*
ウリボウケイソウ属	*Östrupia*

エ
和名	学名
エスジケイソウ属	*Gyrosigma*
エツキクンショウケイソウ属	*Asteromphalus*
エリツキケイソウ属	*Sellaphora*
エリマキモドキケイソウ属	*Raphidodiscus*

オ
和名	学名
オウギケイソウ属	*Licmophora*
オオアナノヒモケイソウ属	*Climacodium*
オオコアミケイソウ属	*Ethmodiscus*
オオシンボウツナギケイソウ属	*Syndetocystis*
オオタルモドキケイソウ属	*Ellerbeckia*
オオツノケイソウ属	*Ceratophora*
オオナガウミハリケイソウ属	*Thalassiothrix*
オオナガケイソウ属	*Climaconeis*
オオハリケイソウ属	*Catacombas*
オオヘラケイソウ属	*Climacosphenia*
オオミツメケイソウ属	*Lampriscus*
オオメダマケイソウ属	*Pseudauliscus*
オオメダマモドキケイソウ属	*Cerataulus*
オオメフタヅノケイソウ属	*Ailuretta*
オカメケイソウ属	*Rhaphoneis*
オカメモドキケイソウ属	*Delphineis*
オシャブリケイソウ属	*Terpsinoe*
オトヒメノクシケイソウ属	*Perrya*
オニクサビケイソウ属	*Didymosphenia*
オニジュウジケイソウ属	*Staurosirella*
オビケイソウ属	*Fragilaria*
オビモドキケイソウ属	*Fragilariforma*
オビササノハケイソウ属	*Fragilariopsis*
オビジュウジケイソウ属	*Staurosira*
オビジュウジモドキケイソウ属	*Pseudostaurosira*
オビダマシケイソウ属	*Cymatosira*
オビフネケイソウ属	*Diadesmis*
オワンケイソウ属	*Porosira*

カ
和名	学名
カイコガタケイソウ属	*Eunotogramma*
ガイコツケイソウ属	*Craticula*
カイコマメケイソウ属	*Reimeria*
カイゾクケイソウ属	*Vikingea*
カキノタネケイソウ属	*Lunella*
カクオカメケイソウ属	*Perissonoe*
カクダコケイソウ属	*Attheya*
カザグルマケイソウ属	*Actinoptychus*
カサケイソウ属	*Planktoniella*
カサネモチケイソウ属	*Druridgea*
カザリグルマケイソウ属	*Lepidodiscus*
カザリコメツブケイソウ属	*Campyloneis*
カタアキフネケイソウ属	*Kurpiszia*

和名 → 学名

カワツキケイソウ属	*Epipellis*	サンカクケイソウ属	*Sheshukovia*
キ		サンカクチョウチンケイソウ属	*Lithodesmium*
キクノハナケイソウ属	*Chrysanthemodiscus*		
キクノモンケイソウ属	*Trossulus*	サンカクチョウチンモドキケイソウ属	*Lithodesmioides*
キテレツケイソウ属	*Undatella*		
キョクジツケイソウ属	*Rhaphidophora*	サンカクノヒモケイソウ属	*Trinacria*
ギョロメケイソウ属	*Auliscus*	**シ**	
ク		シオクサビケイソウ属	*Gomphoseptatum*
クケイケイソウ属	*Cistula*	シオジュウジケイソウ属	*Staurophora*
クサビクチビルケイソウ属	*Gomphocymbella*	シオツメワカレケイソウ属	*Pauliella*
クサビケイソウ属	*Gomphonema*	シオヌサガタケイソウ属	*Pteroncola*
クサビササノハケイソウ属	*Gomphonitzschia*	シオハリケイソウ属	*Tabularia*
クサビノオビケイソウ属	*Martyana*	シオホシガタケイソウ属	*Asterionellopsis*
クサビハリケイソウ属	*Opephora*	シオマガタマケイソウ属	*Falcula*
クサビハリモドキケイソウ属	*Trachysphenia*	シオマガリケイソウ属	*Campylopyxis*
		ジグザグオオメダマケイソウ属	*Pleurosira*
クサビフネケイソウ属	*Gomphoneis*		
クサリケイソウ属	*Bacillaria*	ジクビロダエンケイソウ属	*Austariella*
クシガタケイソウ属	*Rhopalodia*	シグレケイソウ属	*Fogedia*
クシダンゴケイソウ属	*Stephanopyxis*	シマツメワカレケイソウ属	*Lemnicola*
クシノハサンカクチョウチンケイソウ属	*Ditylum*	シマヒモケイソウ属	*Hemiaulus*
		ジャバラケイソウ属	*Acanthoceras*
クジラツキケイソウ属	*Bennettella*	シャモジケイソウ属	*Podocystis*
クジラツキダマシケイソウ属	*Epiphalaina*	ジュウジケイソウ属	*Stauroneis*
		ジュウジハナラビケイソウ属	*Crucidenticula*
クダズミケイソウ属	*Parlibellus*		
クチビルケイソウ属	*Cymbella*	ジュウジモドキケイソウ属	*Stauropsis*
クチビルササノハケイソウ属	*Cymbellonitzschia*	ジュウモンジケイソウ属	*Hydrosilicon*
		ジンゾウガタケイソウ属	*Plagiodiscus*
クチビルダマシケイソウ属	*Campylosira*	シンツキケイソウ属	*Cyclophora*
クチビルマガイケイソウ属	*Seminavis*	シンネジケイソウ属	*Scoliopleura*
クノジケイソウ属	*Hannaea*	シンネジモドキケイソウ属	*Scoliotropis*
クビレスジタルケイソウ属	*Strangulata*	シンボウツナギケイソウ属	*Rutilaria*
クモノスケイソウ属	*Arachnoidiscus*	シンマルニセハネケイソウ属	*Glyphodesmis*
クラガタケイソウ属	*Campylodiscus*		
クンショウケイソウ属	*Asterolampra*	シンヨリケイソウ属	*Biremis*
コ		**ス**	
コアミケイソウ属	*Coscinodiscus*	スカシグルマケイソウ属	*Glorioptychus*
コアミダマシケイソウ属	*Azpeitia*	スカシノマツバケイソウ属	*Adlafia*
コウガイケイソウ属	*Sceptroneis*	スカシヨウジケイソウ属	*Pravifusus*
コウサフネケイソウ属	*Decussata*	スジカクレケイソウ属	*Nupela*
ゴウリキケイソウ属	*Aneumastus*	スジタルケイソウ属	*Aulacoseira*
コウロケイソウ属	*Aulacodiscus*	スナジノコメツブケイソウ属	*Psammococconeis*
コブケイソウ属	*Eolimna*		
コバンケイソウ属	*Surirella*	スナツメワカレケイソウ属	*Psammothidium*
コバンモドキケイソウ属	*Petrodictyon*	スナマルケイソウ属	*Psammodiscus*
コメツブケイソウ属	*Cocconeis*	**ソ**	
サ		ゾウノハナケイソウ属	*Proboscia*
ササノハケイソウ属	*Nitzschia*	**タ**	
サミダレケイソウ属	*Anomoeoneis*	ダイガタケイソウ属	*Isthmia*
サミダレモドキケイソウ属	*Brachysira*	タイコケイソウ属	*Cyclotella*
ザラメフネケイソウ属	*Trachyneis*	タイコトゲカサケイソウ属	*Cyclostephanos*
サンカクガサネケイソウ属	*Hydrosera*	ダエンケイソウ属	*Cocconeiopsis*

和名──→学名

和名	学名
ダエンフネケイソウ属	Placoneis
タコアシケイソウ属	Bacteriastrum
タテゴトモヨウケイソウ属	Lyrella
タテジュウジケイソウ属	Tetracyclus
タテスジフネケイソウ属	Pachyneis
タテナミケイソウ属	Tryblionella
タテミゾケイソウ属	Psammodictyon
タマスジケイソウ属	Luticola
タルケイソウ属	Melosira
タルモドキケイソウ属	Paralia

チ

和名	学名
チクビレツケイソウ属	Mastogloia
チクビレツモドキケイソウ属	Diadema
チチマルケイソウ属	Mastogonia
チャブダイケイソウ属	Praetriceratium
チョウガタイロイタケイソウ属	Petroneis
チョクレツケイソウ属	Bacteriosira

ツ

和名	学名
ツクエノアシケイソウ属	Fontigonium
ツツガタケイソウ属	Rhizosolenia
ツノケイソウ属	Chaetoceros
ツハナシツツガタケイソウ属	Guinardia
ツブスジツメワカレケイソウ属	Karayevia
ツマヨウジケイソウ属	Peronia
ツメケイソウ属	Achnanthes
ツメダマシケイソウ属	Leyanella
ツメワカレケイソウ属	Achnanthidium
ツルギケイソウ属	Actinella

テ

和名	学名
デカハリケイソウ属	Ardissonea
デキソコナイケイソウ属	Phaeodactylum
テンモンタルケイソウ属	Brevisira

ト

和名	学名
ドウナガケイソウ属	Rhabdonema
トゲガサケイソウ属	Stephanodiscus
ドラヤキケイソウ属	Hyalodiscus
トンガリボウシケイソウ属	Pyxilla

ナ

和名	学名
ナガクサビササノハケイソウ属	Gomphotheca
ナガツノナシツツガタケイソウ属	Dactyliosolen
ナガヒモケイソウ属	Subsilicea
ナガフタツノケイソウ属	Sphynctolethus
ナポレオンケイソウ属	Toxonidea
ナンキョクハリケイソウ属	Thalassioneis

ニ

和名	学名
ニセイタケイソウ属	Dimeregrammopsis
ニセイチモンジケイソウ属	Catenula
ニセオニジュウジケイソウ属	Punctastriata
ニセクサビケイソウ属	Pseudogomphonema
ニセクジラツキダマシケイソウ属	Tursiocola
ニセクチビルケイソウ属	Amphora
ニセコアミケイソウ属	Thalassiosira
ニセコメツブケイソウ属	Cavinula
ニセササノハケイソウ属	Pseudonitzschia
ニセチクビレツケイソウ属	Dictyoneis
ニセトゲカサケイソウ属	Pleurocyclus
ニセニレンキュウケイソウ属	Pseudopodosira
ニセハスフネケイソウ属	Neidiopsis
ニセハネケイソウ属	Plagiogramma
ニセハネダマシケイソウ属	Brockmanniella
ニセハラミケイソウ属	Encynopsis
ニセヒトツメケイソウ属	Stictocyclus
ニセフネケイソウ属	Caloneis
ニセマツバケイソウ属	Veigaludwigia
ニレンキュウケイソウ属	Podosira

ヌ

和名	学名
ヌサガタケイソウ属	Tabellaria

ネ

和名	学名
ネコヒゲケイソウ属	Sieminskia
ネジリコノハケイソウ属	Corbellia
ネジリヒモケイソウ属	Abas
ネジレオビケイソウ属	Streptotheca
ネジレコメツブケイソウ属	Eucocconeis
ネジレツケイソウ属	Ceratoneis (Cylindrotheca)
ネジレフネケイソウ属	Scolioneis

ノ

和名	学名
ノコギリケイソウ属	Pseudorutilaria

ハ

和名	学名
ハイザラケイソウ属	Tumulopsis
ハクアノカナエケイソウ属	Medlinia
ハクアノトッキケイソウ属	Archaegladiopsis
ハグルマケイソウ属	Rocella
ハシツブフネケイソウ属	Geissleria
ハジメノミゾモドキケイソウ属	Protoraphis
ハスノミケイソウ属	Stictodiscus
ハスフネケイソウ属	Neidium
ハスフネモドキケイソウ属	Rossia
ハダナミケイソウ属	Cymatopleura
バナナケイソウ属	Pseudohimantidium
ハナビタイコケイソウ属	Puncticulata
ハナラビケイソウ属	Denticula
ハナラビモドキケイソウ属	Denticulopsis
ハネガタモドキケイソウ属	Dimeregramma
ハネケイソウ属	Pinnularia
ハフケイソウ属	Epithemia

和名 ⟶ 学名

和名	学名
ハマキガタケイソウ属	Frickea
ハラスジケイソウ属	Striatella
ハラミクチビルケイソウ属	Encyonema
ハリオカメケイソウ属	Neodelphineis
ハリケイソウ属	Ulnaria
ハンカケイソウ属	Thalassiophysa
ハンマルケイソウ属	Hemidiscus

ヒ

和名	学名
ヒゲツキツノケイソウ属	Gonioceros
ヒシガタケイソウ属	Frustulia
ヒシノバシケイソウ属	Baxteriopsis
ヒトエダユメケイソウ属	Monocladia
ヒトツメケイソウ属	Actinocyclus
ヒトツメコアミケイソウ属	Roperia
ヒノヒカリケイソウ属	Corona
ヒメクダズミケイソウ属	Berkeleya
ヒメツノケイソウ属	Kittonia
ヒメツブケイソウ属	Mayamaea
ヒメナガケイソウ属	Stenoneis
ヒメハネケイソウ属	Chamaepinnularia
ヒメホネツギケイソウ属	Detonula
ヒメホネツギモドキケイソウ属	Lauderia
ヒメヨコスジササノハケイソウ属	Simonsenia
ヒモガタオニジュウジケイソウ属	Belonastrum
ヒモケイソウ属	Bellerochea
ビンノフタケイソウ属	Poretzkia

フ

和名	学名
フシアナケイソウ属	Diatomella
フジツボオウギケイソウ属	Licmosoma
フタエダユメケイソウ属	Dicladia
フタゴケイソウ属	Palmeria
フタツマクケイソウ属	Caponea
フチアナマメケイソウ属	Amicula
フトスジツメワカレケイソウ属	Planothidium
フトスジマメビシケイソウ属	Andrewsiella
フトハリケイソウ属	Hyalosynedra
フナガタクチビルケイソウ属	Cymbopleura
フナガタダマシケイソウ属	Plagiogrammopsis
フナガタモドキケイソウ属	Navigiolum
フネケイソウ属	Navicula
フルイノメケイソウ属	Cosmioneis

ヘ

和名	学名
ヘラガタハリケイソウ属	Synedrosphenia
ヘラケイソウ属	Meridion

ホ

和名	学名
ボウシケイソウ属	Brassierea
ボウスイガタツメダマシケイソウ属	Arcocellulus
ホカツキケイソウ属	Synedrella
ホシガタケイソウ属	Asterionella
ホシズラケイソウ属	Stellarima
ホシツキツメケイソウ属	Astartiella
ホシノタイコケイソウ属	Discostella
ホソイタツナギケイソウ属	Distrionella
ホソオウギケイソウ属	Licmosphenia
ホソジュウジケイソウ属	Stauronella
ホソスジツメワカレケイソウ属	Kolbesia
ホソハリモドキケイソウ属	Synedropsis
ホソミドロケイソウ属	Leptocylindrus
ホソミノクチビルケイソウ属	Delicata
ホソミノマツバケイソウ属	Kobayasiella
ホネカサケイソウ属	Gossleriella
ホネツギケイソウ属	Skeletonema
ボンガタケイソウ属	Ehrenbergia
ボンサイユメケイソウ属	Syndendrium

マ

和名	学名
マガタマケイソウ属	Semiorbis
マガリエスジケイソウ属	Rhoicosigma
マガリクサビケイソウ属	Rhoicosphenia
マガリフネケイソウ属	Rhoikoneis
マクハリタテゴトケイソウ属	Fallacia
マミズツツガタケイソウ属	Urosolenia
マメオビダマシケイソウ属	Nanofrustulum
マメガタツメダマシケイソウ属	Papiliocellulus
マメセンタクイタケイソウ属	Microcostatus
マメハネダマシケイソウ属	Minutocellus
マメホシガタケイソウ属	Bleakeleya
マユケイソウ属	Diploneis
マルオカメケイソウ属	Diplomenora
マルキンツバケイソウ属	Vulcanella

ミ

和名	学名
ミズマクラケイソウ属	Anaulus
ミスミケイソウ属	Trigonium
ミゾツキオビケイソウ属	Frankophila
ミゾナシツメケイソウ属	Entopyla
ミゾマガリナシケイソウ属	Lucohuia
ミツカドケイソウ属	Triceratium
ミツカドダマシケイソウ属	Pseudotriceratium
ミツマタケイソウ属	Centronella
ミツメコゾウケイソウ属	Minidiscus
ミナミノコブネケイソウ属	Diprora
ミネエスケイソウ属	Donkinia
ミバエハリケイソウ属	Ctenophora
ミバエフネケイソウ属	Brebissonia

和名 ⟶ 学名

和名	学名	和名	学名
ミミズケイソウ属	*Stenopterobia*	ムカシノユキジルシケイソウ属	*Actinodictyon*
ミミタブケイソウ属	*Auricula*		
ミミナシツツガタケイソウ属	*Pseudosolenia*	**メ**	
		メガネケイソウ属	*Pleurosigma*
ム		**ヤ**	
ムカシノアナヒモケイソウ属	*Briggera*	ヤッコケイソウ属	*Odontella*
		ユ	
ムカシノイトマキケイソウ属	*Euodiella*	ユミケイソウ属	*Hantzschia*
		ヨ	
ムカシノオカメケイソウ属	*Dickensoniaforma*	ヨコアナヨウジケイソウ属	*Hyalinella*
ムカシノシオイチモンジケイソウ属	*Colliculoamphora*	ヨコスジササノハケイソウ属	*Cymatonitzschia*
ムカシノタイコケイソウ属	*Pliocaenicus*	ヨコヒゲフネケイソウ属	*Muelleria*
ムカシノタテアナケイソウ属	*Dimidiata*	ヨジレケイソウ属	*Entomoneis*
		ヨツカドケイソウ属	*Amphitetras*
ムカシノックエケイソウ属	*Petrochiscus*	ヨツメケイソウ属	*Eupodiscus*
ムカシノフネケイソウ属	*Protokeelia*	**ワ**	
ムカシノメノワケイソウ属	*Sextiputeus*	ワラジガタケイソウ属	*Oxyneis*

引 用 文 献

安達六郎・高野秀昭・入江春彦 1982．赤潮マニュアル．Ⅲ．珪藻類，赤潮研究会分類班．光出版，三重．121 pp.

Anonymous 1975. Proposals for a standardization of diatom terminology and diagnoses. Nova Hedwigia Beih. **53**: 323-354.

Bixby, R. J. and Jahn, R. 2005. *Hannaea arcus* (Ehrenberg) R.M. Patrick: lectotypification and nomenclatural history. Diatom Research **20**: 210-226.

Boyle, J. A., Pickett-Heaps, J. D. and Czarnecki, B. D. 1984. Valve morphogenesis in the pennate diatom *Achnanthes coarctata*. J. Phycol. **20**: 563-573.

Brummitt, R. K. and Powell, C. E. (eds) 1992. Authors of plant names. A list of authors of scientific names of plants, with recommended standard forms of their names, including abbreviations. Royal Botanic Gardens, Kew. 732 pp.

Cholnoky, B. von 1933. Die Kernteilung der *Melosira arenaria* nebst einigen Bemerkungen über ihre Auxosporenbildung. Zeit. Zellforsch. Mikrosk. Anat. **19**: 698-719.

Cholnoky, G. J. 1953. Diatomeenassoziationen aus dem Hennops-rivier bei Pretoria. Verh. Zool.-bot. Ges. Wien **93**: 134-149.

Cleve, P. T. and Müller, J. D. 1877-1882. Collection of 324 diatom slides with accompanying analysis of A. Grunow. Parts 1-6. 9 pp. Upsala.

Cox, E. J. 1975. A reappraisal of the diatom genus *Amphipleura* Kütz. using light and electron microscopy. Br. Phycol. J. **10**: 1-12.

Cox, E. J. 1978. Taxonomic studies on the diatom genus *Navicula* Bory. *Navicula grevillii* (C. A. Ag.) Heiberg and *N. comides* (Dillwyn) H. & M. Pellagallo. Bot. J. Linn. Soc. **76**: 127-143.

Cox, E. J. 2006. *Achnathes* sensu stricto belongs with genera of the Mastogloiales rather than with other monoraphid diatoms (Bacillariophyta). Eur. J. Phycol. **41**: 67-81.

Cox, E. J. and Ross, R. 1981. The striae of pennate diatoms. In: Ross, R. (ed.) Proceedings of the sixth Symposium on recent and fossil diatoms. Otto Koeltz, Koenigstein. pp. 267-278.

Cox. E. J. 1996. Identification of freshwater diatoms from live material. Chapman & Hall, London. 158 pp.

Crawford, R. M. 1978. The taxonomy and classification of the diatom genus *Melosira* C. A. Agardh. III. *Melosira lineata* (Dillw.) C. A. Ag. and *M. varians* C. A. Ag. Phycologia **17**: 237-250.

Crawford, R. M. 1981. The diatom genus *Aulacoseira* Thwaites: its structure and taxonomy. Phycologia **20**: 174-192.

Crawford, R. M. 1988. A reconstruction of *Melosira arenaria* and *M. teres* resulting in a proposed new genus *Ellerbeckia*. In: Round, F. E. (ed.) Algae and the aquatic environment. Biopress, Bristol. pp. 413-433.

Crawford, R. M., Likhoshway, Y. V. and Jahn, R. 2003. Morphology and identity of

Aulacoseira italica and typification of *Aulacoseira* (Bacillariophyta). Diatom Research **18**: 1-19.

Dawson, P. A. 1972. Observations on the structure of some forms *Gomphonema parvulum* Kütz. I. Morphology based on light microscopy, transmission and scanning electron microscopy. Br. Phycol. J. **7**: 255-271.

Dawson, P. A. 1974. Observations on diatom species transferred from *Gomphonema* C. A. Agardh to *Gomphoneis* Cleve. Br. Phycol. J. **9**: 75-82.

Flower, R. J. and Battarbee, R. W. 1985. The morphology and biostratigraphy of *Tabellaria quadriseptata* (Bacillariophyceae) in acid waters and lake sediments in Galloway, southwest Scotland. Br. Phycol. J. **20**: 69-79.

Geitler, L. 1927. Somatische Teilung, Reductionsteilung, Kopulation und Parthenogenese bei *Cocconeis placentula*. Arch. Protistenk. **59**: 506-549.

Geitler, L. 1932. Der Formwechsel der pennaten Diatomeen. Arch. Protistenk. **78**: 1-226.

Geitler, L. 1939a. Die Auxosporenbildung von *Synedra ulna*. Ber. dt. Bot. Ges. **57**: 432-436.

Geitler, L. 1939b. Gameten- und Auxosporenbildung von *Synedra ulna* im Verleich mit anderen pennaten Diatomeen. Planta **30**: 551-566.

Geitler, L. 1940. Die Auxosporenbildung von *Meridion circulare*. Arch. Protistenk. **94**: 288-294.

Geitler, L. 1951. Die Stellungen der Kopulationspartner und der Auxosporen bei *Gomphonema*-Arten. Abh. Ak. Wiss. Lit. Mainz, math.-nat. Kl. Nr. **6**: 217-225.

Geitler, L. 1952. Untersuchungen über Kopulation und Auxosporenbildung pennater Diatomeen. I. Automixis bei *Gomphonema constrictum* var. *capitata*. Österr. Bot. Z. **99**: 376-384.

Geitler, L. 1958. Notizen über Rassenbildung, Fortpflanzung, Formwechsel und morphologische Eigentümlichkeiten bei pennaten Diatomeen. Österr. Bot. Z. **105**: 408-442.

Geitler, L. 1966. Anomalien der Auxosporen und Korrektionen während der Weiterentwicklung bei *Meridion circulare*. Österr. Bot. Z. **113**: 273-282.

Geitler, L. 1969. Notizen über die Auxosporenbildung eniger pennater Diatomeen. Österr. Bot. Z. **117**: 265-275.

Geitler, L. 1971. Die inäquale Teilung bei der Bildung der Innenschalen von *Meridion circulare*. Österr. Bot. Z. **119**: 442-446.

Geitler, L. 1973a. Zur Lebensgeschichte und Morphologie pennater Diatomeen. I. Allogamie bei *Gomphonema constrictum* var. *capitatum* (Ehr.) Cleve. Österr. Bot. Z. **122**: 35-49.

Geitler, L. 1973b. Auxosporenbildung und Systematik bei pennaten Diatomeen und die Cytologie von *Cocconeis*-Sippen. Österr. Bot. Z. **122**: 299-321.

Genkal, S. I. and Kiss, K.T. 1993. Morphological variability of *Cyclotella atomus* Hustedt var. *atomus* and *C. atomus* var. *gracilis* var. nov. In: van Dam, H. (ed.) Twelfth International Diatom Symposium. Kluwer Academic Publishers. Hydrobiologia **269/270**: 39-48.

Håkansson, H. 1982. Taxonomical discussion on four diatom taxa from an ancient lagoon in Spjälko, South Sweden, Univ. of Lund., Deport **22**: 65-81.

Håkansson, H. 2002. A compilation and evaluation of species in the general *Stephanodiscus, Cyclostephanos* and *Cyclotella* with a new genus in the family Stephanodiscaceae. Diatom Research **17**: 1-139.

Håkansson, H. and Locker, S. 1981. *Stephanodiscus* Ehrenberg 1846, a revision of the species described by Ehrenberg. Nova Hedwigia **35**: 117-150.

Hartmann, A. A. 1967. Le genre Diatoma aux Pays-Bas. Phycologia **6**: 240-247,

Hargraves, P. E. 1986. The relationship of some fossil diatom genera to resting spores. In: Ricard, M. (ed.) Proceedings of the eighth International Diatom Symposium. Otto Koeltz, Koenigstein. pp. 67-80.

Hasle, G. R. 1975. Some living marine species of the diatoms family Rhizosoleniaceae. Nova Hedwigia Beih. **53**: 99-151.

Hasle, G. R. and Lange, C. B. 1989. Freshwater and brackish water *Thalassiosira* (Bacillariophyceae): Taxa with tangentially undulated valves. Phycologia **28**: 120-135.

Helmcke, J.-G. and Krieger, W. 1953. Diatomeenschalen im elektronenmikroskopischen Bild. Vol. 1. Berlin: Bild und Forschung.

Hirose, H. and Yamagishi, T. (eds) 1977. Illustrations of the Japanese fresh-water algae. Uchidarokakuho, Tokyo. 933 pp. (in Japanese)

Holmes, R. W., Crawford, R. M. and Round, F. E. 1982. Variability in the structure of the genus *Cocconeis* Ehr. (Bacillariophyta) with special reference to the cingulum. Phycologia **21**: 370-381.

Houk, V. and Klee, R. 2004. The stelligeroid taxa of the genus *Cyclotella* (Kützing) Brébisson (Bacillariophyceae) and their transfer into the new genus *Discostella* gen. nov. Diatom Research **16**(2): 203-228.

Hustedt, F. 1927-1930. Die Kieselalgen Deutschlands, Österreichs und der Schweitz. Rhabenhorst, Kryptogamenflora. vol. 7(1). Akademische Verlagsgesellschaft, Leipzig. 920 pp.

出井雅彦 1993. *Melosira moniliformis* (O. F. Müller) Agardh. 堀輝三（編）藻類の生活史集成 第3巻, 単細胞性・鞭毛藻類. 内田老鶴圃, 東京. pp. 236, 237.

Jahn, R. and Kusber, W.-H. 2005. Reinstatement of the genus *Ceratoneis* Ehrenberg and lectotypification of its type specimen: *C. closterium* Ehrenberg. Diatom Research **20**: 295-304.

河島綾子・小林弘 1995. 阿寒湖の珪藻（3. 羽状類−広義の *Fragilaria* を除く無縦溝類）. 自然環境科学研究 **8**: 35-49.

河島綾子・真山茂樹 2001. 阿寒湖の珪藻（8. 羽状類−縦溝類：*Cymbella, Encyonema, Gomphoneis, Gomphonema, Gomphosphenia, Reimeria*）. 自然環境科学研究 **14**: 89-109.

Kobayasi, H. 1965. Notes on the new diatoms from River Arakawa (Diatoms from River Arakawa -4). J. Jap. Bot. **40**: 347-351.

小林弘 1993. *Gomphonema truncatum* Ehrenberg. 堀輝三（編）藻類の生活史集成 第3巻, 単細胞性・鞭毛藻類. 内田老鶴圃, 東京. pp. 290, 291.

Kobayasi, H. 1997. Comparative studies among four linear-lanceolate *Achnanthidium* species (Bacillariophyceae) with curved terminal raphe endings. Nova Hedwigia **65**: 147-163.

Kobayasi, H., Idei, M., Kobori, S. and Tanaka, H. 1987. Observations on the two rheophilic species of the genus *Synedra* (Bacillariophyceae): *S. inaequalis* H. Kob. and *S. lanceolata* Kütz. Diatom **3**: 9-16.

小林弘・井上弘喜 1985. 日本産小形ステファノディスクス属（ケイソウ類）の微細構造と分類 1. *Stephanodiscus invisitatus* Hohn & Hell. 藻類 **33**: 149-154.

小林弘・井上弘喜・小林秀明 1985. 日本産小形ステファノディスクス属（ケイソウ類）の微細構造と分類 1. *Stephanodiscus hantzschii* Grun. Form. *tenuis* (Hust.) Håk. et Stoerm. 藻類 **33**: 233-138.

小林弘・石田典子 1996. 日本新産珪藻3種の員弁川および多摩川上流部への出現について. Diatom **12**: 27-33.

Kobayasi, H. and Mayama, S. 1989. Evaluation of river water quality by diatoms. Korean J. Phycol. **4**: 121-133.

Kobayasi, H., Nagumo, T. and Mayama, S. 1986. Observations on the two rheophilic species of the genus *Achnanthes* (Bacillariophyceae), *A. convergens* H. Kob. and *A. japonica* H. Kob. Diatom **2**: 83-93.

小林弘・野沢美智子 1981. 淡水産中心類ケイソウ *Aulacosira ambigua* (Grun.) Sim. の微細構造について. 藻類 **29**: 121-128.

小林弘・野沢美智子 1982. 淡水産中心類ケイソウ *Aulacosira italica* (Ehr.) Sim. の微細構造について. 藻類 **30**: 139-146.

Kobayasi, H. and Sawatari, A. 1986. Some small and elliptic *Achnanthes*. In: Ricard, M. (ed.) Proceedings of the 8th International Diatom Symposium. Koeltz Scientific Books, Koenigstein. pp. 259-269.

小林弘・吉田稔 1984. コンクリート池のケイソウとその優れた教材性について. 東京学芸大学紀要4部門 **36**: 115-143.

Körner, H. 1970. Morphologie und Taxonomie der Diatomeengattung *Asterionella*. Nova Hedwigia **20**: 557-724.

Krammer, K. 2001. Taxonomic und Morphologie von *Brebisira arentii* (Kolbe) Krammer gen. nov., comb. nov. In: Jahn, R., Kociolek, J. P., Witkowski, A. and Compère, P. (eds) Lange-Bertalot-Festschrift. Studies on diatoms. Gantner Verag, Ruggell. pp. 9-20.

Krammer, K. and Lange-Bertalot, H. 1986-1991. Bacillariophyceae. 1. Teil. Naviculaceae. 876 pp.; 2. Teil, Bacillariaceae, Epithemiaceae, Surirellaceae. 596 pp.; 3. Teil, Centrales, Fragilariaceae, Eunotiaceae. 576 pp.; 4. Teil, Achnanthaceae, Kritishe Ergänzungen zu *Navicula* (Lineolatae) und *Gomphonema*. 437 pp. In: Ettl, J., Gerloff, H. and Mollenhauer, D. (eds) Die Süsswasserflora von Miteleuropa. Bd 2/1-4. Gustav Fischer, Stuttgart.

Krasske, G. 1923. Die Diatomeen des Casseler Beckens und seiner Randgebirge, nebst einigen wichtigen funden aus Niederhessen. Bot. Arch. **3**: 185-209.

Lange-Bertalot, H. 1993. 85 Neue Taxa und über 100 weitere neu definierte Taxa ergänzend zur Süsswasserflora von Mitteleuropa Vol. 2/1-4. Bibliotheca Diatomologica vol. 27. J. Cramer, Berlin. 454 pp.

Lange-Bertalot, H. 1999. Neue Kombinationen von Taxa aus *Achnanthes* Bory (sensu lato). In: Lange-Bertalot, H. (ed.) Iconographia Diatomologica vol. 6. Phytogeography-Diversity-Taxonomy. Koeltz Scientific Books, Koenigstein. pp. 276-289.

Lange-Bertalot, H. and Krammer, K. 1989. *Achnanthes*, eine Monographie der Gattung mith Difinition der Gattung *Cocconeis* und Nachträgen zu den Naviculaceae. Bibliotheca Diatomorogica vol.18. 393 pp.

Lange-Bertalot, H. and Ruppel, M. 1980. Zur Revision taxonomisch problematischer, Ökologisch jedoch wichtiger Sippen der Gattung *Achnanthes* Bory. Arch. Hydrobiol. Suppl. **60**: 1-31.

Lange-Bertalot, H. and Simonsen, R. 1978. A taxonomic revision of the *Nitzschia lanceolata* Grunow. 2. European and related extra-European freshwater and brackish water taxa. Bacillaria **1**: 11-111.

Locker, F. 1950. Beiträge Zur kenntnis des Formwechsels der Diatomeen an hand von Kulturversuchen. Österr. Bot. Z. **98**: 322-332.

Makita, N. and Shihira-Ishikawa, I. 1997. Chloroplast assemblage by mechanical stimulation and its intercellular transmission in diatom cells. Protoplasma **197**: 86-95.

Mann, D. G. 1981. Sieves and flaps: siliceous minutiae in the pores of raphid diatoms. In: Ross, R. (ed.) Proceedings of the sixth Symposium on recent and fossil diatoms. Otto Koeltz, Koenigstein. pp. 279-300.

Mann, D. G. 1982a. Structure, life history and systematics of *Rhoicosphenia* (Bacillariophyta). I. The vegetative cell of Rh. curvata. J. Phycol. **18**: 162-176.

Mann, D. G. 1982b. Structure, life history and systematics of *Rhoicosphenia* (Bacillariophyta). II. Auxospore formation and perizonium structure of *Rh. curvata*. J. Phycol. **18**: 264-274.

Mann, D. G. 1984. Structure, life history and systematics of *Rhoicosphenia* (Bacillariophyta). V. Initial cell and size reduction in *Rh. curvata* and a description of the Rhoicospheniaceae fam. nov. J. Phycol. **20**: 544-555.

真山茂樹 1993. *Actinella brasiliensis* Grunow. 堀輝三（編）藻類の生活史集成 第3巻, 単細胞性・鞭毛藻類. 内田老鶴圃, 東京. pp. 252, 253.

Mayama, S., Idei, M., Osada, K. and Nagumo, T. 2002. Nomenclatural changes for 20 diatom taxa occurring in Japan. Diatom **18**: 89-91.

Mayama, S. and Kobayasi, H. 1984. The separated distribution of the two varieties of *Achnanthes minutissima* Kuetz. according to the degree of river water pollution. Jap. J. Limnol. **45**: 304-312.

Mayama, S. and Kobayasi, H. 1986. Observations of *Navicula mobiliensis* var. *minor* Patr. and *N. goeppertiana* (Bleisch) H. L. Sm. In: Ricard, M. (ed.) Proceedings of the eighth International Diatom Symposium. Koeltz, Koenigstein. pp. 173-182.

Mayama, S. and Kobayasi, H. 1989. Sequential valve development in the monoraphid diatom *Achnanthes minutissima* var. *saprophila*. Diatom Research **4**: 111-117.

Mayama, S. and Kobayasi, H. 1991. Observations of *Eunotia arcus* Ehr., type species of the genus *Eunotia* (Bacillariophyceae). Jpn. J. Phycol. **39**: 131-141.

Mayama, S. and Kuriyama, A. 2002. Diversity of mineral cell coverings and their formation processes: a review focused on the siliceous cell coverings. J. Plant Res. **115**: 289-295.

Mayama, S., Mayama, N. and Shihira-Ishikawa, I. 2004. Characterization of linear-oblong pyrenoids with cp-DNA along their sides in *Nitzschia sigmoidea* (Bacillariophyceae). Phycol. Res. **52**: 129-139.

Medlin, L. K. and Kaczmarska, I. 2004. Evolution of the diatoms: V. Morphological and cytological support for the major clades and a taxonomic revision. Phycologia **43**: 245-270.

Medlin, L. K. and Priddle, J. 1990. Polar marine diatoms. British Antarctic Survey, Natural Environmental Research Council, Cambridge. 214 pp.

Mizuno, M. 1987. Morphological variation of the attached diatom *Cocconeis scutellum* var. *scutellum*. J. Phycol. **23**: 591-597.

文部省・日本植物学会 1990. 学術用語集・植物学編（増訂版），丸善.

南雲保 1982. *Asterionella formosa* Hasall. 赤潮研究会分類班編，赤潮生物シート No. 102. 水産庁，東京.

南雲保・小林弘 1977. 光顕及び電顕的研究に基く *Melosira arentii* (Kolbe) comb. nov. について. 藻類 **25**: 128-183.

南雲保・小林弘 1985. 淡・汽水産珪藻 *Cyclotella* 属の3種 *C. atomus*, *C. caspia*, *C. meduanae* の微細構造. Bull. Plankt. Soc. Jap. **32**: 101-109.

Nipkow, F. 1953. Die Auxosporenbildung bei *Fragilaria crotonensis* Kitton im Plankton des Zürichsees. Schweiz. Z. Hydrol. **15**: 302-310.

Okuno, H. 1957. Electron-microscopical study on fine structure of diatom frustules XVI. Bot. Mag. Tokyo **70**: 216-224.

Osada, K. and Kobayasi, H. 1985. Fine structure of the brackish water pennate diatom *Entomoneis alata* (Ehr.) Ehr. var. *japonica* (Cl.) comb. nov. Jpn. J. Phycol. **33**: 215-224.

Osada, K. and Kobayasi, H. 1990. Observations on the forms of the diatom *Entomoneis paludosa* and related taxa. In: Simola, H. (ed.) Proceedings of the tenth International Diatom Symposium. Sven Koeltz, Koenigstein. pp. 161-172.

Paddock, T. B. B. and Sims, P. A. 1977. A preliminary survey of the raphe structure of some advanced group of diatoms (Epithemiaceae-Surirellaceae). In: Simonsen, R. (ed.) Fourth Symposium on recent and fossil marine diatoms. J. Cramer, Vaduz. pp. 291-322.

Patrick, R. and Reimer, C. W. 1966. The diatoms of the United States I. Acad. Nat. Sci. Phild. Monogr. **13**. 688 pp.

Potapova, M. G. and Ponader, K. C. 2004. Two common north American diatoms, *Achnanthidium rivulare* sp. nov. and *A. deflexum* (Reimer) Kingston: morphology, ecology and comparison with related species. Diatom Research **19**: 33-57.

Potapova, M. and Snoeijs, P. 1997. The natural life cycle in wild populations of *Diatoma moniliformis* (Bacillariophyceae) and its disruption in an aberrant environment. J. Phycol. **33**: 924-937.

Reichardt, E. 1999. Zur Revision der Gattung *Gomphonema*. Die Arten um *G. affine/insigne*, *G. angustatum/micropus*, *G. acuminatum* sowie gomphonemoide Diatomeen aus dem Oberoligozän in Böhmen. In: Lange-Betalot, H. (ed.) Iconographia Diatomologica Vol. 8. Taxonomy. Koeltz Scientific Books, Koenigstein. 203 pp.

Reichardt, E. 2001. Revision der Arten um *Gomphonema truncatum* und *G. capitatum*. In: Jahn, R., Kociolek, J. P., Witkowski, A. and Compère, P. (eds) Lange-Bertalot-Festschrift. Studies on diatoms. A.R.G. Gantner Verlag, Ruggell. pp. 187-224.

Reichardt, E. and Lange-Betalot, H. 1991. Taxonomische Revision des Artenkomplexes um *Gomphonema angustum-G. dichotomum-G. intricatum-G. vibrio* und ähnliche Taxa (Bacillariophyceae). Nova Hedwigia **53**: 519-544.

Roemer, S. C. and Rosowski, J. R. 1980. Valve and band morphology of some freshwater diatoms. III. Pre- and post -auxospore frustules and the initial cell of *Melosira roeseana*. J. Phycol. **16**: 399-411.

Ross, R. 1963. The diatom genus *Capartogramma* and the identity of *Schizostauron*. Bull. Brit. Mus. (Nat. Hist.) Bot. **3**: 47-92.

Ross, R., Cox, E. J., Karayeva, N. I., Mann, D. G., Paddock, T. B. B., Simonsen, R. and Sims, P. A. 1979. An amended terminology for the siliceous components of the diatom cell. Nova Hedwigia Beih. **64**: 513-533.

Ross, R. and Sims, P. A. 1972. The fine structure of the frustule in centric diatoms: a suggested terminology. Br. Phycol. J. **7**: 139-163.

Round, F. E. 1970. The Delineation of the genera *Cyclotella* and *Stephanodiscus* by light microscopy, transmission and reflecting electron microscopy. Nova Hedwigia Beih. **31**: 591-604.

Round, F. E. 1981. The diatom genus *Stephanodiscus*: An electronmicroscopic view of the classical species. Arch. Protistenk. **124**: 447-465.

Round, F. E. and Basson, P. W. 1997. A new monoraphid diatom genus (*Pogoneis*) from Bahrain and the transfer of previously described species *A. hungarica* and *A. taeniata* to new genera. Diatom Research **12**: 71-81.

Round, F. E. and Bukhtiyarova, L. 1996. Four new genera based on *Achnanthes* (*Achnanthidium*) togerther with a re-definition of *Achnanthidium*. Diatom Research **11**: 345-361.

Round, F. E., Crawford, R. M. and Mann, D. G. 1990. The Diatoms. Biology and morphology of the genera. Cambridge University Press, Cambridge. 747 pp.

Round, F. E. and Håkansson, H. 1992. Cyclotelloid species from a diatomite in the Hartz Mountains, Germany, including *Pliocaenicus* gen. nov. Diatom Research **7**: 109-125.

Round, F. E. and Maidana, N. I. 2001. Two problematic freshwater arahid taxa re-classified in new genera. Diatom **17**: 21-28.

佐竹俊子・小林弘 1991. 淡水産中心類珪藻 *Aulacoseira valida* (Grunow in Van Heurck) Krammer の微細構造. 自然環境科学研究 **4**: 45-57.

Schmid, A. M. 1979. Influence of environmental factors on the development of the valve in diatoms. Protoplasma **99**: 99-115.

Schrader, H.-J. 1973. Types of raphe structures in the diatoms. Nova Hedwigia Beih. **45**: 195-230.

Schütt, F. 1896. Bacillariales. In: Engler, A. and Prantl, K. (eds) Die natülichen Pflanzenfamilien. 1(1b). W. Engelmann, Leipzig. pp. 31-153.

Simonsen, R. 1979. The diatom system: Ideas on phylogeny. Bacillaria **2**: 9-71.

Simonsen, R. 1987. Atlas and catalogue of the diatom types of Friedrich Hustedt. Vol.1. Catalogue 525 pp.; Vol.2. Atlas. 597 pp.; Vol.3. Atlas 772 pp. J. Cramer, Berlin.

Sims, P. A. and Paddock, T. B. B. 1979. Observations and comments on some prominent morphological features of Naviculoid genera. Nova Hedwigia Beih. **64**: 169-191.

Spaulding, S. A. and Kociolek, J. P. 1998. The diatom genus *Orthoseira*: Ultrastructure and morphological variation in two species from Madagascar with comments on nomenclature in the genus. Diatom Research **13**: 133-147.

Stosch, H. A. von 1951. Entwicklungsgeschichtliche Untersuchungen an zentrischen Diatomeen. I. Die Auxsporenbildung von *Melosira varians*. Arch. Mikrobiol. **16**: 101-135.

Stosch, H. A. von 1977. Observations on *Bellerochea* and *Streptotheca*, including descriptions of three new planktonic diatom species. Nova Hedwigia Beih. **54**: 113-166.

Stosch, H. A. von and Fecher, K. 1979. "Internal thecae" of *Eunotia soleirolii* (Bacillariophyceae): development, structure and function as resting spores. J. Phycol. **15**: 233-243.

鈴木秀和・田中次郎・南雲保 1999. 伊豆半島式根島産の紅藻ユカリに着生する珪藻類. 日本歯科大学紀要（一般教育系）**28**: 147-160.

Takano, H. 1961. Epiphytic diatoms upon Japanese agar sea-weeds. Bull. Tokai Reg. Fish. Res. Lab. **1961**: 269-274.

Tanaka, H. and Nagumo, T. 2005. *Puncticula ozensis* sp. nov., a new freshwater diatom in Lake Oze, Japan. Diatom **21**: 47-55.

Tschermak-Wess, E. 1973. Über die bisher vergeblich gesuchte Auxosporenbildung von *Diatoma*. Öterr. Bot. Z. **121**: 23-27.

辻彰洋・伯耆晶子 2001. 琵琶湖の中心目珪藻. 琵琶湖研究モノグラフ no. 7. 滋賀県琵琶湖研究所, 大津. 90 pp.

Tuji, A. 2002. Observations on *Aulacoseira nipponica* from Lake Biwa, Japan and *Aulacoseira solida* from North America (Bacillariophyceae). Phycol. Res. **50**: 313-316.

Van der Werff, A. and Huls, H. 1957-1974. Diatomeënflora van Nederland. Abcoude, Den Haag.

Van Heurck, H. 1880-1885. Synopsis des diatomées de Belgique. Atlas & Text. Anvers. 120+235 pp., 135 pls.

Voigt, M. 1956. Sur certaines irrégularités dans la structure des Diatomées. Rev. Algol. nouv. sér. **2**: 85-87.

Vyverman, W. and Compère, P. 1991. *Nupela giluwensis* gen. & spec. nov. A new genus of naviculoid diatoms. Diatom Research **6**: 175-179.

Williams, D. M. 1985. Morphology, taxonomy and inter-relationships of the ribbed araphid diatoms from the genera *Diatoma* and *Meridion* (Diatomaceae: Bacillariophyta). Bibl. Diatomol. vol. 8. J. Cramer, Vaduz.

Williams, D. M. 1986. Comparative morphology of some species *Synedra* Ehrenb. with a new definition of the genus. Diatom Research **1**: 313-339.

Xie, Shu-Qi and Qi, Yu-Zao 1984. Light, scanning and transmission electron microscopic studies on the morphorogy and taxonomy of *Cyclotella shanxiensis* sp. nov. In: Mann, D. G. (ed.) Proceedings of the seventh International Diatom Symposium. Otto Koeltz, Koenigstein. pp. 185-196.

学名索引

太字は本文解説記載種

A

Achnantheiopsis frequentissima ……… 131
Achnantheiopsis septentrionalis ……… 134
Achnanthes alteragracillima ……… 123
Achnanthes austriaca ……… 135
Achnanthes austriaca var. *helvetica* ……… 135
Achnanthes biasolettiana ……… 127
Achnanthes coarctata ……… 111, 418, 420
Achnanthes convergens ……… 121
Achnanthes crenulata ……… 112, 422, 424
Achnanthes delicatula ……… 134
Achnanthes detha ……… 138
Achnanthes exigua ……… 122
Achnanthes frequentissima ……… 131
Achnanthes gibberula ……… 127
Achnanthes helvetica ……… 135
Achnanthes hungarica ……… 130
Achnanthes hustedtii ……… 132
Achnanthes inflata ……… 113, 426, 428
Achnanthes japonica ……… 124
Achnanthes krasskei ……… 132
Achnanthes kuwaitensis ……… 114, 430
Achnanthes lanceolata ……… 133
Achnanthes lanceolata spp. *frequentissima* ……… 131
Achnanthes lapidosa ……… 139
Achnanthes lineariformis ……… 125, 458
Achnanthes marginulata ……… 136
Achnanthes microcephala ……… 123
Achnanthes minutissima ……… 125
Achnanthes minutissima var. *cryptocephala* ……… 123
Achnanthes minutissima var. *gracillima* ……… 123
Achnanthes minutissima var. *jackii* ……… 123
Achnanthes minutissima var. *macrocephala* ……… 125, 458
Achnanthes minutissima var. *saprophila* ……… 128
Achnanthes montana ……… 137
Achnanthes pusilla ……… 126
Achnanthes pyrenaica ……… 127
Achnanthes septentrionalis ……… 134
Achnanthes subatomoides ……… 138
Achnanthes subhudsonis ……… 129
Achnanthidium alteragracillimum ……… 123
Achnanthidium coarctatum ……… 111
Achnanthidium convergens ……… 121, 124, 450
Achnanthidium crassum ……… 121
Achnanthidium delicatulum ……… 134, 484
Achnanthidium exiguum ……… 122, 452
Achnanthidium gracillimum ……… 123, 454
Achnanthidium hungaricum ……… 130
Achnanthidium japonica ……… 121
Achnanthidium japonicum ……… 121, 124, 126, 456
Achnanthidium lanceolatum ……… 133
Achnanthidium lapidosum ……… 139
Achnanthidium latecephalum ……… 121
Achnanthidium microcephalum ……… 125, 458
Achnanthidium minutissimum ……… 125, 458, 460
Achnanthidium pirenaicum ……… 121
Achnanthidium pusillum ……… 126, 462
Achnanthidium pyrenaicum ……… 127, 464, 466
Achnanthidium rivalare ……… 121
Achnanthidium saplophilum ……… 125, 128, 468
Achnanthidium subhudsonis ……… 129, 470, 472
Actinella brasiliensis ……… 93, 372
Actinella punctata ……… 93
Amphiprora alata var. *japonica* ……… 144
Amphiprora paludosa ……… 145
Asterionella formosa ……… 47–49, 272
Asterionella glacialis ……… 50
Asterionella gracillima ……… 47, 48, 272
Asterionella japonica ……… 50
Asterionella ralfsii ……… 49, 274
Asterionella ralfsii var. *americana* ……… 49
Asterionella ralfsii var. *hustedtiana* ……… 49
Asterionella ralfsii var. *ralfsii* ……… 49
Asterionellopsis glacialis ……… 50, 276
Aulacoseira ambigua ……… 6, 9, 162, 164
Aulacoseira crenulata ……… 9
Aulacoseira distans ……… 7, 166
Aulacoseira epidendron ……… 16
Aulacoseira granulata ……… 8, 168, 170
Aulacoseira italica ……… 9, 172, 174
Aulacoseira longispina ……… 10, 176, 178
Aulacoseira longispina var. *tenuis* ……… 13
Aulacoseira nipponica ……… 11, 180, 182
Aulacoseira solida ……… 11

Aulacoseira subarctica ⋯⋯ 12, 184, 186
Aulacoseira tenuis ⋯⋯ 13, 188, 190
Aulacoseira valida ⋯⋯ 14, 192, 194

B

Bacillaria ulna ⋯⋯ 89
Biblarium lapponica ⋯⋯ 79
Brebisira arentii ⋯⋯ 17, 200

C

Catacombas obtusa ⋯⋯ 51, 278
Cocconeis diminuta ⋯⋯ 115, 432
Cocconeis hustedtii ⋯⋯ 132
Cocconeis lineata ⋯⋯ 118
Cocconeis pediculus ⋯⋯ 115, 116, 434, 436
Cocconeis placentula ⋯⋯ 119
Cocconeis placentula var. *euglypta* ⋯⋯ 115, 118
Cocconeis placentula var. lineata
⋯⋯ 118, 442, 444
Cocconeis placentula var. placentula
⋯⋯ 117, 118, 438, 440
Cocconeis scutellum ⋯⋯ 119, 446
Cocconeis scutellum var. *stauroneiformis* ⋯⋯ 120
Cocconeis stauroneiformis ⋯⋯ 120, 448
Conferva flocculosa ⋯⋯ 91
Conferva moniliformis ⋯⋯ 2
Coscinodiscus arentii ⋯⋯ 17
Coscinodiscus eccentricus ⋯⋯ 19
Coscinodiscus faurii ⋯⋯ 20
Coscinodiscus lacustris ⋯⋯ 22
Coscinodiscus striata ⋯⋯ 37
Ctenophora pulchella ⋯⋯ 52, 280
Cyclostephanos dubius ⋯⋯ 27, 220, 222
Cyclostephanos invisitatus ⋯⋯ 28, 224, 226
Cyclotella arentii ⋯⋯ 17
Cyclotella asterocostata ⋯⋯ 38
Cyclotella atomus ⋯⋯ 29, 228
Cyclotella atomus var. gracilis ⋯⋯ 30, 230
Cyclotella caspia ⋯⋯ 30
Cyclotella criptica ⋯⋯ 31, 232
Cyclotella dubius ⋯⋯ 27
Cyclotella litoralis ⋯⋯ 32, 234, 236
Cyclotella meduanae ⋯⋯ 33, 238
Cyclotella meneghiniana ⋯⋯ 34, 240
Cyclotella meneghiniana var. *stelligera* ⋯⋯ 40
Cyclotella minutula ⋯⋯ 44
Cyclotella ocellata ⋯⋯ 35, 242
Cyclotella pantaneliana ⋯⋯ 36, 244
Cyclotella praetermissa ⋯⋯ 41
Cyclotella pseudostelligera ⋯⋯ 39
Cyclotella radiosa ⋯⋯ 41

Cyclotella rotula ⋯⋯ 45
Cyclotella shanxiensis ⋯⋯ 42
Cyclotella stelligera ⋯⋯ 40
Cyclotella striata ⋯⋯ 37, 246

D

Diatoma elongatum ⋯⋯ 54
Diatoma fenestratum ⋯⋯ 90
Diatoma gracillimum ⋯⋯ 48
Diatoma hiemalis ⋯⋯ 53
Diatoma hiemalis var. *mesodon* ⋯⋯ 53
Diatoma mesodon ⋯⋯ 53, 282
Diatoma moniliformis ⋯⋯ 54
Diatoma tenuis ⋯⋯ 54, 284
Diatoma vulgaris ⋯⋯ 55, 286
Diatomella balfouriana ⋯⋯ 140, 496
Diploneis elliptica ⋯⋯ 141, 142, 498
Diploneis ovalis ⋯⋯ 142, 500
Diploneis smithii ⋯⋯ 143, 502
Discoplea graeca var. *stelligera* ⋯⋯ 40
Discostella asterocostata ⋯⋯ 38, 248
Discostella pseudostelligera ⋯⋯ 39, 250
Discostella stelligera ⋯⋯ 40, 252, 254

E

Echinella circularis ⋯⋯ 66
Ellerbeckia arenaria f. *arenaria* ⋯⋯ 5
Ellerbeckia arenaria f. teres ⋯⋯ 5, 160
Entomoneis alata ⋯⋯ 144
Entomoneis alata var. *japonica* ⋯⋯ 144
Entomoneis japonica ⋯⋯ 144, 504
Entomoneis paludosa ⋯⋯ 145, 506
Eucampia zodiacus ⋯⋯ 46, 270
Exilaria vaucheriae ⋯⋯ 61

F

Fragilaria arcus var. *recta* ⋯⋯ 64
Fragilaria bicapitata ⋯⋯ 62
Fragilaria binodis ⋯⋯ 72
Fragilaria brevistriata ⋯⋯ 67
Fragilaria brevistriata var. *nipponica* ⋯⋯ 68
Fragilaria capitellata ⋯⋯ 56, 288, 290
Fragilaria capucina ⋯⋯ 58
Fragilaria construens ⋯⋯ 71
Fragilaria construens var. *binodis* f. *robusta*
⋯⋯ 69
Fragilaria construens var. *triundulata* ⋯⋯ 74
Fragilaria crotonensis ⋯⋯ 57, 292, 294
Fragilaria elliptica ⋯⋯ 75
Fragilaria lapponica ⋯⋯ 78
Fragilaria mesodon ⋯⋯ 53

Fragilaria mesolepta ……58, 296
Fragilaria neoproducta ……59, 298
Fragilaria parasitica ……81
Fragilaria perminuta ……60, 300
Fragilaria pinnata ……70, 80
Fragilaria pseudogaillonii ……88
Fragilaria robusta ……69
Fragilaria vaucheriae ……56, 61, 302
Fragilaria venter ……76, 77
Fragilariforma virescens ……59
Fragilariforma bicapitata ……62, 304

G

Gallionella crenulata ……9
Gallionella distans ……7
Gallionella granulata ……8
Gallionella italica ……9
Gomphoneis herculeana var. minuta ……96
Gomphoneis heterominuta ……96, 382
Gomphoneis minuta ……96
Gomphoneis olivacea ……96
Gomphoneis rhombica ……97, 384, 386
Gomphonema abbreviatum ……95
Gomphonema capitatum ……109
Gomphonema clevei ……98, 99, 102, 388
Gomphonema clevei var. inaequilongum ……101
Gomphonema constrictum ……109
Gomphonema constrictum var. capitatum ……109
Gomphonema constrictum var. constrictum
…… 109
Gomphonema curvatum ……95
Gomphonema curvipedatum ……99, 102, 390
Gomphonema fibula ……94
Gomphonema gracile ……100, 392
Gomphonema inaequilongum
…… 101, 102, 394
Gomphonema kinokawaensis ……102, 396
Gomphonema lagenula ……105
Gomphonema micropus ……103, 107, 398
Gomphonema minutum ……96
Gomphonema nipponicum ……104, 400
Gomphonema parvulum var. **lagenula**
…… 105, 107, 402
Gomphonema parvulum var. micropus
…… 103, 107
Gomphonema parvulum var.
neosaprophilum ……106, 404, 406
Gomphonema parvulum var. **parvulum**
…… 107, 408, 410
Gomphonema procerum ……110
Gomphonema pseudoaugur ……108, 412

Gomphonema pumilum ……110
Gomphonema rhombicum ……97
Gomphonema schweickerdii ……98
Gomphonema sumatrense ……97
Gomphonema truncatum ……109, 414
Gomphonema turgidum ……109
Gomphonema vibrio ……104
Gomphonema yamatoensis ……110, 416
Grammatophora balfouriana ……140

H

Hannaea arcus ……63, 306
Hannaea arcus var. arcus ……64
Hannaea arcus var. *recta* ……64, 308

L

Lemnicola hungarica ……130, 474, 476
Licmophora minuta ……96

M

Martyana martyi ……65, 310, 312
Melosira ambigua ……6
Melosira arenaria ……5
Melosira arentii ……17
Melosira crenulata ……9
Melosira crenulata var. ambigua ……6
Melosira crenulata var. valida ……14
Melosira distans ……7
Melosira granulata ……8
Melosira italica ……9
Melosira italica subsp. subarctica ……12
Melosira longipes var. tenuis ……13
Melosira longispina ……10
Melosira moniliformis ……2, 148, 150
Melosira nummuloides ……3, 152, 154
Melosira roseana var. asiatica ……15
Melosira roseana var. epidendron ……16
Melosira solida var. nipponica ……11
Melosira teres ……5
Melosira varians ……4, 156, 158
Meridion circulare ……66, 314
Microneis gracillima ……123
Micropodiscus weissflogii ……26

N

Navicula arcus ……63
Navicula elliptica ……141, 502
Navicula smithii ……143
Navicula subatomoides ……138
Nitzschia sigmoidea ……81, 127
Nitzschia vermicularis ……127

Nupela giluwensis ··············· 139
Nupela jahniae-reginae ··············· 139
Nupela lapidosum ··············· 139, 494
Nupela rumrichorm ··············· 139
Nupela tenuistriata ··············· 139

O

Odontidium parasiticum ··············· 81
Opephora martyi ··············· 65
Orthoseira asiatica ··············· 15, 196
Orthoseira epidendron ··············· 15, 16, 198

P

Peronia fibula ··············· 94, 374
Pinnularia ovalis ··············· 142
Planothidium delicatulum ··············· 484
Planothidium frequentissimum ··············· 131, 478
Planothidium lanceolatum ··············· 132, 480
Planothidium septentrionale ··············· 133, 482
Psammothidium helveticum ··············· 134, 484
Psammothidium hustedtii ··············· 135, 486
Psammothidium marginulata ··············· 136, 488
Psammothidium montanum ··············· 137, 490
Psammothidium subatomoides ··············· 138, 492
Pseudostaurosira brevistriata ··············· 67, 316
Pseudostaurosira brevistriata var. nipponica ··············· 68, 318
Pseudostaurosira robusta ··············· 69, 320
Punctastriata linearis ··············· 70, 80, 322
Puncticulata ozensis ··············· 41
Puncticulata praetermissa ··············· 41, 256
Puncticulata shanxiensis ··············· 42, 258

R

Rhoicosphenia abbreviata ··············· 95, 376, 378, 380
Rhoicosphenia curvata ··············· 95
Rossithidium pusillum ··············· 126

S

Sphenella parvula ··············· 107
Stauroneis exilis ··············· 122
Stauroneis inflata ··············· 113
Staurosira construens ··············· 74, 79
Staurosira construens var. binodis ··············· 69, 72, 326
Staurosira construens var. construens ··············· 324
Staurosira construens var. exigua ··············· 73, 328
Staurosira construens var. triundulata ··············· 74, 330

Staurosira elliptica ··············· 75, 332
Staurosira venter ··············· 76, 334
Staurosira venter var. binodis ··············· 77, 336
Staurosirella lapponica ··············· 78, 338
Staurosirella leptostauron ··············· 79, 340
Staurosirella pinnata ··············· 70, 80, 342
Stephanodiscus dubius ··············· 27
Stephanodiscus hantzschii f. tenuis ··············· 28, 43, 44, 260, 262
Stephanodiscus invisitatus ··············· 28, 43
Stephanodiscus minutulus ··············· 43, 44, 264, 266
Stephanodiscus rotula ··············· 45, 268
Stephanodiscus tenuis ··············· 43
Stephanosira epidendron ··············· 16
Synedra acus ··············· 83
Synedra affinis ··············· 82
Synedra biceps ··············· 84
Synedra capitata ··············· 85
Synedra capitellata ··············· 56
Synedra fasciculata var. obtusa ··············· 51
Synedra inaequalis ··············· 86
Synedra lanceolata ··············· 87
Synedra parasitica ··············· 81
Synedra perminuta ··············· 60
Synedra pulchella ··············· 52
Synedra ungeriana ··············· 88
Synedrella parasitica ··············· 81, 344

T

Tabellaria fenestrata ··············· 90, 92, 362, 364
Tabellaria flocculosa ··············· 91, 92, 366, 368
Tabellaria pseudoflocculosa ··············· 92, 370
Tabellaria quadriseptata ··············· 92
Tabularia affinis ··············· 82, 346
Tabularia fasciculata ··············· 51
Thalassiosira allenii ··············· 18, 202
Thalassiosira bramaputrae ··············· 22
Thalassiosira eccentrica ··············· 19, 204
Thalassiosira faurii ··············· 20, 206
Thalassiosira guillardii ··············· 21, 208
Thalassiosira lacustris ··············· 22, 210
Thalassiosira nordenskioeldii ··············· 23, 212
Thalassiosira pseudonana ··············· 24, 214
Thalassiosira tenera ··············· 25, 216
Thalassiosira weissflogii ··············· 26, 218
Triceratium exigua ··············· 73

U

Ulnaria acus ··············· 83, 348
Ulnaria biceps ··············· 84, 350

Ulnaria capitata 85, 352
Ulnaria inaequalis 86, 354
Ulnaria lanceolata 87, 356
Ulnaria pseudogaillonii 88, 358
Ulnaria ulna 86–89, 360
Ulnaria ungeriana 88

著者紹介

小林　弘　Hiromu KOBAYASI
1926年　三重県上野市に出生
1948年　東京文理科大学生物学科修了
1949年　東京教育大学・東京文理科大学助手
1962年　理学博士（東京教育大学・東京文理科大学）
1968年　東京教育大学助教授
1974年　東京学芸大学助教授
1975年　東京学芸大学教授
1986年　日本藻類学会会長・日本珪藻学会会長・国際珪藻学会評議員
1989年　東京学芸大学定年退官・東京珪藻研究所設立
1994年　国際珪藻学会副会長
1996年　7月　逝去

出井　雅彦　Masahiko IDEI
文教大学　教育学部　教授　理学博士

真山　茂樹　Shigeki MAYAMA
東京学芸大学　教育学部　生物学教室　助教授
理学博士

南雲　保　Tamotsu NAGUMO
日本歯科大学　生命歯学部　生物学教室
教授　水産学博士

長田　敬五　Keigo OSADA
日本歯科大学　新潟生命歯学部　生物学教室
助教授　水産学博士

H. Kobayasi's Atlas of Japanese Diatoms based on electron microscopy

2006年11月25日　第1版発行

著者の了解により検印を省略いたします

小林弘 珪藻図鑑
第1巻

著　者 © 小　林　　　弘
　　　　出　井　雅　彦
　　　　真　山　茂　樹
　　　　南　雲　　　保
　　　　長　田　敬　五
発行者　内　田　　　悟
印刷者　山　岡　景　仁

発行所　株式会社　内田老鶴圃　〒112-0012　東京都文京区大塚3丁目34-3
　　　　電話（03）3945-6781(代)・FAX（03）3945-6782
　　　　印刷/三美印刷 K.K.・製本/榎本製本 K.K.

Published by UCHIDA ROKAKUHO PUBLISHING CO., LTD.
3-34-3 Otsuka, Bunkyo-ku, Tokyo 112-0012, Japan

U.R. No. 551-1

ISBN 4-7536-4046-9 C3045

本書の全部あるいは一部を断わりなく転載または複写（コピー）することは，著作権および出版権の侵害となる場合がありますのでご注意下さい．

日本淡水藻図鑑

廣瀬 弘幸　編集
山岸 高旺

B5判・上製・総頁 960 頁
定価 39,900 円（本体 38,000 円＋税 5%）

**創案以来 10 余年を費やして完成した労作.
斯学の権威がそれぞれの専門分野を担当し，網羅的にまとめた貴重な図鑑.**

〔本書の特徴〕

・全巻にわたり原則として図を左ページに，それらの記載を右ページに配し，図と記載とが一見して対照できるよう工夫配列.

・各綱について，その中の目・科・属の分類学的特徴を図などを用いて，わかりやすく説明し，詳細な検索表を付す.

・各属内の種の検索表には，図や記載のページを示し，また図や記載ページには検索表ページを付して，どちらからでも利用できる.

・種については図とそれぞれの特徴，国内の産地，世界的な分布を記し，特記すべき事項は備考として付す.

・随所に淡水藻の特別な種についての特徴，生殖法，または稀産種の産地，産状，分布，淡水藻の利用法，諸外国の著名な藻学者の小伝などをノートとして挿入.

・明治初期以来の日本淡水藻の分類学的研究を中心とする文献を網羅する．採集と研究法，略史も付した.

・巻末には学名，和名，術語の詳細な索引を付しているが，それらの記載ページ，検索表ページなどはそれぞれ字体を変えて示し，利用の便をはかっている.

藻類多様性の生物学

千原 光雄　編著

B5判・上製・総頁 400 頁
定価 9,450 円（本体 9,000 円＋税 5%）

　藻類の複雑な多様性・異質性は藻学およびその周辺領域の多くの学問による長い年月の成果に裏付けられたものであり，その全貌を理解することは容易なことではない.
　本書は，かねてより最近の知識を盛った藻類の教科書の必要性を痛感していた編者が，それぞれの藻群を得意とする専門家の参加を得て，膨大な知識の蓄積を整理するとともに，次々と発表される新しい成果を取り入れつつ編んだもので，藻類の世界をひと通り理解するに最適の書である.

淡水藻類入門

淡水藻類の形質・種類・観察と研究

山岸 高旺　編著

B5判・上製・総頁 700 頁（口絵カラー）
定価 26,250 円（本体 25,000 円＋税 5%）

　「日本淡水藻図鑑」の編者である著者がまとめた，初心者・入門者のための書．多種多様な淡水藻類群を，平易な言葉で誰にも分かるよう，丁寧に解説する．I編，II編で形質と分類の概説を行い，III編では各分野の専門家による具体的事例 20 編をあげ，実際にどのように観察・研究を進めたらよいかを理解できるように構成する.

新日本海藻誌

日本産海藻類総覧

吉田 忠生 著

B5判・上製・総頁1248頁
定価48,300円（本体46,000円＋税5％）

　本書は古典的になった岡村金太郎の歴史的大著「日本海藻誌」(1936) を全面的に書き直したものである．「日本海藻誌」刊行以後の研究の進歩を要約し，日本産の報告のある海藻（緑藻，褐藻，紅藻）約1400種について，形態的な特徴を現代の言葉で記述する．

　編集にあたっては，各種類の学名の原典にさかのぼって検討し，国際植物命名規約に厳密に従って命名法上の正確さを期し，関連する文献を詳しく引用，また命名規約に基づいて多くの種のタイプ標本を確定し，その所在を明らかにするとともに，北海道大学，国立科学博物館などに所蔵されているタイプ標本の写真を多数掲載した．

〔本書の特徴〕

　綱，目，科，属，種の分類階級ごとに形質の特徴および他との比較などを詳細に記述する．綱から目へ，目から科へといったように，わかりやすい検索表を掲げ，目，科，属にはそれぞれタイプ科，タイプ属，タイプ種が記されている．また，各種ごとに極めて詳細，細緻な文献リストを付す．さらに種ごとにタイプ産地，タイプ標本，分布地域名を示す．

　学名は国際命名規約に沿う．和名は現在使われているものを出来る限り網羅した．取り上げた和名の多さも本書の大きな特徴である．学名，和名の由来や生育地の特徴など関連した話題も豊富に収録する．

有用海藻誌

海藻の資源開発と利用に向けて

大野 正夫 編著

B5判・上製・総頁596頁
定価21,000円（本体20,000円＋税5％）

　本書の構成は，「生物学編」，「利用編」，「機能性成分編」に分かれるが，どの項目も独立しており，必要なところから読むことができる．生物学編では，利用分野ごとに分けて，種名の査定に必要な形態，生活史，分布生態が記述されており，さらに，これらの水産，食用などへの利用や産業的背景，利用の歴史についても詳しく記述されている．利用編は，海藻産業の歴史的背景，加工技術から化学構造，品質などにふれており，将来への展望も示される．機能性成分編では，あまり知られていない海藻の成分とその利用範囲が幅広く記述されている．

〔本書の特徴〕

　生物学編，利用編，機能性成分編の3編31章から構成．それぞれの分野で長く研究にかかわってきた各執筆者が，専門の分野を詳述する．海藻の生物学の解説にはじまり応用の具体的事例を数多く紹介するとともに，今後期待される新分野である機能性成分についても現在知られている知見を盛り込む．本文とともに，充実した和名索引，学名索引，事項索引を付し読者が求めている目的，事柄をすぐに探せるように配慮する．

淡水珪藻生態図鑑

群集解析に基づく汚濁指数 DAIpo，pH 耐性能

渡辺 仁治 編著　　浅井一視・大塚泰介　著
　　　　　　　　　辻 彰洋・伯耆晶子

B 5 判・上製・総頁 784 頁
定価 34,650 円（本体 33,000 円＋税 5%）

日本のみならず世界各地から約 1500 のサンプルを採集．膨大なサンプルの生態情報を処理検討し，約 1000 種の珪藻についてその結果を分かり易くまとめる．

好清水か好汚濁か＝きれいな水を好むのか，汚れた水を好むのかを判断する環境指標としての珪藻群集の適性を，多くの図版で具体的に示す．

生態情報の妥当性を期するため，すべてのサンプルを統一条件下で採集．

1. 分類・同定について

珪藻の研究は現在，電子顕微鏡を用いた形態学的研究による微細構造の解明をはじめ，細胞分裂，生活史，光合成に関わる葉緑体，ミトコンドリアの機能など，生理学的分野においても，新しい手法による研究が積極的に進められている．これらの研究は，一方においては新属，新種の設定，あるいは統合とも関わりあい，そのために，分類学はかなり流動的であるといわざるをえない．この点について，本書の編集にあたっては，新しい情報に配慮しながら，基本的には従来からの分類体系を重視し，新しい分類群名はできる限り併記することとした．種名には，Synonym および最近の分類学的情報を文献と併記して，生態学的情報の混乱を避けることにつとめた．

2. 生態学的情報について

陸水の有機汚濁と pH に対する生態学的情報，および生態分布図は，著者らが特定の統一条件のもとで採集した付着珪藻群集の試料を，数理統計的に処理して得た情報を示したものである．なお，プランクトンとして出現する淡水産珪藻も，ほとんど総て含まれる．他の環境要因との関係については，内外の研究者による生態学的情報も精選して紹介する．

DAIpo（Diatom Assemblage Index to organic water pollution）は，止水域，流水域共通の有機汚濁に対する生物学的指数である．0 は最も強い汚濁，100 は最も清浄であることを示す．本書では，この DAIpo（有機汚濁）と pH の変動に伴う，付着珪藻群集中における相対頻度の変動を示す図をできるだけ多くの分類群（種）に添えた．それらの図から，それぞれの種の適応能の特性を知る手がかりが得られる．

〔目 次〕

総 論

- 第 1 章　珪藻研究の歴史
- 第 2 章　環境指標としての珪藻群集
- 第 3 章　湖沼，河川共通の水質汚濁指数 DAIpo
- 第 4 章　珪藻の生活様式
- 第 5 章　試料の採集
- 第 6 章　試料の処理と検鏡
- 第 7 章　形態（種の同定に関わる特性要素）

参考文献

写真編

- I 　中心目 (Centrales) の分類
- II 　羽状目 (Pennales) の分類
- II A 　無縦溝亜目 (Araphidineae) の分類
- II A 　ディアトマ科 (Diatomaceae)
- II B 　有縦溝亜目 (Raphidineae) の分類
- II B_1 　ユーノチア科 (Eunotiaceae)
- II B_2 　アクナンテス科 (Achnanthaceae)
- II B_3 　ナビクラ科 (Naviculaceae)
- II B_4 　エピテミア科 (Epithemiaceae)
- II B_4 　ニチア科 (Nitzschiaceae)
- II B_5 　スリレラ科 (Surirellaceae)

学名総索引
事項索引